COMPUTER AND COMPUTING TECHNOLOGIES IN AGRICULTURE, VOLUME II

IFIP – The International Federation for Information Processing

IFIP was founded in 1960 under the auspices of UNESCO, following the First World Computer Congress held in Paris the previous year. An umbrella organization for societies working in information processing, IFIP's aim is two-fold: to support information processing within its member countries and to encourage technology transfer to developing nations. As its mission statement clearly states,

> *IFIP's mission is to be the leading, truly international, apolitical organization which encourages and assists in the development, exploitation and application of information technology for the benefit of all people.*

IFIP is a non-profitmaking organization, run almost solely by 2500 volunteers. It operates through a number of technical committees, which organize events and publications. IFIP's events range from an international congress to local seminars, but the most important are:

- The IFIP World Computer Congress, held every second year;
- Open conferences;
- Working conferences.

The flagship event is the IFIP World Computer Congress, at which both invited and contributed papers are presented. Contributed papers are rigorously refereed and the rejection rate is high.

As with the Congress, participation in the open conferences is open to all and papers may be invited or submitted. Again, submitted papers are stringently refereed.

The working conferences are structured differently. They are usually run by a working group and attendance is small and by invitation only. Their purpose is to create an atmosphere conducive to innovation and development. Refereeing is less rigorous and papers are subjected to extensive group discussion.

Publications arising from IFIP events vary. The papers presented at the IFIP World Computer Congress and at open conferences are published as conference proceedings, while the results of the working conferences are often published as collections of selected and edited papers.

Any national society whose primary activity is in information may apply to become a full member of IFIP, although full membership is restricted to one society per country. Full members are entitled to vote at the annual General Assembly, National societies preferring a less committed involvement may apply for associate or corresponding membership. Associate members enjoy the same benefits as full members, but without voting rights. Corresponding members are not represented in IFIP bodies. Affiliated membership is open to non-national societies, and individual and honorary membership schemes are also offered.

COMPUTER AND COMPUTING TECHNOLOGIES IN AGRICULTURE, VOLUME II

First IFIP TC 12 International Conference on Computer and Computing Technologies in Agriculture (CCTA 2007), Wuyishan, China, August 18-20, 2007

Edited by

DAOLIANG LI
China Agricultural University

Computer and Computing Technologies in Agriculture, Vol. 2

Edited by Daoliang Li

p. cm. (IFIP International Federation for Information Processing, a Springer Series in Computer Science)

ISSN: 1571-5736 / 1861-2288 (Internet)

ISBN: 978-1-4419-4584-6 e-ISBN: 978-0-387-77253-0

Printed on acid-free paper

9 8 7 6 5 4 3 2 1

springer.com

CONTENTS

FOREWORD

The papers in this volume comprise the refereed proceedings of the First International Conference on Computer and Computing Technologies in Agriculture (CCTA 2007), in Wuyishan, China, 2007.

This conference is organized by China Agricultural University, Chinese Society of Agricultural Engineering and the Beijing Society for Information Technology in Agriculture. The purpose of this conference is to facilitate the communication and cooperation between institutions and researchers on theories, methods and implementation of computer science and information technology. By researching information technology development and the resources integration in rural areas in China, an innovative and effective approach is expected to be explored to promote the technology application to the development of modern agriculture and contribute to the construction of new countryside.

The rapid development of information technology has induced substantial changes and impact on the development of China's rural areas. Western thoughts have exerted great impact on studies of Chinese information technology development and it helps more Chinese and western scholars to expand their studies in this academic and application area. Thus, this conference, with works by many prominent scholars, has covered computer science and technology and information development in China's rural areas; and probed into all the important issues and the newest research topics, such as Agricultural Decision Support System and Expert System, GIS, GPS, RS and Precision Farming, CT applications in Rural Area, Agricultural System Simulation, Evolutionary Computing, etc. In the following sessions, this conference could hopefully set up several meeting rooms to provide an opportunity and arena for discussing these issues and exchanging ideas more effectively. We are also expecting to communicate and have dialogues on certain hot topics with some foreign scholars.

With the support of participants and hard working of preparatory committee, the conference achieved great success on the participation. We received around 427 submitted papers and 180 accepted papers will be published in the Springer Press. It has evidenced our remarkable achievements made in our studies of the New Period, forming a necessary step in the development of the research theory in China and a worthy legacy for information technology studies in the new century. The conference is planned to be organized annually. We believe that it can provide a platform for exchanging ideas and sharing outcomes and also contribute to China's agricultural development.

Finally, I would like to extend the most earnest gratitude to our co-sponsors, Chinese Society of Agricultural Engineering and the Beijing Society for Information Technology in Agriculture, also to Nongdaxingtong Technology Ltd., all members and colleagues of our preparatory committee, for their generous efforts, hard work and precious time! On behalf of all conference committee members and participants, I also would like to express our genuine appreciation to Fujian Provincial Agriculture Department, Nanping City Bureau of Agriculture and Wuyishan City government. Without their support, we can not meet in such a beautiful city.

This is the first in a new series of conferences dedicated to real-world applications of computer and computing technologies in agriculture around the world. The wide range and importance of these applications are clearly indicated by the papers in this volume. Both are likely to increase still further as time goes by and we intend to reflect these developments in our future conferences.

Daoliang Li

Chair of programme committee, organizing committee

ORGANIZING COMMITTEE

Chair

Prof. Daoliang Li, China Agricultural University, China
Director of EU-China Center for Information & Communication technologies in Agriculture

Members [in alpha order]

Mr. Chunjiang Zhao, Director, Beijing Agricultural Informatization Academy, China

Mr. Ju Ming, Vice Section Chief, Foundation Section, Science & Technology Department, Ministry of Education of the People's Republic of China

Prof. Haijian Ye, China Agricultural University, China

Prof. Jinguang Qin, Chinese Society of Agricultural Engineering, China

Prof. Qingshui Liu, China Agricultural University, China

Prof. Rengang Yang, China Agricultural University, China

Prof. Renjie Dong, China Agricultural University, China

Prof. Songhuai Du, China Agricultural University, China

Prof. Wanlin Gao, China Agricultural University, China

Prof. Weizhe Feng, China Agricultural University, China

PROGRAM COMMITTEE

Chair

Prof. Daoliang Li, China Agricultural University, China
Director of EU-China Center for Information & Communication technologies in Agriculture

Members [in alpha order]

Dr. Alex Abramovich, Maverick Defense Technologies Ltd., Israel
Dr. Boonyong Lohwongwatana, Asian Society for Environmental Protection (ASEP), Thailand
Dr. Feng Liu, Mercer University, GA, USA
Dr. Haresh A. Śuthar, Industry (Masibus Automation & Instrumentation (p)) Ltd., India
Dr. Javad Khazaei, University of Tehran, Iranian
Dr. Jinsheng Ni, Beijing Oriental TITAN Technology. Co. Ltd.
Dr. Joanna Kulczycka, Polish Academy of Sciences Mineral and Energy Economy Research Institute, Poland
Dr. John Martin, University of Plymouth, Plymouth, UK
Dr. Kostas Komnitsas, Technical University of Crete, Greece
Dr. M. Anjaneya Prasad, College of Engg. Osmania University, India
Dr. Pralay Pal, Engineering Automation Deputy General Manager (DGM), India
Dr. Shi Zhou, Zhejiang University, China
Dr. Sijal Aziz, Executive Director WELDO, Pakistan
Dr. Soo Kar Leow, Monash University, Malaysia
Dr. Wenjiang Huang, National Engineering Research Center for Information Technology in Agriculture, China
Dr. Yong Yue, University of Bedfordshire, UK
Dr. Yuanzhu Zhang, Suzhou Center of Aquatic Animals Diseases Control, China
Mr. Weiping Song, DABEINONG Group
Prof. A.B. Sideridis, Informatics Laboratory of the Agricultural University of Athens, Greece
Prof. Andrew Hursthouse, University of Paisley, UK

Prof. Apostolos Sarris, Institute for Mediterranean Studies, Greece

Prof. Chunjiang Zhao, China National Engineering Center for Information Technology in Agriculture, China

Prof. Dehai Zhu, China Agricultural University, China

Prof. Fangquan Mei, Institute of Information, China Agricultural Science, China

Prof. Gang Liu, China Agricultural University, China

Prof. Guohui Gan, Institute of Geographic Sciences and Natural Resources

Prof. Guomin Zhou, Institute of Information, China Agricultural Science, China

Prof. Iain Muse, Development into Community Cooperation Policies and International Research Areas, Belgium

Prof. Jacques Ajenstat, University of Quebec at Montreal, Canada

Prof. K.C. Ting, Department of Agricultural and Biological Engineering, University of Illinois at Urbana-Champaign

Prof. Kostas Fytas, Laval University, Canada

Prof. Liangyu Chen, Countryside Center, Ministry of Science & Technology, China

Prof. Linnan Yang, Yunnan Agricultural University, China

Prof. Liyuan He, Huazhong Agricultural University (HZAU), China

Prof. Maohua Wang, member of Chinese Academy of Engineering, China Agricultural University, China

Prof. Maria-Ioanna Salaha, Wine Institute-National Agricultural Research Foundation, Greece

Prof. Michael Petrakis, National Observatory of Athens, Greece

Prof. Michele Genovese, Unit Specific International Cooperation Activities, International Cooperation Directorate, DG Research, UK

Prof. Minzan Li, China Agricultural University, China

Prof. Nigel Hall, Harper Adams University College, England

Prof. Raphael Linker, Civil and Environmental Engineering Dept., Technion

Prof. Rohani J. Widodo, Maranatha Christian University, Indonesia

Prof. Xiwen Luo, South China Agricultural University, China

Prof. Yanqing Duan, University of Bedfordshire, UK

Prof. Yeping Zhu, Institute of Information, China Agricultural Science, China

Prof. Yiming Wang, China Agricultural University, China

Prof. Yu Fang, Information Center, Ministry of Agriculture, China

Prof. Yuguo Kang, Chinese cotton association, China

Prof. Zetian Fu, China Agricultural University, China

Prof. Zuoyu Guo, Information Center, Ministry of Agriculture, China

SECRETARIAT

Secretary-general

Baoji Wang (China Agricultural University, China)
Liwei Zhang (China Agricultural University, China)

Secretaries

Xiuna Zhu (China Agricultural University, China)
Yanjun Zhang (China Agricultural University, China)
Xiang Zhu (China Agricultural University, China)
Liying Xu (China Agricultural University, China)
Bin Xing (China Agricultural University, China)
Xin Qiang (China Agricultural University, China)
Yingyi Chen (China Agricultural University, China)
Chengxian Yu (China Agricultural University, China)
Jie Yang (China Agricultural University, China)
Jing Du (China Agricultural University, China)

RESEARCH ON THE SHORT-TERM AGRICULTURAL ELECTRIC LOAD FORECASTING OF WAVELET NEURAL NETWORK

Qian Zhang
Department of Economic Management, North China Electric Power University, Baoding 071003, China, Email: hdzhq@yeah.net

Abstract: This paper proposes a new method for load forecasting—the wavelet neural network model for daily load forecasting. The neural call function is basis of nonlinear wavelets. A wavelet network is composed by the wavelet basis function. The global optimum solution is got. We overcome the intrinsic defects of a artificial neural network that its learning speed is slow, its network structure is difficult to determine rationally and it produces local minimum points. It can be seen from the example this method can improve effectively the forecast accuracy and speed. It can be applied to the daily agricultural electric load forecasting.

Keywords: Artificial Neural Network, Wavelet Neural Network, Agricultural electric Load Forecasting

1. INTRODUCTION

A daily agricultural electric load forecasting means the load forecasting that time units are the hour, day or month. Because its tendency has a strong randomness, the determination of mathematical models is difficult. The improvement of forecasting accuracy is difficult.

At present, one of the effective forecast methods is the artificial neural network. It can express a complex nonlinear function. But then it has some intrinsic defects that its learning speed is slow, its network structure is difficult to determine rationally and it produces local minimum points.

Zhang, Q., 2008, in IFIP International Federation for Information Processing, Volume 259; Computer and Computing Technologies in Agriculture, Vol. 2; Daoliang Li; (Boston: Springer), pp. 737–745.

In order to overcome these questions, we propose to forecast short-term load with the wavelet neural network. It combines a wavelet transformation with an artificial neural network. It is a new mathematical model method. First a wavelet series is got through an expansion and contraction factor and a translation factor. Then a wavelet function network is formed. Because two new parameters that are an expansion and contraction factor and a translation factor are used, the wavelet neural network has more degree of freedom than an artificial neural network (Karaki et al., 2005). Thus it has a better function approximation ability. The wavelet neural network that it is combined by the series of less term can get excellent approximation effect. The network structure is combined with wavelet basis functions; the subjective determination is avoided (C.N. Lu et al., 2003). Because its network weight number is linear and learning objective function is convex, the global optimum solution is got. Because the network structure is a single implicit layer, the learning speed is faster than general network (Niu et al., 1994).

2. THE WAVELET CONCEPT AND THE WAVELET TRANSFORM

For setting up the load forecast model of wavelet neural network, first we introduce some basic concepts of wavelet and wavelet transform.

We call the square integral function $\varphi(t)\in L2(R)$ that it is asked to satisfy the admissible condition:

$$\int_{-\infty}^{+\infty}[\hat{\varphi}(\omega)]^2[\omega]^{-1}d\omega<+\infty \quad (1)$$

basic wavelet or mother wavelet. $\hat{\varphi}(\omega)$ is the Fourier Transform of the $\varphi(t)$. Let

$$\varphi_{ab}(t)=\frac{1}{\sqrt{|a|}}\varphi(\frac{t-b}{a}) \quad (2)$$

Among them, a, b are real number, and $a\neq 0$, φab is called the wavelet basis that it is generated by mother wavelet and it depends on the parameter a, b.

Let $f(t)\in L2(R)$, The f(t) is tendency function that it shows the variance law of the agricultural electric load.

Let the wavelet transform of the f(t)

$$W_f(a,b) = (f, \varphi_{ab}) = \frac{1}{\sqrt{|a|}} \int_{-\infty}^{+\infty} f(t)\varphi(\frac{t-b}{a})dt \tag{3}$$

Due to the specificity of the load, this transform is discussed only in the real numbers. From (3), the variance of the parameter b in the wavelet basis has the use of translation. The parameter a not only changes the frequency spectrum structure but also changes the length of its window. Thus, a, b are called respectively the expansion and contraction factor and the translation factor of φab (t). The similarity of the Fourier analysis, the wavelet analysis resolves the signal function into the wavelet normal orthogonal basis. It constructs a series to approximate the signal function. This linear combination has an optimum recognition capacity (Srinivasan et al., 2005) .

The mother wavelet is asked to satisfy condition (1), Thus we get

$$\int |\varphi(t)|^2 dt < \infty \tag{4}$$

The condition (1) determines the locality behavior of the wavelet. It equals 0 without a limited interval or approximates 0 fast. The formula (4) determines that the wavelet has the limited energy and is an oscillation (the positive number part equals the negative number part). The wavelet's name is produced from here.

The mother wavelets with the better locality property and smooth property have the spline wavelet and Morlet wavelet usually. Its system of the expansion and contraction and translation composes the normal orthogonal basis of L2(R). The wavelet series generated can approximate f(t) optimally (Rutkowskil et al., 2004).

The similarity of the Fourier transform, the wavelet transform has also a inversion formula.

$$f(t) = \frac{1}{C_\varphi} \int_{-\infty}^{+\infty} \int_{-\infty}^{+\infty} W_f(a,b)\varphi_{ab}(t) \frac{dadb}{a^2} \tag{5}$$

Among them,

$$C_\varphi = \int_{-\infty}^{+\infty} |\varphi(\omega)|^2 |\omega|^{-1} d\omega \tag{6}$$

3. THE LOAD FORECASTING MODEL OF THE WAVELET NEURAL NETWORK

In the wavelet neural network, we replaces the nonlinear sigmoid function with the nonlinear wavelet basis. The fitting of a load historical sequence is completed with the linear superposition of the nonlinear wavelet basis. The limited terms of a wavelet series can approximate the load historical curve. The load curve can be fitted with the wavelet basis φab(t)

$$\hat{y}(t) = \sum_{k=1}^{L} \omega_k \varphi\left(\frac{t - b_k}{a_k}\right) \tag{7}$$

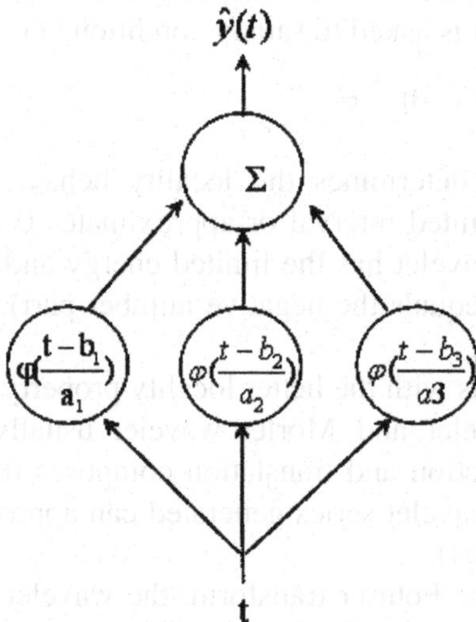

Figure 1. The wavelet neural network structure

In (7), ŷ (t) shows the forecast values sequence of load curve y(t). ωk are weight numbers. bk are translation factors. ak are expansion and contraction factors. L is wavelet basis number. In Figure 1, we give the structure of the wavelet neural network. (see Figure 1) The network is a single implicit layer. It only contains an importation node and a exportation node. We need to determine the network parameters ωk ak bk and L. Let the sequence y(t) and sequence ŷ (t) are fitted optimally. The parameter ωk ak and bk can be optimized on the basis of the minimum square error energy function E_L.

$$E_L = \frac{1}{2}\sum_{t=1}^{L}[y(t) - \hat{y}(t)]^2 \tag{8}$$

We determine L with the method of stepwise test, Thus the network structure is determined also. to determined every L, we compute optimal parameters ωk ak and bk by (8) as follows.

First, we use Morlet mother wavelet in (7)

$$\varphi(t) = \cos(1.75t)\exp(-\frac{t^2}{2}) \tag{9}$$

Let $T = \frac{t - b_k}{a_k}$, then the gradient of (8) is showed respectively as follows.

$$g(\omega_k) = \frac{\partial E}{\partial \omega_k} = -\sum_{t=1}^{L}[y(t) - \hat{y}(t)]$$

$$\cos(1.75T)\exp(-\frac{T^2}{2})$$

$$g(b_k) = \frac{\partial E}{\partial b_k} = -\sum_{t=1}^{L}[y(t) - \hat{y}(t)]\omega_k$$

$$[1.75\sin(1.75T)\exp(-\frac{T^2}{2})$$

$$+\cos(1.75T)\exp(-\frac{T^2}{2}T)]/a_k$$

$$g(a_k) = \frac{\partial E}{\partial a_k} = -\sum_{t=1}^{L}[y(t) - \hat{y}(t)]\omega_k$$

$$[1.75\sin(1.75T)\exp(-\frac{T^2}{2})T$$

$$+\cos(1.75T)\exp(-\frac{T^2}{2})T]/a_k = Tg(b_k)$$

Network parameters ωk bk and ak are optimized with the conjugate gradient method. Let vector ω=(ω1, ω2,…ωk…ωL)), g(ω)=(g(ω1), g(ω2)… g(ωk)… g(ωL)). S(ω)i shows the ith cyclic search direction of the function of w. Then

$$S(\omega)^i = \begin{cases} -g(\omega)^i & i=1 \\ -g(\omega)^i + \\ \dfrac{g(\omega)^{iL} g(\omega)^i}{g(\omega)^{(i-1)L} g(\omega)^{i-1}} \cdot s(\omega)^{i-1} & i \neq 1 \end{cases}$$

The weight vector is regulated as follows.

$$\omega^{i+1} = \omega^i + \alpha_w^i S(\omega)^i \tag{10}$$

4. THE APPLIED STUDY OF SHORT-TERM DAILY LOAD FORECASTING IN CHINA LIAONING POWER

Applying the new method put forward in the paper, we study the daily load forecasting in Liaoning Power Network. The selected forecasting data is the history data and weather factors in October 2006. The agricultural electric load of 24 o'clock of October, 20, 2006 is to be forecasted. To compare the two models, forecast methods are elected by wavelet neural network model (WNNM) and artificial neural network model (ANNM) respectively. Through the imitation computation, we know that the accuracy and speed of the WNNM are raised obviously. (see Figure 2, 3, 4)

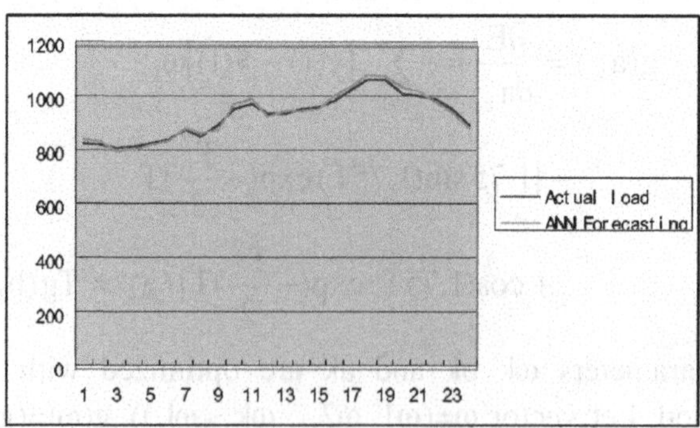

Figure 2. The ANN analysis of the load in China Liaoning power system

Table 1. The new method forecasting analysis of the load in China Liaoning power system

Time	Actual load	ANN Forecasting	ANN Relative error%	WANN Forecasting	WANN Relative error%
1	841.40	854.53	1.56	851.84	1.24
2	835.37	846.06	1.28	845.31	1.19
3	823.13	814.40	-1.06	816.71	-0.78
4	830.54	820.41	-1.22	824.81	-0.69
5	845.34	837.06	-0.98	837.56	-0.92
6	856.02	849.43	-0.77	847.55	-0.99
7	888.97	898.22	1.04	894.13	0.58
8	864.13	880.64	1.91	873.38	1.07
9	907.97	891.35	-1.83	895.80	-1.34
10	968.16	988.59	2.11	985.10	1.75
11	985.78	1005.98	2.05	1002.14	1.66
12	953.11	939.10	-1.47	938.62	-1.52
13	950.53	963.65	1.38	960.23	1.02
14	971.63	962.98	-0.89	965.12	-0.67
15	977.39	968.01	-0.96	969.76	-0.78
16	1002.71	1017.05	1.43	1009.43	0.67
17	1043.51	1055.93	1.19	1053.32	0.94
18	1078.58	1096.60	1.67	1087.97	0.87
19	1077.64	1090.24	1.17	1088.84	1.04
20	1025.51	1046.12	2.01	1039.35	1.35
21	1016.83	1036.86	1.97	1033.81	1.67
22	1009.01	993.37	-1.55	991.35	-1.75
23	971.72	954.62	-1.76	959.48	-1.26
24	908.24	898.80	-1.04	901.34	-0.76

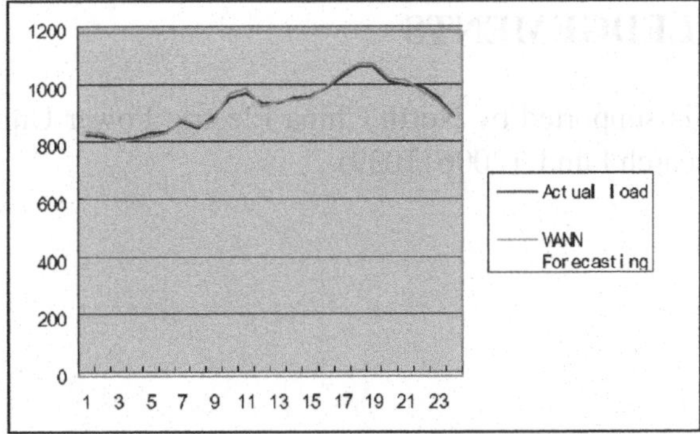

Figure 3. The WANN analysis of the load in China Liaoning power system

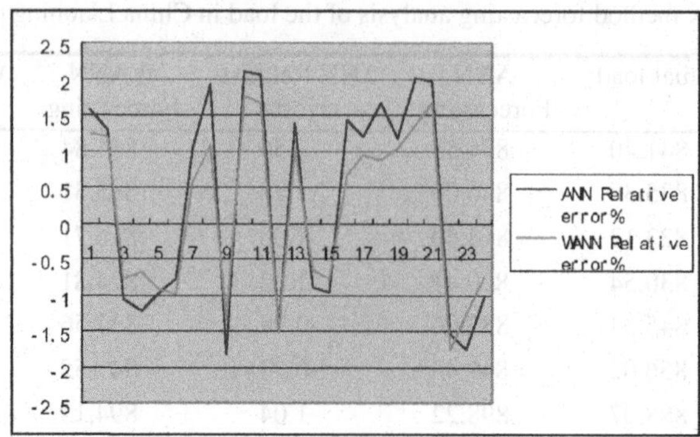

Figure 4. Relative error analysis of the load in China Liaoning power system

5. CONCLUSIONS

(1) In this paper, we propose the wavelet neural network forecast model of agricultural electric load. It overcomes the intrinsic defects of an artificial neural network that its learning speed is slow, its network structure is difficult to determine rationally and it produces local minimum points.

(2) Its nervous cell function is basis of nonlinear wavelets. We get the global optimum fitting effect. Then the accuracy is improved. The network structure is determined rationally with the stepwise test method, Because the network is a single implicit layer structure, Its speed is improved obviously It can be used to forecast daily agricultural electric load.

(3) Through the imitation computation, we prove that the accuracy and speed are improved obviously. This is a new and effective method of agricultural electric load forecasting.

ACKNOWLEDGEMENTS

This paper is supported by North China Electric Power University Youth Scientific Research Fund. (200611039).

REFERENCES

Niu Dongxiao: Adjustment Gray Load Forecasting Model of Power System. The Journal of Gray System, 1994, 6(2):127-134.

Karaki S.H.: Weather Sensitive Short-term Load Forecasting Using Artificial Neural Networks and Time Series. International Journal of Power and Energy Systems, 2005, 2005, 19(3):25 1-256.

XieHong, ChengZhiye, NiuDong xiao: The Research of Daily Load Forecasting Model Based on Wavelet Decomposing and Climatic Influence. Proceedings of the Csee, 2004, 21(5):5-10.

Srinivasan D., Chang C.S., Liew A.C.: Demand Forecasting Using Fuzzy Neural Computation, with Special Emphasis on Weekend and Public Holiday Forecasting. IEEE Transactions on Power Systems, 2005, 10(4):1897-1903.

Rutkowskil L.: Generalized Regression Neural Network in Time Varying Environment. IEEE Transaction on Neural Network, 2004, 15(3):576-596.

Tomandl D., Schober A.: A Modified General Regression Neural Network with New, Efficient Training Algorithms as a Robust Black Box Tool for Data Analysis. Neural Networks, 2001, 14(8):1023-1034.

He S.D., Qu T., Huang X.H.: An Improved Neural Network Method for Short Term Load Forecasting. Automation of Electric Power System, 2000, 21(11):13-14

Charytoniuk W., Chen M.H.: Very Short-term Load Forecasting Using Artificial Networks. IEEE Trans. on PWS, 2000, 15(1):263-268.

T. Haida and S. Muto, "Regression based peak load forecasting using a transformation technique," IEEE Trans. Power Syst., vol. 9, pp. 1788-1794, Nov. 1999.

S. Rahman and O. Hazim, "A generalized knowledge-based short term load-forecasting technique," IEEE Trans. Power Syst., vol. 8, pp. 508-514, May 1998.

S. Rahman and G. Shrestha, "A priory vector based technique for load forecasting," IEEE Trans. Power Syst., vol. 6, pp. 1459-1464, Nov. 2003.

T. Senjyu, S. Higa, T. Yue-Jin, and K. Uezato, "Future load curve shaping based on similarity using Fuzzy logic approach," in Proc. Int. Power Eng. Conf (IPEC), vol. II, 2004, pp. 483-488.

C.N. Lu and S. Vemuri, "Neural network based short term load forecasting," IEEE Trans. Power Syst., vol. 8, pp. 336-342, Feb. 2003.

R. Lamedica, A. Prudenzi, M. Sforna, M. Caciotta, and V.O. Cencellli, "A neural network based technique for short-term forecasting of anomalous load periods," IEEE Trans. Power Syst., pp. 1749-1756, Nov. 1996.

S. AlFuhaid, M.A. El-Sayed, and M.S. Mahmoud, "Cascaded artificial neural networks for short-term load forecasting," IEEE Trans. Power Syst., vol. 12, pp. 1524-1529, Nov. 1997.

D. Papalexopoulos and T.C. Hesterberg, "A regression-based approach to short-term load forecasting," IEEE Trans. on Power Syst. vol. 5, no. 4, pp. 1535-1550, 1990.

H.S. Hippert, C.E. Pedreira, and R.C. Souza, "Neural networks for short-term load forecasting: A review and evaluation," IEEE Trans. on Power Syst., vol. 16, no. 1, pp. 44-55, 2001.

AGRICULTURAL FLOOD LOSSES PREDICTION BASED ON DIGITAL ELEVATION MODEL

Lei Zhu
Information School, Central University of Finance and Economics, Beijing, China, 100081

Abstract: A new agricultural flood losses prediction method using digital elevation data combined with agricultural economic data is presented. The method is based on dynamically simulating the procedure of flood out. An example in Huangdunhu flood detention area, Jiangshu province of China has been given to describe how to use it to predict the incoming agricultural losses. Compared with prior methods, the method is more intuitive and accurate, and can be used to provide a scientific and effective reference for taking flood control and disaster mitigation measures.

Keywords: Flood disaster, Agricultural losses prediction, DEM

1. INTRODUCTION

The Huangdunhu flood detention area in Jiangsu province of China is a temporary flood detention area for Yi-River, Shu-River and Si-River. Its main function is to clip flood peak in order to keep the Luomahu reservoir safety. Huangdunhu covers an area of 335km^2 and contains 20,700 hm^2 of arable land, with a general ground elevation of 21 meters and its maximum flood detention capacity is 1.47 billion m^3 with average water depth of 5–7 meters. The flood disaster in this region is mainly caused by flood discharging of sluices located at Caodianzi and Shuangheqiao. Since the area is planning for flood detention, agricultural losses is the mainly losses when flood comes. How to forecast flood disaster caused by the discharging, built

Zhu, L., 2008, in IFIP International Federation for Information Processing, Volume 259; Computer and Computing Technologies in Agriculture, Vol. 2; Daoliang Li; (Boston: Springer), pp. 747–754.

higher dam in time and diver flood to reduce agricultural losses is an import work for water management department and agricultural bureau of Jiangsu province.

2. BACKGROUND

The prior agricultural flood losses prediction methods mainly include two types: statistic data are used to deduct the incoming losses, and satellite images in inundated region are used to analysis the losses problem (Fen Ping, 1995). In the first type of method, the detailed losses statistic data in the flood area is very hard to acquire for short of time. Hence, only those losses data in typical areas are collected, then total losses statistic data are deducted from these typical data. Obviously, the result data is not accurate and the losses prediction is unfaithful. In the second type of method, the submerged areas can be clearly seen in the remote sensing satellite images, which can be an accurate guidance for the forecasting procedures. This type of methods has been widely used in recent years for its accuracy and convenience. (Badji M.S., 1997).

However, the flood flow directions, flood volume and total discharging time is different in each time, it is very difficult to find an accurate method to deduct incoming losses from the past flood losses data. Some scholars have tried different methods, such as the fuzzy theory and neural network (Huang C.F., 1996; Wang Q.J., 2001). Nevertheless, these results are not very satisfactory.

3. FLOOD AGRICULTURAL LOSSES PREDICTION BASED ON DEM

In order to accurately simulate the flood and forecast waterlogging losses, the paper used a method based on Grid DEM data. DEM, shorted for Digital Elevation Model. Grid DEM consists of a matrix data structure with the topographic elevation of each pixel stored in a matrix node. It is readily available and simple to use and using the DEM data is convenient to accurately simulate the whole inundating process, get the submerged region and access the submerged depth. Furthermore, based on these calculated flood data, it can accurately predicate the agricultural losses.

The whole losses forecasting process includes three steps:

Step 1: Fundamental data preparation.
Step 2: Inundating process simulation
Step 3: Agricultural flood losses prediction

3.1 Fundamental data preparation

Three types of critical data are needed to forecast the agricultural flood losses: the first type is the DEM data in research area. The second one is the vector data geographically matched with the DEM data, including the administrative divisions, road traffic, farmland plots, fish ponds block, economic crops block, orchards block, dams, sewers, culverts, and other location-based information. The last type of data is agriculture, forestry, animal husbandry, fishery, and other socio-economic data, matched with those vector data, as well as the submerged statistical losses data in the past years. The DEM data is used to simulate the whole inundating process, the vector format data is used to calculate the concrete inundated block, and socio-economic data is used to calculate the losses of submerged.

3.2 Inundating process simulation

Inundating process also including three steps:

Step 1: Preprocessing the DEM data.
Step 2: Specifying flow path and pooled start point.
Step 3: Specifying inundated region.

3.2.1 Preprocessing the Grid DEM data

The normal depression is a low-lying plain area with lower in center and higher around, in which water can be stored and may be drowned. In Grid DEM data, the elevation in the depression is less than the elevation of the spots around it. However there always exist some spots in the real Grid DEM data with these characters because of data precision during data acquisition. They are pseudo-depression, not real depression, and should be modified. Or else it will disturb the analysis processing of flood flow directions.

To eliminate the impact of pseudo-depression, threshold judgment is a simple and valid method. The value of threshold can be set to 1–2 grid's area in practice. When the area of the depression region is less than this threshold, it can be considered as a pseudo-depression region. Then the average elevation value of spots around this pseudo-depression region is used to replace the current elevation values in the pseudo-depression region in order to eliminate the impact of these pseudo-depression regions.

3.2.2 Specifying flood flow path and pooled start point

The flood will not be pooled immediately after discharging, but flow downstream region along the floodways, until it reaches relatively opened low areas when pooling began. In this process, the crops in the region of flowing path will subject to certain losses, but it is completely different with those subject to waterlogging, thus it is necessary to analyze the flood flow directions, flow path according to surface topographic information and the pooled starting position.

Surface water is always along the steepest downward slope. Hence the simplest method for specifying flow directions is to assign flow from each spots in the grid DEM to one of its eight neighbors, either adjacent or diagonally, in the direction with steepest downward slope. This method, designated D8 (8 flow directions), was introduced by O'Callaghan and Mark (O'Callaghan, 1984) and has been widely used. In this method the finial flow direction can be decided according to the formulation 1:

$$\text{Direction} = (H - H1) - (H - H2)/1.414 \tag{1}$$

Where Direction represent the flow direction; H is the elevation of spot need to study; H1 is the lowest elevation spot of its eight neighbors in the vertical and heretical direction. H2 is the lowest elevation spot of its eight neighbors in the diagonal direction. If the Direction is negative, the flood will flow to the spot with H2 as its elevation, otherwise the flood will flow to the spot with H1 as its elevation. From the starting point, the similar calculation procedure should be taken for each spot along the flow direction, until the Direction is equal to zero.

In Huangdunhu detention area, because flooding disaster dues to the flood discharge in flood sluice, the starting point of runoff is the location of the sluice. Starting from this point, the flood flows from higher position to the lower position along to the elevation of terrain, which is the flood runoff. At the end point of flood runoff, the flood begins to pool, and then the point becomes the pooled start point of the total detention region.

3.2.3 Specifying inundated region

When inundate process begins, floods in low-lying areas begin pooling. As water continues to pool, the water level in the region gradually raises; eventually it ran across the brink of this low-lying region and flow to other low-lying areas, and then the water pools in the new regions once more. This process will repeat until no more water or no more regions accommodating for more water. Inundated simulation process requires large amounts of water mechanical model, which is a very complicated task, but for a plain region, we can use the following method to simplify the simulation process.

Starting from the pooling point, the flood diffuses to its 8 neighbors. The following code describes this diffusion procedure:

```
Do while no more water and no more area for water
    Presentwaterdepth = Presentwaterdepth + 1cm
    Push the present point into a stack
    Do
        Popup a point from the stack
        If the elevations of the point's 8 neighbors < Presentwaterdepth then
            Marks them with -999 'These points are submerged.
            Save them into a stack.
        Else
            Marks them with 999 'The point is a border point
        End if
    Loop until stack is empty.
Loop
```

When the simulation process begins, the program judges the submerged regions grids by grids until no more water or no more area for water. The storage capacity in submerged regions can be calculated by formulation 2.

$$SC = \sum_{1}^{N} waterdepth \times gridsize \tag{2}$$

SC is the storage capacity. ***waterdepth*** is the depth of water in inundated grid. ***gridsize*** is grid size of DEM data. The inundated region contains ***N*** grids.

When the inundate calculation process finished, all of the grids with marks ***-999*** will form a connected region, which is a candidate submerged region. Water come together in the new area is a relative slow and the water level raise also slowly. Hence we should check weather the candidate submerged area contains another depression or not, and do the same simulation in the new area if flood runs across the low-lying area's border. The following code describes this procedure:

```
    'Check the candidate region
If there has no more depression then
    Accept the result
Else
    Presentwaterdepth = the elevation of the new depression start point
    Restore the submerge situation under the last water level.
    Do the inundating process in the new depression
End if
```

In the flood discharge process, the flood discharge time and the flood flow are identified, and then the total flood volume can be deducted. Next the submerged area can also be deducted by this method. Furthermore, the time arrived in a position can be calculated by the flood flow and the flood runoff.

3.3 Agricultural flood losses prediction

Agricultural losses prediction process includes three parts: pre-disaster evaluation, floods direct losses rate determination and agricultural losses prediction. Determining the pre-disaster value is relatively simple, only the farming season and the crops types factors are important, which are easy to acquire. Flood mainly causes two types of loss to the agriculture, the impact and the waterlogging, while the impact losses could be neglected in the plain area compared with the waterlogging losses. The disaster losses are also different in different land used situation even in the same waterlogging situation. For example: waterlogging in a short time with low water level gives little damage to the trees, however it will create a great damage to the fishery. Hence the floods direct losses rate is not only close related with land use factor but also related with flood area, submerged depth, submerged time and some other factors.

Which farmland plots, fish ponds blocks, economic crops blocks, orchards blocks are drowned can be calculated by overlap analysis, a GIS method, in which several data layers are overlapped and new data layers or modification will make to the present layers. In practice, the drowned area in the DEM data are used to overlay with the farmland plots, the fish pond blocks and so on…, then new areas can be acquired, which means the drowned farmland plots, the drowned fish pond and so on…, and we also can get the submerged depth of a certain farmland block because the DEM data is geographically matched with these farmland.

The total flood agricultural loss can be get with the formulation 3

$$Loss = \sum_{1}^{N} LossRate \times Value \tag{3}$$

Loss is the total agricultural losses in this flood. ***LossRate*** is the floods direct losses rate of the special type crops. ***Value*** is the pre-disaster value of crops. The inundated region contains ***N*** grids.

4. EXAMPLE

In order to make flood losses prediction process more convenient, we developed a prediction system on the basis of ArcGIS. The entire implement and use of this system includes three stages:

In the first stage: Its main target is to build a socio-economic database and accomplish a prediction system using the gathered socio-economic survey data, spatial geographic information and flood direct losses rate data, which shows in a tabulation style.

Figure 1 is the main interface of the prediction system, in which the DEM data and socio-economic survey data are represented in three-dimensional style.

Fig. 1. Main interface of the prediction system

In the second stage: The user can set the parameter of the flood out process (including flood flow, the volume of flood) and watch the inundating process and finally get the submerged information.

Figure 2 shows the finial inundated areas in a full map model. The inundated area shows in yellow color and different color depth represents different water depth. The original terrain color is kept if it is not submerged.

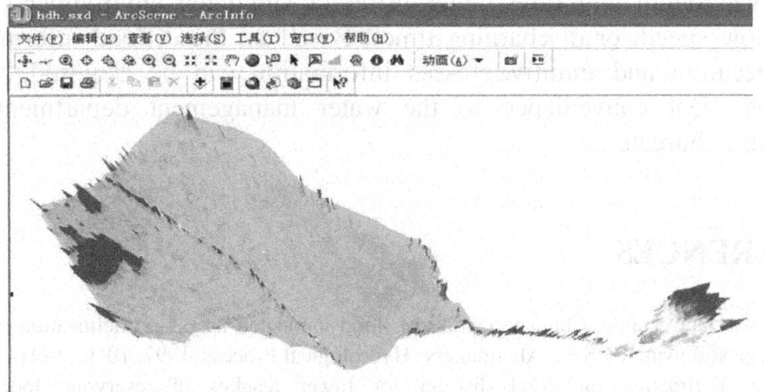

Fig. 2. The inundated area

In the third stage: The total flood agricultural loss can be calculated by using formulation 3, in which the floods direct losses rate is selected automatically from the direct losses rate tables according to the water depth and water logging time.

Figure 3 shows one of the flood losses prediction tables. In this table, which is sorted by the family, the losses of different crops (rice, cotton, corn, kaoliang and so on…) are clearly shown.

居民财产损失补偿统计

请选择查询年份 2003 格式：|2003|

请选择统计方式 户 查询

编 号	户名	水稻(亩)	玉米(亩)	高粱(亩)	豆类(亩)	小麦(亩)	红薯(亩)	棉花(亩)	芝麻(亩)	花生(亩)	蔬菜
2603	0	738	0	0	0	0	0	0	0	0	
2604	0	0	0	0	0	0	0	0	0	0	
2605	0	0	0	0	0	0	0	0	0	0	
2606	0	1107	0	0	0	0	0	0	0	0	
2607	0	369	0	0	0	0	0	0	0	0	
2608	0	369	0	0	0	0	0	0	0	0	
2609	0	738	0	0	0	0	0	0	0	0	
2610	0	369	0	0	0	0	0	0	0	0	
2611	0	369	0	0	0	0	0	0	0	0	
2612	0	0	0	0	0	0	0	0	0	0	
2613	0	738	0	0	0	0	0	0	0	0	
2614	0	738	0	0	0	0	0	0	0	0	
2615	0	0	0	0	0	0	0	0	0	0	
2616	0	738	0	0	0	0	0	0	0	0	
小计		152403.9	0	0	0	863155.98	0	0	0	0	

退 出

Fig. 3. Flood losses prediction statistics

5. CONCLUSION

The paper describes an agricultural losses prediction method based on digital elevation data used in Jiangsu province of China. The most important characteristic of this method is its capability of dynamic simulation. It can be used to not only simulate the flood inundating process but also acquire the submerged depth and time under different simulated environments (e.g. flood flow speed, or discharging time). Based on these flood information, more accurate and intuitive losses information can be acquired, which provides great convenience to the water management department and agricultural bureau.

REFERENCES

Badji M.S., Dautrebande, Characterization of flood inundated areas and delineation of poor drainage soil using ERS-1 SAR Imagery, Hydrological Process. 1997, 10(1): 1441–1450.

Fen Ping, Estimation on flood disaster for lower reaches of reservoir, Journal of Catastrophology, 1995, 10(1): 8–12.

Huang C.F. Fuzzy risk assessment of urban natural hazards, Fuzzy Sets and Systems, 1996, 83(1): 271–282.

O'Callaghan, J.F. and D.M. Mark, The Extraction of Drainage Networks From Digital Elevation Data, Computer Vision, Graphics and Image Processing, 1984, 28(2): 328–344.

Wang Q.J. Global Optimum Approximation of Feed Forward Artificial Neural Network for Taihu Flood Forecast, Journal of Hehai University, 2001, 30(2): 84–900.

EFFECT OF IMAGE PROCESSING OF A LEAF PHOTOGRAPH ON THE CALCULATED FRACTAL DIMENSION OF LEAF VEINS

Yun Kong [1], Shaohui Wang [1], Chengwei Ma [2], Baoming Li [2], Yuncong Yao [1,*]

[1] *Department of Plant Science & Technology, Beijing Agricultural College, Beijing, 102206, China*

[2] *Key Laboratory of Bioenvironmental Engineering of Ministry of Agriculture, College of Hydraulic and Civil Engineering, China Agricultural University, Beijing, 100083, China*

[*] *Corresponding Author, Address: Department of Plant Science & Technology, Beijing Agricultural College, Beijing, 102206, China, Tel: +86-10-80799000, Fax: +86-10-80799000, Email: ky0257@126.com*

Abstract: Digital photography is a promised method for estimating the fractal characteristics of leaf veins. In this study, the effects of different threshold levels and image processing methods using Adobe Photoshop software on the fractal dimension values were examined from a digital photo of nectarine leaf. The results showed that the nectarine leaf vein has typical fractal characteristics and its fractal dimension increased linearly with the increasing levels of threshold. A larger value of fractal dimension was calculated from the image processed by dark strokes, and a smaller value by accented edges or desaturate. A positive relationship was found between the calculated values of fractal dimension and the black coverage in the image. Therefore, it should be cautious to choose threshold levels and image processing methods when processing a digital image for calculating the fractal characteristics of leaf veins.

Keywords: image processing, threshold, leaf vein, fractal dimension

1. INTRODUCTION

Veins of some leaves have typical fractal characteristics. Their box-counting dimension calculated by some fractal software can be used for

Kong, Y., Wang, S., Ma, C., Li, B. and Yao, Y., 2008, in IFIP International Federation for Information Processing, Volume 259; Computer and Computing Technologies in Agriculture, Vol. 2; Daoliang Li; (Boston: Springer), pp. 755–760.

plant classification (Liu *et al.*, 2005; Sasaki *et al.*, 1994). But it took much time to prepare the leaf samples used for scanning photos of leaf veins, because of the difficulty of retaining only leaf veins and removing other leaf parts. With the development of digital camera and image processing, a clear digital photo of leaf can be non-destructively obtained by some image processing software such as Adobe Photoshop after capturing it in the filed. Then the processed photo can be directly used for fractal analysis of leaf veins. However, the information is meager for the effect of image processing method on the calculated result of fractal dimension of leaf vein.

The objective of the paper is to study how fractal dimension values change with the different threshold levels, and the different image processing methods.

2. MATERIAL AND METHOD

A color photograph of lower side of leaf in a nectarine tree (Fig. 1), which was grown inside a chinese type lean-to greenhouse in the horticultural experiment station at Beijing Agricultural College, was taken in a still and overcast sky condition using a digital camera (DSC-F717, Sony Corporation, Japan) on May 12th, 2006. The effective pixel count was 5.0 million pixels. An automatic setting for the aperture width and shutter speed was used. The characteristics of the image were 2048×1536 Pixel in resolution, 1.20M in size and JPG file in type.

The photograph was downloaded directly from the digital camera to a personal computer and then was analyzed by the following steps of image-processing using Adobe Photoshop software for Windows (Adobe Systems Inc., San José, CA, USA). Step 1, with the newly opened image visible in a Adobe Photoshop window, select **Image>Adjust> Invert,** and then choose **Image>Adjust> Threshold** from the menu bar. The threshold was set at seven values ranging from 85–115 every five levels. Then, click on OK to produce 7 different quality binary images and saved them as BMP files (Fig. 2). Step 2, after opening one of the above saved files with the threshold

Figure 1. A color digital photograph of lower side of nectarine leaf

level of 100, select **Filter> Brush Strokes > Accented Edges**, or **> Dark Strokes** and the save them as two different processed images (Fig. 3). Step 3, after opening and inverting the photo, follow the Step 1, choose **Image>Adjust>Desaturate or Filter> Brush Strokes> Dark Strokes**, and then select **Image>Adjust> Threshold,** and set the level at 100. Click on OK to produce 2 different-quality binary images (Fig. 4).

All of the above black-white photos were then be used for calculating fractal dimension of leaf veins by a Fractal Analysis Software (Sasaki *et al.*, 1994), developed by Hiroyuki Sasaki (Grassland Subteam for Integrated Soil Fertility Management, National Institute of Livestock and Grassland Science, Nasushiobara, Tochigi, Japan). The detailed procedure is described in its user manual.

3. RESULT

As found in Fig. 2, with the increment in the levels of threshold, the minor leaf veins became clear and at the same time the edges of some leaf veins became obscure.

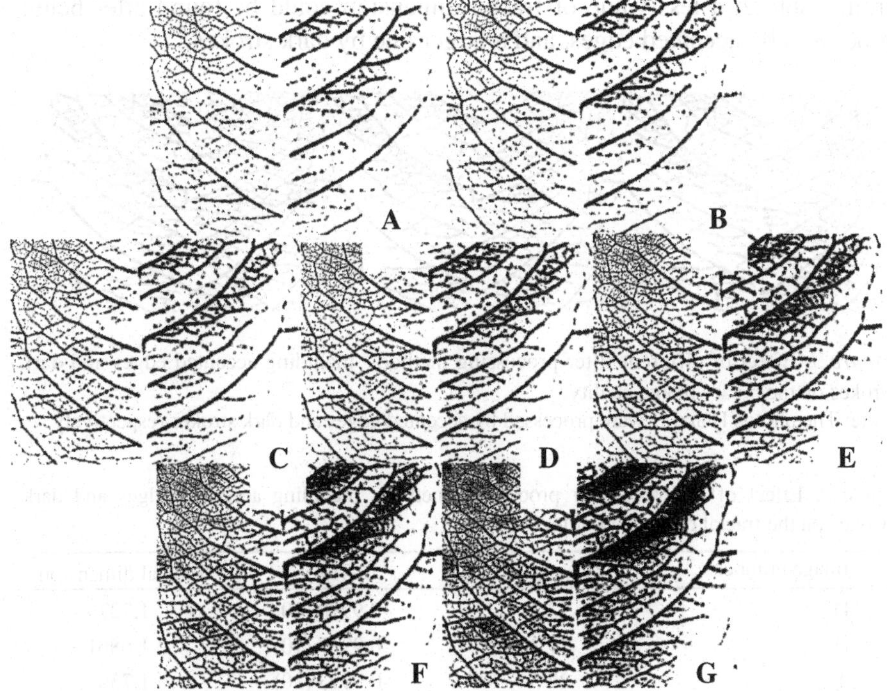

Figure 2. Effect of different threshold levels on the binary image quality
Note: The threshold levels of photo A-D were set at 85, 90, 95, 100, 105, 110, 115 respectively

The nectarine leaf veins have typical fractal characteristics due to a large correlation coefficient of above 0.99(Table 1). In addition, the fractal dimension presented a linear increment with the increasing levels of threshold. The regressed equation was obtained as follows: y=0.0086x + 0.8617 (R^2 = 1).

Table 1. Effect of different threshold levels on the fractal dimension of leaf veins

Threshold level	Cover (%)	R(n)	Fractal dimension
85	11.7	-0.9983 (10)	1.5966
90	14.8	-0.9985 (10)	1.6394
95	18.6	-0.9988 (10)	1.6821
100	23.3	-0.9990 (10)	1.727
105	29	-0.9993 (10)	1.7695
110	36	-0.9995 (10)	1.8124
115	45	-0.9997 (10)	1.8555

Note: cover is the percent of black coverage in the processed image, R, the correlation coefficient, and n, the number of data to calculate.

After the binary images were processed by accented edges and dark strokes, it was difficult to find the significant changes by eye (Fig. 3). But from Table 2, a less value of fractal dimension could be found after being processed by accented edges, and a larger one by dark strokes.

Figure 3. Effect of different filter processing methods including accented edges and dark strokes on the binary image quality

Note: The photos H and D were processed by accented edges and dark strokes respectively

Table 2. Effect of different filter processing methods including accented edges and dark strokes on the fractal dimension of leaf veins

Image number	Cover (%)	R(n)	Fractal dimension
D	23.3	-0.9990 (10)	1.727
H	19.2	-0.9987 (10)	1.6861
I	27.3	-0.9993 (10)	1.756

Before being converted into a binary image, the photo was processed by desaturate and dark strokes produced two different-quality white-black images respectively although both threshold levels were set at 100 (Fig. 4). The former had a remarkably less black coverage, while the latter had a significantly more one. Associated with this, as seen in Table 3, the photo was processed by desaturate before being converted into a binary image had a less value of fractal dimension, and it had been done by dark strokes with a larger value of that.

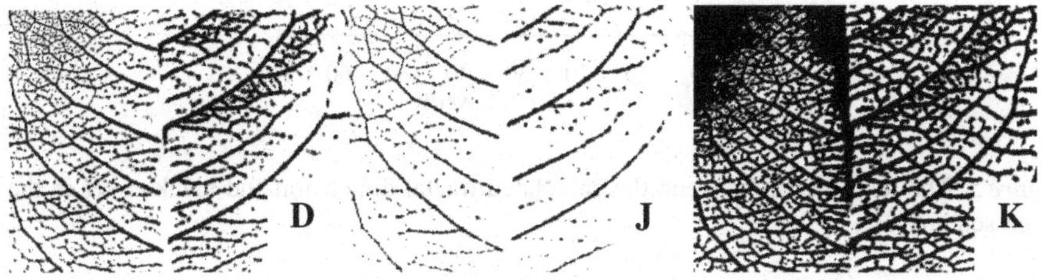

Figure 4. Effect of processing methods of desaturate and dark strokes before converting into binary image on the ultimate image quality
Note: The photos J and K were processed by desaturate and dark strokes respectively

Table 3. Effect of processing methods of desaturate and dark strokes before converting into binary image on the fractal dimension of leaf vein

Image number	Cover (%)	R(n)	Fractal dimension
D	23.3	-0.9990 (10)	1.727
J	7.2	-0.9977 (10)	1.5007
K	57.9	-0.9998 (10)	1.8987

4. DISCUSSION

From all the tables above, it could be found that larger values of fractal dimension were associated with the increment of black coverage in the image. There was a logarithmic relationship between the fractal dimension value and the percentage of the black coverage (Fig. 5).

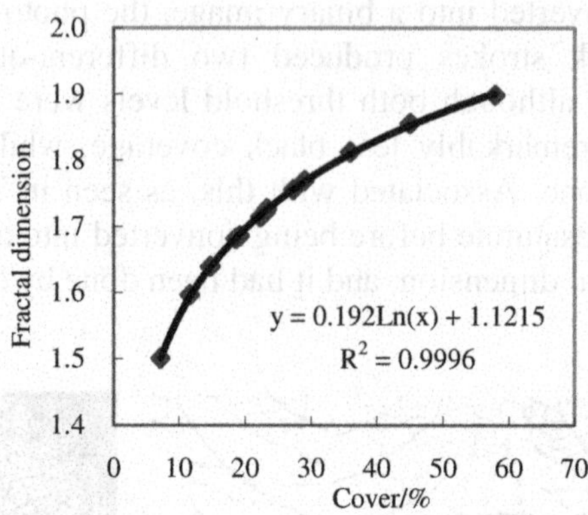

Figure 5. The relationship between the calculated fractal dimension and black cover in the processed images

5. CONCLUSION

Different threshold levels and different image processing methods produced different black coverage in the image and then resulted in different values of fractal dimension of leaf veins. Therefore, when calculating the fractal dimension, the images should be processed by the same methods and the same threshold levels in order to compare the fractal dimension of different images.

ACKNOWLEDGEMENTS

The authors are indebted to the cooperation program between Beijing Education Committee and China Agricultural University (XK100190553) for the financial support.

REFERENCES

Adobe Systems Inc. Adobe Photoshop version 6.0. San Jose, CA. 2001.

Liu T., Zou X., Zhao T., *et al.* Fractal characteristics of leaves of *Osynanthus fragrans* lour. Journal of Wuhan Botanical Research, 2005, 23(2):199–202.

Sasaki H., S. Shibata, T. Hatanaka. An evaluation method of ecotypes of Japanese lawn grass (*Zoysia japonica* STEUD.) for three different ecological functions. Bull. Natl. Grassl. Res. Inst. 1994, 49:17–24.

CYLINDER HEAD FEM ANALYSIS AND ITS IMPROVEMENT

Shixiong Li[1,*], Jinlong Mao[1], Shumao Wang[1]
[1] *College of Engineering, China Agricultural University, Beijing, China, 100083*
* *Corresponding author, Address: P.O. Box 114, College of Engineering, China Agricultural University, 17 Tsinghua East Road, Beijing, 100083, P. R. China, Tel: +86-10-62736460, Fax: +86-10-62736385, Email: li_shixiong@263.net, lsx@cau.edu.cn*

Abstract: The nose bridge between the inlet valve and exhaust valve is the most fragile part on the cylinder head. Fatigue crack is the major failure mode. It is necessary to analyze the structure strength and reliability (SSR) of this zone. In this paper, a three-dimensional (3D) model for a engine cylinder head is built up using Pro/Engineer wildfire 2.0, and the 3D FEM analysis of the SSR is carried out. The temperature field and stress field of the nose bridge are also calculated based on the model. Several improvement designs are given and compared as the result of the above analysis, which lays the foundation for the cylinder head production.

Keywords: engine, cylinder head, nose bridge zone, SSR, FEM, design improvement

1. INTRODUCTION

Computer technology has been widely used in the agricultural area, especially computer simulations, which greatly promote engineering analysis. With the improvement of the computer's capability, the finite element method (FEM) is considered to be one of the most powerful design tools in Computer-Aided Engineering (CAE) (Liu, 2004). Using this method, the complex structural configuration can be modeled and the response at any point of the structure can be easily determined (Cyuan, 2000). Now, the finite element method has been evolving to be a widely accepted tool for the solution of pragmatic engineering problems (Liang, 2004).

Li, S., Mao, J. and Wang, S., 2008, in IFIP International Federation for Information Processing, Volume 259; Computer and Computing Technologies in Agriculture, Vol. 2; Daoliang Li; (Boston: Springer), pp. 761–768.

The cylinder head is one of the most complex parts in the engine, and it endures high thermal and mechanical load during working cycles (Zhang, 2002). The nose bridge on the head between the inlet valve and exhaust valve is subjected to the hot air through the exhaust valve and the cool air through the inlet valve simultaneously, and it is the weakest part in the cylinder head. The fatigue in the nose zone is one of the most important factors that shorten the life of a cylinder head. So it is very meaningful to find a method to weaken restriction of the heat distortion in the nose zone, especially in the preliminary design stage, the using of computer simulation to analyze the strength and fatigue life of a cylinder head is very important and necessary. In this paper, a successful and feasible numerical model was configured first and analyzed thoroughly, and then with the result got from the analysis, several ameliorative models are analyzed based on finite element method.

2. MODELLING AND COMPARISION ANALYSIS

Due to the complexity of cylinder head in geometrical structure and the heat load, the Pro/Engineer Wildfire 2.0 is selected as the three-dimension solid modeling tool. One cylinder area on the cylinder head of diesel engine 4105Q was modeled with it and the model was transferred to the three-dimensional finite element analysis software ANSYS 9.0 to do the further analysis.

2.1 The model of the cylinder head

The three-dimensional model created by Pro/Engineer Wildfire 2.0 (Fig. 1) made some simplify for some structures which have no large effect on the temperature and stress field distribution, such as bolt holes, pin holes and ribs etc. But there are no simplify on the major structure dimensions and shapes in order to have an accuracy correct analysis result.

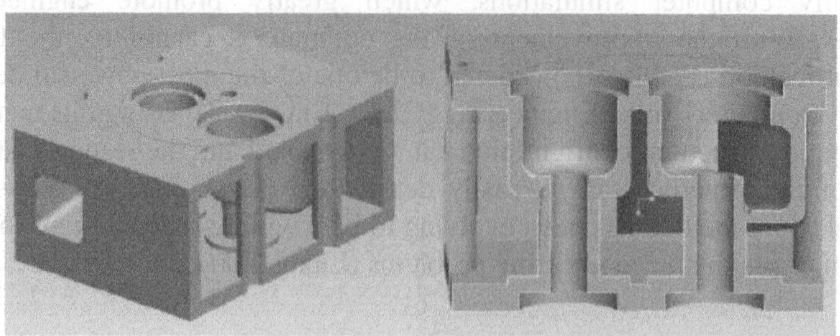

Fig. 1. 3D cylinder head model

When the model is completed, save it to the file of .igs and transfer it to the finite element analysis (FEA) software ANSYS 9.0. In the FEA software, after defining the properties of the material and the elements, the model is meshed by using automatic mesh tool. Fig. 2 show the meshed model, which contains 47310 elements and 84498 nodes.

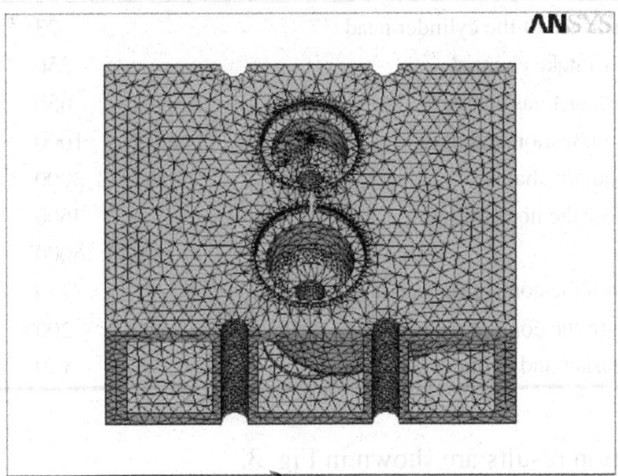

Fig. 2. The meshed model of the cylinder head

2.2 The comparison analysis of the original and the improved designs

The main objective of this study is to provide information on the structure and thermal stress of the cylinder head nose bridge and provide feasible solutions on the structure design improvement.

2.2.1 The temperature and stress field calculation of the original design

In order to calculate the temperature and stress distributions in the cylinder head, the third-type boundary condition should be defined first and then as the thermal load applied on the FEA model.

The third-type boundary condition:

$$k_x \frac{\partial T}{\partial x} n_x + k_y \frac{\partial T}{\partial y} n_y + k_z \frac{\partial T}{\partial x} n_z = \alpha \left(T_\infty - T\right) \tag{1}$$

where:

α: convection coefficient;

k: heat conduction coefficient;

T∞: the liquid temperature;

T: the temperature on the part surface;

nx, ny, nz: the direction cosine of the normal to the boundary.

The boundaries are defined through many modification and calculation using the established experiential equation and values shown in Table 1.

Table 1. The Experiential Values of α and T

Location	*α -W/(m^2.K)*	T
Upper and side surface of the cylinder head	23	293K
The surface of air intake channel	350	335K
The surface of exhaust gas channel	650	973K
The surface of combustion chamber	1000	1200K
The surface of coolant channel	3000	353K
Immediately below the nose bridge	3600	383K
Forced cooling	6000	383K
Cylinder head with the contact surface of the inlet valve seat	150	665K
Cylinder head with the contact surface of the exhaust valve seat	200	803K
Cylinder head, gasket and cylinder block	100	503K

The calculation results are shown in Fig. 3.

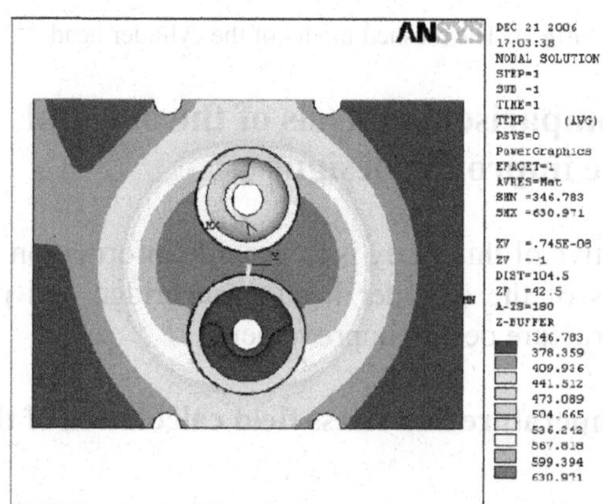

Fig. 3. Temperature distribution of the original design

In order to examine the plausibility of the boundary selection, the calculation results of the temperature distribution and the experimental results of Type Z6110 diesel engine are compared, as shown in Table 2.

Table 2. The Comparison between Calculation and Experimental Temperatures

Points	1	2	3	4	5	6	7	8	9
Measurement	483	424	518	507	558	605	613	471	430
Calculation	496	430	567	552	595	620	631	536	433
Errors (%)	2.7	1.4	9.5	8.9	6.6	2.5	2.8	13.8	0.7

The units for Temperatures are K.

Table 2 show the overall temperature distribution is coherent, which means the boundary condition selection is reasonable.

The thermal stress distribution is achieved through automatic data coupling calculation based on the temperature distribution calculation results. The calculation results of the thermal stress distribution are shown in Fig. 4.

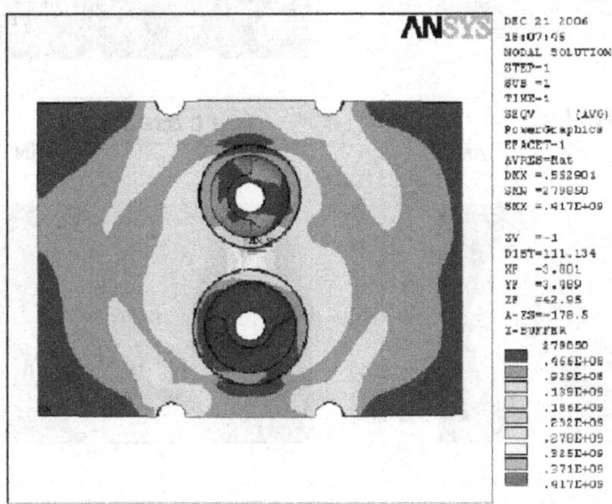

Fig. 4. Thermal stress distribution of the original design

Fig. 3 & 4 show that the max data occurs on the exhaust valve side of the nose bridge which need to find ways to lower its temperature and thermal stress.

2.2.2 The temperature and stress field calculation of the improved designs

In order to lower the temperature and thermal stress of the nose bridge zone, decrease the restriction of heat distortion, several models with different structure are provided.

Case1. Reduce the thickness of the nose bridge to 8 mm;

Case2. Machining a deep and narrow arc groove (unload groove) with about 0.5mm width in the nose bridge;

Case3. Machining a 3 mm diameter cooling water hole to improve the cooling condition in the nose bridge;

Case4. Casting a coolant channel whose outlet is opposite to the nose bridge.

The temperature distribution calculation results are shown in Fig. 5

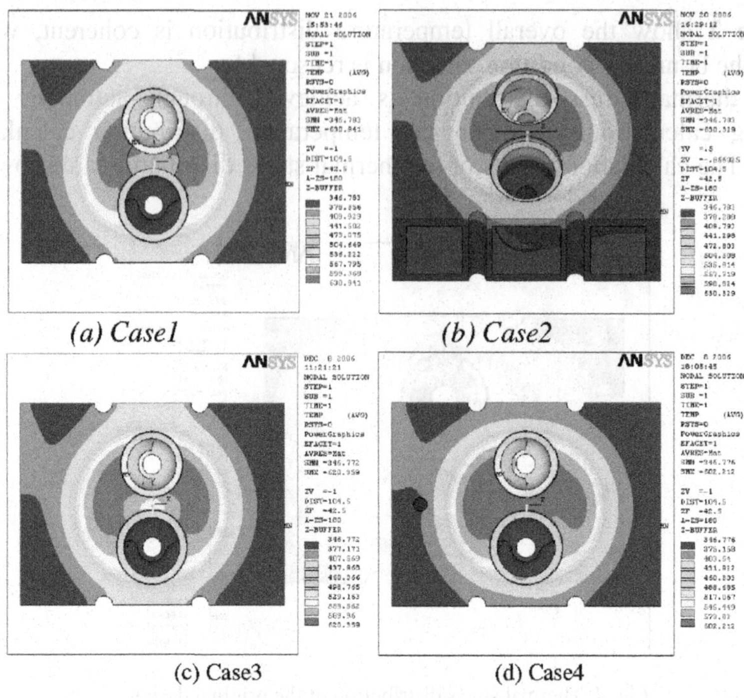

(a) Case1 (b) Case2

(c) Case3 (d) Case4

Fig. 5. The temperature distribution

The thermal stress distribution calculation results are shown in Fig. 6.

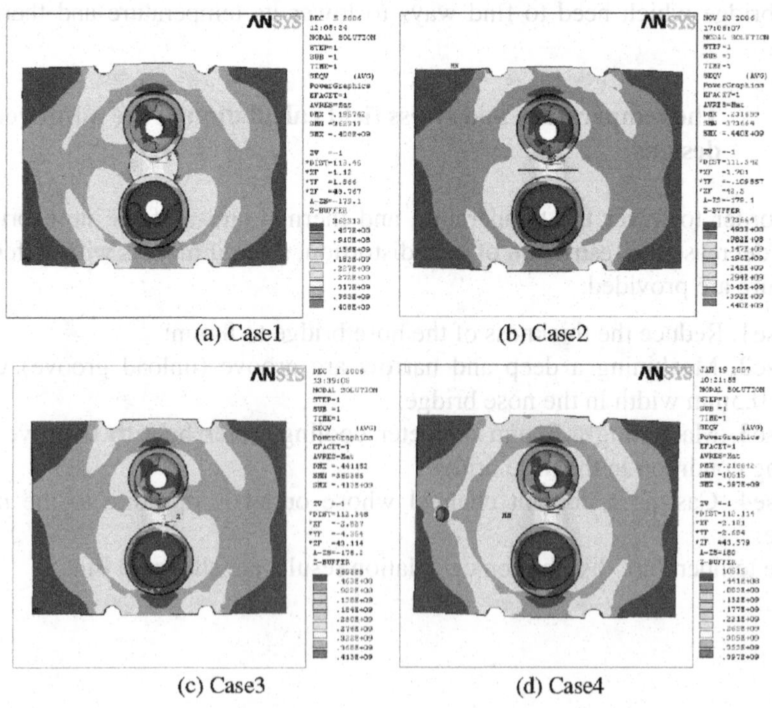

(a) Case1 (b) Case2

(c) Case3 (d) Case4

Fig. 6. The thermal stress distribution

2.3 Comparison

The comparison of the four improved design solutions' temperature and thermal stress distribution calculation results are shown in table 3 and table 4. (Note: A is the original case)

Table 3. The Highest Temperature Distribution Comparisons in Nose Bridge Zone

Cases	A	1	2	3	4
The Highest Temperature (K)	630	610.4	630	590	602

Table 4. The Highest Von Mises Thermal Stress Distribution Comparisons in Nose Bridge Zone

Cases	A	1	2	3	4
The Highest Thermal Stress (MPa)	417	408	440	413	397

3. RESULTS AND DISCUSSION

Case1. Reducing the thickness of nose bridge improves the heat exchange condition in the nose zone, lowering the nose temperature about 20K and the thermal stress about 9Mpa.

Case2. Machining a unload groove have no change to the temperature distribution, while make the thermal stress increased about 23Mpa.

Case3. Machining a coolant holes in the nose bridge effectively lower the nose temperature 40K, and at the same time lower the thermal stress about 4Mpa.

Case4. Connect a coolant injector against the nose bridge is the best way to improve the temperature and thermal stress distributions in these solutions. It not only makes the temperature lower 28K, but also makes the thermal stress lower 20Mpa. It has been already found its application on some engines.

4. CONCLUSIONS AND FUTURE WORKS

From the above analysis, the main outcomes can be outlined as follows:

(1) For the nose bridge of cylinder head, an elaborate temperature field and thermal stress analysis was carried out using ANSYS 9.0. Results from the research could be provided to the designers to be as the reference in their work of improving the fatigue life of the cylinder head.

(2) Just as what have been calculated, in all the cases the nose bridge between the inlet valve and exhaust valve are found to endure the highest temperature and highest thermal load. And the thermal load is dynamic and has a very high rate of repetition, causing the nose bridge to be the most fragile part in the cylinder head. We can know from the results that remove some metal and cool this zone with coolant will effectively lower the temperature and thermal stress.

(3) Because the configuration of cylinder head is too complex to predict the critical area and failure mode in the design phase, the finite element simulation have become the best method to obtain the stress distribution with different structures, this can greatly reduce project span time and total project cost.

ACKNOWLEDGEMENTS

This study is supported by China National 11th 5-Year Planned Project 'The Research on Rice and Wheat Cross-area Harvest Mechanization Technology' (Contract Number: 2006BAD28B03).

REFERENCES

Liu Jinxiang, LiaoRi dong, Zhang You. Finite Element Analysis for Cylinder Head of 6114 Diesel Engine. Transaction of CSICE, 2004, 22 (4): 367–372

Shiang-Woei Cyuan. Finite element simulation of a twin-cam 16-valve cylinder structure. Finite elements in Analysis and Design, 2000, (35): 199–212

S.L. Liang, X.H. Dai, H.M. Yao. 3D FE Analysis of Cylinder Head for Diesel Engine. Agricultural Mechanical Paper, 2004, 35 (3): 45–48

Zhang Weizheng, Zhang Guohua, Guo Liangping. The Thermal Fatigue Test and Modification of High-Temperature Creep for cast Iron Cylinder Heads. Transaction of CSICE, 2002, 23 (6): 67–69

AUTOMATIC NAVIGATION SYSTEM WITH MULTIPLE SENSORS

Xiaojun Wan, Gang Liu *

Key Laboratory of Modern Precision Agriculture System Integration Research, China Agricultural University, P.O. Box 125, Qinghuadonglu 17, Haidian District, Beijing, 100083, P. R. China

** Corresponding author, Address: P.O. Box 125, China Agricultural University, Qinghua Donglu 17, Haidian District, Beijing, 100083, P. R. China, Tel: +86-10-62736741, Fax: +86-10-62736746, Email: pac@cau.edu.cn*

Abstract: An automatic navigation system with multiple sensors including a vision sensor, RTK GPS and an electronic-compass was developed. Data from the sensors were fused by a feedback algorithm. And the steering angle was calculated based on the fuzzy logic module. A steering controller was developed to control steering using guidance information obtained from the sensors. A golf-car was taken as the research platform which was equipped with an angle sensor, an accelerometer and a steering controller. The row-crop was selected as the interested crop. Navigation of the vehicle in the farmland could be realized by the system.

Keywords: GPS, machine vision, navigation, sensor fusion, PID, steering controller

1. INTRODUCTION

Over the past centuries, many changes have occurred on agricultural field production machinery. With the increasing working speed and vehicle size, an operator has to devote more energy to operate the vehicle (Wilson, 2000).

Wan, X. and Liu, G., 2008, in IFIP International Federation for Information Processing, Volume 259; Computer and Computing Technologies in Agriculture, Vol. 2; Daoliang Li; (Boston: Springer), pp. 769–776.

Fatigue impedes the operator to operate the vehicle, especially during the tight constraints of harvest. The operator becomes one of the greatest limitations to the increasing vehicle performance. Therefore, developing automatic navigation system for agricultural vehicles is essential.

Automatic navigation systems attempt to relieve the operator from many, if not all, of the tasks involved in navigating agricultural vehicles. Automatic navigation has potential to reduce operator fatigue and improve positioning accuracy of the vehicle (Gerrish et al., 1997). A number of agricultural navigation techniques have been developed or demonstrated in recent years, such as machine vision, GPS, inertial positioning system, radar, ultrosonic, etc. But generally, a single sensor system can supply only parts of the environment information. The precision of the navigation is sometimes unbelievable. Therefore the navigation system with sensor fusing has become more and more popular in recent years. And the combination of machine vision and GPS becomes most popular (Wilson, 2000).

Automatic navigation system should be able to detect posture of the vehicle, calculate proper steering angle, and steer the vehicle according to the angle. The posture is the position and orientation of the vehicle (Kanayama and Hartman, 1989). Different automatic navigation systems have been developed for agricultural vehicles during the past several decades. Although each system uses different techniques to navigate the vehicle, most of the systems generally need the same navigation parameters, including heading angle and offset, to control steering. Heading angle is the angle between the vehicle centerline and the desired path, and offset is the displacement of the vehicle central mass off the desired path. More precise navigation parameters could be obtained by fusing the data acquired from different sensors. The fuzzy logic module was used to calculate a steering angle. Finally, a steering controller was installed on the vehicle to steer the vehicle following the desired path automatically.

This paper presented a solution to challenging problems for automatically navigated agricultural vehicles with multiple sensors. This solution included obtaining navigation parameters with CCD, GPS and electronic-compass, fusing multi-sensor data for real-time vehicle navigation, calculating the steering angle and steering the vehicle following the desired path with a steering controller. A feedforward-plus-PID steering control algorithm was used to compensate the variation in the vehicle dynamics for ensuring a satisfactory steering performance. A golf-car installed with an angle sensor, an accelerometer and a steering controller was taken as the research platform.

The configuration of the automatic navigation system was shown in Fig. 1.

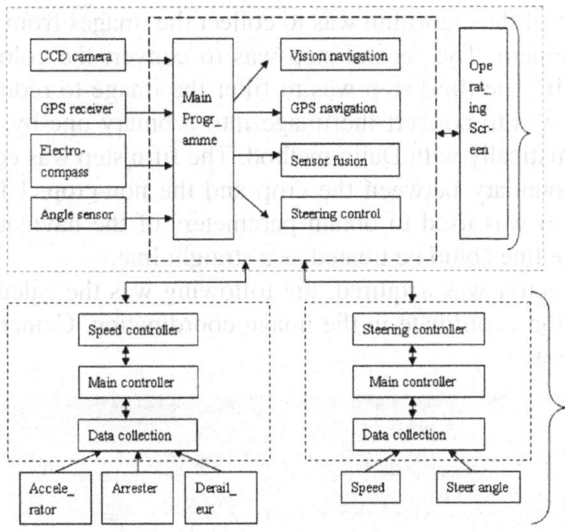

Fig. 1. The configuration of the automatic navigation system

2. MULTIPLE SENSORS AND SENSOR FUSION

2.1 Obtaining navigation parameters based on machine vision

Machine vision can be used to navigate agricultural vehicles automatically when the row structure is distinguishable in the farmland. It has the technological characteristics closely resembling those possessed by a human operator, and thus has great potential for implementation of the vehicle navigation system (Wilson, 2000). The flowchart to obtain the navigation directrix was stepped in Fig. 2.

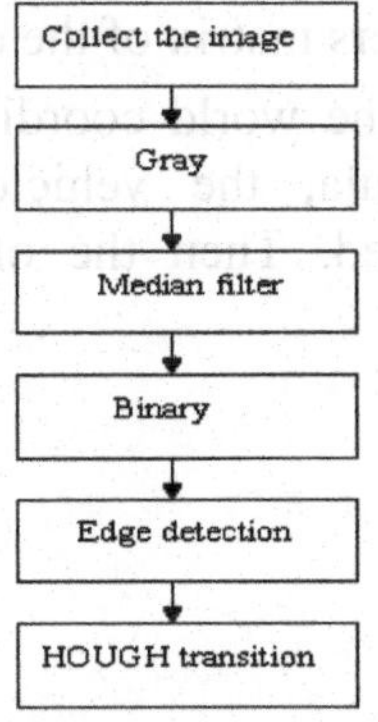

Fig. 2. The flowchart to obtain navigation directrix

The first step of the algorithm was to collect the images from the farmland by the CCD camera. The second step was to convert the color image into gray by 2G-R-B. The third step was to filter the image to reduce the noise. The forth step was to convert the image into a binary one by selecting the threshold automatically with Dajin method. The fifth step was edge detection to detect the boundary between the crop and the non-crop. Ultimately, the Hough transition was used to obtain parameters of the navigation directrix for the guidance line could be treated as a straight line.

Once the directrix was acquired, the following was the calculation of the vehicle projection coordinate in the image coordination. Camera calibration was shown in Fig. 3.

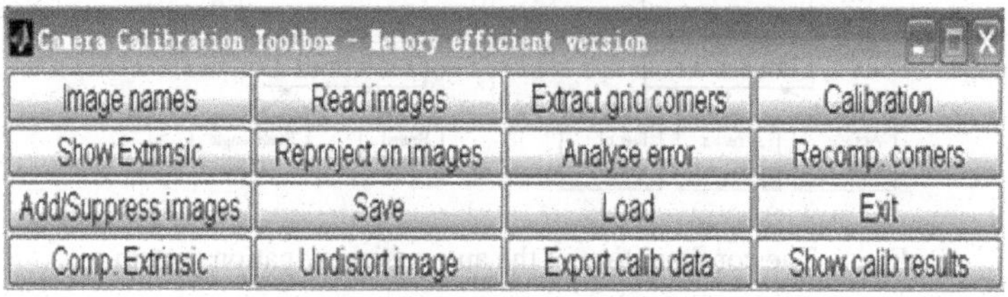

Camera Calibration Toolbox - Memory efficient version

Image names	Read images	Extract grid corners	Calibration
Show Extrinsic	Reproject on images	Analyse error	Recomp. corners
Add/Suppress images	Save	Load	Exit
Comp. Extrinsic	Undistort image	Export calib data	Show calib results

Fig. 3. The calibration toolbox of the camera

The projection relationship between the world coordination and the image coordination could be found.

$$s\begin{bmatrix} u \\ v \\ 1 \end{bmatrix} = M_1 \times M_2 \times \begin{bmatrix} X_W \\ Y_W \\ Z_W \\ 1 \end{bmatrix} \tag{1}$$

where $\begin{bmatrix} u & v \end{bmatrix}^T$ is the coordinate in the image coordination.

$\begin{bmatrix} X_W & Y_W & Z_W \end{bmatrix}^T$ is the coordinate in the world coordination.

M_1 is the internal parameters matrix of the camera.

M_2 is the external parameters matrix of the camera.

The vehicle coordinate in the world coordination is (0, 0, 0), using the projection relationship formula, the vehicle coordinate in the image coordination can be calculated. Then the offset and heading angle are obtained.

$$d = \frac{|Ax_0 + By_0 + C|}{\sqrt{A^2 + B^2}} \tag{2}$$

$$\theta = a\tan\frac{(x_2 - x_1)}{(y_2 - y_1)} \tag{3}$$

where (x_0, y_0) is the vehicle coordinate in the image coordination.

A, B, C are the beeline equation coefficient of the directrix.
(x_1, y_1) is the begin point coordinate of the directrix.
(x_2, y_2) is the end point coordinate of the directrix.

2.2 Obtaining navigation parameters based on GPS

GPS-based navigation technique can be used in many farmland operations such as tillage, planting, cultivating, harvesting, etc. It has the potential to achieve completely autonomic navigation. The flowchart to obtain navigation parameters by GPS was shown in Fig. 4.

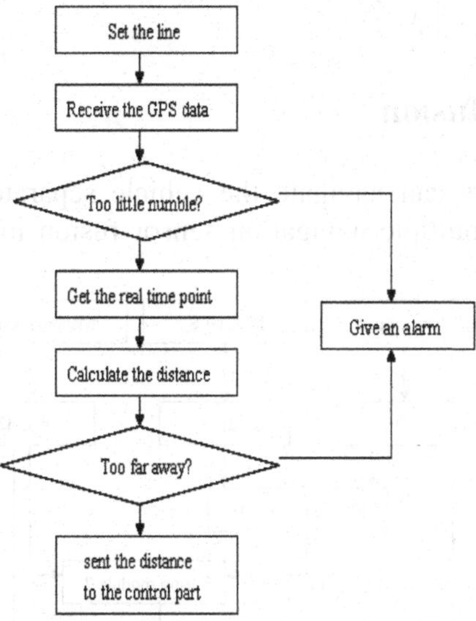

Fig. 4. The flowchart to obtain navigation parameters

Firstly, to set the end point and the begin point of the row. Based on the method that two points determine one line, the row line was got. GPS was connected to the computer through RS-232, received the associated data from satellites and abstracted the longitude and latitude from the data. By coordination transition, X and Y coordinates of the vehicle in the 54-coordination were got. Once the real time position of the vehicle was got, the distance between the row and the vehicle could be calculated. An alarm signal was set when the vehicle going far away from the row, and the alarm distance could be set manually.

To integrate the sensors into a sensor fusion module required unifying coordinate systems of all the sensors. The GPS provided position information of the vehicle referring to the global coordination. The machine vision provided information in the image coordination. To make it compatible to the directrix information obtained from the vision sensor, it was necessary to convert the global coordinates to the image coordinates as well. However, it was necessary to correct the error caused by the vehicle inclination before mapping it to the vehicle coordination.

$$\begin{bmatrix} x_c \\ y_c \\ z_c \end{bmatrix} = \begin{bmatrix} x_a \\ y_a \\ z_a \end{bmatrix} + \begin{bmatrix} a_{11} & a_{12} & a_{13} \\ a_{21} & a_{22} & a_{23} \\ a_{31} & a_{32} & a_{33} \end{bmatrix} \begin{bmatrix} a \\ b \\ h \end{bmatrix} \tag{4}$$

When the variation in the farmland elevation was limited, the 2-D GPS coordinates could be converted into the image coordinates.

$$\begin{bmatrix} x^g{}_v \\ y^g{}_v \end{bmatrix} = \begin{bmatrix} b_{11} & b_{12} \\ b_{21} & b_{22} \end{bmatrix} \begin{bmatrix} x_c \\ y_c \end{bmatrix} \tag{5}$$

2.3 Sensor fusion

The three sensors can navigate the vehicle separately or together. The block diagram of multiple navigation sensor fusion module was shown in Fig. 5.

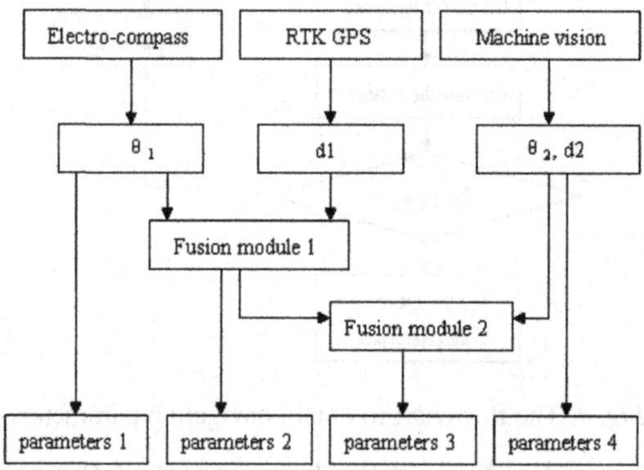

Fig. 5. The block diagram of multiple navigation sensor fuision module

The sensor fusion module evaluated the information acquired from different sensors and selected the chief navigation sensor using a rule-based method. With a corresponding fusion algorithm, more appropriate heading angle and offset information could be obtained in real-time.

Fusion algorithm based on feedback was presented in fusion module 2 and the precision still needed to be validated. The following was the fusion formula, which was mainly based on the fuzzy logic fusion.

$$O_c = M_v \times E_v + M_g \times E_g \tag{6}$$

$$M_v + M_g = I \tag{7}$$

$$M_v / M_g = C_g / C_v \tag{8}$$

where O_C is the output of the fusion module, $[D \quad \theta]^T$.

M_v is the coefficient matrix of machine vision parameters.

M_g is the coefficient matrix of fusion module 1 parameters.

E_v is the matrix of machine vision parameters.

E_g is the matrix of fusion module 1 parameters.

I is a identify matrix.

C_v is the average error of navigation with only vision sensor.

C_g is the average error of navigation with GPS and electronic-compass.

3. STEERING CONTROL OF THE VEHICLE

The steering of the vehicle was performed on front-wheel. Therefore the front-wheel angle should be acquired to steer the vehicle. Agricultural vehicles were complicated objects with the parameter variation, environment uncertainty, nonlinear model and so on. However, the fuzzy logic method was presented to decide the front-wheel angle. The steering control algorithm was shown in Fig. 6.

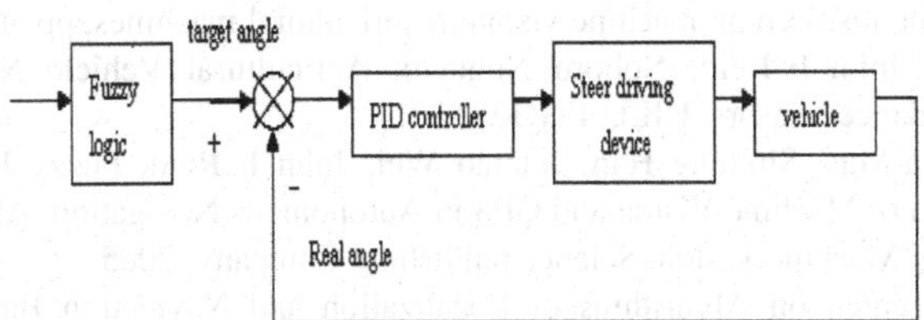

Fig. 6. The steering control algorithm

The input of the fuzzy logic module was the heading angle and offset calculated by the fusion module, the output was the target angle to steer the vehicle. The error between the target angle and the real angle detected by the angle sensor, which was installed on the wheel, was the input of the PID controller. Finally, the steer driving device controlled the vehicle navigation.

4. CONCLUSION

An automatic navigation system based on vision sensor, GPS and electronic-compass was developed for agricultural vehicles. The system offered an algorithm to track the directrix between the farm-crop and the non-farm crop. The fusion of the three sensors, vision sensor, GPS and electronic-compass could offer the posture of the vehicle in real-time. The fuzzy logic module was a component to calculate the steering angle and the steering controller realized the function of steering with the PID algorithm. The camera was mounted on the cab of the vehicle at operator eye level, and this was proved to be a better algorithm now. The system could navigate the vehicle in a straight line, and through access database the result of different navigation algorithm could be recorded. The redisplaying of the trace could compare the precision of different navigation methods and was used to modify the coefficient of the fusion module to achieve a higher navigation precision. The system could implement the function of automatic navigation in the farmland.

ACKNOWLEDGEMENTS

This paper is supported by the national 863 projects: Control Technique and Product Development of Intelligent Navigation of Farming Machines (2006AA10A304).

REFERENCES

Weimin yang, The research on machine vision in agricultural machines, pp.36-48

Qin Zhang and John F. Reid, Noboru Noguchi, Agricultural Vehicle Navigation Using Multiple Guidance Sensors, UILU-ENG-99-7013

Francisco Rovira-Mas, Shufeng Han, Jiantao Wei, John F. Reid, Fuzzy Logic Model for Sensor Fusion of Machine Vision and GPS in Autonomous Navigation, ASAE No.051156

Guangjun Zhang, Machine Vision, Science publishing Company, 2005

Xiaolan He, Research on Algorithms of Localization and Navigation Based on Machine Vision, pp.22-48

Zhigang Zhang, Research of the DGPS Automatic Navigation Control System on Tranplanter, pp.46-59

Zhou Jun, Ji Changying, Liu Chengling, Visual navigation system of agricultural wheeled-mobile robot. pp.47-59

Benson, E.R., J.F. Reid, and Q. Zhang. Machine vision-based guidance system for agricultural grain harverters using cut-edge detection .2003BE86 (4), pp.389-398

Benson, E.R., J.F. Reid, and Q. Zhang. Machine vision-based guidance system for an agricultural small-grain harvester 2003 TransASAE46(4), pp.1255-1264

Jiaming Ye, Digital Image Processing and Pattern Reconnition, pp.6-22

Xiaoling Gong, Theory and Simulation Study on Low-Cost Integrated Positioning System of GPS and SINS, pp.35-39

Zhiyan Zhou, Research of multiple sensor combination for agricultural wheel-vehicles navigation pp.13-28

KINEMATICS MODEL AND SIMULATION OF 5-DOF FINGER BASED ON FLEXIBLE PNEUMATIC ACTUATOR

Libin Zhang, Zhiheng Wang, Qinghua Yang, Tiefeng Shao, Guanjun Bao*
The MOE Key Laboratory of Mechanical Manufacture and Automation, Zhejiang University of Technology, Hangzhou, Zhejiang, China, 310032
* *Corresponding author, Address: the MOE Key Laboratory of Mechanical Manufacture and Automation, Zhejiang University of Technology, Hangzhou, Zhejiang, 310032, P. R. China, Tel: +86-571-88320819, Fax: +86-571-88320819, Email: robot@zjut.edu.cn*

Abstract: Based on flexible pneumatic bending joint and flexible pneumatic spherical joint, a kind of flexible pneumatic 5-DOF finger is proposed, which is composed of two bending joints and a spherical joint. The mathematic model of flexible pneumatic spherical joint is further analyzed. On the researching foundation of flexible pneumatic bending joint and flexible pneumatic spherical joint, the kinematics equation of the 5-DOF finger is deduced. And the redundancy problem of the inverse kinematics solution is resolved by genetic algorithm. The simulation experiment illustrates that genetic algorithm for solving of inverse kinematics is feasible and effective.

Keywords: flexible pneumatic finger, flexible pneumatic actuator, flexible pneumatic bending joint, flexible pneumatic spherical joint

1. INTRODUCTION

The complex structure of human hand enables it to grab objects of different size and shape easily, so researching on dexterous hand similar to human hand is many researchers' goal. Many different kinds of multi-fingered dexterous hand have been developed, such as: the Okada dexterous hand developed by Japanese Electronic Technology Lab in 1974 (Okada,

Zhang, L., Wang, Z., Yang, Q., Shao, T. and Bao, G., 2008, in IFIP International Federation for Information Processing, Volume 259; Computer and Computing Technologies in Agriculture, Vol. 2; Daoliang Li; (Boston: Springer), pp. 777–789.

1982), the Utah/MIT hand similar to human hand developed by MIT and the University of Utah in 1980 (Mason et al., 1985), the Hitachi hand proposed by Japan in 1984 (Nakano et al., 1984), the DIST hand and UB hand presented by Italy in 1990s (Caffaz et al., 1998), the DLR-1 and DLR-2 Hand developed by German Aerospace Center (Hirzinger et al., 1999; Lovchik et al., 1999), the NASA multi-fingered dexterous hand developed by United States National Aeronautics and NASA Johnson Space Center in 1999, the dexterous hand based on the pneumatic artificial muscles developed by Shadow robot company in England.

Some research institutions in China have also launched the study on dexterous hand in 1980s. The Robot Research Institute of Beijing University of Aeronautics & Astronautics has developed BH-1, BH-2, BH-3 and BH-4 dexterous hands (Wang et al., 1997; Shang et al., 2000). Harbin University of Technology and German Aerospace Center have jointly developed a new generation of multi-fingered humanoid robot hand HIT/DLR Hand.

The executive components of dexterous hands introduced above are rigid structure, so the adaptability and security for environment and grasp goal are poor and they need more accurate control system. Flexible dexterous hand based on Flexible Pneumatic Actuator FPA has virtues of good flexibility and adaptability, particularly suitable for agriculture fruit picking, service robots and medical rehabilitation apparatuses, which need higher security and adaptability. This paper presents a new kind of 5-DOF finger based on flexible pneumatic bending joint and flexible pneumatic spherical joint. The Kinematics model of this finger is established. The inverse Kinematics equation is also analyzed and simulated. This flexible pneumatic 5-DOF finger can be applied in multi-fingered dexterous hand design, applicable to agricultural harvesting robot, service robots, and other fields.

2. STRUCTURE AND PRINCIPLE OF FLEXIBLE PNEUMATIC 5-DOF FINGER

2.1 Flexible pneumatic actuator FPA

Yang has proposed the Flexible Pneumatic Actuator FPA, and analyzed its static and dynamic characteristics (Yang et al., 2005). The static model of FPA is:

$$\Delta L = \frac{(P - P_{atm})R}{2E_b t_b - (P - P_{atm})R} L_b \tag{1}$$

where, ΔL: linear deformation of rubber tube, t_b: original thickness of shell, E_b: elastic module of FPA, P_{atm}: atmospheric pressure, P: air pressure in actuator, R: average radius of actuator, L_b: original length of rubber tube.

FPA is the foundation of flexible pneumatic bending joint and flexible pneumatic spherical joint.

2.2 Flexible pneumatic bending joint

Based on FPA, Zhang designed the flexible pneumatic bending joint (Zhang et al., 2006). Fig. 1 shows the structure of the bending joint. Adding constraining wire fixed at both ends in one side of FPA rubber tube results in the flexible pneumatic bending joint.

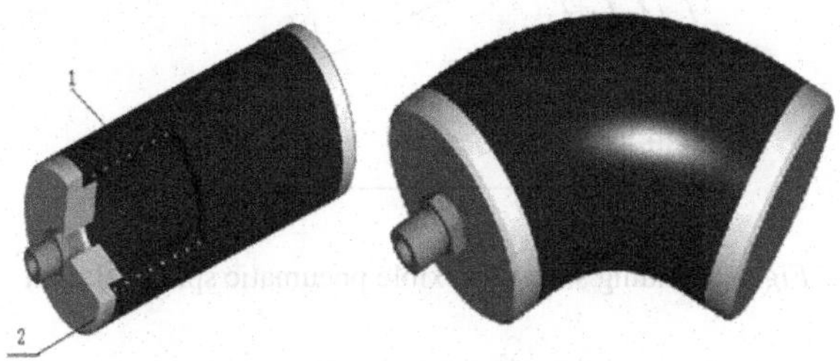

1 Spring; 2 Constraining wire

(a) Original state (b) Bending state

Fig. 1. Structure of flexible pneumatic bending joint

This kind of joint can realize bending movement in a plane, and the bending angle can be reached at 90°. The static model of the flexible pneumatic bending joint is:

$$\theta = \frac{L_b}{4r_b} \times \frac{\pi r_b^3 \Delta P + 6\pi E_b r_b^2 t_b - \sqrt{\pi^2 r_b^6 \Delta P^2 - 20\pi^2 \Delta P E_b^2 r_b^5 t_b + 36\pi^2 E_b^2 r_b^4 t_b^2}}{2\pi E_b r_b^2 t_b - \pi r_b^3 \Delta P} \quad (2)$$

where, θ: curving angle, L_b: original length of rubber tube, r_b: average radius of rubber tube, t_b: original thickness of shell, ΔP: the difference between air pressure in the joint and atmospheric pressure, E_b: elastic module of the bending joint.

2.3 Flexible pneumatic spherical joint

Fig. 2 shows the structure of the flexible pneumatic spherical joint which is composed with three FPAs uniformly distributed by 120°. The working principle of the spherical joint is: if the pressure of compressed air in the three FPAs is adjusted properly, the three FPAs have different elongations.

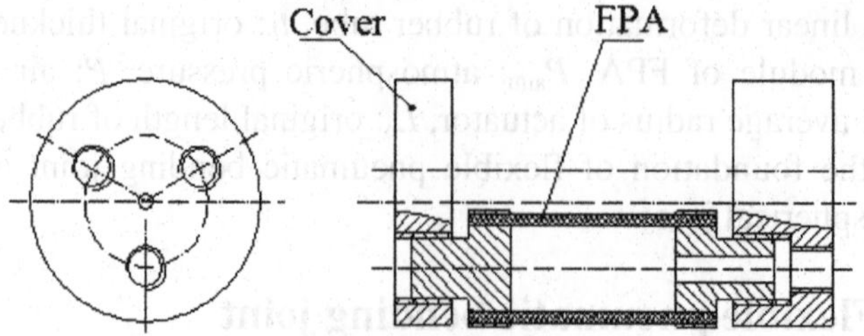

Fig. 2. Structure of flexible pneumatic spherical joint

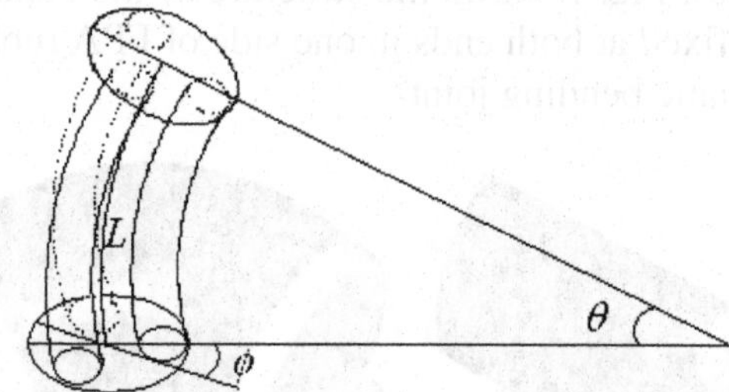

Fig. 3. Bending state of flexible pneumatic spherical joint

The following mathematical model is built:

$$\begin{pmatrix} P_1 - P_{atm} \\ P_2 - P_{atm} \\ P_3 - P_{atm} \end{pmatrix} = \begin{pmatrix} 2 & -1-\sqrt{3}tg\phi & -1+\sqrt{3}tg\phi \\ 1 & 1 & 1 \\ \cos\phi_1 & \cos\phi_2 & \cos\phi_3 \end{pmatrix}^{-1} \begin{pmatrix} 0 \\ \dfrac{3E_b A_w (L_1 - L_0)}{A_l L_0} \\ -\dfrac{E_b R A_w \theta}{A_l L_0} \sum_{i=1}^{3} \cos^2\phi_i \end{pmatrix} \tag{3}$$

In practical application, P_i-$P_{atm} \geq 0$, so each matrix item is greater than 0, when $\phi = 0$, the following equation can be obtained:

$$\begin{pmatrix} 2 & -1-\sqrt{3}tg\phi & -1+\sqrt{3}tg\phi \\ 1 & 1 & 1 \\ \cos\phi_1 & \cos\phi_2 & \cos\phi_3 \end{pmatrix}^{-1} \begin{pmatrix} 0 \\ \dfrac{3E_b A_w (L - L_0)}{A_l L_0} \\ -\dfrac{E_b R A_w \theta}{A_l L_0} \sum_{i=1}^{3} \cos^2\phi_i \end{pmatrix} \tag{4}$$

$$= \frac{E_b A_w}{4 A_l L_0} \begin{bmatrix} 2 & 2 & 0 \\ -1 & 2 & -2\sqrt{3} \\ -1 & 2 & 2\sqrt{3} \end{bmatrix} \begin{pmatrix} 0 \\ 2(L_1 - L_0) \\ -R\theta \end{pmatrix}$$

From equation (4), the following inequalities can be obtained:

$$\begin{cases} L_1 - L_0 \geq 0 \\ (L_1 - L_0) + \frac{\sqrt{3}}{2} R\theta \geq 0 \\ (L_1 - L_0) - \frac{\sqrt{3}}{2} R\theta \geq 0 \end{cases}$$, so $L_1 - L_0 \geq \dfrac{\sqrt{3}}{2} R\theta$. When the largest spherical joint bending angle is $\theta = \dfrac{\pi}{2}$, the length of the joint's center line L_1 can be calculated from the equation:

$$L_1 = \frac{\pi\sqrt{3}}{4} R + L_0 \tag{5}$$

According to equation (5) and (1), the size of initialization importation air pressure is:

$$P_{in} = \frac{\pi E_b t_0 \sqrt{3}}{2R(\dfrac{\pi\sqrt{3}}{4} R + L_0)} + P_{atm} \tag{6}$$

where, ϕ_i : angle between shell i and rotation axis n, i=1,2,3, ϕ : abduction angle of the flexible pneumatic finger, L_0: original length of the spherical joint's center line, d: distance between the axis of FPA and the axis of the spherical joint, A_w: cross-sectional area of FPA shell, A_1: cross-sectional area of FPA gas room, P_i: air pressure in FPA i, P_{atm} : atmospheric pressure, θ: bending angle.

2.4 Flexible pneumatic 5-DOF finger

According the anatomical structure of human hand, flexible pneumatic 5-DOF finger is designed based on flexible pneumatic bending joint and flexible pneumatic spherical joint. Fig. 4 shows the structure of the finger.

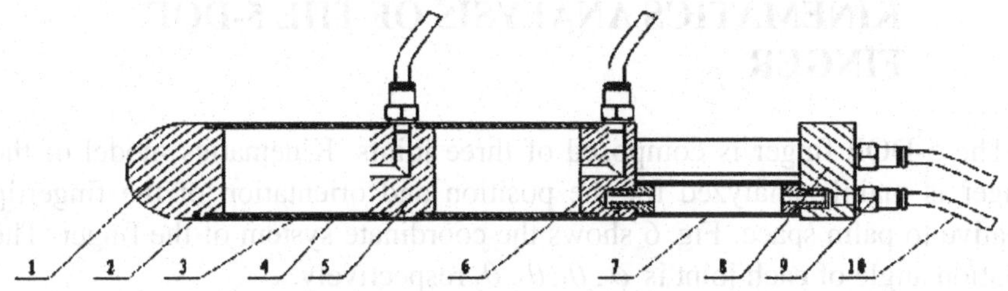

1 Far finger knuckle; 2 Bolt; 3 Constraining wire; 4, 7 FPA;
5 Near finger knuckle; 6 Palm finger knuckle;
8 Carpometacarpal joint; 9 Connector; 10 Air tube

Fig. 4. Structure of flexible pneumatic 5-DOF finger

Two flexible pneumatic bending joints and one flexible pneumatic spherical joint are connected in series by rigid knuckles. The two constraining wires embedded in the flexible pneumatic bending joints should be maintained in a straight line, and the spherical joint is the first joint near the digital root. The length of the far finger knuckle, near finger knuckle and palm finger knuckle are $l_3 = 15\text{mm}$, $l_2 = 10\text{mm}$, $l_1 = 25\text{mm}$ respectively. And the length of the finger joint, metacarpophalangeal joint and carpometacarpal joint are $l_6 = 15\text{mm}$, $l_5 = 30\text{mm}$, $l_1 = 25\text{mm}$ respectively.

The flexible pneumatic 5-DOF finger designed in this paper is shown in Fig. 5. Compressed air fills into three joints by five windpipes. When the air pressure inside two bending joints adjusted properly, they can realize different angle bending movement. Controlling the air pressure in three FPAs of the pneumatic spherical joint, the spherical joint can move within 360° of rotation and bending. When the pressure of air inside the joints gradually reduces, the pneumatic finger reverts to the original state. The flexible pneumatic finger with five DOFs has virtues of excellent flexibility, simple structure, and does not need the complex transmission device and variable speed mechanism. For the use of rubber material as actuators, the finger has good flexibility. When grasping objects, elastics rubber tube can adapt to the shape of the object, and does not cause unnecessary damage to the target.

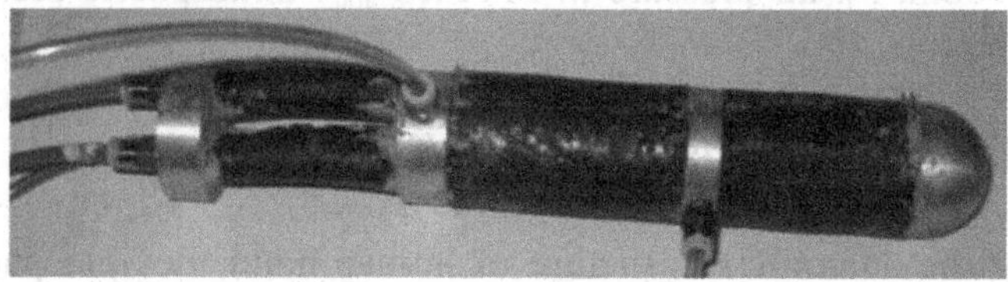

Fig. 5. The photo of the 5-DOF finger

3. KINEMATICS ANALYSIS OF THE 5-DOF FINGER

The 5-DOF finger is composed of three joints. Kinematics model of the finger is mainly analyzed for the position and orientation of the fingertip relative to palm space. Fig. 6 shows the coordinate system of the finger. The rotation angle of each joint is ϕ, θ_1, θ_2, θ_3 respectively.

Coordinate system 0 is fixed on the carpalmetacarpal; its origin O_0 is the crossing point between axis of the spherical joint and carpalmetacarpal, when the thumb and the other four fingers are in the same plane. The Z_0-axis is perpendicular to the palm outward, while the X_0-axis lies along the axis of

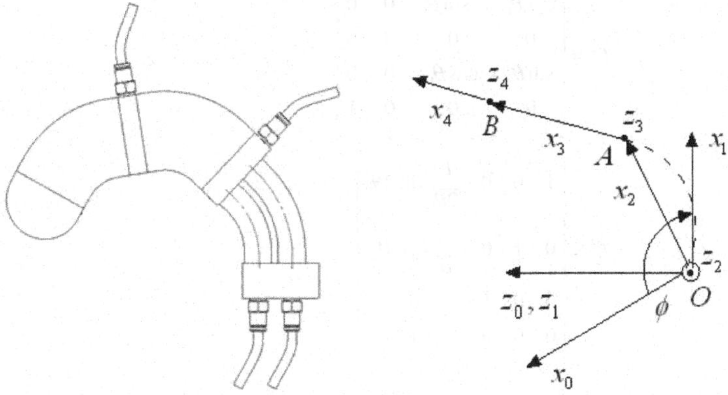

(a) Flexible pneumatic 5-DOF finger (b) Moving analysis of the finger
Fig. 6. Coordinate map of flexible pneumatic 5-DOF finger

the joint, and the Y_0-axis is determined by the right-hand rule. Located in the spherical joint's end connecting with carpalmetacarpal, the coordinate system 1 rotates relative to the coordinate system 0 along with the spherical joint's abduction movement. The origin O_1 coincides with O_0, while the Z_1-axis is perpendicular to the palm outward. Initially, X_1-axis mergers with X_0-axis, and then rotates around the Z_1-axis; while the rotation angle is ϕ. The Y_1-axis is determined by the right-hand rule. The origin of the coordinate system 2 coincides with origin O_0. The Z_2-axis is perpendicular to bending finger plane, while the X_2-axis lies along the axis of the joint upward, and the Y_2-axis is also determined by the right-hand rule. Coordinate systems 3, 4, 5, 6, 7, 8 are defined by the method described above. Coordinate 8 is located at the fingertips.

When the abduction angle is ϕ, Fig.7 shows the simplified model of the finger movement under the coordinate system 1. In Fig. 7, the chord lengths of the three joints 1, 2, 3 are denoted by the symbols O_1A, BC, DE, while the three knuckles' lengths are denoted by the symbols AB, CD, DE respectively. Three symbols θ_1, θ_2, θ_3 stand for angles between three knuckles and three chords respectively. According to the geometric relationship, the bending angles of three joints can be calculated, and they are $2\theta_1$, $2\theta_2$ and $2\theta_3$. The homogeneous transformation matrixes for the thumb movement are as follows:

$$T_0^1 = \begin{bmatrix} \cos\phi & -\sin\phi & 0 & 0 \\ \sin\phi & \cos\phi & 0 & 0 \\ 0 & 0 & 1 & 0 \\ 0 & 0 & 0 & 1 \end{bmatrix} \tag{7}$$

$$T_1^2 = \begin{bmatrix} \cos\theta_1 & -\sin\theta_1 & 0 & 0 \\ 0 & 0 & -1 & 0 \\ \sin\theta_1 & \cos\theta_1 & 0 & 0 \\ 0 & 0 & 0 & 1 \end{bmatrix} \tag{8}$$

$$T_2^3 = \begin{bmatrix} 1 & 0 & 0 & \frac{l_4}{2\theta_1}\sin 2\theta_1 \\ 0 & 1 & 0 & \frac{l_4}{\theta_1}\sin^2\theta_1 \\ 0 & 0 & 1 & 0 \\ 0 & 0 & 0 & 1 \end{bmatrix} \tag{9}$$

$$T_3^4 = \begin{bmatrix} \cos\theta_1 & -\sin\theta_1 & 0 & l_1\cos\theta_1 \\ \sin\theta_1 & \cos\theta_1 & 0 & l_1\sin\theta_1 \\ 0 & 0 & 1 & 0 \\ 0 & 0 & 0 & 1 \end{bmatrix} \tag{10}$$

$$T_4^5 = \begin{bmatrix} \cos\theta_2 & -\sin\theta_2 & 0 & \frac{l_5}{2\theta_2}\sin 2\theta_2 \\ \sin\theta_2 & \cos\theta_2 & 0 & \frac{l_5}{\theta_2}\sin^2\theta_2 \\ 0 & 0 & 1 & 0 \\ 0 & 0 & 0 & 1 \end{bmatrix} \tag{11}$$

$$T_5^6 = \begin{bmatrix} \cos\theta_2 & -\sin\theta_2 & 0 & l_2\cos\theta_2 \\ \sin\theta_2 & \cos\theta_2 & 0 & l_2\sin\theta_2 \\ 0 & 0 & 1 & 0 \\ 0 & 0 & 0 & 1 \end{bmatrix} \tag{12}$$

$$T_6^7 = \begin{bmatrix} \cos\theta_3 & -\sin\theta_3 & 0 & \frac{l_6}{2\theta_3}\sin 2\theta_3 \\ \sin\theta_3 & \cos\theta_3 & 0 & \frac{l_6}{\theta_3}\sin^2\theta_3 \\ 0 & 0 & 1 & 0 \\ 0 & 0 & 0 & 1 \end{bmatrix} \tag{13}$$

$$T_7^8 = \begin{bmatrix} 1 & 0 & 0 & l_3\cos\theta_3 \\ 0 & 1 & 0 & l_3\sin\theta_3 \\ 0 & 0 & 1 & 0 \\ 0 & 0 & 0 & 1 \end{bmatrix} \tag{14}$$

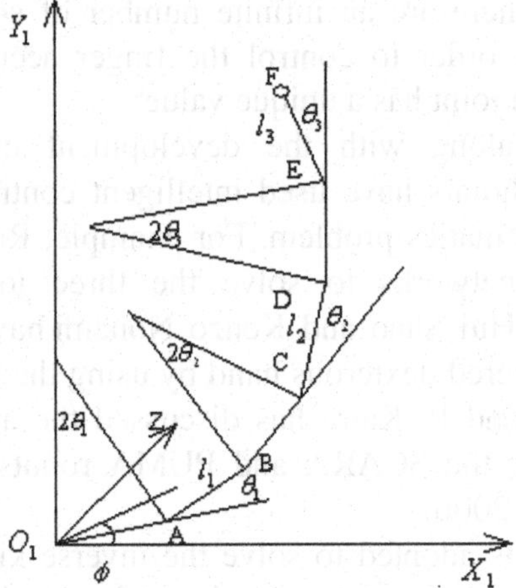

Fig. 7. Simplified model of the finger movement under the coordinate 1

According to the analysis above, we can get the fingertip's position and orientation under the fixed coordinate system 0:

$$T_0^8 = T_0^1 T_1^2 T_2^3 T_3^4 T_4^5 T_5^6 T_6^7 T_7^8 \tag{15}$$

4. INVERSE KINEMATICS ANALYSIS OF THE 5-DOF FINGER BASED ON GENETIC ALGORITHM

4.1 Description of genetic algorithm

In inverse kinematics analysis, the fingertip positions are given to compute the bending angles of each joint. The bending angles ϕ, θ_1, θ_2 and θ_3 of the three joints can be compute from the equation:

$\begin{bmatrix} x \\ y \\ z \\ 1 \end{bmatrix} = T_0^8$, so we can get the abduction angle: $\phi = actg\frac{z}{x}$, θ_1, θ_2 and θ_3 are determined by the liner equation which has two variable x_1, and y_1 in the coordinate system 1. Where, $x_1 = \dfrac{x}{\cos(actg\frac{z}{x})}$, $y_1 = y$. According to the previous equations, there are an infinite number of solutions for a unique fingertip position. In order to control the finger accurately, we hope the bending angle of each joint has a unique value.

In recent years, along with the development of intelligent control technology, many scholars have used intelligent control solutions to solve the robot inverse kinematics problem. For example, Rasit and Kokera have applied the neural networks to solve the three joints robot's inverse kinematics problem, Hui Shao and Kenzo Nonami have solved the control problem of multi-fingered dexterous hand by using the fuzzy neural network (Shao et al., 2006), and P. Kalra has discussed the application of genetic algorithm for solving the SCARA and PUMA robots' inverse kinematics problem (Kalra et al., 2006).

Genetic algorithm is adopted to solve the inverse kinematics problem of the finger in this paper. The main steps of genetic algorithm are as follows:

(1) Code: The mapping between the genotype and the phenotype is called coding. Binary coding is applied in this paper, and the binary variable median is 20.

(2) Initial population producing: N data strings are randomly generated, while each data string is called an individual, and then N individuals constitute a group. Initially, genetic algorithm iterates from the N data strings.

(3) Fitness evaluation testing: The fitness function illustrates the individual or solution is superior or interior. The fitness function in this paper is:

$$\begin{cases} \min \ f_1 = \frac{l_4}{2\theta_1}\sin 2\theta_1 + l_1\cos\theta_1 + \frac{l_5}{2\theta_2}\sin 2\theta_2 + l_2\cos\theta_2 + \frac{l_6}{2\theta_3}\sin 2\theta_3 + l_3\cos\theta_3 - X_1 \quad f_1 \geq 0 \\ \min \ f_2 = \frac{l_4}{\theta_1}\sin^2\theta_1 + l_1\sin\theta_1 + \frac{l_5}{\theta_2}\sin^2\theta_2 + l_2\sin\theta_2 + \frac{l_6}{\theta_3}\sin^2\theta_3 + l_3\sin\theta_3 - Y_1 \quad f_2 \geq 0 \\ 0 \leq \theta_1 \leq \frac{\pi}{3} \quad 0 \leq \theta_2 \leq \frac{4\pi}{9} \quad 0 \leq \theta_3 \leq \frac{\pi}{2} \end{cases} \tag{16}$$

where, (X_1, Y_1) is the target coordinates.

(4) Selection: The individuals in the group which have strong vitality are kept down to the next generation groups by the select operator. According to the fitness value of the individual, genetic algorithm does the operation based on the rule of the survival of the fittest.

(5) Cross: The group is treated by the crossover operator. Two individuals are selected from the group with the greater probability. The crossover operation exchanges partial parts of the two individuals.

(6) Mutation: Changing one or several bits of the coding string, we obtain the new individuals. The mutation operation can improve the local search ability of genetic algorithm and maintain the genetic diversity of the population.

(7) Stop judgment.

4.2 Simulation analysis of the inverse kinematics

Matlab are used to do simulation experiments on genetic algorithm. The number of the individuals is 100, the number of the maximal genetic generations is 80, and the generation gap is 0.9. Fig. 8 and Fig. 9 show the optimal solutions and tracking performance of the first and the second objective function respectively. From Fig. 8 and Fig. 9, it can be seen that the 10th generation function solutions tend to be 0, which are the expected values. So the No.15 generation group can be the inverse kinematics solutions of θ_1, θ_2, and θ_3. Fig. 10 shows the optimal solutions tracking of the two objective functions.

Through the simulation, the redundancy problem on the inverse kinematics of the 5-DOF finger can be solved by the genetic algorithm. The method is simple and clear, no redundant constraints.

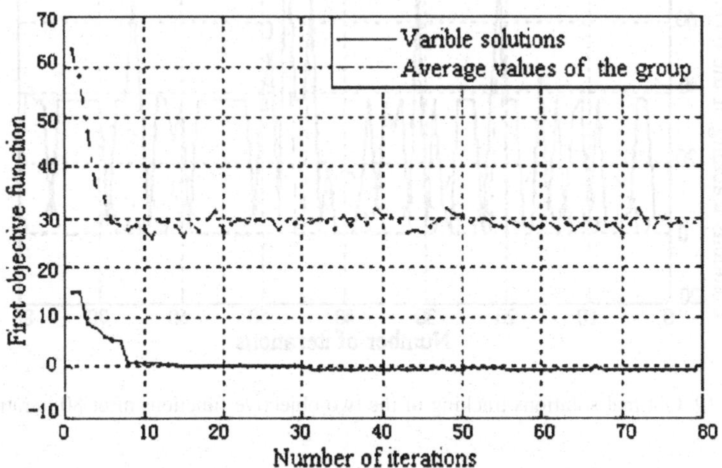

Fig. 8. Optimal solutions and tracking performance of the first

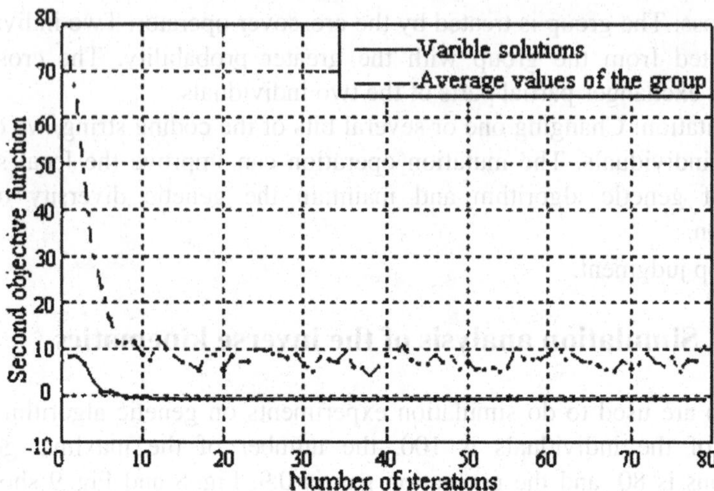

Fig. 9. Optimal solutions and tracking performance of the second objective function after 80 iterations

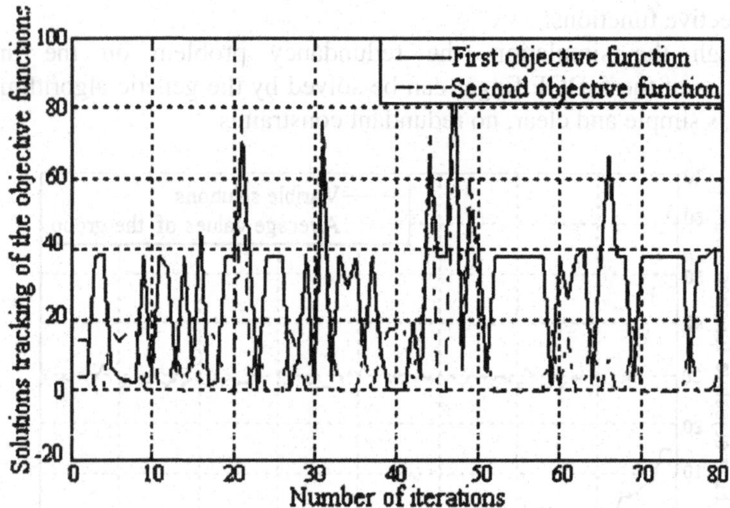

Fig. 10. Optimal solutions tracking of the two objective functions after 80 iterations

5. CONCLUSION

(1) Based on the flexible pneumatic bending joint and flexible pneumatic spherical joint, flexible pneumatic 5-DOF finger is proposed. It has many advantages such as good flexibility, simple structure and good adaptability.

(2) Kinematics characteristic of the 5-DOF finger is profoundly studied and the kinematics model is built.

(3) Inverse kinematics characteristic of the flexible pneumatic 5-DOF finger is analyzed, and the genetic algorithm is applied to solve the inverse kinematics equation. The simulation experiments show that applying genetic algorithm to solve the inverse kinematics problem is feasible and effective.

REFERENCES

Caffaz, Cannata G. The Design and Develoopment of the DIST-Hand Dextrous Gripper [J]. Proceedings of the IEEE International Conference on Robotics and Automation, Leuven, Belgium. 1998:2075-2080.

Hirzinger G., Fischer M., Brunne B. Advances in Robotics: The DLR Experience [J]. The International Journal of Robotics Research, 1999, 18(1): 1064-1087.

http://robonaut.jsc.nasa.gov/robonaut.html

http://www.dlr.de/rm/en/desktopdefault.aspx/tabid-117/

http://www.hitrobot.net/

http://www.shadow.org.uk/projects/openhardware.shtml

http://wwwrobot.gmc.ulaval.ca/liens/liens_a.html

Hui Shao, Kenzo Nonamib, Tytus Wojtara, etc. Neuro-fuzzy position control of demining tele-operation system based on RNN modeling [J]. Robotics and Computer-Integrated Manufacturing, 2006, 22: 25-32.

Libin Zhang, Guanjun Bao, Qinghua Yang, Jian Ruan and Liyong Qi. Static Model of Flexible Pneumatic Bending Joint, Proceeding of the 2006 9th Int. Conf. Control, Automation, Robotics and Vision, Singapore, 2006.12: 1749-1753.

Lovchik C. S., MDifler M. A. The Robonaut Hand: A Dextrous Robotic Hand for Space [J]. Proceedings for the IEEE International Conference on Robotics and Automation. Detroit, Michigan, 1999: 907-912.

Mason M. T., Salisbury J. K. Robot Hands and the Mechanics of Manipulation [J]. MIT Press, Cambridge, USA, 1985: 3-93.

Nakano Y., Fujie M., Hosada Y. Hitachi's Robot Hand [J]. Robotics Age, 1984, 6(7): 18-20.

Okada T. Computer control of multi-jointed finger system for precise object-handing [J]. IEEE Transactions on Systems, Man and Cybernetics, 1982, 12(3): 289-299.

P. Kalra, P.B. Mahapatra, D.K. Aggarwal. An evolutionary approach for solving the multimodal inverse kinematics problem of industrial robots [J]. Mechanism and Machine Theory, 2006, 41: 1213-1229.

Rasit Koker, CemilOza, Tarık, etc. A study of neural network based inverse kinematics solution for a three-joint robot [J], Robotics and Autonomous Systems 2004, 49: 227-234.

Shang Xisheng, Guo Weidong, Zhan Hao, etc, Grasp Planning and Realization of BH-4 Dexterous Hand [J], Robot, 2000, 22(7): 608-612 (in Chinese).

Wang Guoqing, Zhang Qixian, Li Dazhai, He Yongqiang, Grasping Control of the Dexterous Hand Based on the Degree of Stability of Grasping [J]. Acta Aeronautica Et Astronautica Sinica, 1997, 18(3): 294-298 (in Chinese).

Yang Qinghua, Research on Flexible Pneumatic Joints and Their Application Based on Flexible Pneumatic Actuator, Dissertation for Doctor's Degree in Engineering of Zhangjiang University of technology, 2005 (in Chinese).

Yang Qinghua, Zhang Libin, Bao Guanjun, Ruan Jian, Pneumatic Squirming Robot Based on Flexible Pneumatic Actuator, Proceedings of SPIE ICMIT2005: Control Systems and Robotics, Sept. 2005, Vol. 6042: 60422w-1-60422w-5.

Zhang Peiyan, Lv Tiansheng, Song Libo, Study on BP Networks-based Inverse Kinematics of Motoman Manipulator [J], Mechanical & Electrical Engineering Magazine, 2004, 20(2): 56-58 (in Chinese).

DIGITAL DESIGN AND IMPLEMENTATION OF SOYBEAN GROWTH PROCESS BASED ON L-SYSTEM

Hongmin Sun [1,*], Leqiang Ai [1], Xinzhong Tang [1]

1Department of computing, Engineering College, Northeast Agricultural University, Harbin 150030, China

** Corresponding author, Address: Department of Computer Science and Technology, Engineer College, Northeast Agriculture University, Harbin, China, 150030, Tel: +86-0451-55191749, Fax: +86-0451-55190170, Email: sunhongmin111@126.com*

Abstract: Aiming at the Present Condition of Research on soybean growth model, in order to recur the overall dynamic process of soybean growth, we parameterized the main factors that can affect the plants' physical development and built the soybean plant growth model by adopting the classical biologic plant growth logistic equation and the Self-similarity of plant morphology. In VC6.0 environment, we applied OpenGL technology and L-system to simulate the soybean plant type topological structure and the plant leaves in computer to simulate the whole process of soybean plant type growth more factually which creates advantaged conditions to research on high yield soybean plant shape.

Keywords: Virtual soybean, growth model, topological structure, L-system

1. INTRODUCTION

Virtual plant is a simulation to growth and development status of plant in 3D space which is based on a large number of data of individual plant. It can create plant with three-dimensional effects and visualizing function, also obtain the common results of plant physiological ecology processes and morphological structure process.

Soybean is the one of the world's major food crops, the study on virtual soybeans mainly focus on two research direction, one direction focus on the simulation of plant structure, the other focus on the simulation of physiology

Sun, H., Ai, L. and Tang, X., 2008, in IFIP International Federation for Information Processing, Volume 259; Computer and Computing Technologies in Agriculture, Vol. 2; Daoliang Li; (Boston: Springer), pp. 791–797.

function recently, the next hot research will be the simulation of common-effect that integrated the two factors above (Zheng et al., 2006). Soybean physiological function can influence the form of soybeans; the other way round, soybean pattern of soybean can also affect the efficiency and process of physiological function. By controlling the environment factors that can affect the soybean growth to simulate the form of soybean, then according to the simulation of soybean production patterns to determine efficiency of the soybean biological function (such as photosynthesis efficiency), finally selecting the high-yielding soybean plant type and transforming soybean plant type through the biological genetic engineering. This research can make important impact to increase soybean yield and have great practical significance (He et al., 2004).

L-system simulation to herbs can easily be understood and realized (Jin et al., 2002), so the obvious and practical approach of virtual plant is to model by using L-system and OpenGL technology with combining the plant growth equation in the VC++ MFC framework, digital designing and implementing the simulation of soybean growth process.

2. MATIERIALS AND METHOD

2.1 Geography and Meteorology

Variety selection: DongNong45. Main stem type, limited growth, round leaf, early maturing, The period of growth was about 87 days.

Planting place: The virtual plant laboratory of northeast agricultural university, the Average light were 8.6 hour, The temperature in laboratory 18°C–25°C, humidity 30% –70%.

Measurement tool: 15-cm ruler (precision 1mm) and spiral micrometer (precision 0.01mm), electronic balance, CI-203 laser leaf area meter, vernier caliper.

Shoot tool: Canon Power Shot S45.

2.2 Data Collection

Observation period: From November 23, 2005 to February 10, 2006 (sowing in November 17, 2005)

Experiment content: In this experiment, we select four representative soybean plant, measure and record a individual soybean plant at the same time every two days. The measurement results include the internode distance, internode thickness, the length and thickness of petiole and little petiole, the length and width of every leaf, leaf weight, petiole weight, internode weight etc, 1745 Data were presented. We also take photography for the individual soybean plant when it's measured.

Table1 shows the partial data:

Table 1. Soybean growth process parameters

Week	Internode			Petiole			Leaf area (cm^2)/ fresh weight (g)		
	Cross sectional diameter (mm)	Length (cm)	Fresh weight (g)	Cross sectional diameter (mm)	Length (cm)	Fresh weight (g)	Leaf 1	Leaf 2	Leaf 3
1	1.60	0.90	0.04	1.36	0.40	0.02	—/0.06	—/0.06	—/0.06
2	2.40	1.55	0.10	1.88	3.20	0.11	8.364/0.12	10.679/0.14	8.116/0.15
3	3.12	1.90	0.44	1.36	6.20	0.24	6.890/0.09	5.257/0.13	9.490/0.17
4	3.70	3.4	0.37	1.58	10.3	0.27	14.915/0.28	16.711/0.30	13.179/0.23
5	4.72	2.30	0.38	2.06	9.40	0.22	9.577/0.13	13.394/0.18	11.865/0.14

3. RESULTS AND ANALYSIS

3.1 Soybean Topological Structure L System Design

Soybean topological structure has certain characteristics, For example: Relatively simple structure, Hierarchy, Self-similarity etc, so we use symbol instead of each organ which makes it is easy to design L-System, Ultimate design of the L-system are as follows:

//Iteration Times

Iteration = N

//According to observation, set space rotation angle

Angle=45

//the initial Character Set

V={I, i, Pl, Pr, P, p, A, L, l, z, [,], +, -, &, ^, \, /, |}

The initial expression: // Soybean topological structure which only has cotyledon

ω: I[+z][-z]A

//production expression

(N = 1) //Structural Iterative Expression of the soybean univalent period

P1: A→I[+iL][-iL]A

(N > 1) //Structural Iterative Expression in different growth periods

N is odd P2: A→I[+Pl]A

N is even P3: A→I[-Pr]A

//The final iteration expression

//Description of the petiole and trifoliate leaf

Pl→ p[\iL][/iL][-iL]

Pr→ p[\iL][/iL][+iL]

Various symbols which represent the meaning indicated in the following table 2.

Table 2. Soybean framework represented by some symbols

Symbol	Meanings	Symbol	Meanings
I	Internode	i	Cotyledon, compound leaf ramastrum
Pl	Left branching	Pr	Right branching
P	Compound leaf part	p	Compound leaf petiole
z	Cotyledon	L	Compound leaf
l	Univalent	A	Apical meristem
Iteration	Iteration Times	Angle	The included angle of main stem and ramificate
[,], +, -, &, ^, \, /,			The meaning of L-system

3.2 Construction of Soybean Growth Equation

During the process of the experiment, integrant data of the soybean previous growth was recorded due to the environmental restriction of measurement. The measuring cycle is two days, and the function trend reflected from the data can be shown in table 3.

Table 3. Data of internodes length from 1st to 13th days

Day N	1	3	5	7	9	11	13
Internodes length/mm	9.0	11.20	13.30	15.50	16.70	18.10	19.0

These data roughly show the growth process of internodes length. If these data is directly adopted to construct a growth equation, the accuracy of it will be greatly discounted. To make up its inexact resulting from the inadequacy of data, we adopt a method of data combination. Here six degree polynomial is used to describe these data. Constructing a .mmf file through 1stOpt software to answer, so as to get interpolation polynomial as followed:

$$y=-0.0017x^6+0.073x^5-1.215x^4+9.769x^3-39.14x^2+77.08x-35.479 \quad (1)$$

Then a number of interpolation data are estimated from this interpolation polynomial as showed in table 4.

Plant growth follows the classical logistic equation, therefore making use of the mathematical software—1stopt to construct .mmf program to fitting logistic equation by using interpolation data, at last confirming the growth equation parameter of each organ of the individual soybean plant. We can get the result as follow:

W=54.336683977435
k=6.40478984980516
n=0.506995309056227

Thus a more accurate curve equation of the soybean internodes length can be got as followed. W(t) indicates the length of certain growth time and t indicates the parameter of the growth time.

$$W(t) = \frac{54.3367}{1+6.4.48e^{-0.5070t}} \tag{2}$$

Table 4. Data of internodes length after polynomial interpolation (mm)

Day N	1	2	3	4	5	6	7	8	9	10	11	12	13
Internodes length (mm)	9.0	10.10	11.20	12.00	13.30	14.42	15.50	16.20	16.70	17.46	18.10	18.70	19.0

3.3 Construction of the Main Organ Model

The paper takes soybean leaf as the main object to discusses the construction method of the soybean organ model. By means of the computer image processing, L system is applied to realize the simulation of leaf configuration and texture. The setting of light source and leaf texture can lead to the visual model of it. The construction of the leaf visual model is in the same plane of the whole leaf. But in fact, the axis of the soybean leaf configuration is the main nerve and the left and right leaves shape a certain angle. Therefore the left and right leafs respectively need to be processed in terms of rotation and introduced parameter α to realize the control of rotation. α refers to the plane model of the leaf and the value range of the rotating angle is from 0 to 90. (Kang et al., 2006)

The translation and rotation of the whole leaf also makes a parameter design. By means of regulating the translation and rotation parameter, the three-dimensional visual model of the leaf can show different angles and conditions. In imitating the duplicating leaf, the leaf moves the leaf handle along the axis Y. Fig. 1 simulates the duplicating leaf.

Fig. 1. Three-dimensional visual model of soybean leaves

The position processing method of a three-leaf leaf is that the leaf on the left and right moves the length of the leaf handle lt along the axis Y and then respectively rotates the corresponding the angle z along the axis Z. The length of the leaf handle is larger than that of the leaf in the middle.

3.4 The Realization of the Visual Model

After confirming the topological structure and growth equation, we can apply the equation into the topological structure and simulate the soybean growth process in the three-dimensional environment by means of the integration development environment VC and 3D function library OpenGL (Song et al., 2003). In simulating the soybean growth process, the method is the technique of s mapping in OpenGL (Cai et al., 2002) and the final simulating effect is shown as Fig. 2.

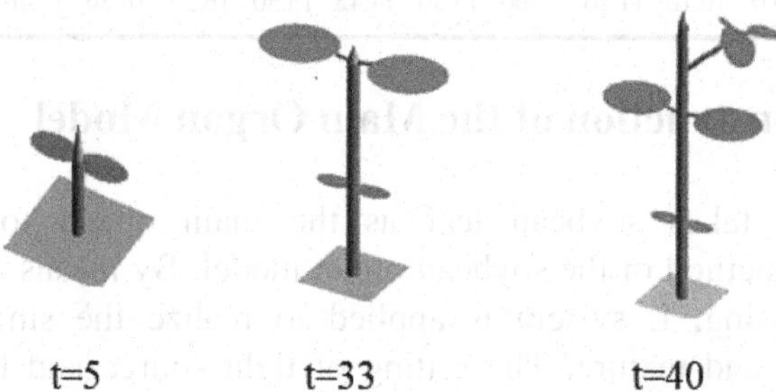

Fig. 2. Simulation graphics of soybean growth

4. CONCLUSION

The paper combines L system and the plant growth model to realize the imitation of the soybean growth process. In terms of configuration and function, the effect of environment on the plant growth is fully revealed. However, the environment element involved in the model only refers to the time element and other environment elements closely related with the plant growth such as light, water and temperature need to be further perfected after collecting the experiment data to be used in the model. At the same time the real sense process of the soybean vein and leaf surface still need to be further researched into and dealt with.

ACKNOWLEDGEMENTS

This study has been funded by Natural Science Foundation of Heilongjiang province (Contract Number: C200607).It is also supported by Science and Technology Department of Heilongjiang province, China (Contract Number: GC04B712), and program for Innovative Research Team of Northeast Agricultural University, "IRTNEAU". we also thanks to the Northeast Agricultural University Virtual Agricultural Research Center for the experimental equipment and technical support.

REFERENCES

Zheng P, Su Z B, Kang L. Modeling of virtual soybean topology based on growth function. Journal of Agricultural Mechanization Research, 2006, 7:193-195.

He S W, Chang S H, Wu D L. Biological Characteristics of four different soybean varieties. Journal of Acta Pratacultural Science, 2004, 13:70-75.

Song Y H, Guo Y, Li B G, Philippe de Reffye. Virtual maize model I. biomass partitioning based on plant topological structure, Acta Ecologica Sinica, 2003, 23:2333-2341.

Cai Z J. The application of L-system in plant imitation and visualization on computer, Journal of Agricultural University of Hebei, 2003, 26: 98-101.

Jin R Z, Wang Z Y. Essential Concept of L-system and Examples. Journal of Tianjin Agricultural College, 2002, 9: 49-54.

[Kang L, Su Z B, Zheng P, Li Y F. Research on modeling leaf venation based on L-system Journal of agricultural mechanization research, 2006, 7:180-181.

DEVELOPMENT OF A WEB-BASED WIRELESS TELEMONITORING SYSTEM FOR AGRO-ENVIRONMENT

Keming Du[1], Zhongfu Sun[1,*], Huafeng Han[1], Shuang Liu[1]

[1] *Institute of Environment and Sustainable Development in Agriculture (IEDA), Chinese Academy of Agricultural Sciences (CAAS), No. 12, Zhong-guan-cun South Street, Beijing, 100081, P. R. China*

[*] *Correspondence author: Zhongfu Sun, Tel/Fax: 86-10-62119558, E-mail: sunzf@263.net*

Abstract: In order to satisfy the requirements of agro-environmental data acquisition, a project scheme of wireless telemonitoring system was plotted out according to the agricultural characteristics of scattered-sites far from developed community, multiple environmental factors, mutable conditions disturbed by natural disasters, and so forth. Integrated with modern information technology (IT), an environmental data acquisition system, named WITSYMOR V1.0, was developed, which was designed to be compatible with most of common sensors, such as temperature and moisture for air and soil, CO_2, air pressure, PAR, total solar radiation, soil pH and EC, etc. The kernel techniques were to realize a seamless connection between wireless mobile network (GPRS/CDMA) and Internet with TCP/IP software design, the data from the remote agricultural sites were real-timely acquired and transmitted to the central database servers, and could be browsed, applied and downloaded by authorized users at anytime and in anywhere. At present, it has been set up in some agricultural stations, horticultural greenhouses, methane-gas pools, as well as in animal shelters. The results show that the system works stably, and it is much adaptive to monitor various agricultural environmental factors in the sites far away.

Keywords: Data acquisition, Telemonitoring, Sensors, GPRS/CDMA, Agro-environment

Du, K., Sun, Z., Han, H. and Liu, S., 2008, in IFIP International Federation for Information Processing, Volume 259; Computer and Computing Technologies in Agriculture, Vol. 2; Daoliang Li; (Boston: Springer), pp. 799–807.

1. INTRODUCTION

For the requirement of precision management and control technology in modern agriculture, the dependence on production field information has become more and more obvious. With the rapid development of modern science and technology, the application of information technology has become an important mark of agricultural modernization. Application of modern IT began in the late 1970s, in spite of that, its development is quite rapid (Lu et al., 2004). For the recent thirty years, making full use of intelligent information technology to promote agricultural modernization, lots of countries have made great achievements. with support of using such technologies, people could exchange and share information resources rapidly, and acquire convenient services about weather, markets, production, crop planting, and so on, which could make the most of the demand for the market with minimum input and maximum benefit simultaneously.

Controlled environment agriculture has become one of the most representative examples of information technology application in agriculture, which integrates with technologies of sensor, network, automatic control, etc (Cox, 2002). Nevertheless, how to develop a telemonitoring system for agricultural data acquisition is still one of the important problems that should be solved in modern agriculture (Sun, 2005; Wang et al., 2006; McKinion et al, 2004; Sobeih, 2005) for accurate data acquisition of crop growth and environmental factors, timely prediction of crop growth and crop diseases, as well as efficient realization of remote diagnosis and management.

At present, two traditional methods are still prevailing in lots of existing environmental monitoring systems (Sun et al., 2005). One of which is so called onsite closed system that uses a field computer with linkage of a series of sensors. Another one is to keep long-distance communication with leased cable. It is well known that most of monitoring sites in agriculture are far away from communities and administrators, such as typical sites like greenhouses, agricultural and forestry ecosystems, meteorological and hydrological stations, etc, for which it is not easy to get environmental information automatically. By contrary, extra input cost would much increase if the leased cable communication network was adopted. What is more, the position of observed sites changes frequently, and the administrators, who often go out for business from place to place but want to know the information of the observed sites far away from themselves. What should they do? Aimed at all mentioned above, a remote monitoring technical scheme is proposed with seamless connection between wireless communication network and Internet, which could provide a suitable solution for efficiently resolving the problems of remote communication in data acquisition and transmission. It is well-known that the research object is in quite accordance with the development trend of current monitoring system for agricultural environment (Messer et al., 2006; Serodio et al., 2001).

2. STRUCTURAL DESIGN AND KERNEL TECHNOLOGY OF THE SYSTEM

According to the agricultural characteristics mentioned above, integrating with the former research results (Sun et al., 2006), a structural scheme of wireless telemonitoring system for agricultural environment was optimized, upgraded and validated by practical application. And the wireless telemonitoring system for agricultural environment (WITSYMOR V1.0) was realized. This multi-function modern monitoring system was made combined with environmental factor measuring technology, modern transducer technology, wireless communication technology, and computer network information technology, etc. The GPRS/CDMA wireless communication technology was adopted to resolve the difficult problem of field data transmission field, which is a convenient way to solve the so-called 'final one kilometer bottle-neck problem of information freeway' for information transmission from agricultural field to the users.

In structural design (Figure 1), this system consists of three modules: remote monitoring terminal module in observed site for data acquisition and transmission, database server module for data receiving and saving in central database server, and Web server module for data management and process.

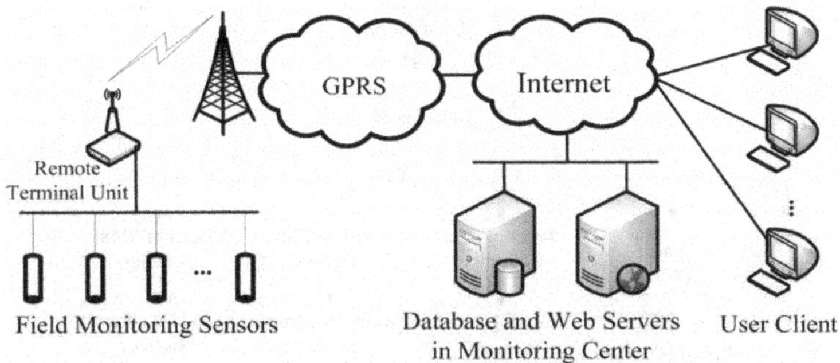

Figure 1. Structure of the system

The following contents are focused on the detailed description about system structure, function, key technology and realization of the three modules.

2.1 Remote Monitoring Terminal Module of Data Acquisition and Transmission

The terminal module is placed in an agricultural observed site, which is responsible for achieving real-time data and sending them to remote central database server through the seamless connection between GPRS/CDMA wireless communication network and Internet.

The hardware device of the module consists of programmable RTU (Remote Terminal Unit), sensors for environmental factors, A/D transform device, RS-485 field bus and mobile SIM card. RS-485 acts as a bridge to connect the programmable RTU and the output signal of sensors through A/D transform devices. The A/D module with multi-channel analog signal input could be connected with various types of analog sensors with standard signal output, from 0 – 5.0V voltage or 0 – 20mA electric current. In addition, digital sensors could be directly connected to RTU only through RS-485 cable without A/D transform.

RTU is widely used in the industry for measurement and control. The programmable RTU of the system provides embedded programming interface. Using integrated development environment of Dynamic C, a specific C language, the program for data acquisition and process, wireless transmission and regulation have been developed. The program execution flow is shown in Figure 2.

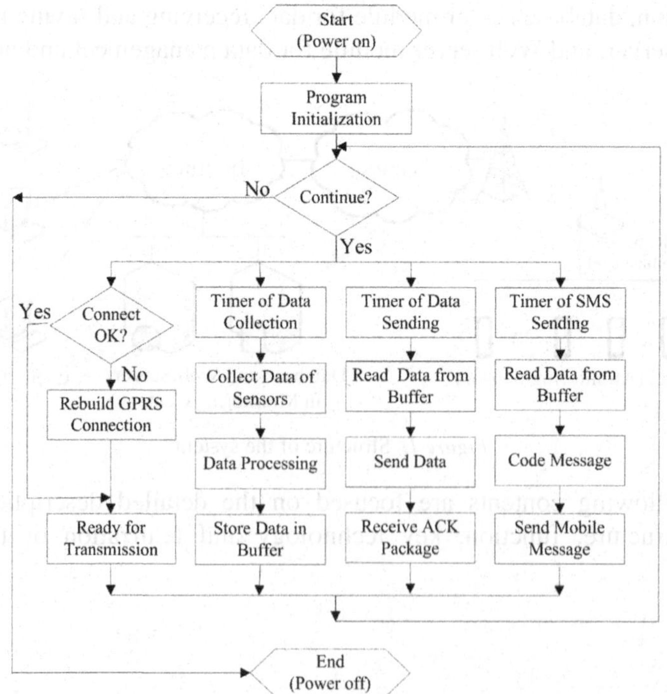

Figure 2. Flow chart of the program for data acquisition and transmission

2.2 Central Database Server Module of Data Receiving and Saving

The database server module acts as a control and management platform consisting of server computer, Microsoft SQL Server database and server application program for data receiving and saving.

DELL PowerEdge 4600 was selected as server computer, and Microsoft SQL Server 2000 as database management. The server application program, installed on the server computer, takes charge of the following functions such as to identify and accept TCP Socket connection request from remote monitoring terminal module, and to analyze and save the data into corresponding database table.

The server application program of the system adopts network programming with Winsock Control Component, using development tool of Microsoft Visual Studio.NET 2003 and program language of VB.NET. The program execution flow is shown in Figure 3.

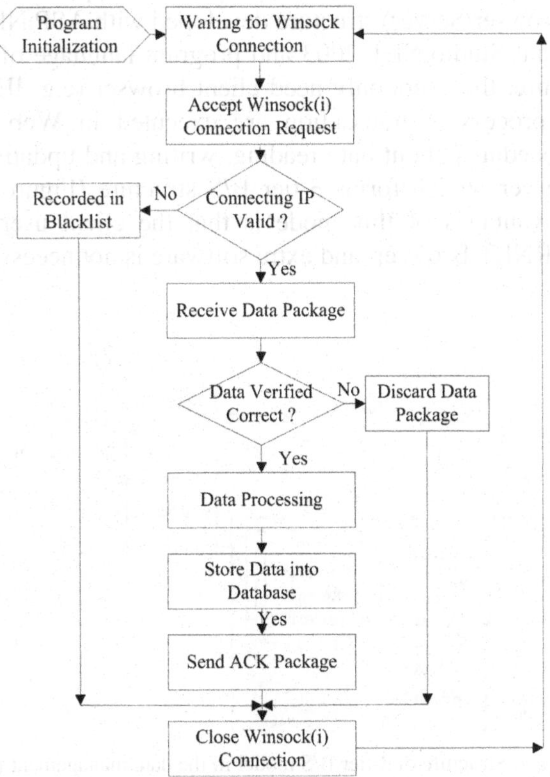

Figure 3. Flow chart of server program for data process

2.3 Web Server Module of Data Management

The Web server module consists of Web server platform and Web server application. The hardware of the Web server adopts enterprise-level server computer (DELL PowerEdge 4600), and for the software of the Web server, Windows 2003 server operation system and IIS6.0 (Internet Information Server) were used. IIS, integrated with Windows operation system, is one of the most popular Web server software, which has advantages of easy configuration and friendly interface.

In order for promoting security and reliability of the database, the system provides a safe scheme that the data could be saved into more than one physical independent servers, which means that the real-time data from the field sites would be sent to two or more different database servers simultaneously.

Authorized users could conveniently access to the following websites e.g. www.sinong.com and www.cea.net.cn, which represent two independent server computer placed in different sites. The Web application of the system adopts B/S (Browser/Server) structure developed with ASP.NET technology, Microsoft Visual Studio.NET 2003 and program language of VB.NET. For the B/S structure, the users only need client browser (e.g. IE or Netscape); Most of the processing transaction are executed in Web Server; while processing procedures about data reading, writing and updating are finished in database server, which forms 3-tier B/S structure (Figure 4). Evidently, one of the advantages of this mode is that the client users only need a popular INTERNET Browser, and extra software is not necessary.

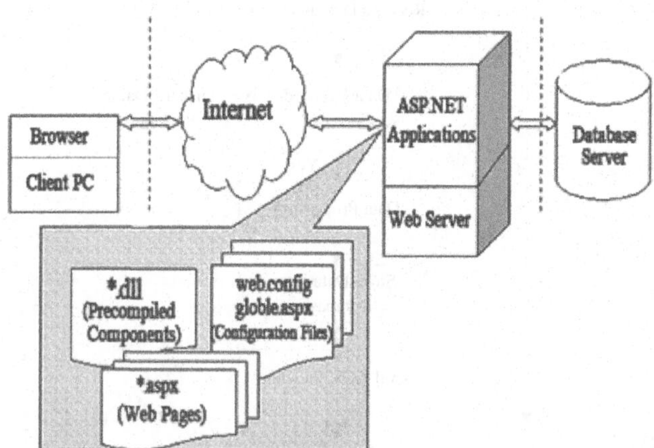

Figure 4. Structure of 3-tier B/S model in the data management module

The execution procedures of the Web server application is as follows: Firstly users have to access to the main Web page for registration through client's browser; Secondly, after achieving authorization, the users could do a lot of request operations on the page including real-time data display, historical data inquiries, dynamic data analysis and data file download; And finally, the above requests are submitted to Web server, then the Web server starts to deal with the operations requested and return the results to client's browser. The primary Web pages of data management are shown in Figure 5.

Figure 5. Main page of the system (in Chinese interface)

The Web server application program provides a friendly human-computer interface, by which registered users could browse and analyze the data from remote sites according to a specific object, and similarly, server administrators and production managers also could simultaneously manage information coming from scattered sites at anytime and anywhere.

The most useful functions of the system are summarized as follows:

1) Real-time data display. As soon as the new data arrive to the central server, the Web page refreshes itself automatically.

2) Historical data inquiries. Selecting an observed station, and indicating the starting and ending time, users could make data inquiry of the given period for the selected site.

3) Dynamic analysis. In order to visualize the trend of the data changes with time, the Web server application program provides the function of drawing dynamic curves both for real-time and historical data.

4) Data extraction and download. In order to be convenient for users' download, The Web server provides the function of data transform from MS SQL Server to MS Excel format.

3. APPLICATION AND SUMMARY OF THE SYSTEM

In this paper, an integrated technical solution of the 'field data acquisition – data transmission – Web based data management' was introduced. This system carries out all-weather and real-time data monitoring for various environmental factors. Any user logged in the Web site could use and download the data by accessing to the server. By further analysis of the data, experts could make remote diagnosis and decision support for common farmers.

Up to now, this system has been put into applications in some sites both for practical production and for validation. The observed stations distribute in Liao-ning, He-bei, Guang-dong Provinces, and areas around Beijing, etc. for the data acquisition of greenhouses, methane gas pool and animal farms, the filed scenes of several typical sites equipped with the system are shown in Figure 6. The results show that this system is suitable for the majority of

(a) Greenhouse in the garden of CAAS's scientific technology demonstration, Lang-Fang, He-bei Province

(b) Greenhouse in Daxing District, Beijing, P.R.C

Figure 6. Application scene of the system

agricultural environment monitoring applications both for research and for practical production.

Although the system has been put into practical applications, and it has been proved to be useful, there still exist some weaknesses to be improved. For example, the system should be further integrated and optimized, the amount of sensors should be expanded to meet more needs, and especially, it is very important to perfect a web-based software that could provide multiple functions, such as data analysis, intelligent control and management for crop cultivation, fertilization and irrigation, as well as for remote pest diagnosis, according to the field data and the combination of expert experiences.

ACKNOWLEDGEMENTS

This project were jointly supported by National Program-Instrument upgrade and reconstruction (2006JG003500), National Scientific Support Program of Eleventh Five Year Plan (2006BAD04B08), the National Natural Science Foundation of China (30671211)

REFERENCES

Cox S. Information technology: the global key to precision agriculture and sustainability [J]. Computers and Electronics in Agriculture. 2002, 36: 93–111.

Lu L, Sun J. Agricultural development and modern agricultural construction of China in the new period [J]. Engineering Sciences. 2004, 6(1): 22–29. (in Chinese)

McKinion JM, Willers JL, Jenkins JN. Wireless Local Area Networking for Farm Operations and Farm Management [A]. St. Joseph. ASAE Annual Meeting Papers [C]. Michigan, USA: The American Society of Agriculture Engineers, 2004: 1–16.

Messer H, Zinevich A, Alpert P. Environmental Monitoring by Wireless Communication Networks [J]. Science. 2006, 312: 713–714.

Serodio C, Cunha JB, Morais R, et al. A networked platform for agricultural management systems [J]. Computers and Electronics in Agriculture. 2001, 31: 75–90.

Sobeih A, Chen W, Hou J, et al. J-Sim: A Simulation Environment for Wireless Sensor Networks [A]. Proceedings of the 38th annual Symposium on Simulation [C]. Washington, USA: IEEE Computer Society. 2005: 175–187.

Sun Z, Qiao X, Wang Y. Research and application of IT on controlled environment agriculture. Progress and development in digital agriculture [C]. China Agricultural Science Press, Beijing, 2005: 75–78. (in Chinese)

Sun Z, Cao H, Li H, et al. GPRS and WEB based data acquisition system for greenhouse environment [J]. Transactions of CSAE, 2006, 22(6): 131–134. (in Chinese)

Sun Z, Influence of controlled environment agriculture on modern agriculture of China [J]. Greenhouse & Horticulture, 2005, (01): 15–17. (in Chinese)

Wang N, Zhang N, Wang M, et al. Wireless sensors in agriculture and food industry-Recent development and future perspective [J]. Computers and Electronics in Agriculture. 2006, 50(1): 1–14.

THE APPLICATION OF WAP AND WEBGIS IN THE SYSTEM OF INSECTS' CLASSIFICATION RETRIEVAL

Qing Zou [1], Lin Huang [1,*], Xuexia Wang [1], Dongmei Lang [2]

[1] *College of Information Engineering, Northwest A & F University, Yangling, Shaanxi, China, 712100*

[2] *College of Changzhi, Changzhi, Shanxi, China, 046011*

* *Corresponding author, Address: College of Information Engineering, Northwest A & F University, Yangling, Shaanxi, China, 22 Xinong Road, 712100, P. R. China, Tel: +86-29-87091546, Fax: +86-29-87092353, Email: hl@nwsuaf.edu.cn*

Abstract: This paper mainly introduced how WAP, Google API and 51ditu API were applied to the System of Insects' Classification Retrieval. The introduction of Google API makes the query interface more visualized, convenient and humane. The purpose of 51ditu API is to provide support for decision-making when open country investigators are considering investigation routes. The application of WAP enables to extend the scope of the functions of the retrieval system from computer to cell phone, which renders more conveniences to open country investigators to access this system.

Keywords: Insects' Classification Retrieval, WEBGIS, WAP

1. INTRODUCTION

In these days, as computer technology has rapidly advanced, computer has been applied in every field of science and technology. Especially, the advent of computer network technologies has significantly changed the ways people live and work. However, the application of computer technology in the fields of identification and retrieval of insects still comparatively lags behind as many experts and workers (especially domestic) in the mentioned fields are still using the traditional ways to work and study, which will definitely have a strong negative impact on the advance of the science research. Although many researchers both here and abroad are dedicated to the application of

Zou, Q., Huang, L., Wang, X. and Lang, D., 2008, in IFIP International Federation for Information Processing, Volume 259; Computer and Computing Technologies in Agriculture, Vol. 2; Daoliang Li; (Boston: Springer), pp. 809–816.

the technologies in this field, the introduction and use of them are far from being sufficient and meeting the needs of the societies.

In the past ten more years, entomologist in China has developed many expert systems for assistant identification of insects. For example, Lu Huimeng has developed the *Design and Complementation of Expert System for Classification Retrieval and Diagnosis for China ACRIDOIDEA (ESCA)* (Lu et al., 2003) and *Lucid: A Multi-way Expert System for Identification and Diagnosis* (Sun et al., 2002). Both are only for stand-alones and not available for different persons simultaneously. There is also some web-based query systems for classification such as *Design and Development of Web System for Isoptera Taxonomy* (Xu et al., 2004) developed by the Research Center of City Insects of Zhejiang University. Nevertheless, such a system includes little new technology, has an insufficient function, covers a small scope of the related area, and so is unable to eliminate the inconveniences occurring in fieldwork when using a retrieval system. To solve this problem, this thesis pictures how WAP and WEBGIS are applied and actualized in retrieval system for entomologic classification. The introduction of WEBGIS makes the query interface more visualized, convenient and humane and provide support for decision-making when open country investigators are considering investigation routes. The application of WAP enables to extend the scope of the functions of the retrieval system from computer to cell phone, which renders more conveniences to open country investigators to access this system.

2. THE APPLICATION OF WEBGIS

2.1 Google API

The System of Insects' Classification Retrieval Based on WEBGIS provided users three ways to query: basic query, advanced and regional query. Google API is basically meant for regional query which including two functions: providing the queried information of insect distribution in a certain region and of distribution of a certain insect.

Google maps API can display in map information of a designated region by loading XML files, which are used to describe the regional information. By constructing the XML files describing the regional insect distribution and through Google maps API, the distribution region of insects can be visualized in the map.

2.1.1 Query of the Information of a Regional Insect Distribution

Actualization is demonstrated by Fig. 1 as follow.

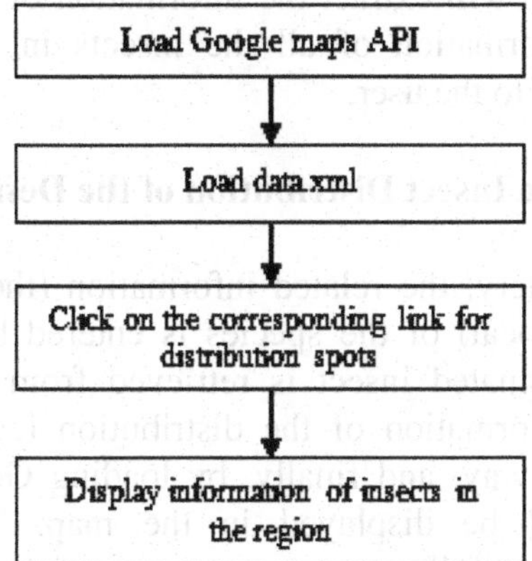

Fig. 1. Query of information of regional insect distribution

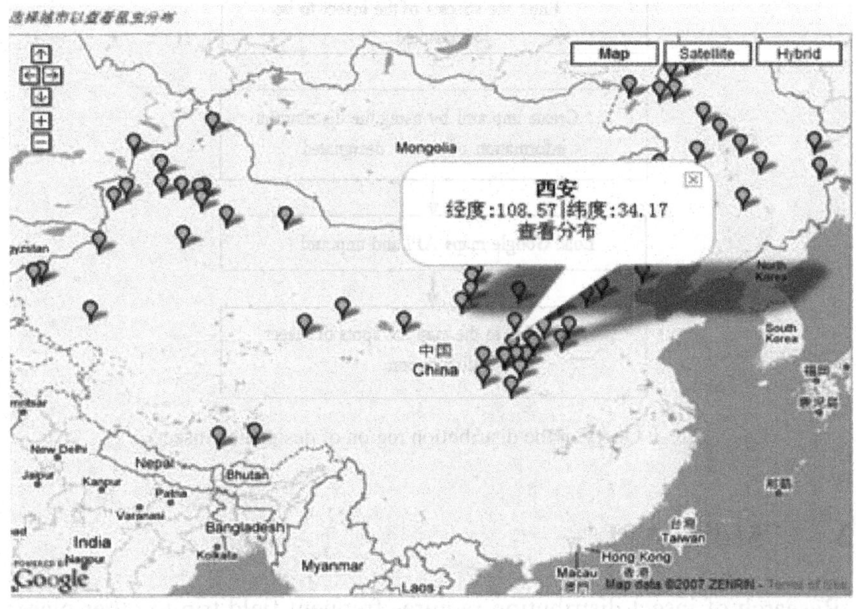

Fig. 2. Display of the region of insect distribution

In the process of query, the region of insect distribution is comparatively stable, so its information should be saved in a fixed XML file (data.xml), or such a file will be created termly from the datasheets containing insect information. System will display the insect distribution spots in the map by data.xml. When the user's mouse is pointed at a certain distribution spot, the system will display the information of the geographical location of that spot, and provide a query link "consult distribution" (Fig. 2); when the user clicks on the link, the system will extract the information of the distribution spot, and then retrieve information of all the insects in the region from the database and display it to the user.

2.1.2 Query of the Insect Distribution of the Designated Species

When using this query, the related information (the scientific names in Chinese or Latin or local) of the species is entered by the user, then the distribution of a designated insect is retrieved from the database, and a tmp.xml including information of the distribution region of the insect is created in a dynamic way, and finally, by loading Google maps API, the distribution spot will be displayed in the map. The actualization is demonstrated in Fig. 3 as follow.

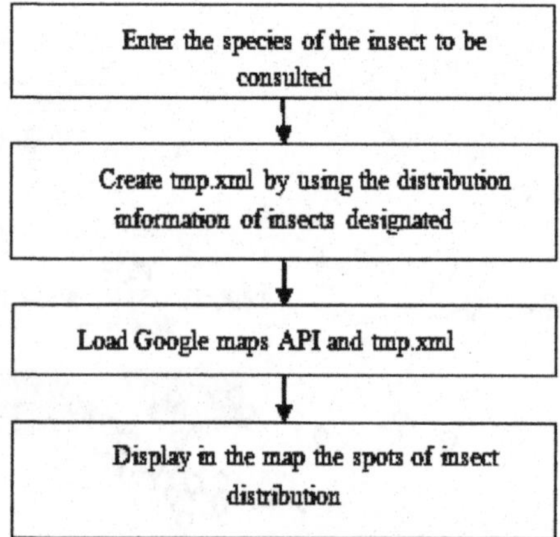

Fig. 3. Query of the distribution region of designated insects

2.2 51ditu API

Research of insect distribution requires frequent field trip to other places in the country. The system has the function of displaying the dynamic trip routes in the map at the user's convenience when the user enters a start point and end point (including midway station).

This function developed through 51ditu API, an interface of WEBGIS. When the user picks a city as the destination of investigation listed in the table of option of cities, red round dots will flash signaling as chosen cities.

When clicking on "displaying the routes", there will be lines linking all the chosen cities. When clicking on any line between any two cities, the total distance (specifically, the length of the railways between cities) of the field trip will be displayed. Besides, for the convenience, the system offers future 72 hour weather forecast for the national major cities. The result is demonstrated in Fig. 4 as follow.

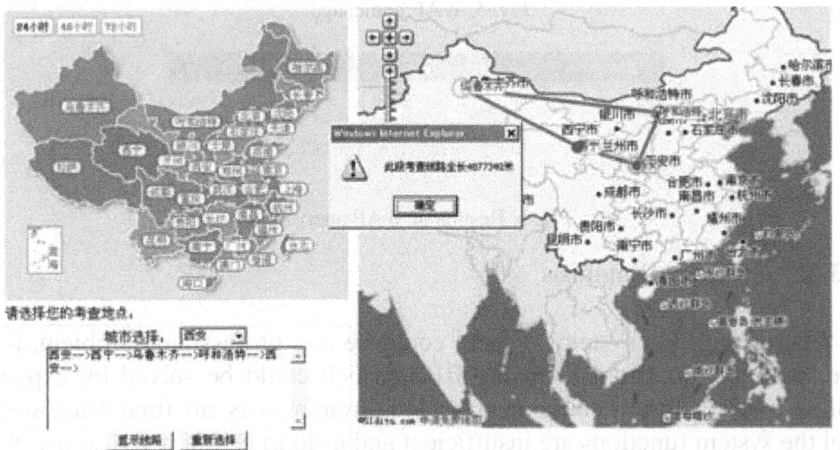

Fig. 4. Field trip route

2.3 The application of WAP

The popularization of WAP phone makes mobile phones possible to share computer network information. In regard with the mass tool-mobile phone, this system adopts WEB and WAP as means for users to share information of insect data on servers, so that users can not only communicate with servers through WWW, but log on WAP with a mobile phone at convenience of those who have no access to WWW. Therefore, constraint of time and space dissolves and complete sharing of information is achieved.

In the Webpage for a mobile phone to log on WAP, the primary functions of query and expert diagnosis on insect information are achieved.

2.3.1 Retrieval on WAP Phone

The mobile phone provides fuzzy query. When using this function, the user may enter a complete insect name in Chinese or Latin, or may enter any characters of a complete name. The homepage of WAP and the results of query are shown as following Fig. 5 and Fig. 6.

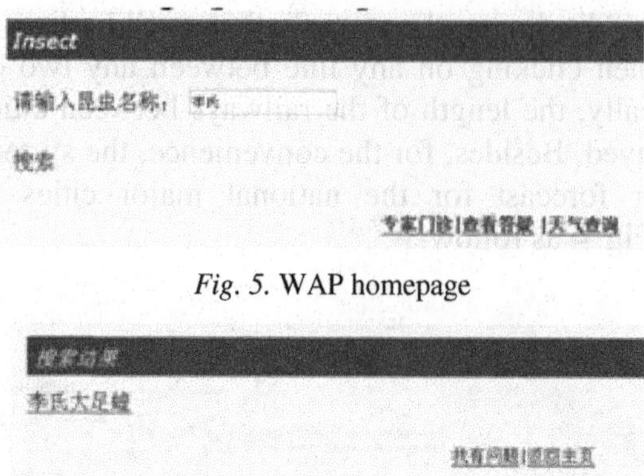

Fig. 5. WAP homepage

Fig. 6. Results of WAP query

2.3.2 Expert Diagnosis

When using this system, if there could be any professional problem, the user may ask questions by using BBS, which could be solved by experts online. And feedbacks could be presented through posts any time when users feel the system functions are insufficient and need to be improved so that the system management committee could be able to consider advice or ideas in time and get the system improved.

The interface of cell phone expert diagnosis is shown in Fig. 7 as following.

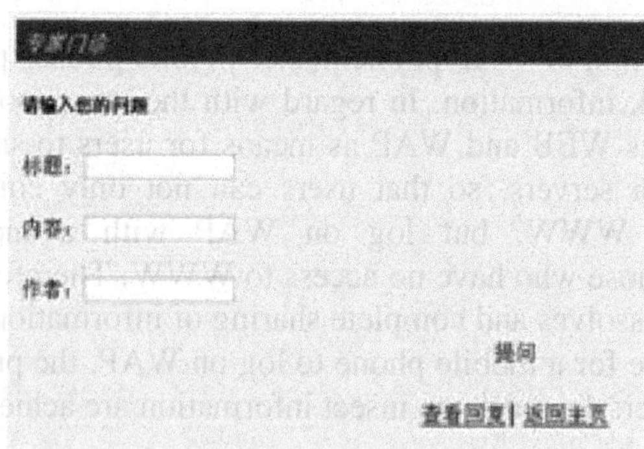

Fig. 7. Expert diagnosis of WAP

3. DISCUSSIONS

This thesis introduces the application of WEBGIS and WAP in the System of Insects' Classification Retrieval and the approaches of actualization of them. However, there are still a couple of problems found in the process of the actualization of the system.

The First problem is the display level of the insect distribution region in the map. Because the data of domestic geographic information and information for insect distribution in hand are not exclusive enough, the displayed spots can only represent regions at a municipal level instead of smaller ones.

The second is a trip route made. When the start point and end point are entered, the system can display in the map the dynamic travel routes. However, the spots entered are confined in cities at least at municipal level. Further study is required when more detailed spots are dealt. Furthermore, since at present 51ditu API interface can only provide addition of polylines connecting two places and displaying the distance of the road, which fails to picture the particular routes for transport vehicles in visualized way in the map.

The last is images displayed on mobile phone. Viewing a picture is related to the internet speed (the actual speed is between 20 to 40K), then cell phone signals and situation of network. The test results show that, to display a bigger, more memory demanding map, a higher WAP speed is required. The existing one is a major problem.

4. CONCLUSION

The application of WEBGIS and WAP in the System of Insects' Classification Retrieval provides the user a new retrieval system and new approach to display results of query, which enables the user to learn insect distribution in a visualized way. Besides, the introduction of WAP renders convenience for those users who work in open country. The user is only required to have a WAP phone to log on the system and could be able to consult needed information. The system provides a function "create a field trip route" which help analyze and decide a field trip route.

ACKNOWLEDGEMENTS

We wish to express our thanks to Prof. Yuan Feng and Yuan Xiangqun, Northwest A & F University, for their valuable suggestions to accomplish this system. The authors thank the reviewers for their kindly comments and suggestions. This research was supported by the special fund of Northwest A & F University (project code: 080807; 08080209).

REFERENCES

Lu Huimeng, Huang Yuan 2003. Design and Complementation of Expert System for Classification Retrieval and Diagnosis for China ACRIDOIDEA (ESCA). Journal of ACTA ZOOTAXONOMICA SINICA, 28: 428-433 (in Chinese).

Sun Guanying, Chen Xinxue, Cheng Jiaan 2002. Lucid: A Multi-way Expert System for Identification and Diagnosis. Journal of ACTA ZOOTAXONOMICA SINICA, 27: 871-875 (in Chinese).

Xu Xiaoguo, Mo Jianchu, Cheng Jiaan 2004. Design and Development of Web System for Isoptera Taxonomy, Journal of ENTOMOTAXONOMIA, 26: 86-90 (in Chinese).

DEVELOPMENT OF INFORMATION SUPPORT SYSTEM FOR THE APPLICATION OF NEW MAIZE VARIETY BASED ON SMARTPHONE

Feng Yang[1], Shaoming Li[1,*]

[1] *Collage of Information and Electrical Engineering, China Agricultural University, Beijing, China, 100083*

* *Corresponding author, Address: P.O. Box 698,17 Tsinghua East Road, Haidian District, Beijing, P. R. China, 100083, Tel: +86-10-82856450-8388, Fax: +86-10-82856430, Email: lshaoming@sohu.com*

Abstract: Assisted by mobile and information technologies, the risk of new maize variety application will be reduced and the efficiency will be enhanced significantly. According to the demands of new maize variety adoption and application, then an information support system which runs on Smartphone has been developed. The system consists of three main parts: information support, field problem diagnosis and business information management. Mobile development, mobile communication, mobile database and expert system technologies were used in the process of this program. The system offers a kind of new technology and approach for the new maize variety application. Moreover, it can provide powerful information support to users, and is applicable and easy-to-use.

Keywords: new maize variety application, Smartphone, mobile database, expert system

1. INTRODUCTION

The seed industry is now in a new period of great change in China (Sun Shixian, 2003). At present, there is a widespread phenomenon in China's seed industry that good variety has always lacked suitable method. On the

Yang, F. and Li, S., 2008, in IFIP International Federation for Information Processing, Volume 259; Computer and Computing Technologies in Agriculture, Vol. 2; Daoliang Li; (Boston: Springer), pp. 817–824.

one hand, it is difficult to select a suitable cropping field for the new variety. Moreover, we often lack scientific cultivation methods which take actions that suit to local circumstances. On the other hand, farmers often are accustomed to cultivate all of the varieties in the same way. So it is difficult for the new variety to have a good performance and higher yield. However, maybe there is a reduction in output (Zhu Jianfang, 2001). For the rapid conversion of new varieties to productivity, it is necessary to cultivate the new variety in a suitable and scientific way. The informationalization of the new variety adoption and application will play a crucial role to the scientific application of new variety.

The major works of the new variety application are carried out in the field in China's rural areas, but computer and Internet is still in the preliminary stage of development there. So portability, simplification and economy are the basic requirements for the information products of new variety application. As a representative of mobile computer system, Smartphone is becoming popular rapidly. Combined with modern wireless communication and WAP, Smartphone has been applied to mobile commerce, mobile office, and mobile life and so on. It changed the mode of thinking in the traditional sense that the computer is only used in the office or fixed place. It also has brought about the informationalization and digitalization developing into personal application which is deep and anywhere. Since Smartphone is easy to carry, powerful, economical and other reasons, the information system based on Smartphone has become the only way of the informationalization of new variety application.

An information support system for new maize variety adoption and application based on Smartphone has been designed and developed in this paper, which used modern information technologies synthetically such as mobile development technology, mobile database technology and expert system. The system runs on the operation system of Windows Mobile 5.0. And it consists of information support, field problem diagnosis and business information management. Because it is small in size, easy to operate and economical, there will be a good prospect.

2. OVERALL STRUCTURE AND FUNCTIONS OF SYSTEM

The information support system for the application of new variety runs on Smartphone. It is designed against more information demands, portability and resource sharing. System is divided into three levels, including the movable terminal, server and database. The overall structure of system is shown in Fig. 1, including: information database, information support module, field problem diagnosis module, business information management module, information upload and download module.

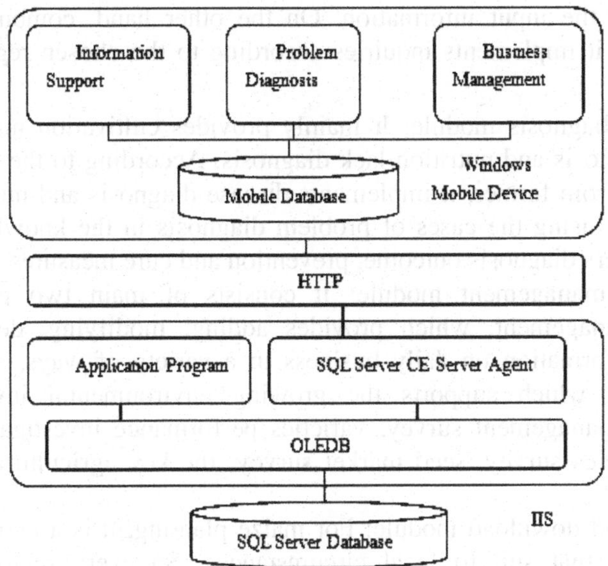

Fig. 1. The overall structure of system

Information Database: There are many kinds of and large quantities of information for the system. But the storage space and data processing capacity of Smartphone are limited, it is necessary to establish a database system including server database and mobile terminal database. When network is connected, Smartphone is used as a client of server through a wireless network, and completes data processing and maintenance by taking advantage of the mature database technology of PC platform; otherwise, Mobile database can ensure that the system is used in the off-line state, and also can record the updated and added information on the mobile terminal. Database mainly stores three kinds of information: one is used for inquiry, including variety information, county's environment information, dealers' information and business information; the other is used for field problem diagnosis, including expert knowledge, maize disease information, nutrition lack information and cultivation information; and another is survey information, including planting growing environmental investigation information, agronomic management survey information, varieties performance investigation information, competitive species survey information, and seed market survey information.

Information support module: It mainly implements adding, modifying, inquiring and deleting information on varieties, county's environment information, cultivation habits, variety preference and dealer information. There are two ways for inquiring: On the one hand, it implements inquiries in line with the input information. On the other hand, combined with a digital map, it implements inquiries according to the chosen region on the digital map.

Problem diagnosis module: It mainly provides cultivation management, disease diagnosis and nutrition lack diagnosis. According to the description of problem from farmer, it implements disease diagnosis and nutrition lack diagnosis by using the cases of problem diagnosis in the knowledge base, and then offers diagnosis outcome, prevention and cure measures.

Business management module: It consists of main two parts: daily business management, which provides adding, modifying, deleting and inquiring information on daily business in a variety of ways, and special investigation which supports the growing environmental investigation, agronomic management survey, varieties performance investigation, competitive species survey, seed market survey, the key agricultural files and so on.

Upload and download module: For maize planting, it is a prerequisite to take actions that suit to local circumstances. So every region requires suitable varieties. Because the storage space of Smartphone is limited, it is necessary to download information according to different region, and update information to the Information Center.

3. KEY TECHNOLOGY RESEARCH

3.1 Problem Diagnosis Technology

3.1.1 Knowledge Representation

It is the basis of problem diagnosis that the experts' knowledge is represented correctly and stored effectively. According to the characters of development platform, this program structures the knowledge of maize diseases and nutrition lack in terms of the object-oriented knowledge representation (Yang Zhong, 2003): First, identifying the object. To identify all objects, the research problem should be analyzed in detail. If some objects have the similar character, a special class, which is defined as lower class, should be created by abstracting these similar objects. Then, if the lower classes have the similar character, a parent class should be created by abstracting these similar objects. It is should be going on until there is no common. Moreover, all objects should be defined and named. Finally, there will be a knowledge base model of the problem field. Taking maize diseases for example, at first, the common attribute of disease place should be classified into root, stem, leaf, fruit four main categories according to the symptoms of the disease. In four major categories, then the common of the objects' symptoms should be abstracted respectively, such as lesion or lesion place, color and so on. The common character should be abstracted and summarized gradually until the final diagnosis of various diseases.

3.1.2 Inference Engine Implementation

Problem diagnosis module is an expert system virtually. The way used for reasoning is forward inference. The basic idea of this reasoning is that, starting from the known facts, a conclusion is reasoned out by using positive rules (Gao Chunming, 2002). To reason, the module requires moralized rules and man-machine dialogue interface which adapted the requirements of Smartphone. Taking maize disease for example, the module diagnoses the maize diseases, nutrition lack by reasoning mechanism, and gives the corresponding prevention and cure methods. As follows:

if (diseaseSite = = "leaf"

and spotsAndModuleStateMembranes = = "moiré, brown bacteria"

and earlySymptoms = = "oval lesion appeared on leaves"

and lateSymptoms = = "lesions expand into moiré; they are grass green in the central of lesion, grayish brown on the edge; and there are white and brown small bacteria tumors; leaves even wither when serious");

According to the expert knowledge, the conclusion is shown in the Fig. 2.

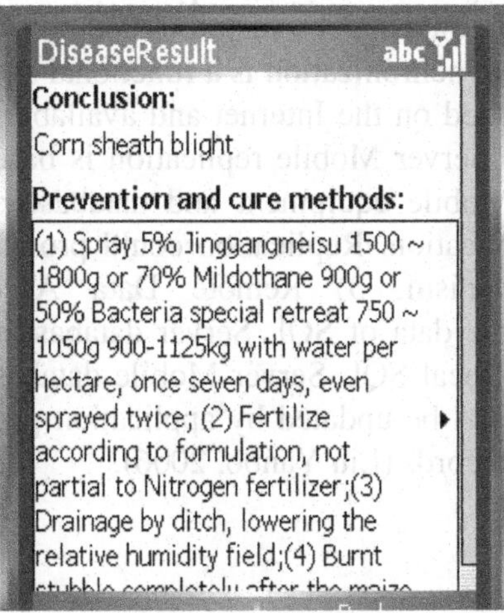

Fig. 2. The diagnosis result

3.1.3 Knowledge Base Maintenance

It is ideal for expert system to learn and preserve knowledge themselves. But the capacity of Smartphone is limited (Ouyang Jianquan, 2002). So it is difficult for the expert system runs on Smartphone to learn and preserve knowledge themselves. However, it is clever to maintain knowledge base by a distributed way. That is, the database used on Smartphone could be created on PC firstly; or the database on PC could be converted to the database that Smartphone requires. By this way, the knowledge database on Smartphone could be updated by the knowledge database on PC when Smartphone could communicate with PC. Advantages are that it could take advantage of the mature database technology on PC and avoid maintaining the knowledge database in a small screen; moreover, the interface is much friendlier. At the same time, the resources of PC and Smartphone are fully exploited. In addition, they could use the resources of each other.

3.2 Mobile Database Technology

With the popularization of mobile terminals, there are higher demands for the real-time processing and management to mobile data. Mobile database is increasingly showing its advantages. For resource constraints, mobile equipments are used as the front-end application together with general application system. The data sets on the mobile equipment may be the subset of the data set on back-end server or the copy of the subset on the server. Generally, mobile database keeps synchronous with the database server through some data replication mode (upload, download or mixed mode). Thus meet the demands that people visit arbitrary data in arbitrary location and at any time (Zhou Shumin, 2007).

Web Service data synchronization, Replication and Remote Data Access (RDA) are three main data synchronization methods (Zhou Shumin, 2007): 1) Web Service data synchronization is a functional set which is packed into a single entity published on the Internet and available for other procedures. 2) Replication, SQL Server Mobile replication is based on Microsoft SQL Server replication. Mobile equipment and database server exchange data automatically by replication. Replication could provide a resolution to the data conflict mechanism. 3) Remote Data Access (RDA) permits applications access the data of SQL Server database from the remote, and store the data in the local SQL Server Mobile database table. The table of SQL Server table could be updated by application procedures according to the changes of local records (Liu Yanbo, 2006).

Data on server are downloaded by SQL Server Mobile applications running on mobile equipment through RDA. In the system, the database, which based on Windows Mobile 5.0, is designed with SQL Mobile and SQL Server 2005 development environment. And data synchronization is implemented through RDA.

4. DEVELOPMENT AND IMPLEMENTATION OF SYSTEM

Smartphone, which supports wireless communication and network connection, is used as hardware. The system runs on the operation system of Microsoft Windows Mobile 5.0, and based on the Microsoft .NET Compact Framework 2.0. And it is implemented with C # and Microsoft Visual Studio 2005. Database management platform is SQL Mobile.

5. CONCLUDING REMARK

In this paper there is a desk study on the design and implementation of the information support system for the application of new maize variety based on Smartphone. According to the business characteristic of the new variety application, the system provides a solution program against the problems of immediate inquiry and resources sharing in the process of new variety application. Mobile development, mobile communication, mobile database and expert system technologies were used in the process of this program. Application of information technology and mobile technology, the program has been designed and implemented to assist strong and updatable information for the application of new maize variety. Thus, the new good variety and suitable cultivation methods are offered together. It is likely that the new variety will have a good performance. So the risk of new crop variety application will be reduced and the efficiency will be enhanced significantly. With the development of Smartphone, mobile communication, the Internet and other technologies, information systems based on Smartphone in the variety industry would have a wider application in the near future.

ACKNOWLEDGEMENTS

This research is funded by the Support of Science and Technology Project of the State, Programmed award No. 2006BAD10A01 from March 2007 to November 2009.

REFERENCES

Gao Chunming, Chen Yuexin, Su Liang, Liu Dongbo. Direct Reason-Machine System Research. Computer Engineering & Application. 2002, (19):78-80.

Guo Yinqiao, Guo Xinyu, Li Cundong, Zhao Chuande, Zhao Chunjiang. Knowledge model-based decision support system for maize management. Transactions of the Chinese Society of Agricultural Engineering. 2006, 22(10):163-166.

Liu Yanbo, Hu Yanyan, Ma Ji. Windows Mobile Application and Development Platform. Beijing: POST & TELECOM PRESS. 2006, 336-339.

Ouyang Jianquan, Qian Queliang, Chu Chengyuan, Li Jintao. The Design and Implement of PDA-Oriented Expert System in Agriculture. Computer Engineering & Application. 2002, (2):30-31.

Peng Hai-yan, Chen Gui-fen, Peng Hai-ying, Yang Lian-hui, Tong Mei. Study and development of expert systemdiagnosis of maize about the scathe for disease and worm. Computer and Agriculture. 2002, (3):16-17.

Sun Shixian. Forecast for the generalization of new varieties during “tenth-five”. Seed Science & Technology. 2003, (3):127-128.

Yang Zhong, Zuo Hongfu, Shen Chunlin. Realization of Object-Oriented Rule-Type Expert System Template. Transactions of Nanjing University of Aeronautics & Astronautics. 2003, 20(2):218-223.

Zhou Shumin, Zhang Tiantai, Xu Zhiwen. The Data Synchronization of Embedded Database and Its Application]. Control & Automation. 2007, 23(5-2):79-80.

Zhu Jianfang, Yang Xianzhong, Wang Yunhua. Accelerate the pace of new varieties, promote the development of seed industry. Seed. 2001, (3):78-79.

GRASS COVER CHANGE MODEL BASED ON CELLULAR AUTOMATA

Shuai Zhang [*], Jingyin Zhao, Linyi Li
Digital Agricultural Engineering Technological Research Center, Shanghai Academy of Agricultural Sciences, Shanghai, China, 201106
** Corresponding author, Address: Digital Agricultural Engineering Technological Research Center, Shanghai Academy of Agricultural Sciences, 2901 Beidi Road, Shanghai, 201106, P. R. China, Tel: +86-21-62208434, Fax: +86-21-62208434, Email: zhangshuai@saas.sh.cn*

Abstract: This research attempt to establish a grass cover change model based on cellular automata. It took the relationship between elevation, slope, aspect, settlement, road, water system and grass cover change as the foundation of the model, the dynamic progress of grass cover change was expressed by using cellular automata, the grass cover change model based on the basic geographical control condition which has been mentioned in above paragraph was constructed. The grass cover state of Madoi County in 2003 was simulated, and has been compared with the actual state of Madoi County in 2003. It was found that the grass cover change model has a high precision. The value of kappa index has reach 0.8801.

Keywords: cellular automata, grass cover change

1. INTRODUCTION

Cellular automata was conceived and advanced by John Von Neumann during the 1950s. Cellular automata is a kind of dynamic model which is discrete, it has not been confirmed by strictly defined physics equation or function, but a set rule of construct model, it is a sum of a kind model. Because it has reciprocity between different cells, cellular automata is fit to simulate complex system, at the same time it is also has spatial conception which can make the change of object visibility, and it is fit to simulate changes in large scale.

Zhang, S., Zhao, J. and Li, L., 2008, in IFIP International Federation for Information Processing, Volume 259; Computer and Computing Technologies in Agriculture, Vol. 2; Daoliang Li; (Boston: Springer), pp. 825–832.

We can characterize cellular automata as follows:

Cell-Cell is the basic element of cellular automata, and each cell has a set of states. The states could be a binary system; it also could be internal variables.

Grid-Grid is a aggregate of cells, it could be 1-dimensional, 2-dimensional or multidimensional.

Neighborhood-The neighborhood of cell is a collection of cells around the cell, they will impact the state of the cell at next time. In 1-dimensional cellular automata the neighborhood is confirmed by radius. In 2-dimensinal cellular automata the neighborhood has three style, they are Von.Neumann, Moore and prolate Moore which are showed in Fig. 1.

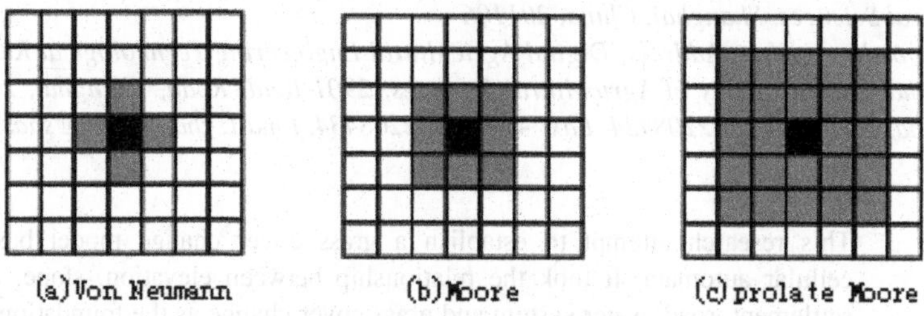

Fig. 1. Neighborhood model of cellular automata

Rule-The state of a cell is determined by a set of rules which depend on the current state of cell and the state o f the neighborhood cells.

Time-Time is a set of internal integers.

Cellular automata based models have been used to study forest fires, soil erosion and so on.

Li use the cellular automata which based on nerve network simulate the complex system of land use and land cover change in zhujiang delta (Li et al., 2005). Chen has used the cellular automata model based on 3s technology to simulate the hungriness change, simulate and forecast the trend of the hungriness change trends in Beijing and adjacent (Chen et al., 2004). Fang has analysed the land cover and land change data in Peoria area from 1993 to 2000 (Fang et al., 2005). Filho has simulated the landscape change in Amazon area with stochastic cellular automata (Filho et al., 2004).

Cellular automata has been used in many fields, but is has not been used to simulate the change of grass cover. This paper presents an cellular automata model to simulate grass cover change in Madoi County.

Madoi County, which is located in the source region of yellow river, is the main part of the source region of yellow river, it is between 96°50′–99°32′ E, 33°52′ – 35°39′ N, total area of Madoi County is $2.45 \times 10^4 km^2$. In recent years, the source region of yellow river has become an important research focus area, the degradation in this region is significantly serious in past decades (Dong et al., 2002). Degradation has a grate impact on the ecosystem health in the source region of yellow river (Feng et al., 2004).

2. MATIERIALS AND METHOD

2.1 Data and pretreatment

We used the DEM data which resolution is 100m. With the spatial analyst in Arcmap we can get the slope and aspect data which resolution are also 100m. Use the reclassify function in spatial analyst model the elevation data was divided by a interval of 100m, then the slope data was reclassified by 3°, 5°, 8°, 12°, 15°, 18°, 21°, 25°, 30°, 35°, the aspect data was reclassified by north, northeast, east, southeast, south, southwest, west, northwest; the distance buffer grid was created with the distance command by use the buffer distance is 100m.

Put the grass cover change data overlap with the elevation, slope, aspect, settlement, road and water system data, the grass cover change statistic of each zonal grid, the ratio of the change area and the zonal grid area, which is short for the rate of change in the below paragraph. We fit each factor with the rate of changes by Origin.

2.2 Model establishment

2.2.1 Basic geographical control condition

We fit the basic geographical control condition with the rate of change by use Origin. The result was showed in Fig. 2–7.

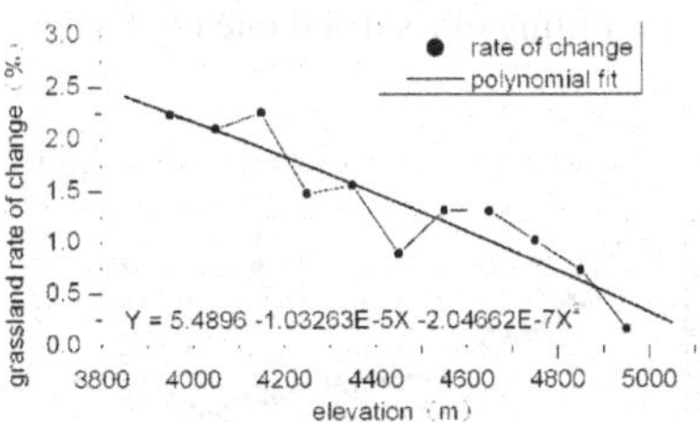

Fig. 2. Fitting of elevation and rate of change

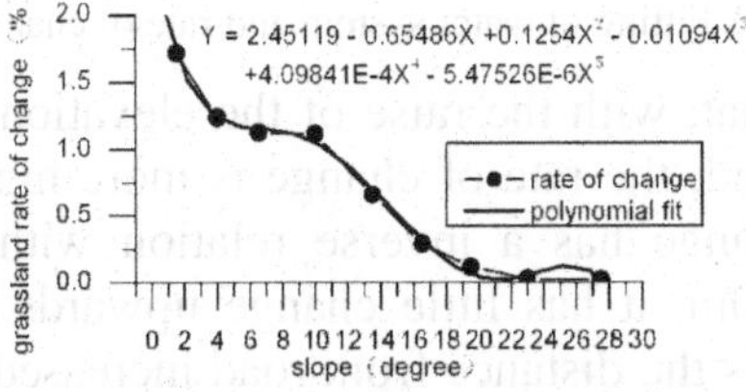

Fig. 3. Fitting of slope and rate of change

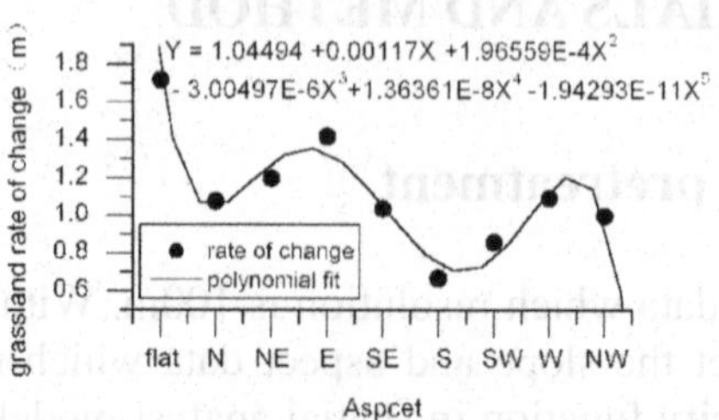

Fig. 4. Fitting of aspect and rate of change

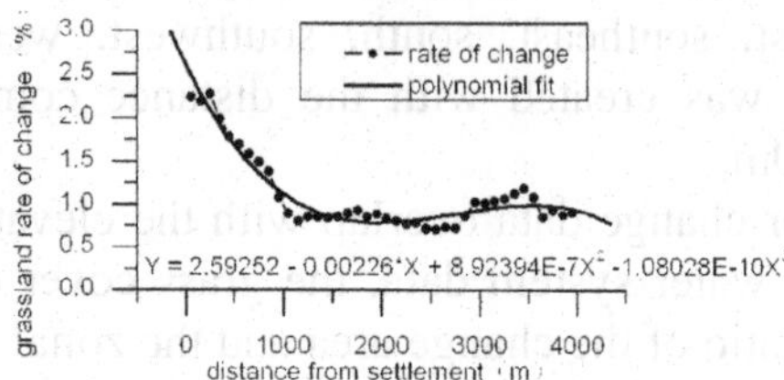

Fig. 5. Fitting of settlement and rate of change

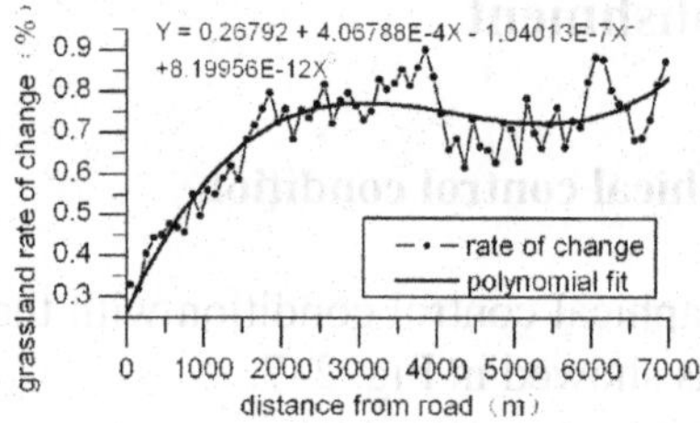

Fig. 6. Fitting of road and rate of change

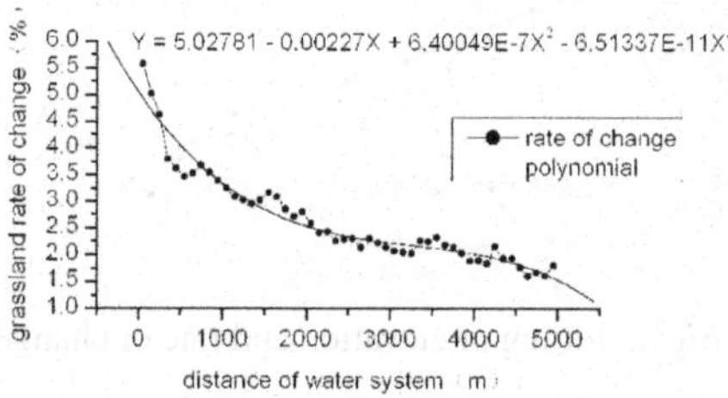

Fig. 7. Fitting of water system and rate of change

It has been found that, with the raise of the elevation, the rate of change has present a downtrend; the rate of change is more high in where slope is lower, the rate of change has a inverse relation with the distance from settlement within 1000m, it has little change upwards 1000m; the rate of change has increased as the distance from road increased; the rate of change has increased rapidly as the distance from water system within 2000m.

2.2.2 Rules of model

Based on the analysis, this paper has established a grass cover change model by use cellular automata, which has make hypothesis before modeling as follows:

(1) The grade of grass cover will reduce when the rate of change has reached the threshold value for 3 years, and the conversion will not span grade.

(2) The conversion of shrubbery to grass will take place only when the rate of change has reached the threshold value for 5 years.

(3) The sandlot expand 100m every 3 years, but only the low cover grassland around sandlot will turn into sandlot, other types will not be influenced by the expand of sandlot.

(4) Each base geographical control condition is independent.

(5) Only the conversion relationship between shrubbery, grass and sandlot were considered, the conversion relationship between other types are not considered for the moment.

(6) The impact of climate and policy are not considered in this paper.

According to the concept and theory of cellular automata, this paper has set the construction of model.

(1) Cell - The shape of the cell is square, and its resolution is 100m.

(2) Cell space - The cell space is the research area.

(3) State - Each cell has two states sometime: 0 and 1. 0 stands for the type of cell will not change, 1 stands for the type of cell will change. Each cell has a change probability, which show the probability of the change.

(4) Time - The time interval is one year.

(5) Neighborhood - Use the prolate Moore model.

(6) Rule - For each cell, the change probability is the average of the probability of neighborhood.

The process of cellular automata is as follows:

(1) Fit the elevation, slope, aspect, settlement, road, water system with grass cover change data, get the chart of influence factor of grass cover change, and make standardization of it.

(2) Delete water body and other type from grass cover map.

(3) Compute the grass cover change probability chart of current year by using the chart of influence factor map.

(4) Simulate the uncertainty of grass cover change by using random function.

(5) The threshold value was confirmed at first, then the finally threshold value was confirmed after simulate and compare, compute the rate of change which big than the threshold value, get the map of grass cover change in current year.

(6) Get the grass cover map in next year by overlap the grass cover change map and grass cover map.

3. RESULTS AND DISCUSSION

Input the grass cover data of year 1990, compute the grass cover data of year 2003, the result was showed in Fig. 9. And the actual state of grass cover in year 2003 was showed in Fig. 8.

Input the compute result in ENVI to compute Kappa index, the Kappa index between the simulate data of year 2003 and the actual data of year 2003 has reached 0.8801.

The compute result was compared with actual data in Table 1, low cover grassland and sandlot has difference quantitatively.

Table 1. The comparison between simulate value and actual value

	Simulate value	Actual value
High cover grassland	570887	556682
Mid cover grassland	426285	423009
Low cover grassland	889516	927951
Shrubbery	13091	12922
Water body	323117	327573
Sandlot	154805	126949
Other type	70519	72636

The units for simulate value and actual value are km^2.

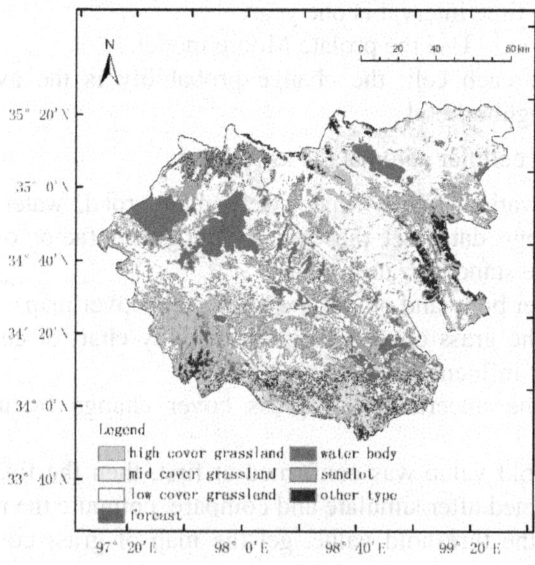

Fig. 8. Actual state of grass cover

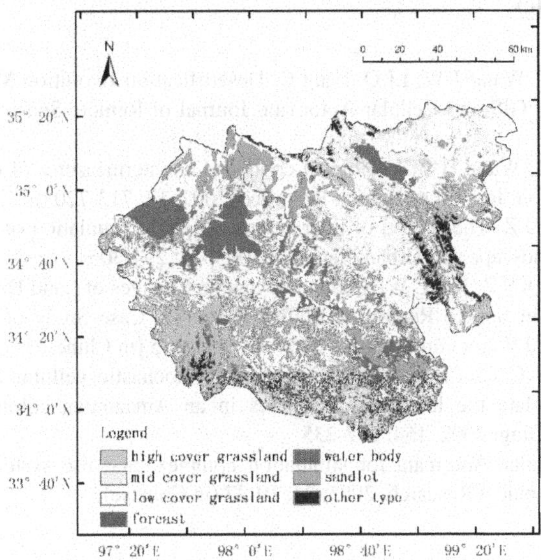

Fig. 9. Simulate state of grass cover

4. CONCLUSIONS AND FUTURE WORKS

This paper has established a set of rules for grass cover change according to the theory of cellular automata, and it was based on the considering of influence of the basic geographic control condition such as elevation, slope, aspect, settlement, road, water system on grass cover change. Then we construct a dynamic model to express the dynamic change of grass cover, the grass cover state of Madoi County in 2003 was simulated by using this grass cover change model, and it was compared with the actual state of grass cover of Madoi County in 2003, it was found that the precision was high, the Kappa index between simulate result and actual value has up to 0.8801, it has show that cellular automata could simulate the grass cover change was predominated adequately.

ACKNOWLEDGEMENTS

We are grateful to colleagues who provide the grass cover data used in the present study – Qinghai sanjiangyuan research group, Institute of Geographic Sciences and Natural Resources Research. This work is supported by Research and Demonstrate for Key Technology of Modern Country Information Project (Contract Number: 2006BAD10A11) of National Science & Technology Pillar Program.

REFERENCES

Chen J P, Ding H P, Wang J W, Li Q, Feng C. Desertification Evolution Modeling through the Integration of GIS and Cellular Automata. Journal of Remote Sensing, 2004, 8: 254-260 (in Chinese)

Dong S C, Zhou C J, Wang H Y. Ecological crisis and countermeasures of the Three River' Headstream Region. Journal of Natural Resource, 2002, 17: 713-720 (in Chinese)

Fang S F, Gertnera G Z. The impact of interactions in spatial simulation of the dynamics of urban sprawl. Landscape and Urban Planning, 2005, 73: 294-306

Feng J M, Wang T, Qi S Z, Xie C W. Study on Dynamic Changes of Land Desertification and Causal Analysis in Source Region of Yellow River – a Case Study of Maduo County. Journal of Soil and Water conservation, 2004, 18: 141-145 (in Chinese)

Filho B S, Gustavo C C, Ca'ssio L P. Dinamica: a stochastic cellular automata model designed to simulate the landscape dynamics in an Amazonian colonization frontier. Ecological Modelling, 2002, 154: 217-235

Li X, Ye J A. Cellular Automata for simulation complex land use systems using neural networks. Geographical Research, 2005, 24: 19-27 (in Chinese)

RESEARCH AND DESIGN OF THE AGRICULTURAL SHORT-MESSAGE MANAGEMENT SYSTEM

Xinlan Jiang[1], Wanlin Gao[1,*], Wei Liu[2], Ganghong Zhang[1], Hongqiang Yang[1], Yang Ping[1]

[1] *College of Information and Electrical Engineering, China Agricultural University, Beijing, China, 100083*

[2] *Zibo energy supervision center, Zibo City, Shandong Province, China, 255031*

[*] *Corresponding author, Address: P.O. Box 105, College of Information and Electrical Engineering, China Agricultural University, 17 Tsinghua East Road, Beijing, 100083, P. R. China, Tel: +86-10-62736755, Fax: +86-10-62736746, Email: gaowlin@cau.edu.cn.*

Abstract: Aiming at the current status of the agricultural information service business, this paper elaborates the characteristics of the agricultural short-message resources, and puts forward a kind of framework and a design method that sets up the agricultural short-message management system, and the key technique of carrying out the system. Especially in the upper-level applications of analyzing, processing, and automatic replying the agricultural short-message, this system adopts a method that separate the short-message application support platform from the agricultural short-message management center. The method would make system more practical and flexible. Such quick, convenient, omnipresent information service would make the application have a good future.

Keywords: agricultural information, SMS, SMSC, short-message management system

1. INTRODUCTION

On the background of new countryside construction, the rural informationization has already become an important mission of promoting the agricultural development and the society improvement of countryside

Jiang, X., Gao, W., Liu, W., Zhang, G., Yang, H. and Ping, Y., 2008, in IFIP International Federation for Information Processing, Volume 259; Computer and Computing Technologies in Agriculture, Vol. 2; Daoliang Li; (Boston: Springer), pp. 833–840.

(Yearly report on agriculture information-based development research in Chinese countryside in 2006-2007). Our country's countryside information-based have been already turned into a new diversified stage of development currently. The first, the requirements of farmer about the market information have already changed from simple releasing supply and demand information, but earnestly hope to gain more large-scale benefits through a more extensive platform. Second, integrate resources. He Yuan and Wu Yuzheng (2007) says, "Now, a lot of the type of science and technology, and living information all come from the literature of ivory tower, there are a lot of technical terms, the information of farmer's real demand still needs someone to arrange, and data mining."

At present, farmers obtain the agriculture information mainly through several methods, such as newspaper, broadcast, television, the telephone consultation, the computer network and mobile, etc. Each improvement of method, all obtained various information to provide more convenient path for farmers. Although the Internet comes ubiquitous, it still falls through at home to get to the Internet in largely village, not flourishing region at present. But the current wireless correspondence network almost overlaid national and each corner (Ma et al., 2006), if use the mobile short-message as the delivering way of agriculture information, can be free from the restriction of time, region and equipments to obtain agriculture information easily.

However, how the numerous and jumbled agricultural information get to the different farmers, aiming at farmers' miscellaneously need and with low cost, through the mobile short-message method, releasing the farmer's information in time, ensuring the information in time and accurately, preventing from the mistaken agriculture information, is still the three greatest hard nut to cracks of agricultural information-based. Thus this paper elaborates the characteristics of the agricultural short-message resources, and puts forward a kind of framework and a design method that sets up the agricultural short-message management system, and the key technique of carrying out the system.

2. CHARACTERISTICS OF THE AGRICULTURAL SHORT-MESSAGE RESOURCES

A suitable agriculture short-message resource is one of the most important conditions of providing characteristic and colorful agriculture short-message service. Because of the specialty that the agricultural short-message resources face to the agriculture and village population, there is higher request on its specialty, easy-understanding, brief-capability, and reliability (He et al., 2006).

1) Information resources should be abundant, overall, accurate, and practical

The agriculture short-message information resources wants to insure its reliability, accuracy and practicality, the information source is also the key aspect, in addition to the process of professional processing and auditing. Currently the practical technique database of being applicable to the farmer and village in the domestic is mostly small scaled and self-use, the large database is particularly less, and the databases that combine the completely tracking of the comprehensive news and the market dynamic state is tightly lack.

2) The management of the information resources needs a norm

The norm management of the information resources contributes to the knowledge-based and deeply mining of the information resources. Accumulate intelligence of numerous agriculture experts, gather together the research strength of the university, college, the research organization, set up an agriculture mobile short-message database which is easy to manage and use, and have a sea of information, the norm management of the information resources is essential.

3) The information service should be professional and characteristic

The characteristic service means a kind of service that specially provides the service method and the service product according to the need characteristics, such as different service environment, service object and the access habit, etc. Want to carry out the characteristic agriculture short-message service with strongly time-limited efficacy and good instruction, the professional process of the information resources and the extensive understanding of farmer's information demanding is an initial condition.

3. FRAMEWORK AND DESIGN METHOD OF THE AGRICULTURAL SHORT-MESSAGE MANAGEMENT SYSTEM

3.1 Requirements analysis

The type of the short-message business can basically divided into the following four major types currently: 1. The type of actively sending out like notify; 2. The type of interaction search through question and answer method; 3. The type of sending out on time which subscripted by customer; 4. The type of replying confirmation after passively received (Ren et al., 2007).

For the sake of the demand of farmers, this system joins above-mentioned ex-service of 3, such as providing some instant information through the notice method to send out actively to the customer; customer and network agricultural information carry on information alternant function with the type of question and answer through the mobile short-message, and categorizing and arranging the agriculture authority information aiming at the actual need, and then converting to the form of short-message, for the purpose of customer can subscribe; Otherwise, the information, such as the farmer self's experience etc., can also send out to the network directly, to provide a basis for enlarging the contents information database later.

3.2 Multi-layer structure of the system

The agricultural short-message management system mainly is absorbed in the upper-level applications of analyzing, processing, and automatic replying the agricultural short-message, carries out the function that customer and network agricultural information carry on the alternant function of information with the question and answer type, through the mobile short-message, and carries on a management to all short-messages. This system adopts a method of separate the short-message application support platform from the agricultural short-message management center, making system more vivid and practical. System framework sketch is following. (Fig. 1).

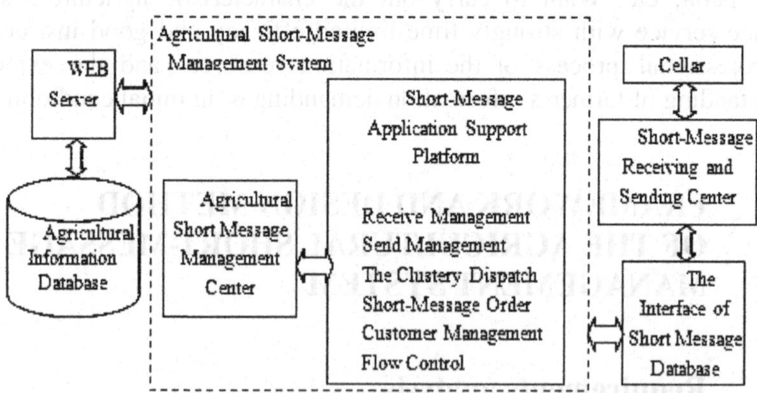

Fig. 1. Agricultural short-message management system framework

The message application support platform mainly carry out the functions of data interaction with the short-message receive and dispatch center through the short-message database interface, it has the data transmission and the confirming process after receiving transmission; Having the stable data comes error processing mechanism, interface data transmission control strategy dependable and perfect; Having a stable contents control and filter mechanism.

The information filter module: Provide the information filter mechanism, complete the filter of information through frequently adjusting main factors as keyword etc.; The information that the application hand over is divided into two types to filter and not to filter, The information that passes the information filter module percolation just can be sent out, or enter into the agricultural short-message management center.

The data management module: Realize the functions of taking out and saving the related technique data of each short-message, and business statistics etc.

3.3 Design of the system function module

The service is divided into four parts that the agriculture short-message management system provide, namely the information search function, short-message receive and dispatch management, short-message order module and information release function, its function module structure sketch is following. (Fig. 2).

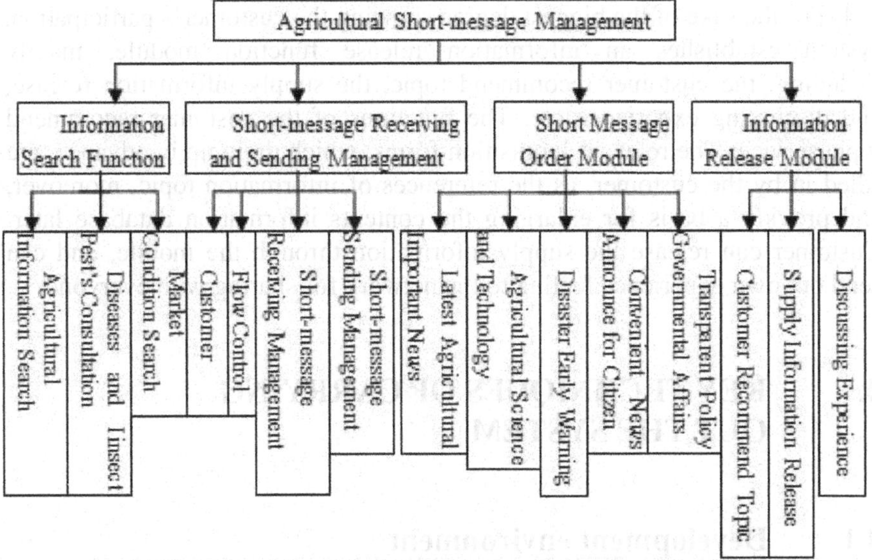

Fig. 2. Agricultural short-message management system function module

1) The information search function module

This module mainly includes the services, such as the market condition search, the agriculture information search, and plant diseases and insect pest's consultation etc. The customer sends out a message to server through a mobile, the server then carries on match according to the message contents and the keyword, and then filter and sieved the information that the customer need from the agriculture information database, according to the keyword. After the operation, the system will send out successful search result or the hint information of failure to the customer's mobile.

2) The short-message receiving and sending management module

The short-message receiving and sending management module is mainly providing the functions, such as receiving and sending out short-message management, customer management, and flow control etc.

3) Short-message order module is to provide the instant information for customer to order, such as the transparent policy governmental affairs, the latest agricultural important news, agricultural science and technology, the disaster early warning, and the convenient news announce for citizen etc. It's the expanding function about the Web agriculture information network. This kind of wireless-network-based special service, change into actively providing an information toward the customer from originally waiting customer passively, extend and enrich the function of the agriculture information network, and provide a better and faster service for the customer.

4) For the sake of the biggest degree showing the customer's participation, system establishes an information release function module, mainly including: the customer recommend topic, the supply information release, and discussing experience etc. The functions of the customer recommend topic service: The relevant inquisition forms, which drew up in advance, are filled in by the customer, as the references of information topic, moreover, and provide a basis for enlarging the contents information database later. Customer can release the supply information through the mobile, and can send out own experience feeling to the network for sharing with everyone.

4. KEY TECHNIQUES OF CARRYING OUT THE SYSTEM

4.1 Development environment

This system takes the Visual Studio.NET 2003 as the developing platform, and uses C# programming language to develop the application. The server software is installed as follows: The operating system is the Windows 2003 Server, the server software is an IIS 6.0, and the database is SQL Server 2000.

4.2 ADO.NET data access techniques

The database management uses the advanced ADO.NET technique. The ADO.NET model has two very important characteristics: the first is programmable. It adopts the operation object of type, the object defined by oneself can be programmed by the programmer, can describe business object better and make other customers more easily comprehend. The second is interoperability, this kind of characteristic is shown by using DataSet (the data gather), one DataSet can include arbitrarily many data tables, the DataSet constitute a non-linking database view and this kind of non-linking structure system makes only to just need to use a database server resources while read and write database. As a result, it provides better flexibility (Chen, 2006).

In the system all the data commutations complete through a database, the agriculture short-message management system acquired the data which need to process from the short-message database, and then write the results into the short-message database after the processing completed. Moreover, some module, which alternate with the information source in the exterior system, should possibly convert the accepted original data (like text file's data, electronic spreadsheet...etc.) to the data which is deposited in database, reduce the process task of non- database format data in system (Tang, 2005), so can availably manage data.

4.3 Debug

At initial stage of debug can carry on program test in interior network, finally carry on function debug with real mobile, main contents include: information release, test of the function on the receiving short-message→ analysis and processing→replying short-message, ordering short-message→ sending out short-message, clustery dispatching short-message→sending out short-message etc.

5. CONCLUSION

This paper mainly introduced the application of the mobile short-message service at the agriculture information. It has the characteristics of quick, convenience, omnipresent information-based service, and will have a good future. The system can also enlarge a development of which can browse, search, release the agriculture information in the WAP mobile in the future, and will have a greater application in the realm that the Internet can't arrive or can arrive but with high cost, and will be an important complement to the Web agriculture information network.

REFERENCES

Chen Ying, Song Ling. 2006. Implementation of SMS Platform Management System Based on ASP.NET [J]. Computer Knowledge and Technique (Academic Exchanges), (5): 37-38.

Consultant of SAIDI, Consultation Center of City Information-based. 2007. Yearly report on agriculture information-based development research in Chinese countryside in 2006-2007. http://www.qcdz.cn/InfoHtml/2007-3/20073151600.shtml.

He Qiyun, Huang Liang, Wang Zhong. 2006. The mode and expansion counter plan of agriculture information service based on short-message [J]. Journal of the China Society for Scientific and Technical Information, 25(10): 176-179.

He Yuan, Wu Yuzheng. 2007. In 10 CIO eyes of 2007: Surmount the technique Fusion business [OL]. The Computer World. http://e.chinabyte.com/346/3031846.shtml.

Ma Xiaojin, Zhou Yong, Jia Shaorui, etc. 2006. Design and realization of score inquiry sub-system based on .NET and WAP mobile [J]. Journal of Hebei Institute of Architectural Science and Technology, 23(3): 80-82.

Ren Chen, Huang Zhengqian. 2007. Face to gather a group of message customers to unify to connect into terrace directly according to the XML [J]. China New Telecommunications (Technical Edition), (5): 83-87.

Tang Chunsheng. 2005. The research and design water conservancy wireless search system based on WAP [J]. Jiang-su Water Conservancy, (8): 8-9.

THE CONSTRUCTION OF SMALL TOWN INFORMATION PORTAL USING OPEN SOURCE SOFTWARE

Cheng Peng, Jing Li, Jianwei Yue*, Qiaozhen Guo

Institute of Resources Technology and Engineering, Institute of Resources Science, Beijing Normal University, Beijing, China, 100875

* *Corresponding author, Address: Institute of Resource Science, Beijing Normal University, 19 XinJieKouWai Street, Beijing, 100875, P. R. China, Tel: +86-10-58807713, Fax: +86-10-58807163, Email: yuejw@ires.cn*

Abstract: Along with the development of small towns, traditional or common methods of urban informatization construction are not fit for small towns. Therefore it's essential to bring forward an appropriate way. By studying on the latest open source portal software uPortal, the paper discussed the application of personalized service, portal technology and information integration technology in informatization construction of small towns. Finally, the design and realization of the information portal and a portal website of small towns, which achieve the management and sharing of information in small towns, were presented.

Keywords: open source software, portal technology, uPortal, small town information portal

1. INTRODUCTION

In recent years, the means of informatization management has been adopted in increasing small towns of China. However, as a whole, information service system is immature and the level of informatization is not high. A few application cases in demonstration plot are just initial experiments, far below the proper level. These cases have limitations in their extensibility, because most of them are established in metropolitan or developed area, where there are good informatization conditions and large investment, being a sharp contrast to numerous small towns where there are

Peng, C., Li, J., Yue, J. and Guo, Q., 2008, in IFIP International Federation for Information Processing, Volume 259; Computer and Computing Technologies in Agriculture, Vol. 2; Daoliang Li; (Boston: Springer), pp. 841–850.

poor informatization conditions as well as shortage of funds (Li, 2005). Therefore, customized solution to the construction of small town informatization should be proposed to meet the special need of small towns. Besides, reusable models and other methods should be applied to reduce developing cost.

Considering departments and organizations are usually simple in small towns, "centralized" informatization method is considered to be applicable in construction of small town informatization. Thus, a general integrated platform of small towns can be established to enable each industry department to process informatization in the platform, based on which a portal website can be built to implement information integration, process and management of small town with high efficiency. This paper focuses on the application of Open Source Software and portal technology in establishment of integrated informatization platform of small town and proposes a new method of propelling the development of small town informatization.

2. INTRODUCTION OF UPORTAL

2.1 Open source software

Open Source Software (OSS), with features of being free used and open source code, is a new mode of software development and publication in current years. With the continuous development of information technology, OSS has more extensive application and wide influence. From scientific research to business application, from operating systems and application servers to application software of education, office, ERP and CRM, OSS has been extended in various industry fields.

Compared with traditional commercial software, OSS has many advantages which lie in: □Low cost: installation and operation cost is not necessary; □Open source code: codes of OSS are usually distributed with software, which make it possible that the source codes can be modified to add new functions; □Easy to modify: anyone can modify the software to catch and solve errors as he likes (http://www.open08.com, 2006). Due to the modification and improvement of many users, software that are developed in this structure are very flexible and reliable (Paolo, 2005).

In contrast with the informatization of metropolitan such as Beijing and Shanghai, the informatization scale of small town is so small that OSS not only fits the features like small-scale, low-cost and flexibility in the informatization of small town, but also avoids high cost and un-changeability of normal commercial software.

2.2 Portal technology

Portal technology is a kind of integration solution based on application layer and presentation layer. As a web application program and presentation of information system, Portal provides functions of personalized service, uniform login and content integration function. From view of architecture, Portal, an application level service above web server and application server, is essentially a web application in platform level (Huang, 2004). The key technology of Portal model is Portlet. Being a web component based on Java technology, Portlet is a packaging component based on web content and application function, running in Portlet container of Portal server. Portlet can connect different data sources, which may be local or remote web page, data from database, or application program (Huang, 2004). On the other hand, Portal can display Portlet from different sources from view of user interface.

As Fig. 1 shows, the portal page is divided into several Portlet zones, each of which can display content that is independent from each other. Moreover, content can be customized according to specific requirements and each Portlet can be shut down. An excellent Portal has features of friendly interface, self organization, self service, expandability and safe compatibility (Jin et al., 2005). Fig. 2 shows the implementation environment of Portal.

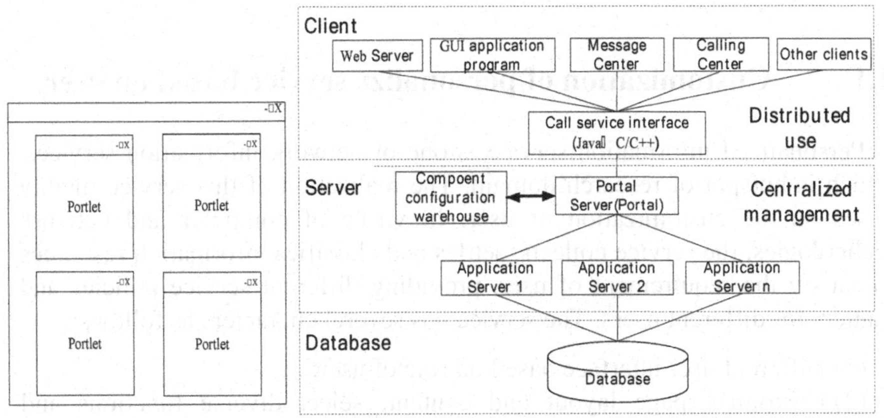

Fig. 1. User interface *Fig. 2.* The implementation environment of portal

2.3 Open source program uPortal

Portal is an important developing tendency of web application. At present, almost all big software companies have their own Portal products, such as WebSphere Portal of IBM, WebLogic Portal of BEA and Oracle Portal of Oracle. Besides commercial products of Portal, there are several open source Portal products, such as Jetspeed, Pluto, uPortal and Enhydra provided by Apache Community. An open source product uPortal, that has been developed and maintained by many US colleges, is discussed here.

uPortal is an open source program which is based on Java, XML and XSL. As followed by OSS development and distribution protocol, uPortal can be freely downloaded, installed and operated. It also can be improved and distributed according to requirements without profit making (Li, 2005).

Focusing on colleges and education institutions, uPortal is also applied for some business organization to construct cooperation community framework. uPortal enables users to add new features and functions in which they interest as well as customize the display style. Current uPortal includes function components such as web service channel, authority and user group management.

uPortal adopts mainstream technologies in open source field and provides a whole set of free and open platform. The installation and operation of uPortal are very easy and it can run in any server that is compatible with JSP. There are two distribution versions: fast version and source code version. Fast version is appropriate for users who have not deployed product environment; source code version suits to those that already have had J2EE and product database environment (http://www.uportal.org, 2004).

3. TECHNIQUE ROUTE

3.1 Customization of personalize service based on user

Personalized information service is one of network information services, which is hotspot of research domain. The realization of this service mainly bases on the customization of user. In virtue of computer and network technologies, the service collects, settles and classifies information resources to satisfy the requirement of user, providing different service policies and matters for different users. The service has several characters as follows:

(1) Different user interface based on role of user;

(2) Customize page layout and content, select diverse functions and contents local or remote to assemble page layout;

(3) Customize the display style of page layout, such as font, color or style;

(4) Control of user authority. Different users have relevant available operation and authorization of accessing content.

Portal provides personalized service of single sign-on access and user can customize page layout, content and style according to his demand and like. After entering into Portal, each kind of service corresponding to the user's role is available. The web pages offer different users different contents and functions to help them obtain information rapidly. For example, the user, who is interested in the weather information, can put the weather website on the customized web page. Therefore, such personalized service certainly makes acquiring information more easily, which increases user's satisfaction, the attraction of the website, and improves the integration of information (Cheng, 2005). At present, such service is already practicable, including the technique of customizing web page, channel or column.

The method of realizing individual service of uPortal is introducing identity authentication center to validate user's login information. After the user logins the system, the identity authentication center saves the whole information of the user; on the other hand, the role and authority of the user are saved in each system, thus the authority of the user is judged by each system. In the identity authentication center, cookie is used to save the user's login information; in each system, session is used.

The process of login of uPortal is as follows:

(1) Validation: Firstly, the user inputs username and password. Secondly, uPortal makes use of AuthenticationServlet to transfer validation service and AuthenticationServlet makes use of Security Provider to dispose data submitted by the user. Then SimpleSecurityContext class is used to validate the encrypted md5 password. If validated and got across, AuthenticationServlet will turn to main interface;

(2) Acquiring the directory information of current user: When the validation finishes, uPortal makes use of PersonDirectory service to acquire directory attribute of validated user. PersonDirectory service gets attribute information by JNDI (Java Naming and Directory Interface) or JDBC source, matches it to the user and acquires the user's directory information. The document of PersionDirs.xml manages the process of matching;

(3) Acquiring the identity of current user: IUserIdentityStore interface creates identity and authority for current user to get the original user interface.

3.2 Integration of multi-source information systems

Another important character of Portal is integration of multi-source information systems. Having considered the requirement of integration of Portlet from multi-source, it's key to operate multi data sources transparently and display content from diverse systems through a portal web application. XML technique is a good solution.

XML (Extensible Markup Language), come of semi-structured mark language from web, is a meta-marker language. As the agency of data exchange, XML has the capability of solving the information exchange problem since XML fulfills two essential demands:

(1) Taking data as the core, XML separates data and expression. In this sense, various data sources will be shielded by XML and showed with uniform XML format for users. The receiver deals with data according to

DTD of XML, such as breaking up data or displaying with different manners. For instance, the weather information can be showed in diverse equipments, like TV, mobile telephone or others.

(2) Transmitting data in different applications, XML provides a format which describes structured data, permits user to create his own tags and endow them with different semantemes. Moreover, tags can be changed dynamically. This capability of self-description makes XML applicable to the integration of data exchange and heterogeneous data among all kinds of applications (Liu et al., 2005).

UPortal offers integration framework for different information sources and effective flexible tools for displaying centrally. In view of different information sources and style-sheets, uPortal framework takes XML as its kernel to manage the relation between XML and last interface document. The architecture of uPortal consists of function component, framework component, theme component and skin component (Fig. 3). Processing requires the following steps (Justin, 2004):

(1) Structure Transformation: Obtain the user interface, an XML document generated from database, and transform it into the structure layout document by XSLT transformation. For example, the interface framework of guest user will be obtained including tables, rows and columns;

(2) Theme Transformation: Add the function component to the structure interface document and convert it to marker language by XSLT transformation. For instance, HTML is the result in Internet Explorer and WML in mobile telephone browser;

(3) Finally, the markup language utilizes CSS (Cascading Style Sheet) to control the display style of page and add logos, pictures of buttons and etc to get the last user layout.

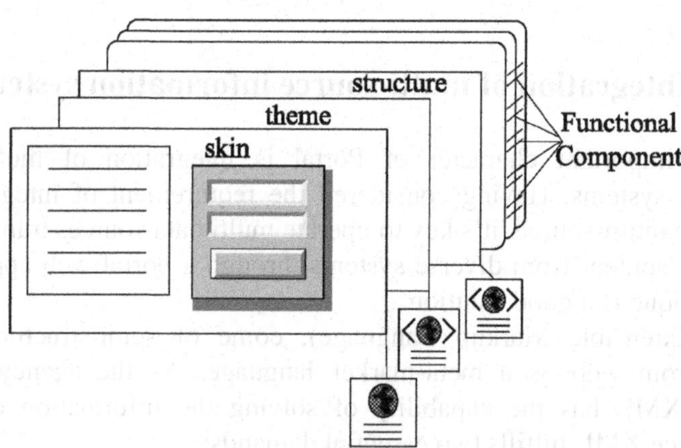

Fig. 3. The architecture of uPortal

The advantages of these multi-stage transformations include:

(1) With the aid of XML, it's simple for uPortal to connect different information sources from different systems, convert XML to HTML by XSLT and display uniformly in the end, which implement the kernel aim of Portal: manage in centralization, use in distribution;

(2) XSLT, a language which is used to describe how XML document transforms into another XML document, can transform one data format into another data format from different systems. The business logic and presentation are separated by XSLT to enhance the flexibility of system;

(3) Multi-stage transformations allow for flexible designs targeting different markup languages and devices (Justin, 2004), just like person computer or mobile telephone, etc;

(4) CSS creates coherent interface for diverse function components and applications.

4. CASE STUDY

4.1 Aims and implementing environment

Based on the above research, the prototype system was constructed. By constructing uniform information portal system, the system not only integrated various inner application systems and business process in existence, published information through Web, provided safety management and personalized service for distributed users, but also allowed cooperating with other business websites and application systems to realize integration.

Main functions in the system included release, subscription and integration of web component; Customization, such as customization of content, format, skin, language and so on; User management, such as group management, role management.

The system was developed using J2EE framework and Open Software tools: Apache, MySQL and JSP.

4.2 Architecture

The whole framework of system was made up of three layers: Browser/ Server/Database.

In database layer, MySQL was used to storage and manage data. MySQL is a small cross-platform relation database. It's excellent in powerful function, fast operation, good performance and high reliability and security.

In server layer, Tomcat was adopted as web server. As an open-source servlet container, Tomcat has stable performance, advanced technology and is completely free.

In application layer, the client was composed of JSP (Java Server Page). JSP is a dynamic web page technique based on Java, which is easy to develop complex web application program and support cross-platform and cross-web server.

4.3 Operation Interface

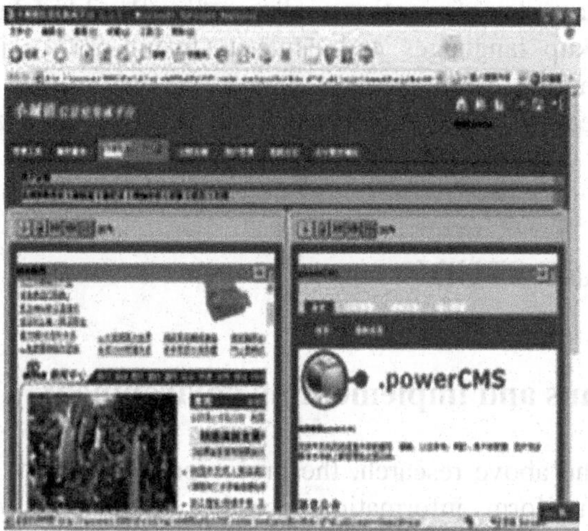

Fig. 4. Customization of web component

Fig. 5. Authority management

4.4 Characters

(1) Constructed by open source software

The small town information portal system is constructed by all open source software such as uPortal, Tomcat and MySQL. It's an ideal solution for constructing information portal to provide excellent, reliable performance and low cost;

(2) Integration of information resource

The advancement of Portal is mostly reflected in the capability of integrating various kinds of application systems, especially in the portal website. The system integrates content management system, email service system, calendar, video-on-demand system and other existing systems to achieve the transferring, sharing of information among diverse systems and efficient operation, management of all types of information in small towns.

(3) Personalized service

Different services are available for different users in the system. It's convenient for the user to customize the layout, content of page as he likes, improving user's efficiency and satisfaction.

5. CONCLUSIONS

This paper analyzed the basic condition of small town informatization development, proposed a new construction method, and adopted the latest portal technology to process information integration and to establish an advanced portal site which can provide integrated, intellectual, shared and open information services, all of which have been applied successfully in a real case. The further work includes:

(1) Adoption of pre-construction of Portlet and declared develop tools can not meet user's complicated need when integrating portal content information (Cheng, 2005). The next step is to develop more general customized Portlet to implement the application integration of portal sites in small towns;

(2) There are still many technical problems such as how to synchronize different accounts in the system and control different authorities among access nodes in portal site.

ACKNOWLEDGEMENTS

This study has been funded by the China National Science Support Plan Program (2006BAJ09B01) [Application of information technologies for the construction of small towns].

REFERENCES

Cheng Hong-bin, Research on personalized service and safety in campus web based on J2EE,Wuhan:Wuhan technology university,2005 (in Chinese).

Huang Xue-wu, Constructing integrated application with Portal and web Service, Journal of Shaoguan University (Natural Science), Vol. 25, No. 3, 2004, pp. 24-26 (in Chinese).

Jin hong, Yang yan-hong, Review of Personalized information service in university library, Modern Information, No. 6, 2004, pp. 17-19 (in Chinese).

Justin Tilton, Chief Executive Officer instructional media + magic, inc. at the JA-SIG Conference Destin, Florida, 2004.

Li Guang-qian, Informatization of Small Town must make clear the idea, http://www.echinagov.com, 2005

Li Xin, Research on open source software in electronic commerce, Fujian Computer, No. 3, 2005, pp. 33-34 (in Chinese).

Liu Ke-yan, Wan Li-rong, Zeng Qing-liang, and etc, Research and realization of information integration system based on XML, Research of Computer Application, No.4, 2005, pp. 149-154 (in Chinese).

Paolo Diviacco, An Open source, web based, simple solution for seismic data dissemination and collaborative research, Computer & Geoscience, Vol. 31, 2005, pp. 599-605

Solution of open source software website, Review of Open source program, http://www.open08.com, 2006

uPortal official Website, uPortal Home. http://www.uportal.org, 2004

SIMULATION STUDY ON CONSTRUCTED RAPID INFILTRATION FOR TREATMENT OF SURFACE WASTEWATER IN TOWN

Zheng Yanxia[1], Feng Shaoyuan[2], Zhao Xuyang[3,*]

[1] *Zheng Yanxia, Doctor. Resource & Environment Department, Shijiazhuang College, Hebei, 050035, P. R. China*

[2] *Feng Shaoyuan, professor. College of Water Conservancy & Civil Engineering, China Agricultural University, Beijing, 100083, P. R. China*

[3] *Zhao Xuyang, Professor. Resource & Environment Department, Shijiazhuang College, Hebei, 050035, P. R. China*

* *Corresponding author, Address: Shijiazhuang College, Shijiazhuang, 050035, Hebei, P. R. China, Tel: 86-13931981959, Email: log2008@163.com*

Abstract: On the base of land disposal, the objective of this study was to use constructed rapid infiltration (Abbreviation: CRI) to improve slightly polluted surface water in town. The data indicated that the average removal rates of BOD_5 and NH_4^+-N were 80% and 65% in summer, respectively. In winter, the removal rates of BOD_5 decreased evidently with the lowering of water temperature, but the change of NH_4^+-N was not obvious with the lowering of water temperature. The average removal rate of NH_4^+-N was 60% in winter. The study results indicated that the constructed rapid infiltration system had better purification effect on the slightly polluted water. The system could still operate normally at low temperature in winter.

Keywords: Constructed Rapid Infiltration; Water temperature; Surface wastewater

1. INTRODUCTION

As an ecological treating technique, the land treatment systems are of good developing prospects because of its advantages, such as simple of facilities, economization of investment, convenience of operation, lower cost

Yanxia, Z., Shaoyuan, F. and Xuyang, Z., 2008, in IFIP International Federation for Information Processing, Volume 259; Computer and Computing Technologies in Agriculture, Vol. 2; Daoliang Li; (Boston: Springer), pp. 851–858.

of energy and better effect of purgation. But it does still have some problems. Among the problems, lower hydraulic loading and lower capability of treatment quantity per area were the most serious one. Constructed rapid infiltration was one kind of land treatment. It enhanced the hydraulic loading by replacing natural medium with artificial medium (Zhang et al., 2001, 2002; He 2001, 2005; Zheng, 2005). Buried sand-filter system and CRI system shared many similarities, such as filter velocity and the removal of particles, microorganisms, and biodegradable substances, but the hydraulic loading was slow, typically 0.2 meters per day (m/d) (Pauel et al., 1990; Markus et al., 1993; Andreas, 1997; Zhang et al., 2001). Some information about buried sand-filter may be helpful. In china, CRI system was developed in recent years, it was necessary to research more.

CRI system was utilized to improve the slightly polluted surface water; especially the application in the northern low-temperature area was studied rarely. For the feasibility of exploring CRI system to improve slightly polluted surface water, this paper developed Simulation experimental study on CRI system for BOD_5 and NH_4^+-N removal effect under different seasons.

2. MATERIALS AND METHODS

2.1 Experimental apparatus

The experimental apparatus had been constructed and operated by the canal under the reservoir. Two replicates were performed per sand. Figure 1 is the sketch of the experimental apparatus of CRI system.

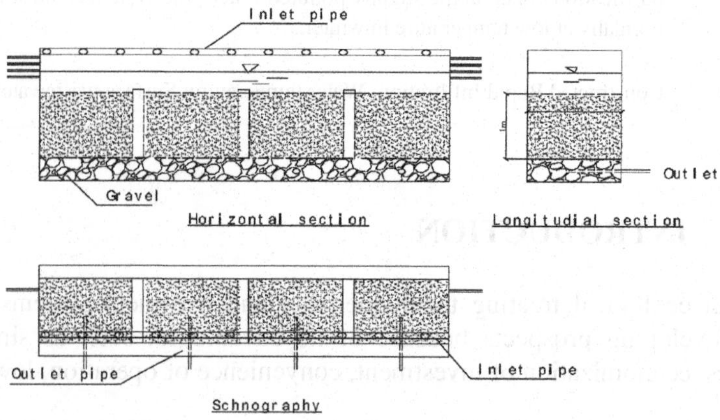

Figure 1. The sketch of experimental CRI system

2.2 Porous media

Sand was used as porous media in the study. Mechanical composition of the two kinds of sand was described in Table 1. Two kinds of sand were not different obviously, were taken from the potential effluent recharge site in a river way.

Table 1. Mechanical composition of porous media

Size/mm					*pH*	Organic matter/%	CEC/ (cmol/kg)
10–2	2–0.5	0.5–0.25	0.25–0.075	<0.075			
5.26%	41.65%	39.91%	10.66%	2.73%	7.95	0.45	2.33
0%	47.26%	41.33%	9.61%	1.8%	7.8	0.52	2.2

2.3 Analytic methods of water quality

Influent water in this study was taken from surface wastewater in town. Parameters were determined by Environmental quality standard for surface water (State Environmental Protection Administration of China, GB 3838 – 2002).

3. RESULTS AND DISCUSSION

3.1 Water temperature on removal efficiency of BOD_5

Figure 2 showed effect of water temperature on removal of BOD_5. Removal of BOD_5 lowered with the decrease of water temperatures. The average removal rate of BOD_5 was 80% in summer with high temperature; in autumn with middle temperature and in winter with low temperature, the average removal rate of BOD_5 was 30%. Water temperature and removal rate of BOD_5 had similar trend of variation with a correlation coefficient of 0.9199.

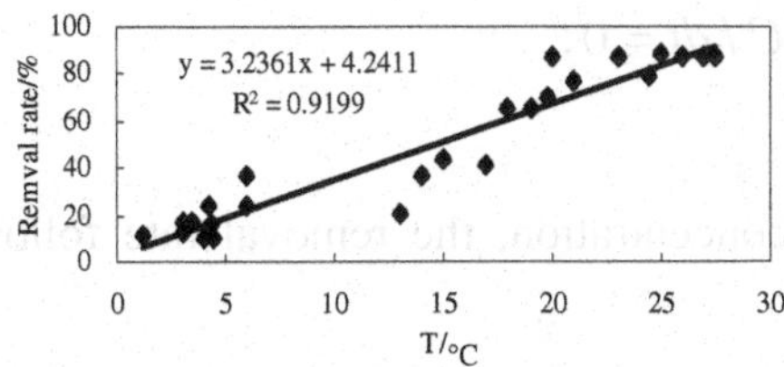

Figure 2. Effect of water temperature on removal of BOD_5

Basis on reaction formula equation, CRI system can be assumed as a plug flow reactor. That is, only in CRI system direction perpendicular to the flow phenomena was mixed, and the flow direction was not mixed. Of course, this was an idealized model, but from a theoretical point of view, the analysis was feasible (Long, 2002). Figure 3 was Plug Flow Reactor Model.

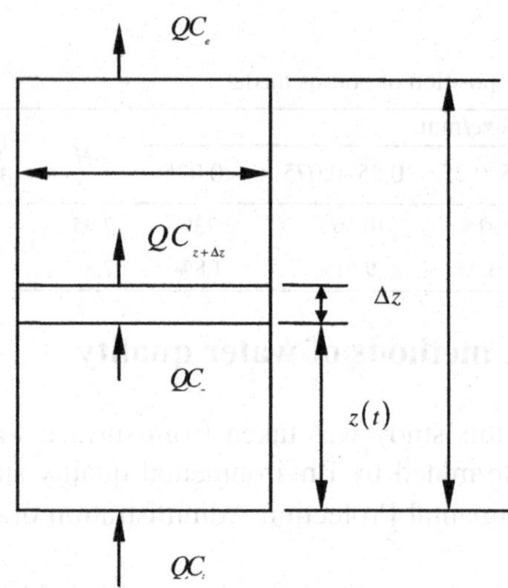

Figure 3. The sketch of Plug Flow Reactor Model

Calculated for energy conservation in $A\Delta z$:

$$QC_z + (A\Delta z)r = QC_{z+\Delta z} + (A\Delta z)(dC_z / dt) \quad (1)$$

Where: Q is flow. C_z is concentration. r is reaction rate

Ordered $\Delta z \to dz$, Equation (1)can be written:

$$-(Q/A)(dC_z/dz) + r = dC_z/dt \quad (2)$$

Where Q/A can be used to replace by U, and z/U can be used to replace by θ, omitted subscript z, equation (2) can be written:

$$-(dC/d\theta) + r = dC/dt \quad (3)$$

In steady state, $dC/dt = 0$:

$$dC/d\theta = r \quad (4)$$

Due to the low concentration, the removal rate followed the first-order reaction formula:

$$r = -kC \quad (5)$$

Where: k is reaction rate constant.

Put equation (5) into equation (4), and integration

$$C = C_i \exp(-k\theta) \tag{6}$$

On both sides of equation (6) with log:

$$\ln(C_i / C) = k\theta \tag{7}$$

Basis on the concentration of BOD_5, the approximation of reaction rate k can be drawn in CRI system. k was a coefficient with water temperature. Such as BOD_5, $K_{20°C}/K_{25°C}$ was 0.94, and $K_{10°C}/K_{15°C}$ was 0.48, and $K_{15°C}/K_{20°C}$ was 0.93. These indicated reaction rate decreased 6%, when water temperature was dropped from 25°C to 20°C; Reaction rate decreased 7%, when water temperature decreased from 20°C to 15°C; reaction rate decreased 52%, when water temperature decreased from 15°C to 10°C. These suggested water temperature more low; reaction rate decreased the proportion more big. This accorded with the experimentation result that the effect of change of water temperature was little in high temperature stage, but in low temperature stage, the effect was obvious.

3.2 Water temperature on removal efficiency of NH_4^+-N

Figure 4 showed effect of water temperature on removal of NH_4^+-N. Removal of NH_4^+-N was not obvious with the change of water temperature. But the trend was increased with the increasing of water temperature. Firstly, NH_4^+-N was adsorbed into the infiltration media and happened nitration. Because NH_4^+-N was biochemical substances, it was totally absorbed by microorganism. So the removal rates of NH_4^+-N were more constant than on the removal of organic matter. Even in the low temperature close to 0°C, the ammonia removal rate remained at 45%. Therefore, CRI system had a certain removal effect on NH_4^+-N in slightly polluted water under low temperatures.

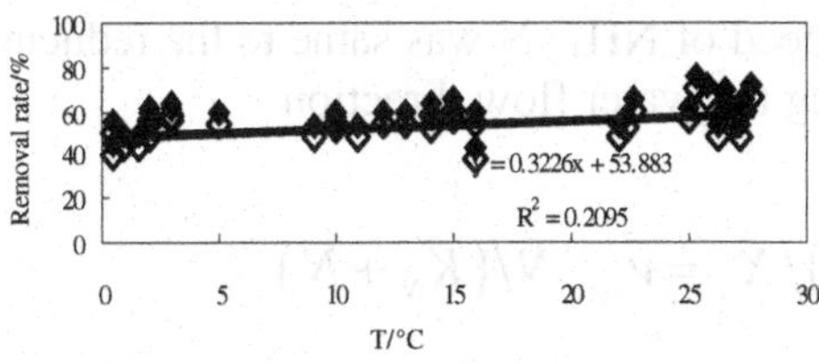

Figure 4. Effect of water temperature on removal of NH_4^+-N

Basis on reaction formula equation, the lower of NH_4 -N removal rate indicated the lower of NH_4^+-N removal speed in CRI system. Therefore, Monod kinetic equation was used to note ammonia oxidation reaction (Gu, 1993).

Basis on Monod equation:

$$\mu_{NS} = (\mu_{max})_{NS} N/(K_N + N) \tag{8}$$

Where: μ_{NS} is growth rate of nitrosobacteria (min^{-1}). $(\mu_{max})_{NS}$ is max growth rate of nitrosobacteria (min^{-1}). N is concentration of NH_4^+-N(mg/L). K_N is saturation constant (mg/L, $K_N = 10^{0.051T-1.158}$). T is temperature (°C).

$$(dX/dt)/X = -y_0(dN/dt)/X \tag{9}$$

$$\mu_{NS} = (dX/dt)/X \tag{10}$$

$$v = -(dN/dt)/X \tag{11}$$

$$\therefore\ \mu_{NS} = y_0 v \text{或} (\mu_{max})_{NS} = y_0 v_{max} \tag{12}$$

Where: X is concentration of nitrosobacteria (mg/L). y_0 is yield (mg Nitrosobacteria/mg NH_4^+-N). v is removal speed of NH_4^+-N(min^{-1}). v_{max} is max removal speed of NH_4^+-N(min^{-1}).

Put equation (12) into equation(10):

$$v = v_{max} N/(K_N + N) \tag{13}$$

In the wastewater treatment process, removal of the pollutants was major, and microbial growth was the result of pollutants removal. Therefore, equation (13) was more practical.

Assumed:

(1) The nitrosobacteria concentration of entrance was similar to the concentration of exports in CRI system. At this point, the nitrosobacteria concentration can be said that the average concentration X_a.

(2) The using speed of NH_4^+-N was same to the reducing speed of NH_4^+-N concentration along the water flow direction.

So:

$$v = -(dN/dt)/X_a = v_{max} N/(K_N + N) \tag{14}$$

Arranged:

$$K_N dN + NdN = -v_{\max} X_a N dt \tag{15}$$

Integral:

$$(N_i - N_e) + K_N \ln(N_i/N_e) = v_{\max} X_a t \tag{16}$$

Ordered $V_{\max} = v_{\max} X_a$, called the max removal rate (mg/L•min):

$$(N_i - N_e) + K_N \ln(N_i/N_e) = V_{\max} t \tag{17}$$

Where: N_i is concentration of NH_4^+-N of entrance (mg/L). N_e is concentration of NH_4^+-N of exports (mg/L).

Basis on the concentration of NH_4^+-N, the approximation of max removal speed $V_{\max}$ can be drawn in CRI system. $V_{\max}$ was a coefficient with water temperature. Such as NH_4^+-N, $V_{\max(5°C)}/V_{\max(10°C)}$ was 0.65, and $V_{\max(0.5°C)}/V_{\max(5°C)}$ was 0.525. This indicated max removal speed decreased 35%, when water temperature was dropped from 10°C to 5°C; The max removal speed decreased 47.5%, when water temperature decreased from 5°C to 0.5°C. Removal efficiency of NH_4^+-N was not significantly different with the increase of water temperatures. The average removal rate of NH_4^+-N was 65% in summer with high temperature, in autumn with middle temperature and in winter with low temperature. The changes of NH_4^+-N were not obvious with the lowering of water temperature.

CRI system had better removal effect on NH_4^+-N in slightly polluted water, but water temperature had not obvious effect on removal effect. This suggested that nitrobacteria and nitrosomonas adapted to low temperature (Wang et al., 1999; Zhou et al., 2000). Therefore, CRI system had a certain removal effect on NH_4^+-N in slightly polluted water under low temperatures.

4. CONCLUSIONS

(1) CRI system had a significant removal effect on slightly polluted surface water. The simulation experiment indicated that the average removal rates of BOD_5 and NH_4^+-N were 80% and 65% in summer with high temperature, respectively. The average removal rates of BOD_5 and NH_4^+-N were 30% and 60% in winter with low temperature, respectively.

(2) Removal of BOD_5 was lowered with the lowering of water temperature. In winter, the removal rates of BOD_5 lowered evidently with

the lowering of water temperature, but the change of NH_4^+-N was not obvious with the lowering of water temperature.

(3) The removal efficiency decreased with the lowering of water temperature. This suggested that microbial activity decreased with the lowering of water temperature. The lower water temperature, microbial activity was affected greater. In winter, the CRI system should take measures to keep from freeze. The system could operate normally at low temperature in winter.

REFERENCES

Andreas S. & Markus B. Transport phenomena in intermittent filters [J]. Wat. Sci. Tech., 1997, 35(6): 13-20.

Gu X.S. Wastewater mathematical models of biological treatment [M]. Tsinghua University Press, 1993. (In Chinese)

He J.T. A new technique for wastewater reuse: Constructed rapid infiltration system (picture) [J]. Earth Science Frontiers (China University of Geosciences, Beijing; Peking University), 2005, 12 (Suppl.): 12-13. (In Chinese)

He J.T. Experimental study on constructed rapid infiltration system [D]. Beijing: China University of Geosciences, Beijing, 2001. (In Chinese)

Long X.Q. Bioactive filter characteristics and transformation projects [D]. Tsinghua University, 2001. (In Chinese)

Markus B. & Schwager A. Dynamic behavior of intermittent buried filters [J]. Wat. Sci. Tech., 1993, 28(10): 99-107.

Pauel S. & Markus B. Onsite wastewater treatment with intermittent buried filters [J]. Wat. Sci. Tech., 1990, 22(3/4): 93-100.

Wang Z.S. & Liu W.J. Drinking Water Treatment Processes with Micro-polluted Raw Water [M]. Beijing: China Architecture and Building Press, 1999. (In Chinese)

Zhang J.B. Study on constructed rapid infiltration system for wastewater treatment [D]. Beijing: China University of Geosciences, Beijing, 2002. (In Chinese)

Zhang J.B., Chen H.H. & Zhong Z.X. Buried sand-filter system and its environmental significance [J]. Geoscience, 2001, 15(3): 346-350. (In Chinese)

Zhang J.B., Chen H.H. & Zhong Z.X. The experimental study on the treatment of both wastewater by artificial rapid infiltration composite system [J]. Hydrogeology and Engineering Geology, 2001, 28(6): 30-32. (In Chinese)

Zheng Y.X., Feng S.Y. & Cai J.B. Experimental study on removal of organic substance in reservoir water by means of soil aquifer treatment system [J]. Journal of Hydraulic Engineering, 2005, 36(9): 29-32. (In Chinese)

Zhou Q.Y. & Gao T.Y. Microbiology of Environmental Engineering [M]. Beijing: Higher Education Press, 2000. (In Chinese)

I-UDDI4M: IMPROVED UDDI4M PROTOCOL

Xiang Li[1], Lin Li[*], Lei Xue[1]
[1] *College of Information and Electrical Engineering, China Agricultural University, Beijing, China, 100083*
[*] *Corresponding author, Address: P.O. Box 215, College of Information and Electrical Engineering University, China Agricultural University, 17 Qinghuadong Road, Beijing, 100083, P. R. China, Tel: +86-10-62732323, Email: lilincau@sohu.com*

Abstract: Web Service is growing to its peak for static networks, e.g. UDDI, but the improvement of the service for wireless networks, especially for MANETs is still required. At the same time, service discovery is basic technology in the MANETs. So, in this paper we propose an Improved UDDI Protocol used for MANETs (I-UDDI4m thereafter) in obtaining the information of Web Service actively, which is based on the theory of UDDI4m Protocol.

Keywords: I-UDDI4m; MANETs; UDDI; Web Services

1. INTRODUCTION

Mobile ad hoc Network (MANET) is a temporary network that is constituted by some self-regulating wireless mobile nodes, does not have central controls and does not require any infrastructures. Because MANET does not have fixed routers, the nodes appear not only as terminal nodes which need information, but also as routers. Consequently, MANET is a self-organized, non-centre-controlled, multi-hop network system (Martin Mauve, 2001). The particularities of MANET are characterized by self-organization, highly dynamic topology, limited bandwidth, dynamic capability of the link layer, the restriction of the node caused by the equipment environment, multi-hop communication, distributed control, limited physical security, low expansibility of the network, half duplex

Li, X., Li, L. and Xue, L., 2008, in IFIP International Federation for Information Processing, Volume 259; Computer and Computing Technologies in Agriculture, Vol. 2; Daoliang Li; (Boston: Springer), pp. 859–866.

wireless channel, and the short lifetime. More attentions are being paid to MANET. and the discovery and description of Web Services will play an important role in the use of MANETs

Much work has been done on service discovery for MANETs recently, and several excellent protocols, such as Konark (Helal S, 2003), Allia (Ratsimor O, 2002) and MDFNSSDP (Gao Zhen-guo, 2006), were proposed. Cremonese *et al.* (Clement L, 2004) proposed the UDDI4m (UDDI for MANETs) protocol that is similar to but simplified Universal Description, Discovery and Integration protocol (UDDI) widely used in static networks. UDDI4m is a three-layer structure with the Application Layer, the Service Layer, and the Middleware Layer. The Service Layer is further partitioned into a UDDI4m API and an API for lower layer protocols. Therefore, the UDDI4m intensifies in expansibility and the ability of the protocol configuration.

Following the simplified UDDI entity construction proposed by Cremonese *et al.* (Clement L, 2004), we propose the Improved UDDI for MANETs protocol (I-UDDI4m) in which the mechanisms of dynamically Add operation and Delete operation are introduced into UDDI4m in this paper. Additional functions are expected to be included in the proposed protocol: writing available service information into UBRs by listening to the service query request in the network in order to avoid repetitious query, and deleting useless information in the UBRs in time through listening to the changes of the topology of the network, which will lessen the amount of the useless query and reply, decrease the quantity of the data stream, save the bandwidth, and enhance the usability of the network.

2. THE ARCHITECTURE OF I-UDDI PROTOCOL

2.1 The Characteristics of MANET

MANET has its own characteristics compared with a static network as follows:

The topology of the network is changed rapidly and the transformation of the network is unpredictable.

MANET is based on the wireless links, so the bandwidth of a MANET network is narrower than that of the static networks.

The security is restricted by the wireless transformation.

The package lose-rate in MANET is higher, its latency is longer, and its shakiness is stronger compared with the static network, and

MANET saves energy greatly.

Not all of the nodes in MANET have their own UBRs, the nodes which don't have UBRs (referred to Client Nodes) only own I-UIID4m clients. The operations of service searching, publishing and deleting will be done by following procedures: the Client Nodes first find neighboring nodes which have UBRs (referred to Full Nodes), and then send requests to the Full Nodes. The Full Nodes will complete the operations they request.

2.2 The Characteristics of I-UDDI4m Protocol

For the design of I-UDDI4m, we should consider the Characteristics of MANETs. Service searching protocol in MANET must have some characteristics thereinafter:

The information of the UBRs must be updated in time in order to adapt the frequent changes of topology.

Useless packages should be reduced in order to saving bandwidth.

All of the nodes are in the same order, a central control node is forbidden.

Useless information is deleted in time for the sake of reducing the energy consuming resulted by maintaining these information.

Based on these characteristics, I-UDDI4m is designed as a protocol with following specialties:

All of the nodes are in the same statuses, the central control node is forbidden.

For saving energy, not the whole web services information of MANET is kept in the UBRs.

The service request is monitored, the available service information in UBRs is updated in time, and the useless information is deleted based on the network state.

2.3 The Software Architecture of I-UDDI4m Protocol

Based on the analysis above, and following the architecture of the UDDI4m software, the software of I-UDDI4m protocol is divided into three layers: the application layer, the service layer, and the middleware layer. Among the three layers, the application layer is to provide a user interface, the middleware layer is to transform information and route. the service layer is to link the application layer and the middleware layer. Standard UDDI output messages are provided to the application layer, so that users can get information they need, while standard UDDI SOAP messages are enveloped for the purpose of the specific requirement of MANET and transmitted to the middleware layer. We provided a special listener, which is called Topology Changes and Services Available Listener (TCSAL for short), to monitor the topology changes and the status of web services, and update register information in UBRs. The software architecture of I-UDDI4m protocol is shown in Figure 1. We use Pastry protocol as the look up protocol in the middleware layer in order to keep it in consistent and comparable with UDDI4m.

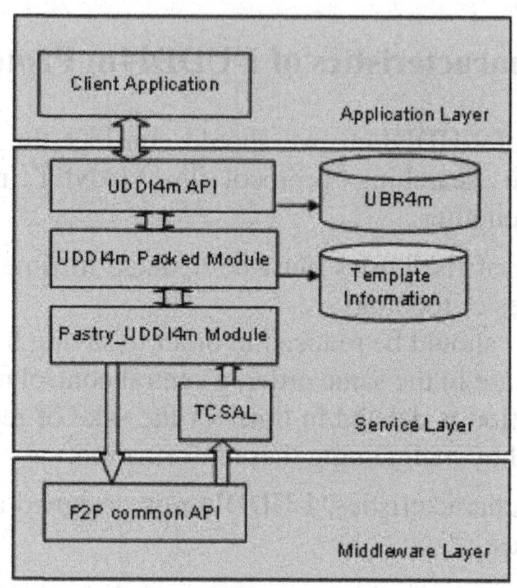

Fig. 1. The software architecture of I-UDDI4m protocol

Cremonese and Vanni (Piergiorgio Cremonese, 2005) simplified the central entities of UDDI (shown in Figure 2), only including Business_Entity, Business_Service, Binding_Template, TModel and Contacts, and realized the main APIs of UDDI (Find_XXX(), Save_XXX(), Delete_XXX()). We keep the same APIs as they are in standard UDDI.

The components of the Service Layer are as follows:

(1) The UDDI4m API is to receiving the requests coming from the application layer and query information from local UBR4m (UBR for MANETs). If no information is found, the UDDI4m API would transfer the messages to the lower layer (only simplified entity is included), and process the messages which are sent back by the UDDI4m packed Module and in which the result of the query is included. Then the UDDI4m API updates the UBR4m, and sent the result back to the Application Layer.

(2) The UDDI4m Packed Module is to accept the standard SOAP messages sent by UDDI4m API, and capsulate these messages with additional information (the node identifier and the life circle information) into a new package which is called Improved UDDI Package (IUP for short), and the information is written into the Template Information database. The structure of IUP is discussed in section three.

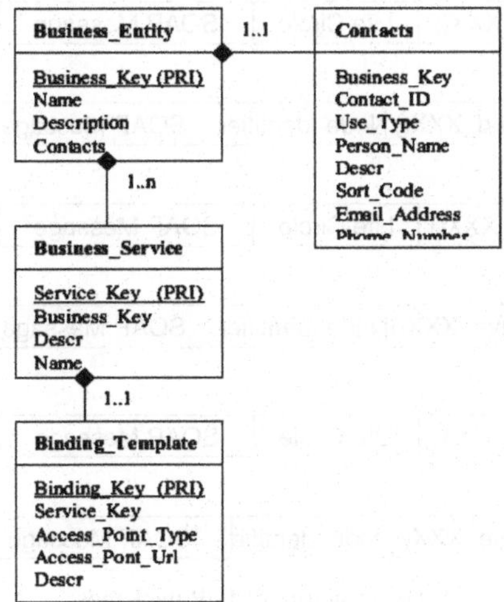

Fig. 2. A part of relational model UDDI4m registry

(3) In order to be consistent with UDDI4m, we use Pastry protocol as the look up protocol in the middleware layer. The function of the Pastry_UDDI4m Module in our architecture is almost the same as the Pastry_UDDI4m Module in the UDDI4m architecture. It is used for pack up the data which are sent by the lower layer route protocol. It will be called by TCSAL and sent the IUP which is independent from the router protocol back to UDDI4m Packed Module.

(4) There is a semi-transparent listener which is called TCSAL under the Pastry_UDDI4m Module. It is used for monitoring the topology changes of the network and the responses of the service requests, It will match the information with the template information, call the upper layer API to execute, and then update the information in UBR4m.

2.4 The Basic Data Package of I-UDDI4m

For achieving the specialty mentioned in section 2.2, the SOAP messages are packed up by the UDDI4m Packed Module before they are sent. The life circle information (such as life time, max hops) is added for the request messages, and the node identifier information is added to the response messages. The packages are shown in Figure 3.

Though some of the fields (such as the life circle information in Delete_XXX() and the node identifier information in Ret_Delete_XXX(). We will have a further discussion at section 3.3) are not in use yet because of its independence from the other nodes. We still keep these fields into the package for the purpose of keeping the ability of consistency and extension.

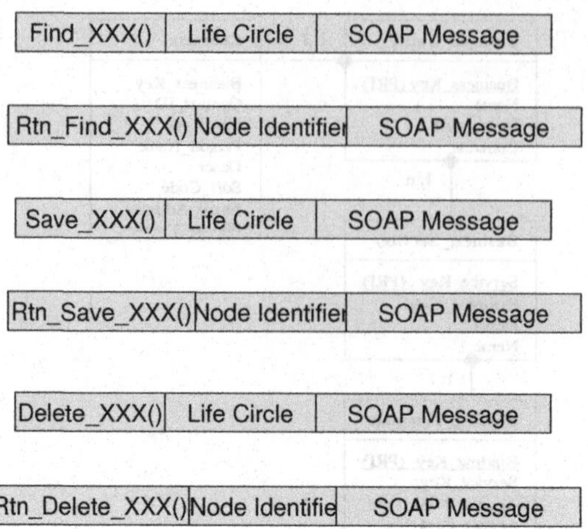

Fig. 3. The structure of IUP packages

3. THE CIRCULATE PROCESS OF I-UDDI4M

The primary operations for services which are included in MANET are listed here: Publish/Update Service information, Query Service according to some fixed characters, Delete Service, and Execute Service. We will discuss them respectively.

3.1 Publishing/Updating

The information is only needed to be written into local UBR4m when a service is published or updated.

3.2 Querying

The querying service is the key point which we pay special attention to, and a significant improvement is made here. When the query is send by node A in Figure 4, (a) firstly the query will fetch the UDDI4m API, and related

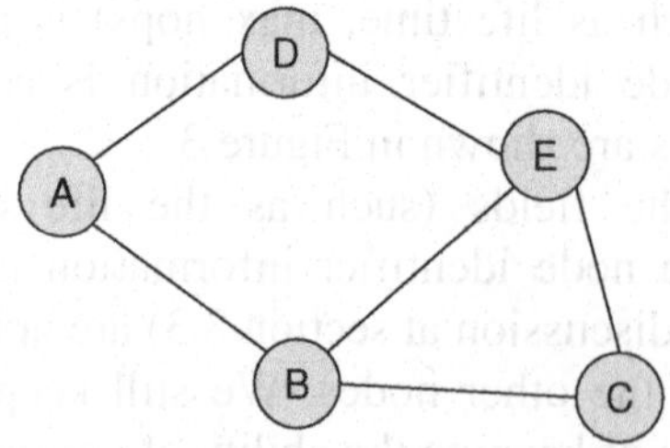

Fig. 4. A topology of a simple MANET

function will be called to query from the local UBR4m; (b) The fallback will be returned while the services are found; otherwise; (3) the query information will be packed up and sent to the middleware layer through the lower module (Pastry_UDDI4m Module in our paper); (4) The middleware layer will broadcast the message to its neighbour nodes. The neighbour node (take B for example) will submit the message to the upper layer when the node gets it. When TCSAL gets the message, it will be sent to the Pastry_UDDI4m Module; the Pastry_UDDI4m Module will unpack the message, and write the life circle information into the template information database, then execute the corresponding UDDI4m API. The UDDI4m API will query in its UBR4m, if the services are found, the result will be sent back, and (5) Otherwise, the process (4) will be continue till the max hops are exceeded or the life time is exhausted.

3.3 Deleting

The deleting operation is triggered by two different reasons, one is the node, the other is the TCSAL.

a) The deleting operation which is triggered by the node. While the node no longer provide the service, it execute the Delete_XXX() of the UDDI4m, and this will delete the information in local UBR4m.

b) The deleting operation which is triggered by the TCSAL. While the TCSAL monitored the topology is changed in the MANET, it will deliver the change to the upper layer, execute the appropriate UDDI4m API to finish the deleting. For example, when node C is not in the overlay of A, it cannot be fetched from A, the Delete_XXX() will be called and the service information which is correlative to C will be deleted.

3.4 Executing a Service

When a service is required, the node will firstly look up the local UBR4m. If the information of the service is found, the related service will execute, else, querying will be done.

If service is unavailable while it is called, TCSALs of the nodes in the path will receive this information, and each Delete_XXX() will be called to deleting information registered in the UBRs. In this way, the validity of the registered information is ensured, and the energy which is used to maintain UBRs is saved.

4. THE COMPARING WITH UDDI4M AND THE CONCLUSION

Comparing with UDDI4m, two aspects in I-UDDI4m are improved:

a) Service query is listened in order to updating the registered information in UBRs.

b) The topology changes are monitored for the sake of deleting the useless information in UBRs.

Due to the improvements mentioned on the above, I-UDDI4m, with the characteristics of UDDI4m, reduce the energy consuming, enhance the availability of the registered information; at the same time, it reduces the quantity of the data stream, saves the bandwidth, and the usability of MANETs is strengthened.

REFERENCES

Clement L, Hately A, Riegen CV, Rogers T. Universal description discovery & integration (UDDI) 3.0.2. 2004. http://uddi.org/pubs/uddi_v3.htm.

Gao Zhen-guo, Wang Ling, Zhao Yun-long, Cai Shao-bin Li Xiang. MDFNSSDP: A Minimum Dominating Forward Node Set Based Service Discovery Protocol for MANETs. [J] ACTA ELECTRONICA SINICA. Vol. 34, No. 11. Nov. 2006. 2030-2036.

Helal S, Desai N, Verma V, Lee C. Konark2a service discovery and delivery protocol for ad hoc networks [A]. Proceedings of the Third IEEE Conference on Wireless Communication Networks [C]. New Orleans, USA, 2003. 2107-2133.

Martin Mauve, Jorg Widmer, Hannes Hartenstein. A Survey on Position-based Routing in Mobile ad hoc Networks [J]. IEEE Network, 2001, 15 (6): 30-39.

Piergiorgio Cremonese, Veronica Vanni. UDDI4m: UDDI in Mobile ad hoc Network. [A] Proceedings of the Second Annual Conference on Wireless On-demand Network Systems and Services (WONS'05) [C]. Vol. 00. 26-31.

Ratsimor O, Chakarborty D, Joshi A, Finin T. Allia: alliance-based service discovery for ad hoc environments [A]. Proceedings of the 2nd ACM International workshop on Mobile commerce [C]. Atlanta, Georgia, USA, 2002. 1-9.

RESEARCH AND DESIGN OF ONLINE DECLARATION AND APPROVAL SYSTEM BASED ON MVC

Wanlin Gao *, Hongqiang Yang, Xin Chen, Yang Ping, Zhen Li, Xinlan Jiang, Ganghong Zhang

College of Information and Electrical Engineering, China Agricultural University, Beijing, China, 100083

** Corresponding author, Address: P.O. Box 105, College of Information and Electrical Engineering, China Agricultural University, 17 Tsinghua East Road, Beijing, 100083, P. R. China, Tel: +86-10-62736755, Fax: +86-10-62736746, Email: gaowlin@cau.edu.cn*

Abstract: This paper presents a development program of the online declaration and approval system. The system is against the government's functional departments-oriented online community to declaration and approval. System using MVC Framework achieve the administrative declaration and approval and external services, thus simplifying declaration and approval procedures, improve functions of the government department office efficiency and achieve administrative declaration and approval information, for promoting the building of e-government, it is a very important significance. MVC-based framework is suitable for the design of the system with a high degree of reliability, scalability and security.

Keywords: declaration and approval, MVC, E-government

1. INTRODUCTION

In recent years, the international and domestic levels of government generally improving office efficiency, simplify administrative procedures, advance online declaration approving the building of e-government as an important content. Using Web-based declaration and approval way, instead of the traditional declaration and approval way in China is an inevitable trend. It is not only conducive to the public and the government to use the

Gao, W., Yang, H., Chen, X., Ping, Y., Li, Z., Jiang, X. and Zhang, G., 2008, in IFIP International Federation for Information Processing, Volume 259; Computer and Computing Technologies in Agriculture, Vol. 2; Daoliang Li; (Boston: Springer), pp. 867–873.

most modern methods of information transfer and establish direct, unified communication channels; reduction of the monopoly of information, but also as a result of reduced cumbersome management areas and intermediate costs, and to a large extent, avoid the breeding of corruption. Therefore research and design of online declaration and approval system based on MVC, promoting the development of e-government is of great significance.

Online declaration and approval system is online functions of the government departments for approval (Zhang Lei, 2007), oversight, and consulting functions and services, can run through the network environment, strengthen the public, mainly corporate users and organizations with the government departments and business through the Internet, greatly reducing the business processing cycle, it will reduce the pressure and work intensity on government departments.

2. SYSTEM REQUIREMENT ANALYSIS

2.1 Functional Requirement

Declaration units and individuals can fill the online declarable items and transfer files through Internet (ShiDong Wang, 2007). The unit or individual has declared the projects and materials information to undertake a comprehensive inquiry. Government managers to be on-line completion of the project information and materials selection, declaration, assessment, approval notices Print operation, all of the projects and state material information integrated query.

2.2 Reliability Requirement

To ensure that the project information and materials declaration process data integrity and validity (ShiDong Wang, 2007), avoiding as a result of system software failure caused data loss, bad data and database damage. We should choose a reliable database management system, and gives a reasonable set of database backup program, and must fully consider the safety system.

2.3 Scalability Requirement

Declaration and approval system is the e-government information technology first step, and with its further development will increase the requirement for more, so in the software development cycle early should fully consider the whole system can be expanded and scalable.

3. SYSTEM DESIGN

3.1 System Framework Design

System provides external services as inquiries (ChuanBao Zhu, 2006), download forms, online declaration, replied feedback information and so on System provide internal management as time supervision management, flow control management and so on. Under the e-government to the actual demand, online declaration and approval system's basic features include: declaration functions, online help functions, published information function, information security, organization and management functions, processing functions, system management can be customized functions, log function declaration and approval flow process control functions. As shown in figure 1.

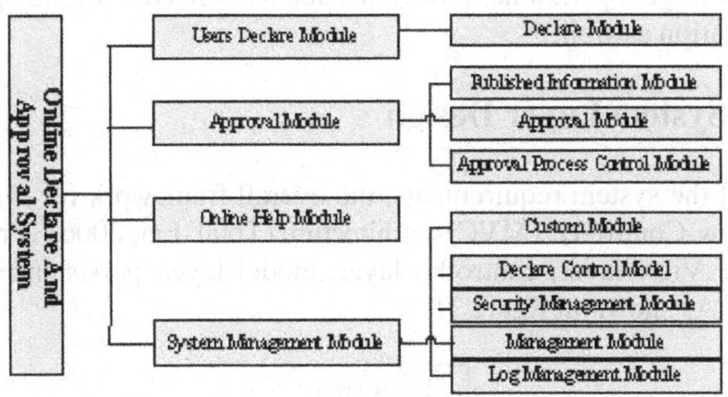

Fig. 1. Online declaration and approval system overall function chart

Online declaration and approval system the following major functions:

Custom functions

The customization process is based on flow algorithm can be customized to the user's own customized process (ChuanBao Zhu, 2006), taking into account the actual needs of the use, can be divided into two customizable: First, the process of specific documents can be customized. This applies to a fixed format and strict vetting procedure audit system with special or unusual approval process; Another is the information reported to the general approval of the customization process. These two customized, users can be flexible on, the realization of the freedom of independent approval process customization (ShiDong Wang, 2007).

Declaration function

Download the appropriate form to fill in, upload form and the corresponding materials.

Approval function

Administrators can query declaration items and materials, approval items.

Process monitoring function

System managers can monitor the implementation of the process of declaration and approval procedures for the implementation of the monitoring, for the implementation of the relevant control commands.

Log function

Maintaining records and logs the basic information, including project personnel of the process and activities of the operation. System operation is detailed log records, guarantee different vetting personnel, and other users of the system in an orderly scheduling will be able to observe their duties, fulfill their duties.

Online Help

System provides friendly on-line help, users can be quickly and convenient to get operational guidelines and associated with the operation of the information as well.

3.2 System Layer Design

To meet the system requirements, the overall framework for systems uses Model-View-Controller (MVC) architecture (Total Lei, 2006). From top to bottom are View layer, controller layer, model layer, persistence layer and data layer. As shown in figure 2.

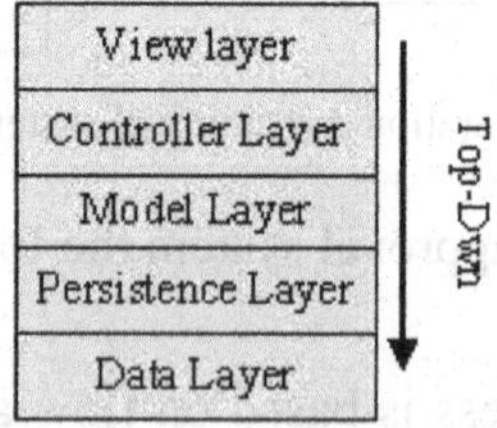

Fig. 2. System structure chart

Layers of dependence between the principles of top-down, the upper layer can rely on the lower layer, and the lower layer should minimize reliance on the upper layer. Meanwhile layer through the interface between interactive and reduce the dependence on implementation.

MVC can be applied to large-scale expansion of the application system development (Total Lei, 2006), application of the input, processing and output separately. Classified as a model layer, layer and View-Controller three different levels, so that they each perform different tasks, any level of change will affect the other two.

MVC three-layer model of the process is very clear, the specific process is as following (Xu Yi, 2006):

Controller receiving user request and then decide which model to call for disposal.

Model for the corresponding logic, then returned to the processed data.

Controller View will be called back to the data model presents to the user.

Change the system, in particular the requirements of the data will change data layers become very cumbersome (Xu Yi, 2006). Because specific data source may be diversified, for example, might be XML or database. In our persistence layer of the design, using shielding data access methods use simple, easy-to-use interface to satisfy the upper layer of data demand, thus very convenient solution to the data requested by the changes brought about by the issue.

The persistence layer used to achieve DAO design pattern. DAO mode is used to reduce core business methods and specific data sources and the coupling achieve data access and data sources unrelated to the purpose. As shown in figure 3.

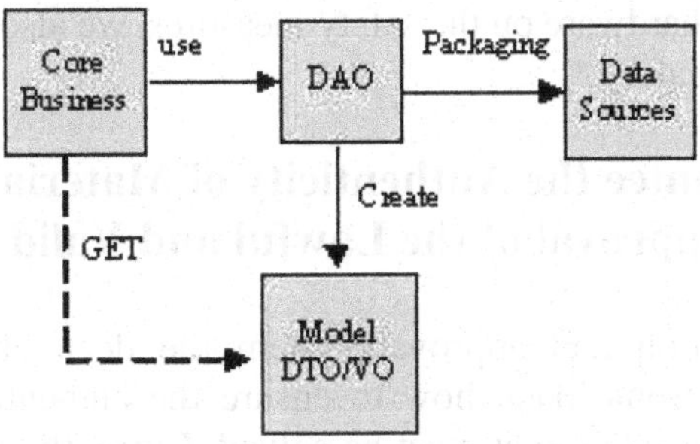

Fig. 3. DAO Mode application chart

Each persistence layer of a DAO corresponding category (Xu Yi, 2006), it achieved a persistence kind of create, query, update and delete methods. When to switch to other persistence or persistence mechanism of middleware, only need to create a new DAO realized, is no need to change the application of business logic code.

4. SYSTEM SECURITY DESIGN

Online declaration and approval system to achieve the office of the Chief of electronic (Liu Yang, 2006), Internet-based, the security of the system will have a direct impact on the normal operation of the Administrative Office of effectiveness, efficiency and stability. System through the network within the network, networks boundaries and physical confinement measures erection of a security network platform. As shown in figure 4.

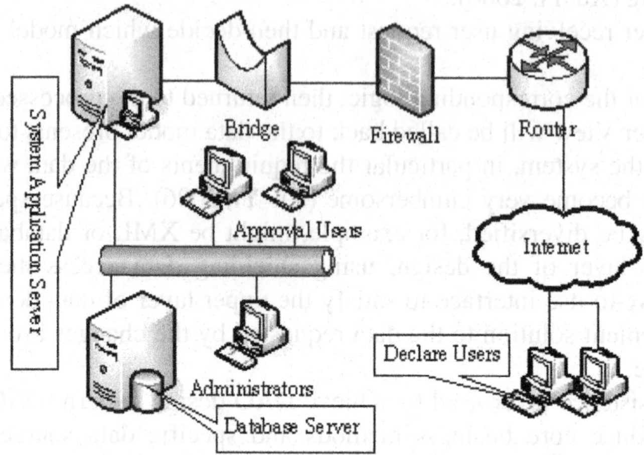

Fig. 4. Declaration and approval system physical topology map

Apart from the hardware on the safety measures, we also need the software system to ensure safety.

4.1 Guarantee the Authenticity of Materials Declaration and Approval of the Lawful and Valid Information

Online declaration and approval system the declaration and approval materials are electronic data, how to ensure the authenticity of electronic data system as a problem that must be solved. Currently, the electronic data taken digital signature technology to ensure its authenticity, digital signatures can effectively solve the problem.

4.2 Ensure Data Transmission in the Process of Secrecy

Because many of the data systems are sensitive information, the transmission process of encryption needs. We used standard SSL protocol to ensure that the process of transmission of confidential data.

5. CONCLUDING REMARKS

Administrative approval online services e-government as a public service focus application projects (Zhang Lei, 2007), is "to further transform government functions, improve the management, the formation of standardized behavior and operational harmony and is fair, transparent, a clean and efficient administrative system" an important component.

Online declaration and approval system not only applies to taxation (Zhang Lei, 2007), and other departments, but also widely used in public security departments household registration management, Land resources management departments, municipal planning declaration and approval, project contracting management. System uses a large number of government departments commonly used standardized document online forms to achieve the declaration approval process. And system support tables and reports generated and modified, the approval authority set up and standardize the process of management, automatic classification of statistical tables and reports archiving management, to facilitate future inquiries statistics and checks. Thus, the system of government departments to improve the working efficiency, and promote the development of e-government has a very important significance.

REFERENCES

ChuanBao Zhu, Lee from the East, 2006. Based on. Net platform electricity infrastructure information management system [J]. computer engineering, 2006, 32 (14): 255-257.

Liu Yang, Wang Jian Hua, Huang, 2006. Web-based framework of the Chief of General System Research and Implementation [J]. computer engineering, 2006, 32 (14):263-265.

ShiDong Wang, Zheng, Zhang Zhihai, and so on, 2007. Web-based document template-processing system [J]. computer application, 2007, 24 (6):289-294.

Total Lei, Li, Zhou Wei, 2006. Proficient J2EE-Eclipse Spring Struts Hibernate Integration Applications [M]. Beijing: People's Posts & Telecommunications Publishing House, 2006, 237-333.

Xu Yi, LiHong Jiang, Dong Li, 2006. A J2EE-based software framework with the application of [J]. Computer Application, 2006, 23 (9):146-148.

Zhang Lei, Hua-Rui Wu, Zhao Chunjiang, and so on, 2007. Based on the Struts framework small towns e-government system [J]. Micro-Computer Information, in 2007, 23 (4-3): 158-160.

APPLICATION ON IOC PATTERN IN INTEGRATION OF WORKFLOW SYSTEM WITH APPLICATION SYSTEMS

Limin Ao[*], Xiaodong Zhu, Wei Zhou

College of Information Engineering, Northeast Dianli University, Jilin, Jilin, China, 132012

[*]*Corresponding author, Address: College of Information Engineering, Northeast Dianli University, Jilin, Jilin, 132012, P. R. China, Tel: +86-432-4807268, Email: aolm@163.com*

Abstract: The integration of the two is a key on introducing workflow technology in the application systems. However, an inappropriate method easily results in invasive system code, structural damage, tight coupling system, and reduces the flexibility of the systems, increases the difficulty to maintain. Then two integration models were presented, and a new integration method was brought forward to solve the problem. Finally, an example was provided to illustrate the implementation, which has been put into practice in the job sheet management system of a power supply enterprise.

Keywords: IoC, workflow, application system, integration

1. INTRODUCTION

Workflow technology is mainly applied to enterprise business process analysis, simulation, definition and the operation implementation, which is the core technology of the realization of business process management and control, process reengineering (WfMC 1995). It can change information processing method of the business and the collaborative approach among the enterprise applications, which has an important influence on the operation and management of enterprises (Zhou et al., 2005). Therefore, the workflow technology has become one of the essential elements on the enterprise information-building program.

Ao, L., Zhu, X. and Zhou, W., 2008, in IFIP International Federation for Information Processing, Volume 259; Computer and Computing Technologies in Agriculture, Vol. 2; Daoliang Li; (Boston: Springer), pp. 875–882.

But, the current workflow products all have to combine with certain application systems, by which the workflow system can play the role only. However, often due to the diversity and differences of the application, and the improper integration approach, their combination easily results in the use of invasive system code, structural damage and coupling system, and reduces the flexibility of the system, increases the difficulty to maintain.

IoC (Inversion of Control) pattern is a design concept, based on object-oriented basis, to solve the dependence, configuration and life cycle between components, and processing the dependence is its core (Lou et al., 2007). IoC can be used to solve integration-coupling degree of the workflow management system with application systems.

2. INTEGRATION MODEL PRESENTATION

According to the relationship of the workflow management system and application systems, workflow system can be divided into the autonomous and the embedded, two categories (Michael et al., 2005). This paper described the two integration models what has been suggested by Li et al., in 2006.

2.1 Autonomous workflow system integration model

On the interactive mode with the business applications, autonomous workflow engine will provide remote procedure calls (RPC) with the WAPI (Workflow API). In addition, the autonomous workflow engine must also provide the solution how to call the business application.

As shown in figure 1 (Li et al., 2006), the application systems access and drive the workflow engine through WAPI, and the workflow engine triggers and calls applications through the agents that are disposed in the

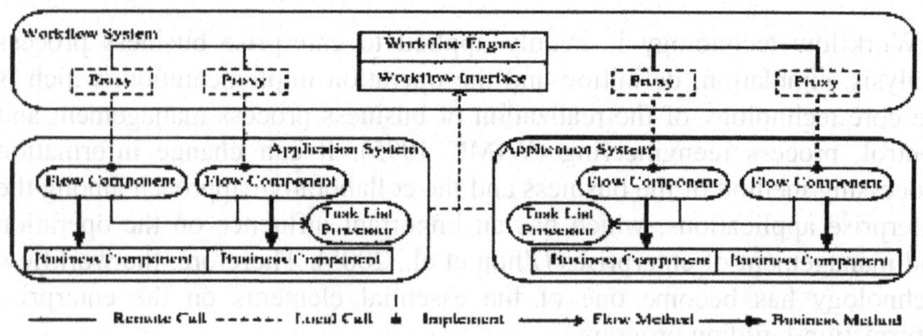

Figure 1. Autonomous workflow system integratiom model

flow components. But, the workflow interface and proxy mode will restrict the integration capacity.

2.2 Embedded workflow system integration model

Embedded workflow engine doesn't run alone. Generally it is deployed in the application system, as a software component. Therefore, local methods can be called to implement the integration which often is implemented through the two kinds of approach: workflow expansion and component expansion respectively, as illustrated in figure 2 (Li et al., 2006).

(1) Workflow expansion Business components are the smallest points of function. The flow components that are used to implement the flow agents are directly driven by the engine; Representation layer components directly access the workflow interface. This kind of approach is suitable for flow-oriented system development.

(2) Component expansion Flow component expands the original business components, and implements the flow driver by calling the workflow interface; Representation layer components access the specific flow components. This kind of approach is suitable for increasing workflow management of the existing application system.

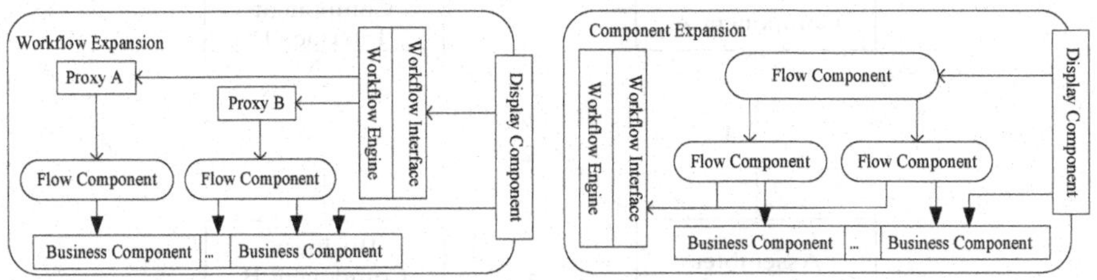

Figure 2. Embedded workflow system integration model

3. IOC BRIEF

IoC was brought out by Stefano Mazzocchi, one of the founders of Apache Avalon project. The aim is to emphasize the safety of the design. Hollywood's famous principle: "Don't call us, we will call you" is the embodiment of this idea.

The principle is: components themselves do not actively call the components that are required to complete certain functions, but only declare their interfaces, thereafter inform the superior manager. The superior manager will provide components that are satisfied to the interfaces. In this way, the dependent relationship will be reversed, and doesn't directly been established any more, instant established in other place. Such, the components won't be hard-coded together any longer each other, then they can obtain the greatest degree of reuse (Tou 2003).

4. INTEGRATION PRINCIPIUM BASED ON IOC

An application is generally composed of a number of components, and those components are interdependent according to a specified rule. In the integration model given hereinbefore, there is the situation of interdependence via calling flow methods or business methods between components. And that relationship of interdependence is represented by hard-coding directly, not through intermediaries. However, such invasive approach will lead to the system tight coupling, lower expandability and maintainability. To a great extent, the components can't be reused, which deviates from software design concept of the "highly cohesive, loosely coupled". Introducing the IoC will provide the solution of this issue, which can achieve good management of the dependence between components.

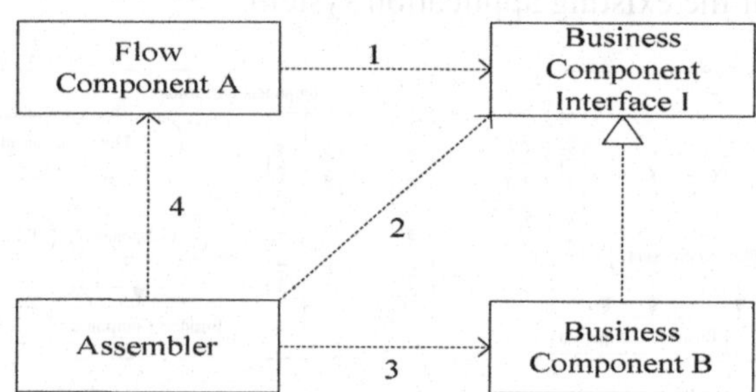

Figure 3. Integration method based on the IoC

According to the principium of IoC pattern, it is known that the essential of solving the component dependence is transferring the dependence. For example, the dependence A--> B between the flow component A and the business component B can be transferred (A->I) + (I<-B) through introducing the inter-protocol and assembler. When the A component calls the method of the B component, at first the I will be declared in the A component, then at runtime the assembler obtains a instance of the interface I according to the configuration file, and injects this instance into the A component, as shown in figure 3. In this way, the component does not actively access the collaboration components or objects no longer, but dynamically inject ones into the caller by the assembler at runtime (Martin 2004). This integration method obviously reduces the coupling, avoids a lot of hard coding, is propitious to reuse components and expand systems.

5. SOFTWARE IMPLEMENTATION AND ILLUSTRATION

IoC is a kind of design pattern, whose implementation generally is a container, such as Spring, PicoContainer, and Jdon etc. So, this paper has adopted Java EE lightweight platform, which was composed of Struts, Spring, and Hibernate, and workflow software OSWorkflow to integrate the workflow with applications, at the same time described the example of job sheet management system of a power supply enterprise.

5.1 Integration of lightweight Java EE framework and OSWorkflow

OSWorkflow is a Java workflow engine that has been issued by the well-known open source community OpenSymphony, which adopted FSM (Finite State Machine) theory based on action.

Java EE provides an enterprise-class computing model and the environment, supports the development and deployment of multi-tier architecture application. OSWorkflow is an embedded workflow engine, which can be deployed in the Java EE framework as the part of business logic, the software hierarchy as shown in figure 4.

In this figure, OSWorkflow engine, process components, and business components are all placed in the IoC container, no direct interaction, and their dependence relations are described in the configuration file. IoC container will dynamically inject those collaboration components according to the configuration file at runtime. Thus the workflow system and application systems can be integrated together. Then the degree of coupling can be reduced, synchronously reusability of components also is increased.

5.2 Illustration

In electric power system, the job sheet management system includes the job sheet of No. 1 and No. 2 of transformer substation etc. Under the above, the dependence relationship of component is incarnated by IoC container. So, the components should be registered into the container at first through

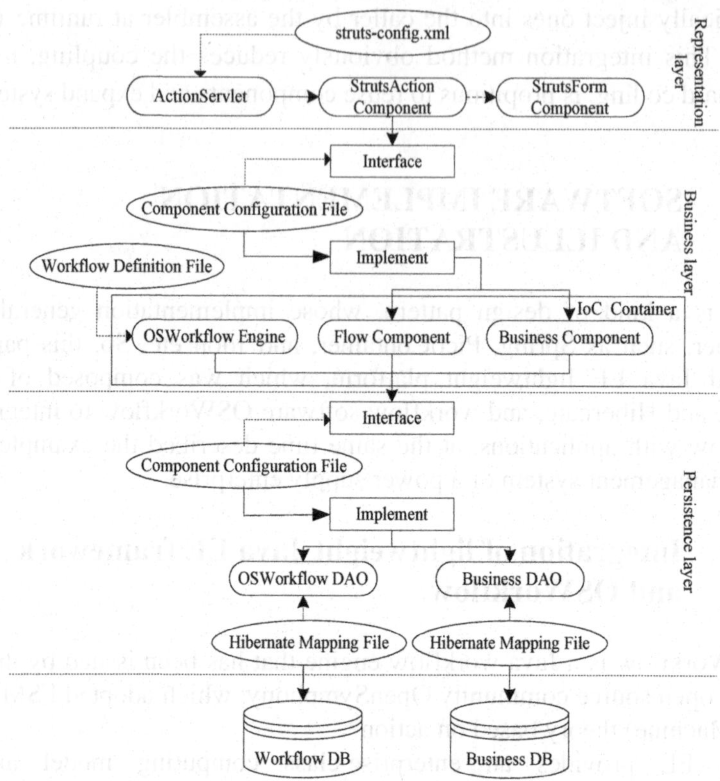

Figure 4. Software hierarchical graph that is embedded workflow engine

XML desperation file. The example based on the No.1 job sheet of transformer substation is described as follows:

```
Registration to Workflow Engine
<bean id="workflow "
        class="org.springframework.aop.framework.ProxyFactoryBean">
    <property name="interceptorNames">
        <list><value>workflowTarget</value>... ...</list>
    </property>... ...</bean>
Registration to Flow Component
<bean id="flowControl" class="com.mycompony.workflow.FlowControlBean">
  <property name="workflow"><ref local="workflow"/></property></bean>
Registration to Display Layer Component
<bean name="/bdfsBasicInfoAction" class="com.mycompany.struts.action.
        bdfirst.BdfsBasicInfoAction">
    <property name="flowControl"><ref local="flowControl" /></property>
</bean>
```

While registering, the depended components should be declared through the element <ref>. Such, the dependence relationship of component is transferred to the description file, when changing the dependence, only modify the description file.

Below are the core codes that the representation layer component calls flow components:

```
public class BdfsBasicInfoAction extends Action { ......
public ActionForward execute (...) {......
    WebApplicationContext wac = WebApplicationContextUtils.
                getRequiredWebApplicationContext (servletContext);
      // injected by container
        flowControlBean=(IFlowControlBean)wac.getBean("flowControlBean");
      ... ...}
  ... ... }
```

In this class, although the representation layer component needs to call the flow components, the latter has not been shown in the former and instant through declaring the interface, and then dynamically injected by the container at runtime, thus avoiding the hard coding.

6. CONCLUDING REMARKS

The key of introducing workflow technology in application systems is to solve the integration coupling degree of the two. Applying IoC pattern in this paper can solve this problem easily. This way cuts down the number of hard coding, greatly improves the reusability of components, makes the whole system flexibility. This solution has been applied to the job sheet management system of a power supply enterprise, which has been showed the availability and practicality.

ACKNOWLEDGEMENTS

This study has been funded by Doctoral Startup Foundation of Northeast Dianli University (BSJXM-200601).

REFERENCES

Li Qing, Wen Jingqian, Zhao Meng 2006. Research on AOP-based integration of workflow system with enterprise information systems [J]. Computer Integrated Manufacturing System,12(3):401-406 (in Chinese).

Lou Feng, Sun Yong 2007. Design and Implementation of Lightweight IoC Container [J]. Computer Technology and Development, 17(1):91-93 (in Chinese).

Martin Fowler 2004. Inversion of Control Containers and the Dependency Injection Pattern [EB/OL]. http://www.martinfowler.com/articles/injection.html.

Michael Zur Muehlen, Rob Allen 2005. Workflow management coalition workflow classification: Embedded & autonomous workflow management systems [EB/OL]. http://www.wfmc.org/standards/docs/MzM_RA_WfMC_WP_Embedded_and_Autonomous_Workflow.pdf.

Tou Ming 2003. Inversion of Control–Components, Containers and IoC [J]. Programmer, (12):92-94 (in Chinese).

WfMC 1995. The workflow reference model [S].

Zhou Jiantao, Shi Meilin, Ye Xinming 2005. State of Arts and Trends Flexible Workflow Technology [J]. Computer Integrated Manufacturing System, 11(11):1501-1510 (in Chinese).

A TEMPORARILY-SPATIALLY CONSTRAINED MODEL BASED ON TRBAC IN WORKFLOW SYSTEM

Limin Ao [*], Wei Zhou, Xiaodong Zhu
College of Information Engineering, Northeast Dianli University, Jilin, Jilin 132012, China
** Corresponding author, Address: College of Information Engineering, Northeast Dianli University, Jilin, Jilin 132012, China, Tel: +86-432-4807268, Email: aolm@163.com*

Abstract: A temporarily-spatially constrained model based on TRBAC (Task-Role Based Access Control) is proposed in workflow system. Temporary and spatial Constraint is that user is not only constrained by temporality, but also constrained by spatiality when user executes the task. The model suggests that a property of security level should be increased in task. The newly increased property can make the workflow more safety and flexibility.

Keywords: TRBAC, Workflow, Constraint, Security

1. INTRODUCTION

Workflow is a kind of business flow entirely or partly disposed by computer (WFMC, 1995). The task can only be executed by user who was authorized. For the sake of the task can not be executed by non-authorized users and make task completed favorably, a safer access control model suitable for workflow management system is needed.

Traditional access control model consist of DAC (Discretionary Access Control) and MAC (Mandatory Access Control) (Shen et al., 2005). DAC and MAC are not suitable for workflow management system. So Ferraiolo and Kuhn proposed the model of RBAC (Role-Based Access Control) in 1992. Then R.Sandhu in University of George Mason described the model of

Ao, L., Zhou, W. and Zhu, X., 2008, in IFIP International Federation for Information Processing, Volume 259; Computer and Computing Technologies in Agriculture, Vol. 2; Daoliang Li; (Boston: Springer), pp. 883–889.

RBAC in 1996 (Sandhu et al., 1996) that is called the model of RBAC96. It has some disadvantages when make it combined with workflow technologies. The model of TRBAC can deal with the problem effectively. This paper adds space-time constraints on TRBAC and proposed authorized model TSC-TRBAC with space-time constraints. The model indicates that execution of workflow is not only constrained by time, but also by spatial, and increases the property of security for task. Therefore, it can keep synchronization between workflow and authorization.

2. TASK-ROLE BASED ACCESS CONTROL (TRBAC)

Task as an individual conception in TRBAC was proposed (Xing et al. 2005). A conceptual model of TRBAC is relevant to this paper in Fig. 1.

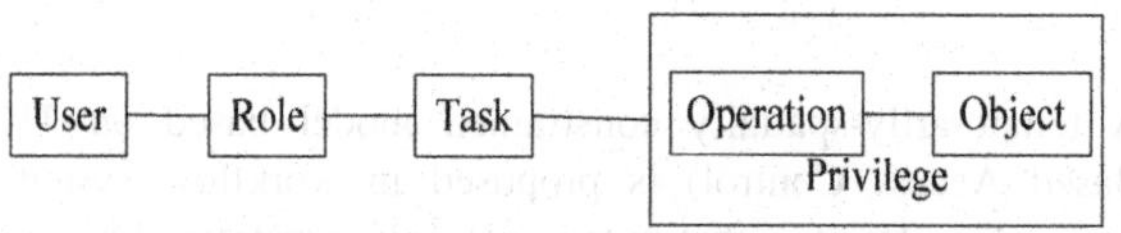

Fig. 1. A conceptual model of TRBAC

User will obtain the corresponding privilege through task that executed by himself. Once task has been executed completely, privilege will be disposed automatically. Privilege will be assigned until user will get another task in the next time. It really achieves assignment according to requirement of user and makes the operation of administrator more convenient. To assure task will be completed on time, temporal constraint is needed in the workflow management system. Particular narration about this point in literature (Xing et al., 2005). Temporal constraint has been recognized by the people who is doing the research in the area of workflow and obtains great success in practical work. To keep the safety of task in the workflow, constraints about physical space to executer of task is proposed in this paper. Although literature [6] (Xu et al., 2006) put forward the conception of spatial constraints, it only talked about logical space to executer of task. However, the problem also exists in the system relatively.

2.1 Definitions

Definition 1. The binary group (TS; $\leq$) is a region of tense, TS=$\{t \in R | t \geq 0\}$ is a set of time, $\leq$ denotes total order in TS.

Definition 2. $[t_s,t_e]$ denotes a temporal interval, $t_s,t_e \in$ TS, and $t_s < t_e$. The temporal interval is bounded by a lower bound t_s and a upper bound t_e. If interval $[t_1,t_2]$ in the interval $[t_3,t_4]$, iff $t_3 \leq t_1$ and $t_4 \geq t_2$. If a temporal point t_1 in the interval $[t_3,t_4]$, iff $t_3 \leq t_1 \leq t_4$.

Definition 3. $IP=\{IP_i|i=1,2,\ldots,n\}$ is a set of IP address.

Definition 4. $MAC=\{C_i|i=1,2,\ldots,n\}$ is a set of MAC address.

Definition 5. $WT=\{wt_i|i=1,2,\ldots,n\}$ is a set of workflow task, task is a static conception in workflow. It is a set of operation, defined by user to complete some function.

Definition 6. $TI = \{ti_i|i=1,2,\ldots,m\}$ is a set of task instance. A task instance is an instance of task. It is a pentad group consist of role, privilege, state of task, temporal and spatial constraints.

Definition 7. M: $WI \rightarrow TI$ is a mapping of instance. It makes each task mapped to the corresponding task instance. To the task instance ti_i, if $M(ti_i)=wt_i$, then ti_i is a instance of wt_i.

To essence of model, task can be operated by users from different department. TRBAC can control operation of task-execution in any department through add the property of IP address in it. If we don not use spatial constraints, users would have the privilege of inter-departmental operation. Administrator of task should consider whether the constraints of IP and MAC address are needed according to practical situation.

2.2 Description of basic elements in TRBAC

$U=\{u_i|i=1,2,\ldots,m\}$ states a set of users,

$R=\{r_i|i=1,2,\ldots,n\}$ states a set of role,

$IP=\{IP_i|i=1,2,\ldots,o\}$ states a set of IP address,

$MAC=\{C_i|i=1,2,\ldots,p\}$ states a set of MAC address,

S={sleep, activity, terminate, hang, abortion} states the state of task. It is a set include five element,

$WT=\{wt_i|i=1,2,\ldots,j\}$ states task in workflow. Task have upper security should be restricted by physical space,

$OP=\{op_i|i=1,2,\ldots,l\}$ states set of operation,

$OBJ=\{obj_i|i=1,2,\ldots,q\}$ states set of object,

$P=\{P_i|i=1,2,\ldots,x\}$ states set of privilege, and $P_i=(op_i, obj_i, [t_{si},t_{ei}], IP_i, C_i)$. IP_i, C_i are selective option,

$TI = \{ti_i|i=1,2,\ldots,k\}$ states set of task instance, and $ti_i = (r_i, P_i, S, [t_{si},t_{ei}], IP_i, C_i)$.

URA (User Role Assignment) indicates that relationship between user and role, $URA \subseteq U \times R$.

RTA (Role Task Assignment) indicates that relationship between role and task, $RTA \subseteq R \times T$.

TPA (Task Privilege Assignment) indicates that relationship between task and privilege, $TPA \subseteq T \times P$.

2.3 Delegation

Delegation is denoted by septenary group in TSC-TRBAC. Like (U, R, T, P, [t_s, t_e], IP, C). IP and C are separated into options and will be option. As an option when security level is higher or otherwise. User u is assigned right of task execution when time in t_s point and revoked in t_e point.

Task T has many states. The state of task can be expressed with a figure of state transition (Song et al., 2005).

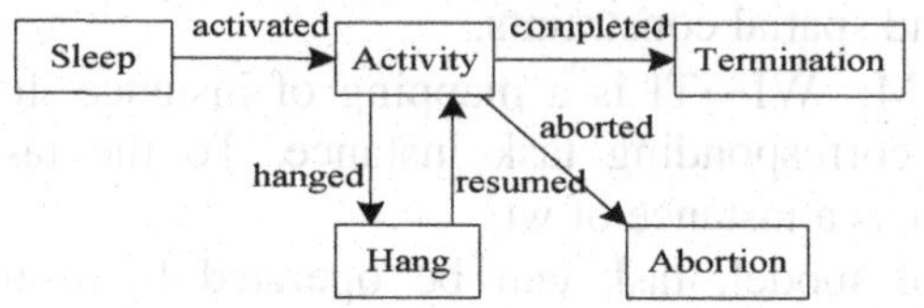

Fig. 2. State transition of task

Signification of every state:

(1) Sleep: It states that there is not have task instance in workflow.

(2) Activity: It states that task instance is created.

(3) Termination: It states that task is completed.

(4) Hang: It states that task is suspended by administrator because of certain reason in the process of task execution.

(5) Abortion: It states that task what could not be executed is terminated forcibly by administrator in the process of task execution.

3. APPLICATION OF TSC-TRBAC

There is very important part about managing the people and controlling the resource in workflow management system. To assure the task can be completed successfully in workflow and enhance the security of workflow management system more, it is necessary of restricting the user and object in system. This paper will apply model of TSC-TRBAC to organization modeling tools in workflow.

3.1 Framework of modeling tool

This modeling tool is based on framework of .NET. It adopts traditional three-layer system framework, namely, expression layer, business logic layer and data layer. The expression layer is regarded as an interface between users and system. This part is carried out by technology of ASP.NET. The business logic layer is a bridge that connects users and database. It is core of system and achieved by language C#. The main function of data layer is to store some relevant model data with database. System adopts Microsoft SQL Server 2000 as a database to store data about organizational model. The framework of system is shown in Fig. 3.

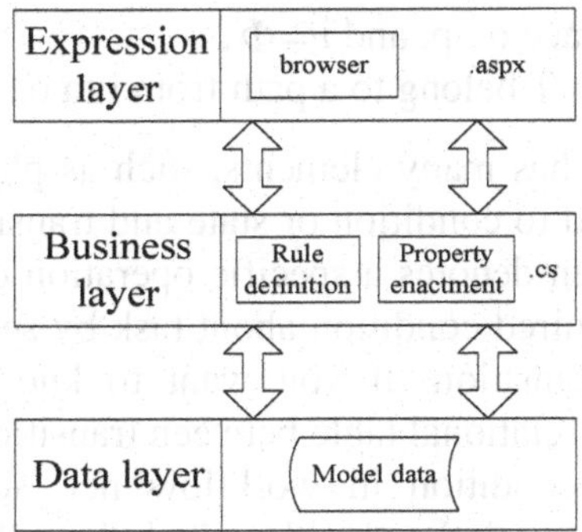

Fig. 3. Framework of system

3.2 Elements of model in organization modeling

According to the main idea of TSC-TRBAC, each element has relationship with others in model. Relationship between them is very tight, so it is very important to design a reasonable database. Based on description on elements in model, the table of department, users, role, task (operation), privilege, relation between users and role (play), relation between role and task (hold), relation between task and privilege (authorize) and so on.

3.3 Process model of Petri Net

Organizational model should be applied by process model. The system adopts Petri Net for process modeling. Van der Alast in university of Holland Eindhoven introduced technology of Petri Net in process of workflow modeling. According to extending Petri Net, he proposed workflow net. The essence of modeling with workflow net is a method that applies Petri Net to process definition in workflow.

Definition 8 (Petri Net): triple group N=(S, T; F) is called Directed Net, for short Net. Sufficient and necessary condition is (Yuan 2005):

(1) $S\cap T=\Phi$;

(2) $S\square T\neq\Phi$;

(3) $F\subseteq S\times T\square T\times S$ ("×" is Cartesian product);

(4) dom(F)□cod(F) = S□T, and, dom(F) = {x| ∃ y : (x,y)□F}, cod(F) = {y|∃x: (x,y)□F}, they are domain and range.

Definition 9 (Workflow Net) (Yuan 2005): Directed Net is Sufficient and necessary condition of WF_net:

(1) PN has a source place i□P, and ·i=Φ.

(2) PN has a sink place o□p, and i·=Φ.

(3) Each node x□P□T belong to a path from i to o.

Model of Petri Net has many elements, such as place, transition, arc and token. Place is mapped to condition or state and transition is mapped to task in workflow net. Token denotes a specific operation object in process flow. We can check the required condition about task by selecting relational table between place and transition. If you want to know the next task, it is necessary of selecting relational table between transition and place. Owing to place is mapped to condition in workflow net, so this paper will put constraint condition into relational table called place. To maintain security of system, constraint condition must be checked when task is executed by users.

3.4 Example

If we have a workflow that students submit thesis. The task consists of four parts in workflow, such as thesis-written, thesis-examined, thesis-revised and thesis-submitted. Two tasks of thesis-written and thesis-revised is completed by the role of student and thesis-examined and thesis-submitted is achieved by the role of mentor. The mentors in college of computer can examine thesis written by students in the same college. Mentors in other college have no privilege to do this job. Although we have achieved the least granularity of privilege through adopting TRBAC to access control safely, TRBAC can not control inter-departmental operation by the same role itself. In this example, in another word, mentors in college of management or others also have right to examine thesis written by students in college of computer. It is fall short of practical situation. The measure we have introduced into this paper is that role which executes task should be restricted by spatial constraints, such as IP address or Mac address according to segment of IP address in different departments. This method can dominate inter-departmental operation of role effectively when keep synchronization between delegation flow and workflow.

4. CONCLUSION

The thought of model is that users who have the same role can carry out inter-departmental operation. It improves fatalness of system in the process of task execution in this way. This paper has an idea to deal with the above problem. Through executer of task implementing the physical space constraints to control inter-departmental operation of role and better guarantee security for the implementation of the tasks.

ACKNOWLEDGEMENTS

This work was supported by Doctoral Startup Foundation of Northeast Dianli University (BSJXM-200601).

REFERENCES

WFMC. The Workflow Reference Model, Doc. No. TC00-1003.http://www.wfmc.org/

Hai-bo Shen, Fan Hong. Survey of Research on Access Control Model [J]. Application Research of computers, 2005, 6:9-11(in Chinese).

Ravi S. Sandhu, Edward J. Coyne, Hal L. Feinstein, Charles E. Youmar. Role-Based Access Control Models [J]. IEEE Computer. 1996, 29(2):38.

Guang-lin Xing, Fan Hong. A Workflow Access Control Model Based on Role and Task [J]. Engineering and Application of computers, 2005(2):210-213 (in Chinese).

Guang-lin Xing, Fan Hong. Workflow Authorization Model Based on RBAC [J]. System of Microcomputer, 2005, 26(3):544-547 (in Chinese).

Hong-xue Xu, Xiu-ying Guo, Yong-xian Liu. Temporarily-Spatially Constrained Workflow Authorization Model Based on RBAC [J]. Transaction of Northeastern University (Edition of Natural Science), 2006, 27(2):217-220 (in Chinese).

Shan-de Song, Wei Liu. Task-Role-Based Access Control Model [J]. Computer Engineering and Science, 2005, 27(6):4-6 (in Chinese).

Chong-yi Yuan. Principle and Application of Petri Net [M]. Beijing: Publishing House of Electronics Industry, 2005.3 (in Chinese).

FREIS: A WEB-BASED RESOURCES AND ENVIRONMENT INFORMATION SYSTEM FOR AGRO-ECOSYSTEM MANAGEMENT

Mingxin Men *, Yuepu Qi, Boyang Du, Hao Xu

College of Agricultural Resources and Environmental Sciences, Hebei Agricultural University, Baoding, China, 071001

** Corresponding author, Address: College of Agricultural Resources and Environmental Sciences, Hebei Agricultural University, 215 Jianshe South Road, Baoding, 071001, P. R. China, Tel: +86-312-7528231, Fax: +86-312-7528218, Email: menmingxin@sina.com*

Abstract: Agro-ecosystem plays a crucial role in conserving biodiversity, sustaining ecosystem functions and processes, and maintaining land productive capacity. In order to effectively manage agro-ecosystem, the planners and government agencies are increasingly seeking better tools and techniques. In this paper, we describe the development of a Web-based resources and environment information system (FREIS), which helps to set up agricultural policy to improve productivity level and resource utility efficiency in terms of yield stability evaluation model. The system design involved four steps, the first was to set up a system platform for FREIS, then a spatial database was developed for analysis, after this the evaluation model was established, and lastly a Web-based interface with analysis tools was developed using client-server technology. FREIS provided a valuable technical scheme of the intelligent and comprehensive agricultural information management. The potential of a Web-based information system for agro-ecosystem management and challenges for its development was discussed.

Keywords: WebGIS; resources and environment information system; agro-ecosystem; productivity stability

Men, M., Qi, Y., Du, B. and Xu, H., 2008, in IFIP International Federation for Information Processing, Volume 259; Computer and Computing Technologies in Agriculture, Vol. 2; Daoliang Li; (Boston: Springer), pp. 891–898.

1. INTRODUCTION

Agro-ecosystem is a man-made ecosystem under human intervention and control. It is a three dimensional ecosystem with the organic combination of nature, society and economic, and have the functions of energy flowing, material cycling, value and information transmission. It plays a crucial role in conserving biodiversity, sustaining ecosystem functions and processes, and maintaining land productive capacity (Zhao et al., 2003; Yang et al., 2005). The stability is one of important characteristics and is also the basis of maintaining higher the agricultural productivity and ensuring the ecosystem health development. In order to effectively manage agro-ecosystem, the planners and government agencies are increasingly seeking better tools and techniques. The appearance of GIS and other information technology makes all of these possible. People can use information technologies to access, store, disseminate different sources and spatio-temporal scale data, and analyze spatial variation (Tony, 1998; Jos et al., 1995; Issolah et al., 2001).

The history of Web-based GIS can be traced back to the development of the Xerox Map Viewer which used a Web Browser via HyperText Markup Language (HTML) format and Common Gateway Interface (CGI) programs to provide interactive mapping functions via the Internet (Putz, 1994). There is a great potential for using WebGIS in the areas of natural resources management and environmental assessment and monitoring. A web-based system was constructed to supply farmers and agricultural advisers with just-in-time information and decision support for crop management (Allan et al., 2000). A web-based system was developed to advise on the relative efficacy of different for mixes of weed and crop species at different times of the year in a forestry of farm forestry setting (Alan et al., 2004). A web-based environmental decision support system was developed for environmental planning and watershed management, which integrated the spatial analysis model into ArcIMS based on the client/server model (Ramanathan et al., 2004). A prototype web site has been developed to provide easy access of geospatial information and to facilitate Web-based image analysis and change detection capabilities for natural resource managers and regional park rangers (Tsou, 2004).

In this paper, the main object was to develop an interactive web-based resources and environment information system using ArcIMS, geodatabase and internet technology in Hebei Province of China. The information system integrated agricultural productivity stability model to analyze the productivity fluctuation trends. The main function of the FREIS is to provide map and data query for the agro-ecosystem manager and the planner, and the technology support to the ecosystem management.

2. FREIS CONSTRUCTION

WebGIS is the realization of GIS functions on Web s through integration of Internet and GIS. It provides the functions of browsing spatial data and thematic maps and analyzing the spatial data on Web with the browser. As one of the ESRI's WebGIS products, ArcIMS which uses Java Applet, Java Servlet and XML application technology can obtain dynamic Map, GIS data and various service items. The system in this study uses ArcIMS as the WWW based GIS server. As a result of the working environment of ArcIMS is in Java, the components of Web Server, Java virtual machine (JVM) and Servlet Engine are required to guarantee ArcIMS operation normally. So the system uses Microsoft Internet Information Server (IIS6.0) as Web Server, J2SDK1.4.2.06 as Java virtual machine (JVM) and ServletExec as Servlet Engine. The FREIS design is based on the browser/server mode in which clients send requests to services running on a server and receive appropriate information in response. The browser/server used in this study has a three-tiered configuration: a WWW client of Microsoft Internet Explorer (IE), a WWW server of Microsoft Internet Information Server (IIS) and a WWW based GIS server of ArcIMS (Fig. 1).

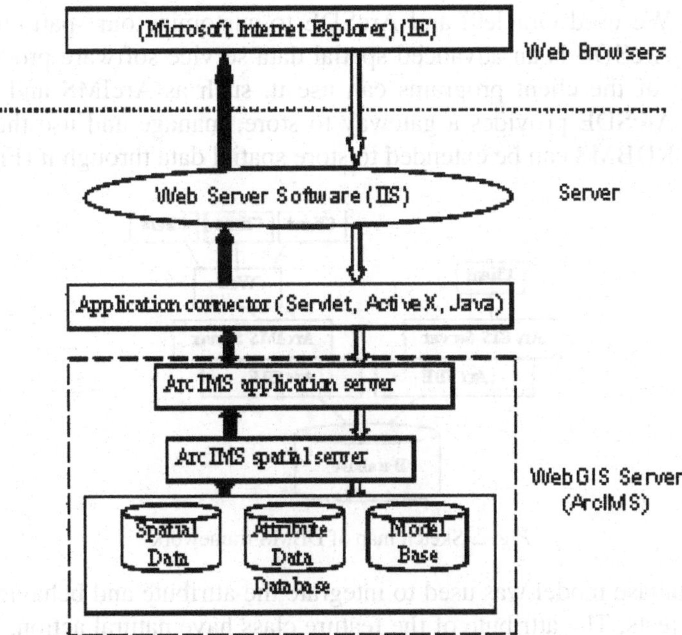

Fig. 1. The process of information transfer between client and server

It shows the flow of information in the client-server transaction. At first, the users initiate a request to the Web Server by manipulating tools and buttons in the browser. The Web server passes the request to the application connector. Then, the application connector translates the request to the

ArcXML and passes the ArcXML to ArcIMS Application Server. ArcIMS Application Server processes loading distribution and passes the request to ArcIMS spatial server. There are seven types of ArcIMS spatial server, such as image, feature, query, geocode, extract, metadata and ArcMap. Among of these images, ArcMap, feature and metadata can access the spatial server. The access isn't direct but through virtual server Tools. At last, ArcIMS spatial server gives a corresponding response after receiving the request, and passes the results to ArcIMS Application Server in opposite direction. The Web server displays the querying outputs on HTML pages. The Web browser on the client machine displays the results and supports further user interaction, which creates additional requests. The whole process makes up of a request/response Cycle.

3. DATABASE DESIGN

3.1 Database Manage System

FREIS uses geodatabase model to create an object-oriented spatial database. We used Oracle9i and ArcSDE to customize our spatio-temporal database. ArcSDE is an advanced spatial data service software produced by ESRI. All of the client programs can use it, such as ArcIMS and ArcGIS Desktop. ArcSDE provides a gateway to store, manage and use the spatial data. The RDBMS can be extended to store spatial data through it (Fig. 2).

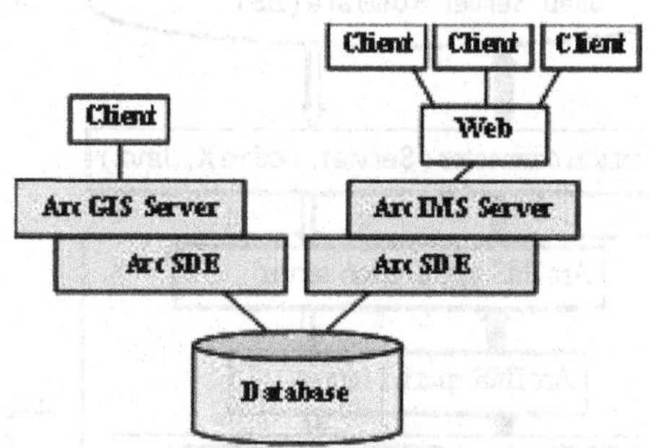

Fig. 2. Sketch map of DBMS framework

Geodatabase model was used to integrate the attribute and behavior of the spatial objects. The attribute of the feature class have natural action, and any type of the relationships can be defined in Geodatabase model. The data is oriented to the users in it. As compared with previous spatial data model, the relation and attribute with behavior had greatly extended the data representation ability. And the new model allows users use the regulations to define the more GIS applications. The excellence of the model is that the feature class completely stored in one database. Therefore the large scale geographic features can be stored smoothly. Using the Geodatabase model, the feature was a record in the database. Thus the physical and logical data model was more adjacency. The data object in Geodatabase, which realized mostly customization without compiling any code, was generally consistent to the object which was defined in logical data model. It uses the field, virtual regulation and other functions to do this.

3.2 Data in FREIS

The data of FREIS includes of the spatial data and attribute data. The spatial data is mainly the vector maps which compose of the basic spatial database and the results maps. The spatial database includes administrative maps, soil type maps, organic matter distribution maps, rainfall distribution maps, topographical maps with different scale etc. Attribute data included natural data and social and economic data and was shown in Table 1.

Table 1. Data of Agricultural Resource and Environment Information System

Data Type	Data Name
Spatial Data	Administrative Map of Hebei
	Soil Types Map of Hebei
	Organic Matter Distribution Map of Hebei
	Rainfall Distribution Map of Hebei
	Topographical Map of Hebei
	Agro-ecosystem Stability Value Map of Hebei
	Crop Production Fluctuation Value Map
	Agro-ecosystem Stability Grade Map
Attribute Data	Meteorological Data
	Soil Properties Data
	Social and Economic Data

4. STABILITY EVALUATION MODEL

At first, agro-ecosystem stability indicators were determined through AHP model. Then, the weight and standardization index was calculated with the Z-Score method. At last the integrated index was calculated. The value of agro-ecosystem productivity stability (API) is a synthetic evaluation results of the multilevel system. According to its characteristic, we choose the linear weighted suming method in the study. The formula is:

$$API = \frac{\sum w_i * x_i}{100\sum w_i} \tag{1}$$

Where: *API* is comprehensive index; *Wi* is weight index; *Xi* is value of single standardization index.

The stability evaluation model was store in model bases of FREIS and was integrated to the user interface. The clients can use ArcIMS server to transfer it.

5. INTERFACE AND APPLICATION

The functions of FREIS include displaying of the map, consulting service and statistics and information outputting etc. The interface is shown in Fig. 3.

Fig. 3. The main function interface of FREIS

5.1 Map Displaying

Users can display and analyze data in the interactive interface. The GIS data browsing tools of FREIS created by ArcIMS is standard, such as zooming out, zooming in, panning, zooming to full extent, refresh, distance and area measuring, etc. These tools provide interactive operations between the users and the maps. The buttons of panning, zooming in or out allow the user to move around in the map displaying and change the maps scale. Full screen button allows the map come back to full screen display at any scale display. The distant measuring button allows user to measure the distant from one point to another on the map. Users also can use the area measure button to draw the outline of polygon to measure its area.

The left of the interface is displaying frame of the map layers. Users can point the button in the check box to show the map layers. The middle of the interface is map displaying window that can show the maps in different scale. The right of the interface is query tools of attribute data and statistic tools. Users can select different area and attribute name to query basic data. The information statistic tools allow users do basic statistical function.

5.2 Statistic Information Output

The functions of information statistic and consultation service were the main function. FREIS can display and output information based on the spatial database. Users can select different regions and attribute names to complete the data query. Users can do basic statistics by using chart tools. (Fig. 4). Output function of FREIS allows users to output the information using digital table and papery documents.

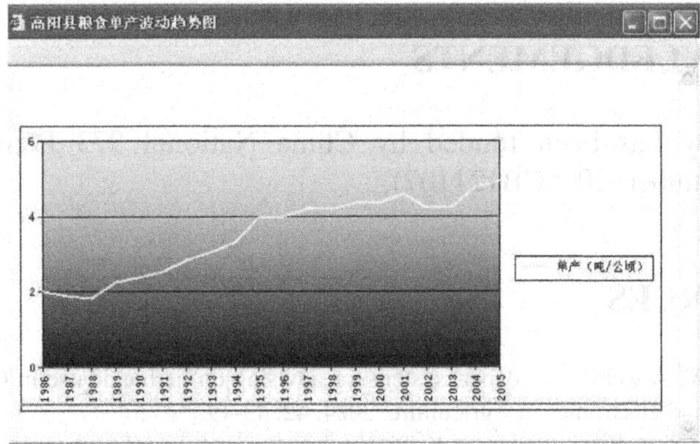

Fig. 4. Statistic chart

5.3 Stability Evaluation

On the right-down of the interface is the stability evaluation tool. Users can select different regions to evaluate the agro-ecosystem productivity stability. After pointing the stability evaluation tools, the results will be shown. This tool can evaluate the productivity stability in different scales.

6. CONCLUSION AND DISCUSSION

FREIS, a Web-based system developed for this project, provides the users with a simple information query tool. This system can be used by planners and mangers within local government, the general public, environmental analysts, farmland researchers and other interested parties. Users can know more to the agro-ecosystem by using FREIS. The FREIS has a lot of limitation now. First, it can not completely describe the agro-ecosystem, because of the data is imperfect and the influence factors of the agro-ecosystem are more complicated. It should be perfected from collected more information. FREIS can evaluate the productivity stability of agro-ecosystem of Hebei Province. But this model has not been validated in practice. In the future, based on feedbacks received from users, we hope to revise and perfect the model constantly. In addition, the data spatial scopes is also limited. It only includes basic information of Hebei Province. We must collect more spatial data to extend it. WebGIS has the potential to share data, provide easy access for users with limited GIS knowledge, and assemble data and information customized for specific topics. The information system based on internet technology provided a customizable interface to users to organize the data and information. The WebGIS will be the development trend of natural resources and environment information system in Future.

ACKNOWLEDGEMENTS

This study has been funded by China National 973 Plans Projects (Contract Number: 2005CB121107).

REFERENCES

Alan J T, Ian W. A web-based expert system for advising on herbicide use in Great Britain. Computers and electronics in Agriculture, 2004, 42: 43-49.

Allan L J, Peter S, Boll I T, Pathak B K. Pl@nteInfo®-a web-based system for personalized decision support in crop management. Computers and Electronics in Agriculture, 2000, 25: 271-293.

Issolah R, Giovannetti J F. The Algerian agricultural information and document system: how does it support national research and training. International Journal of Information Management, 2001, 21: 289-299.

Jos A A, Ruud B, Huirne M, Aalt A D. Economic value of management information systems in agriculture: a review of evaluation approaches. Computers and Electronics in Agriculture, 1995, 13: 273-288.

Putz S. Interactive Information services using World Wide Web Hypertext. In Proceedings of the First International Conference on the World Wide Web, 1994, 25-27.

Ramanathan S, James C M, Jim D. A Web-based environmental decision support system (WEDSS) for environmental planning and watershed management. Geograph Syst, 2004, 6: 307-322.

Tony L. Evolution of farm management information Systems. Computers and Electronics in Agriculture, 1998, 19: 233-248.

Tsou M H. Integrating Web-based GIS and image processing tools for environmental monitoring and natural resource management Geograph Syst, 2004, 6: 155-174.

Yang Z X, Zheng D W, Wen H. Studies on service value evaluation of agricultural ecosystem in beijing region. Journal of Natural Resources, 2005, 20: 564-571 (in Chinese).

Zhao R Q, Huang A M, Qin M Z. Study on farmland ecosystem service and its valuation method. System Sciences and Comprehensive Studies in Agriculture, 2003, 19: 267-270 (in Chinese).

THE MONITORING POPULATION DENSITY OF PESTS BASED ON EDGE-ENHANCING DIFFUSION FILTERING AND IMAGE PROCESSING

Yuehuan Wang, Guirong Weng*

School of Mechanical and Electrical Engineering, Soochow University, 215021, Suzhou, Jiangsu, China

* *Corresponding author, Address: School of Mechanical and Electrical Engineering, Soochow University, 215021, Suzhou, P. R. China, Tel: +86-512-67165761, Fax: +86-512-67165607, Email: wgr@suda.edu.cn*

Abstract: As is known, agriculture is very important in China, but the problem about pests has hampered the further development of Chinese agriculture. Digital image-processing technology and mathematical morphology are referred to as the main research methods, and tiny pets like aphids among field are referred to as the research objects. Image processing technology such as edge-enhancing diffusion filtering, mathematical morphology and watershed segmentation algorithm is used to monitor pest population density, which greatly raises efficiency of pest data acquisition. After the segmentation of the image of the pests, the number of the insect individuals can be obtained from the background by using image processing technology. Computer image processing technology provides a possibility to solve this problem and becomes a very important direction to monitor regional pest population density.

Keywords: Digital Image-processing; Monitoring Population; Mathematical Morphology; Watershed Algorithm; Edge-enhancing filtering

Wang, Y. and Weng, G., 2008, in IFIP International Federation for Information Processing, Volume 259; Computer and Computing Technologies in Agriculture, Vol. 2; Daoliang Li; (Boston: Springer), pp. 899–907.

1. INTRODUCTION

As is known, agriculture is very important in China, but the problem about pest has hampered the further development of Chinese agriculture. It has been long since our country has the ability to forecast plant diseases and insect pests. Nowadays, many foreign countries apply some modern information technology into application of forecasting plant diseases and insect pests (Jason W. Chapman, 2002; Minghua Zhang, 2003). However our country still applies the outdated technology. Mathematical morphology has already been widely used in all fields of image processing with the development of information technology in recent years. With the development of computer network technology, artificial intelligence, images recognition and decision support system, precision farming technology system is used more and more widely (Zhaozhi Lu, 2005; Xiaochao Zhang, 2003). Currently precision farming technology system, the research of fast data acquisition technology is far behind the researches in other fields and becomes the bottleneck of development and practice of precision farming technology system. So it is very urgent for our country to improve the level of pest data acquisition.

The analysis to plant diseases and insect pests has limitations in terms of literal descriptions, while this can be well solved by image identification. The technology of image identification is used in the fields of crop quality monitoring and crop growth state monitoring. There are less application in the field of plant diseases and insect pests monitoring. When the monitor targets are tiny pets like aphid, the survey of sampling among field will take a lot of time and pose a great threat to the veracity and reliability of the data. Using digital camera and other digital devices, we can obtain the image of pest colony, put it into computer, segment the image of the pest and get the number of the insect individuals from the background by using image processing technology. Computer image processing technology provides a possibility to solve this problem and becomes a very important direction to monitor regional pest population density (Xinwen Yu, 2001). So the application of image processing technology in monitoring pest population density will improve the ability of forecasting plant diseases and insect pests and the administration skills, which is of great significance.

In this paper, image information processing technology such as edge-enhancing filtering, mathematical morphology and watershed segmentation algorithm is used to monitor pest population density, which raises efficiency of pest data acquisition greatly.

2. NONLINEAR ISOTROPIC DIFFUSION FILTERS

In the last decade, PDE based models have become very popular in the fields of image processing and computer vision. The basic idea of this theory is to build nonlinear partial differential equations and use the original image as initial condition, and then the solutions in different time are the result of the filtering. Methods of this type have been proposed for the first time by Perona and Malik (Perona, Malik J, 1990).

The anisotropic form of the diffusion equation can be written as:

$$\partial_t u(x,y,t) = div(D \cdot |\nabla u|) = div(g(|\nabla u|) \cdot \nabla u), t > 0$$

$$u(x,y,t) = I(x,y), t = 0 \qquad (1)$$

Where u is the evolving image, div is the divergence operator, D is the diffusion tensor, $\nabla u = [u_x \; u_y]$ is image gradient norm and $g(|\nabla u|)$ is the function of the image gradient norm, it's used to protect the image edge by adjusting the value of $|\nabla u|$, because the image gradient reflects the character of this image in some degree, the normal arithmetic edge detection is also according to image gradient. Perona and Malik method use this arithmetic, so it can remove the noise and protect the image edge information at the same time.

However, Perona and Malik method also has its problems. In theory, one should not expect is that a solution of this type is unique or stable with respect to the initial image. In practice, if the image with large amplitude noise which generates gradient value too, we will not differentiate noise from the image edge information. So the result of filtering is bad.

If there is large amplitude noise on the rim, it will cause large vibration of the image gradient and its amplitude and direction as well. All nonlinear diffusion filters that we have investigated so far utilize a scalar-valued diffusivity $D = g(|\nabla u_\sigma|)$ which is adapted to the underlying image structure. Therefore, they are isotropic and the flux $j = -D \cdot \nabla u$ is always parallel to ∇u, we can not smooth the noise on the image edge so we should rotate the flux. These requirements cannot be satisfied by a scalar diffusivity anymore, a diffusion tensor to anisotropic diffusion filters (J. Weickert, 1997) has to be a matrix D, Weickert proposed a nonlinear anisotropic diffusion equation:

$$\partial_t u = div(D(|\nabla u_\sigma|) \cdot \nabla u) \qquad (2)$$

with D which is constructed by eigenvector and eigenvalue is a positive definite diffusivity. we choose different D to construct Edge-enhancing diffusion (J. Weickert, 1999) or Coherence-enhancing diffusion.

Coherence-enhancing diffusion constructs D as:

$$D = \begin{bmatrix} v_1 & v_2 \end{bmatrix} \begin{bmatrix} \lambda_1 & 0 \\ 0 & \lambda_2 \end{bmatrix} \begin{bmatrix} v_1^T \\ v_2^T \end{bmatrix} \quad (3)$$

eigenvector: $\lambda_1 = g(|\nabla u_\sigma|)$, $\lambda_2 = 1$, $u_\sigma = K_\sigma * u$, eigenvalue: $v_1 \parallel \nabla u_\sigma$, $v_2 \perp \nabla u_\sigma$.

Weickert model (m=2,3, and 4):

$$g(|\nabla u_\sigma|) = \begin{cases} 1 & \text{if } |\nabla u_\sigma| = 0 \\ 1 - \exp\left(-\dfrac{C_m}{\left(|\nabla u_\sigma{}^2 / \lambda^2| \right)^m} \right) & \text{if } |\nabla u_\sigma| > 0 \end{cases} \quad (4)$$

$$\text{Gaussian core: } K_\sigma(x) = \frac{1}{2\pi\sigma^2} \cdot \exp\left(-\frac{|x|^2}{2\sigma^2} \right) \quad (5)$$

The value of C_m can be obtained from the equation: $\exp(-C_m)(1 + 2C_m m) = 1$ (m=2,3, and 4), so C_m =2.33666, 2.9183, and 3.31488.

The solution of polynomial (2) is put into next iterative process:

$$u^{n+1} = u^n + \Delta t \cdot (div(D^* \cdot u^n)), \quad (6)$$

$$D^* = \begin{bmatrix} v_1(u^n) & v_2(u^n) \end{bmatrix} \begin{bmatrix} \lambda_1(u^n) & 0 \\ 0 & \lambda_2(u^n) \end{bmatrix} \begin{bmatrix} v_1^T(u^n) \\ v_2^T(u^n) \end{bmatrix} \quad (7)$$

Where u^n is the result of n steps, Δt is step length, D^* is refreshed according to u^n.

3. TOP-HAT TRANSFORMATION

Top-Hat transformation has the property of enhancing "Gray-scale peaks" or "Gray-scale valleys" of the image signal by applying respectively the opening or closing operator (Yi Cui, 2002). Top-Hat transform can be divided into top-hat arithmetic operators and bottom-hat arithmetic operators obtained from openings or closings. Top-hat arithmetic operators are defined by: top-hat(f)=f-(f ∘ B); Bottom-hat operators are defined by: bottom-hat(f)=(f • B)-f. An application of Top-Hat transform is used to prevent from the effects of environmental conditions and irrelevant structural information and extract isolated targets and noises which have the similar shape as structural element. So the Top-Hat transformation is also a kind of high-pass filter.

4. WATERSHED ALGORITHM OF MORPHOLOGY

The watershed transformation is a powerful Mathematical Morphology tool for segmentation (Rafael C. Gonzalez et al., 2005). The basic principle is to think of an image as a topographic model, and suppose that water is oozing and rising at equal speed from every regional minimum, starting from the lowest one and then from each of the others as soon as the global water level reaches its altitude. Dams are built in the places where water from different minima would merge, separating the watersheds. Watershed divide lines in a gray-level image viewed as a topographic model. The dams rising above the water surface constitute the watershed divide lines, which are composed of closed contours that involve each of the regional minima and correspond to the crest lines of the relief, achieving a good segmentation by the single line.

Since watershed algorithm is sensitive to noises, images will turn out to be successive erosions after the watershed transformation and need further combination of relevant areas. The method to control successive erosions is to bring in the concept of markets. A market is the connected component of an image. Finding the market of every target is the key to solve the problem of successive erosions based on mathematical morphology.

Erosion operation is used in this image. At each step, connected components of pixels can be reduced, separated or even disappear. The residues derived from each component constitute the last erosion of the image, which is often used as marker sets for further processing.

The relevant definitions in watershed algorithm are as follows:

Iterate the image f for k times, $f_k = f\Theta kB$, B is the unit circle, kB is the circle and radius of which is k. Ultimate erosion subset Y_k is the element of f_k. If $l \succ k$, f_k will disappear in f_l. $U_k = (f_{k+1} \oplus \{B\}); f_k$, so $Y_k = f_k - U_k$. If there are many targets in the image, the ultimate erosion image is: $Y = \underset{k=l,m}{U} Y_k$. m is the time of erosion.

5. WATERSHED ALGORITHM BASED ON PRIORI INFORMATION

Direct application of the watershed segmentation algorithm generally leads to over-segmentation due to noise and other local irregularities of the gradient. Over-segmentation means a large number of segmented regions.

There are two methods to solve this problem. The first method is to smooth the gradient image to reduce to effect of noises, which directly reduces the number of segmented regions. We can also improve the morphological smoothing. We can limit the number of allowable regions by incorporating a preprocessing stage designed to bring additional knowledge into the segmentation procedure. People often aid segmentation and high-level tasks in every-day vision by using a priori knowledge. Another practical solution to this problem is to apply the watershed transformation to the image first and then incorporating relevant regions according to certain principles.

In this paper, morphology segmentation algorithm is used to diminish the influence of over-segmentation based on the priori knowledge of the size of the pests.

According to the target, set proper area thresholding a (through the methods of experiment), filter the regional objects whose.

general pixel of the regional area are less than a, eliminate the small targets of background and noises (J. Weickert, 1997).

According to the target, bring in the circular structural element (□) which is similar to the target and modify the gradient image based on the mathematical morphology (Edward R. Dougherty, 2003). The purposes of the modification are to eliminate the irregular details and noises that might lead to over-segmentation and maintain the exact orientation of the regional contour. The sense region of this structural element is the rim and adhesion area of the target.

Some targets have serious problem of adhesion. In terms of the influence of noises and adhesion, conditional extension can prevent over-segmentation effectively. Under the circumstance of K, (K is referred to as a limited set), conditional extension is used based on the mathematical morphology. Structural element B is be used to dilate image f : $f \oplus B; K$. So $f \oplus B; K = f \oplus B \bigcap K$. The experiment turns out adhesion area will tail off and be separated into many tiny areas under the use of appropriate structural element and the right radius of the structural element.

6. APPLICATIONS

Fig. 1 shows the pests image of aphid scanned by CCD. This paper use coherence-enhancing theory as polynomials (3), (4), (5), (6) states where: $\sigma = 5$, $m = 2$, C_m=2.33666, $\rho = 5, \Delta t = 0.1$, $t = 20$. In fig. 3 the gray-scale image fusion can be done as follows: step 1: adding the original image to the image after top-hat transformation; step 2: subtracting the image after bottom-hat transformation from the result; step 3: computing the complement of the residue $k = 125$. Fig. 4 shows the global thresholding segmentation, bwareaopen operation and regional filling operation. The original pests' image is transformed into binary image. we choose regional block, filter the noises whose general pixels of the regional area are less than 200 in the image, eliminate tiny objects in the background by using opening operation and execute regional filling operation based on morphology. Apply the distance transform to fig. 5, then apply the opening operation with '2D''Octagon distance' circular structural element and choose the cross structural element at the same time. Adhesion area will decrease gradually after there times of dilation operation with the radius of 6. Fig. 5 eliminates the irregular details and noises that might lead to over-segmentation and maintains the exact orientation of the regional contour. Fig. 6 shows morphology watershed segmentation. The outcome of the segmentation shows that the purpose of the segmentation is accomplished without over-segmentation according to watershed algorithm based on Priori Information. Fig. 7 shows the result of the edge detection. Fig. 8 is the calculation about the centroids of the individual image block and location of the centroids. It also shows the number of the centroids and array of the centroids (from top to bottom, from left to right). There are 12 aphids in this example and all of them are successively segregated.

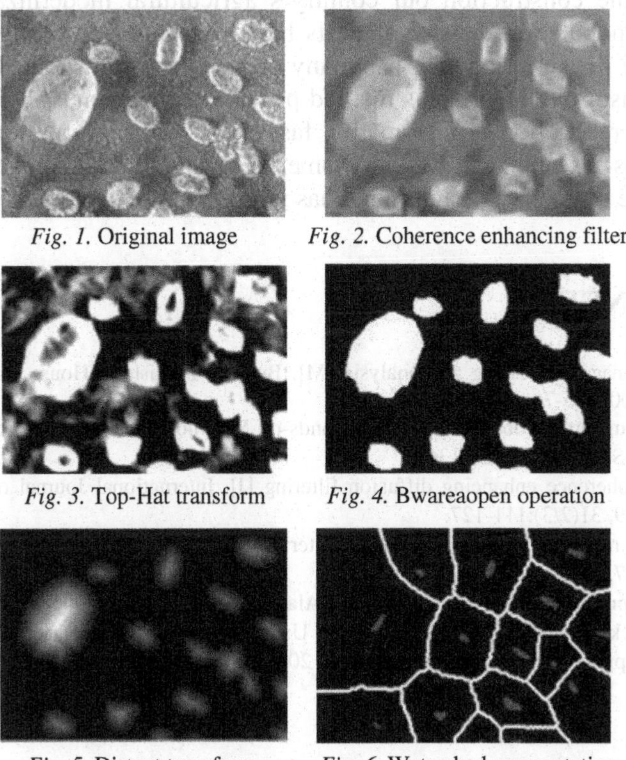

Fig. 1. Original image

Fig. 2. Coherence enhancing filter

Fig. 3. Top-Hat transform

Fig. 4. Bwareaopen operation

Fig. 5. Distant transform

Fig. 6. Watershed segmentation

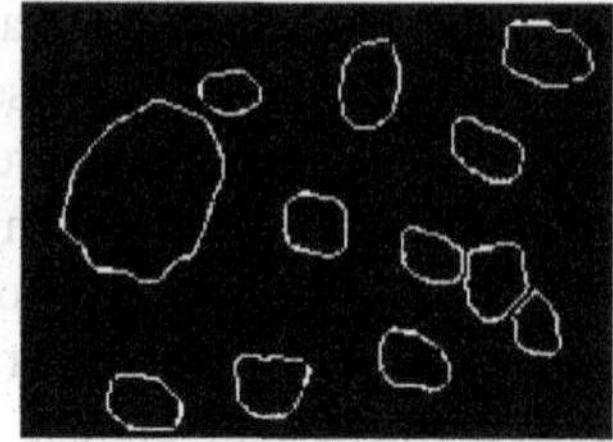

Fig. 7. Edge detection

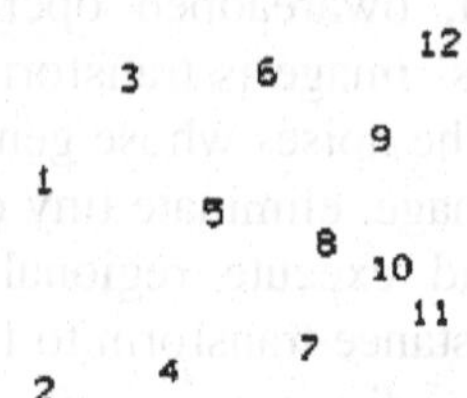

Fig. 8. The result of the statistics

7. CONCLUSION

After the processing given above, we can identify the exact number of the pests; restrain the effect of the similar background and noises and realize the fast collection of information from the original pests' image of aphid scanned by CCD. The efficiency of the information acquisition and the administration level will be improved a lot when edge-enhancing filtering, Top-Hat Transformation, mathematical morphology, watershed segmentation algorithm, edge detection and calculation about the centroids are used in monitoring density of the construction our country's agricultural modernization. The monitoring population density of pests based on edge-enhancing diffusion filtering and image processing has many properties such as expeditiousness and it can also save labor. This method provides convenient and convincing data for the realization of information fast collection, transmission and prevention. This research will provide some theory for the automatic administration of the crop of our country and has important practical meaning.

REFERENCES

Cui Yi. The image processing and analysis [M]. Beijing: Publishing House of Electronics Industry, 2002, 75-76, 146-149.

Edward R. Dougherty, Roberto A. Lotufo. Hands-on Morphological Image Processing [M]. SPIE PRESS July 2003, 129-136.

J. Weickert Coherence enhancing diffusion filtering [J]. International Journal of Computer Vision, 1999, 31(2/3):111-127.

J. Weickert. A review of nonlinear diffusion filtering. [J] Scale Space Theories in Computer Vision, 1997, 1252:3-28.

Jason W. Chapman, Reynolds Don R, Smith Alan D, et al. High-altitude migration of the diamondback moth Plutella xylostella to the U.K., a study using radar, aerial netting, and ground trapping [J]. Ecological Entomology, 2002, 27(26):641-650.

Lu Zhaozhi, Shen Zuorui, Cheng Dengfa, et al. Application of information technologies in monitoring the population density of pests [J]. Transactions of the CSAE, 2005, 21(12): 112-115.

Minghua Zhang, Zhihao Qin, Xue Liu, et al. Detect ion of stress in tomatoes induced by late blight disease in California, USA, using hyperspectral remote sensing [J]. International Journal of Applied Earth Observation and Geo information, 2003, (4):295-310.

Perona, Malik J. Scale space and edge detection using anisotropic diffusion. [J] IEEE Trans Pattern Analysis Machine Intellgence. 1990, 12(7):629-639.

Rafael C. Gonzalez et al. Digital Image Processing Using MATLAB [M]. Beijing: Publishing House of Electronics Industry, 2005.9 P268-P270.

Yu Xinwen, Shen Zuorui. Segmentation Technology for Digital Image of Insects [J]. Transactions of the CSAE, 2001, 17(3):137-141.

Zhang Xiaochao Fang Xianfa Zhao Huaping. Information Acquisition Techniques of Precision Agriculture[J]. Transactions of The Chinese Society of Agricultural Machinery, 2002, 33(6):125-128.

SECURE PRODUCTION OF FARM PRODUCE-ORIENTED MANAGEMENT AND SPATIAL DECISION-MAKING SYSTEM FOR PRODUCING AREA

Yan Wang [1,2], Yuchun Pan [2], Bojie Yan [1,2], Anyun Li [2,3], Jihua Wang [2,*]

[1] *School of Geography, Beijing Normal University, Beijing, 100875, P. R. China*

[2] *National Engineering Research Center for Information Technology in Agriculture, Beijing, 100097, P. R. China*

[3] *School of Information and Electrical Engineering, China Agricultural University, Beijing, 100083, P. R. China*

* *Corresponding author, Address: National Engineering Research Center for Information Technology in Agriculture, BanJing Road, HaiDian District, Beijing, 100097, P. R. China, Tel: +86-10-51503488, Fax: +86-10-51503570, Email: wangjh@nercita.org.cn*

Abstract: To address security of farm produce's quality, using object-oriented technologies to build multi-purpose database, it meets the current and future data needs of a variety of operational systems. Further, base on WebGIS, to establish management and decision-making system for producing area, This system could be implement digital management for the entire agricultural park, and evaluate individual plots momentarily, can also provide accurate results in the decision-making fertilizer.

Keywords: Secure production, spatial decision-making, WebGIS, multi-purpose database, environmental estimation for habitat

1. INTRODUCTION

Appropriate environment of habitats is infrastructure that be ensure the quality and safety of agricultural products, therefore, the level of safety of agricultural products can be fixed according to production environment. At present, with the rapid development of agriculture, agricultural chemicals,

Wang, Y., Pan, Y., Yan, B., Li, A. and Wang, J., 2008, in IFIP International Federation for Information Processing, Volume 259; Computer and Computing Technologies in Agriculture, Vol. 2; Daoliang Li; (Boston: Springer), pp. 909–916.

such as fertilizers, pesticides, feed additives and others play a very important role, and cause agricultural pollution, soil productivity reduction and agro-ecological damage to the environment. Especially, the quality and safety of agricultural products, become the bottlenecks restraining continued efficient development in the new stage of agricultural production (Dai et al., 2002). Using spatial information technology to agricultural production for efficient management, and providing services for scientific decision-making about safe production of agricultural products, is the base to ensure efficient and safe agricultural production. Many scholars have begun to construct some intelligent management and decision-making systems using GIS technology. Moreover, they have applied to the actual production in a small area, and achieved certain results. But these databases only store a single-phase production data, thus making the decision-making is only reflecting status in quo, but the trend analysis of the habitats environment in a specific period, usefulness of the system has been greatly reduced; Furthermore, we only consider the evaluation of large field, divide the entire evaluation of the regional to grid directly, and evaluate by district, but not evaluate in accordance with the actual planting plot, this is detrimental to the overall agricultural park guide for effective operation of the plant. Current problems: database is static and can not reflect the real-time production and environmental conditions; Databases are designed using in only one system, and uniform standards and norms are not viable. Most systems are poor in data sharing and reuse.

This paper describes the design of a unified, scalable, multi-purpose database model. Based on it, we provide a management and decision support system based on WebGIS technology, which is good at updating database online, and using the real-time data for environmental assessment and appropriate decision-making. The entire decision-making is a dynamic process, not just for the historical inquiries of the decision-making result.

2. MATERIALS AND METHODS

2.1 System Framework

2.1.1 Technology truss

The design of system technology truss scheme is playing the significant role in the system design and application. In order to achieve the system target, this system adopts the B/S three development structural model. System collectivity technology truss as shown in Fig. 1. It contains client browser, Web server, GIS application server and database.

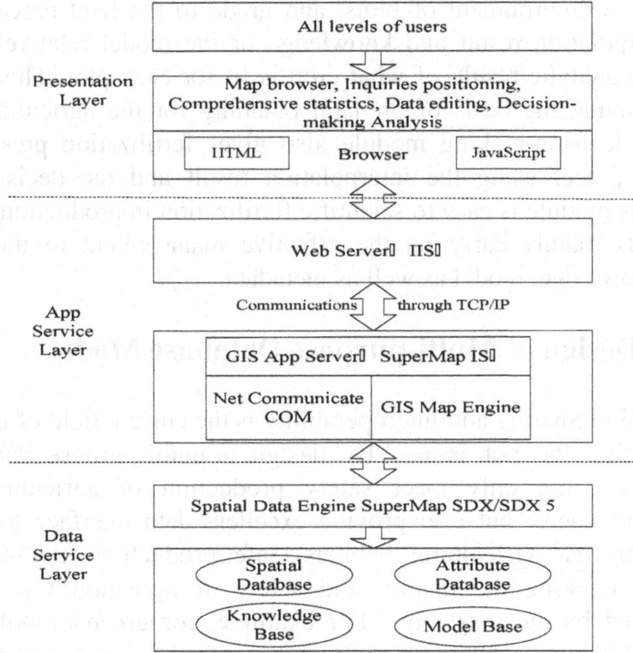

Fig. 1. System technology truss

The client browser supports a variety of data and information display and can communicate with the server. The Web server selects IIS (Internet Information Server), and is responsible for the basic network communication and the coordination. GIS server selects SuperMap IS, and mainly supports the realization of network geographic information system function. The Web server and GIS application server communicate through the Tcp/IP protocol (Ma et al., 2007). Use spatial data engine SuperMap SDX/SDX 5 to manage and operate special database, and to save the spatial data and the attribute data integrated to the database.

2.1.2 Function structure

The function structural design of this system divides into six big function modules. Information management module for plots and sampling points is foundation part of this system. The module mainly realizes map browser, spatial data and attribute data inquiry, statistical analysis as well as online edition. Online browser on spatial distribution of the attribute data is interpolation for the limited sampling point data in the land parcel scope, and obtaining the spatial distribution of soil nutrient or metal content, and the result can display in the form of thematic maps, also can display the value of random point on the map real time. Habitats environment appraisal module estimates the environment of plots, and grade to the land parcel by using space interpolation result and knowledge or the model relatively. Finally, display the analytic result, of what crop is fit for each plot. This module is easy to provide the basis of prenatal planning for the agricultural garden area. The decision-making module also gives fertilization prescription of each land parcel using the interpolation result and the decision-making model. This module is easy to scientific fertilization in production. The other two models mainly carry on the effective management to the decision-making knowledge, model as well as metadata.

2.2 Design of Multi-purpose Database Model

Information Sharing and Interoperability is the current field of information need to solve the hot issues. By design a multi-purpose database, the database will not only meet safety production of agricultural digital management needs, but also provide excellent data interface to the green or pollution-free agricultural habitats, safe production file management (IC card), tracking the quality and safety of agricultural products and other related business systems. For example, for green or pollution-free agricultural habitats declare operational systems, the system can provide the necessary testing of soil, water or nutrients metal elements, and can show grading data in plots of various products in different periods directly. Wherefore, the construction of multi-purpose database effectively improve the utilization of resources is also to avoid the duplication of database construction (Luo, 2006).

For building a multi-purpose database: firstly, it is necessary to standardize the data, in particular from the different departments, a need to formulate a unified standards for the corresponding standardized in order to achieve multi-source data integration and information sharing. The main problems of data standardization in the database design are as follow: the time consistency between space data and attribute data, the consistency of metric units, the consistency of data classification standard, the consistency of each data file for each number and each data field name.

Secondly, the establishment of the database metadata is also an important means of a multi-purpose database. This article metadata, including data collector, access time, content, themes, and the conversion of various operational information description and so on. For example, the users of space data can quickly understand the data names, scale, coordinate system and other describe information by metadata, so that different users, especially in Web environment could use the available data from heterogeneous platform expediently.

2.3 Decision-making Principle and Implementation

2.3.1 Process of extracting decision-making information

Information extracting is a very important part of the decision-making, which need certain data storage in database for the screening, operating, etc. Then, we can use knowledge directly or input into decision-making model (Li et al., 2006). Following is the data flow diagram of the decision-making system (Fig. 2). The figure can be divided into two parts: inquiries sampling point and access plot data distribution.

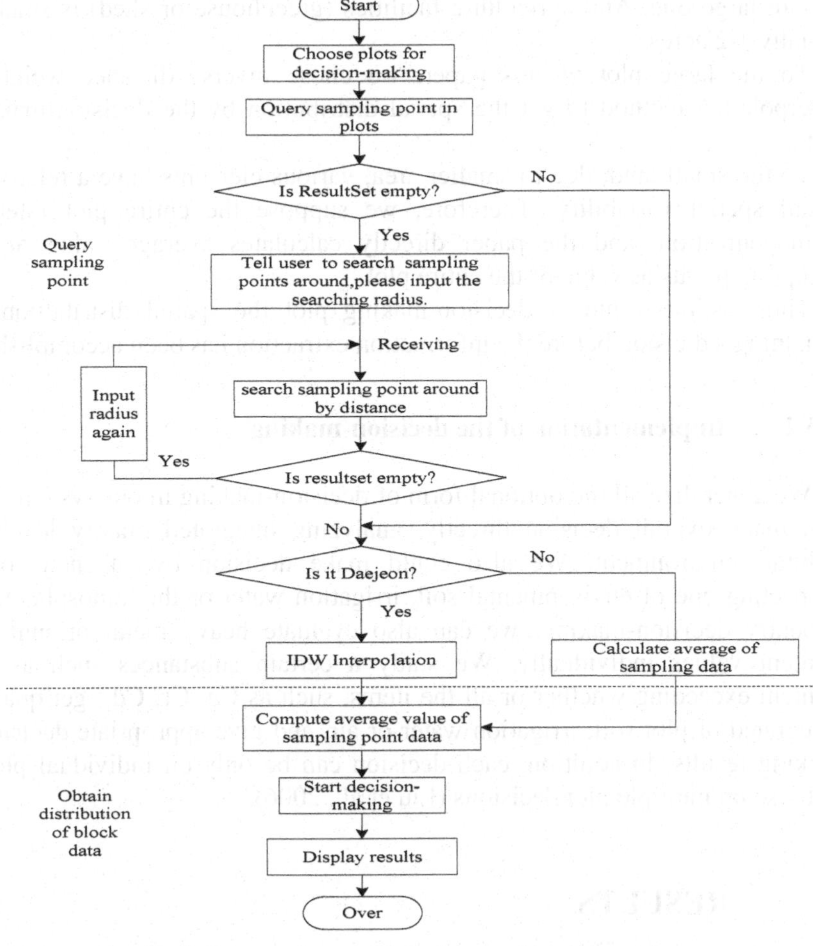

Fig. 2. Process of extracting decision-making information

(1) Inquiries data of sampling points

The decision-making system is targeted at farmland plot, and the decision-making data is sampling data. We need calculate the value of the entire plot through sampling data, and then could get the results of decisions based on the decision-making model. For some objective reasons, it is impossible for every plot has a sampling distribution, and consequently, there are some plots sampling, and some plots were not. For the former, we can use the plots directly within the sampling plot for decision-making, and for the latter, we need to search around the plot from the sampling points for decision-making. The paper describes the Map Control Methods of SuperMap, that finding plot away from the center within a certain range of other sampling sites.

(2) Inquiries data of plot distribution

In the forward step we got the sampling data, and on the following, we need to compute plot data through point data. In the agricultural park, the plot of farmland can be divided into two categories by size. Daejeon and so on are large one. And agriculture facilities (greenhouse or shed) is smaller, usually 1-2 acres.

To the large plot of this paper, we chose inverse distance weighted interpolation method to get the spatial distribution by the decision-making data.

To the small land, due to smaller area, various elements have a relatively small spatial variability. Therefore, we suppose the entire plot internal homogenization, and the paper directly calculates average value of all sampling points as value of the entire plot.

Thus, as the value of decision-making plot the spatial distribution of obtaining a dicision before the information extraction has been accomplished.

2.3.2 Implementation of the decision-making

We generalize all the optional form of decision-making in our system. We can make overall decision directly, analyzing integrated quality level of habitats environment. We also could make decision by subentry, only inspecting one of environmental soil, irrigation water or the atmosphere. In subentry decision-making, we can also evaluate heavy metal or nutrient content values individually. We study a certain substances such as Pb content exceeding whether or all the items, such as Cu, Cr, Cd, get quality integrated of plot soil, irrigation water or air, and give appropriate decision-making results. In addition, each decision can be only on individual plots, but also on multiple plot decisions (Liu et al., 2006).

3. RESULTS

We develop the platform with the aid of Visual Studio.NET 2003, and use the C# language to complete the system development, select the SuperMap

IS.net specifically and the user-defined engine and C# development pattern, and use SQL Server 2000 as database.

This system realizes the main function that has illuminated before. The core function is the environmental estimation for habitat and the suitability decision-making. Fig. 3 is result chart of many land parcel decision analysis.

Fig. 3. Plots environmental assessment

The result of many land parcel decision-making demonstrates in form of the thematic maps. In the chart the green and the blue are two land parcels which be estimated. Rest on legend of the lower right corner. The green express that soil quality of the plot is first level. The first level reaches standard that protect region natural ecology and maintain natural background. It is optimal soil quality grading. Blue is third level. The third level are the minimum standard to guarantee the production of agriculture and forestry, and the normal growth of plant. Moreover, this module may carry on the single item appraisal for certain elements to inspect some elements specifically whether exceeding the allowed range. Thus, the policy-makers of agricultural garden area may carry on planning before the production based on the appraisal result for the land parcel, and consider which land parcels should be plant which kind of crop. Moreover, they may consider adopting what measure to resume the land parcel already had be polluted, and may use fertilizer decision-making function to obtain precise fertilization prescription of each land in concrete production.

4. DISCUSSION

This paper proposes to construct management and decision-making system for farm produce habitat by using the WebGIS technology. First, it adopts the object-oriented technology to establish unified, extensible and the multi-purpose database. This database can not only satisfy the need of Secure Production of Farm Produce-Oriented Management, but also provide the very good data interface for declaration of the green or pollution-free production base, the production secure-oriented records management (the IC card), safe tracking quality of farm produce operational systems correlatively and so on. In the foundation, this database can manage the policy-making knowledge and model through constructing the knowledge library and the model library. Second, using SuperMap IS.NET as GIS development platform, it has realized basic functions, such as map browsing and inquiry, moreover realized real-time remote online data update. Thus it has guaranteed that the habitat decision-making is dynamic and real-time. Finally, the most core of the system is carrying on the environment appraisal and the suitability decision-making to the agricultural product habitats. So the system can provide scientific guidance for pre-natal planning of the agricultural garden area. Function of this system is not very perfect. For example, Historical data can be used for years of trend analysis. These will gradually complete in future. We will seek to establish a more perfect management and decision-making system to guarantee safe production of the farm produce.

ACKNOWLEDGEMENTS

This work was subsidized by Beijing Natural Science Fund (4061002), the program from Ministry of Agriculture (2006-G63) and the National High Tech R&D Program of China (2006AA10Z201, 2007AA10Z203). The authors thank all teachers and students for their helpful cooperation in obtaining data documents and guiding of programming.

REFERENCES

Dai Xiaofeng, Zhao Bingqiang, Present State on Development of Safe Production Technology of China's Agricultural Products and Its Priority Field, Science and Technology Review, No. 3, 2002, pp. 46.

Ma Jin-feng, Pan Yu-chun, Shen Tao, Spatial decision support system for controlling the outbreak and spread of animal epidemics, Computer Applications, Vol. 27, No. 5, 2007, pp. 1289-1292.

Luo Ming-yun, Designing and Applying of a Soil Database System of Nanchong District Based on GIS, Chinese Journal of Soil Science, Vol. 37, No. 1, 2006, pp. 61-63.

Li Weijiang, Wu Yongxing, Mao Guofang, WebGIS-based information system for evaluation of soil environment in prime farmland [J], Transactions of the CSAE, Vol. 22, No. 8, 2006, pp. 59-63.

Liu Xiaojun, Zhu Yan, Yao Xia, et al. WebGIS-based system for agricultural spatial information management and aided decision-making [J],Transactions of the CSAE, Vol. 22, No. 5, 2006, pp. 125-129.

BAYESIAN NETWORK AND ITS APPLICATION IN MAIZE DISEASES DIAGNOSIS

Guifen Chen[1,2], Helong Yu[1,2,*]
[1] *Computer Science and Technology Institute, Jilin University, ChangChun 130021, China*
[2] *Information Technology Institute, Jilin Agricultural University, ChangChun 130118, China*
[*] *Corresponding author, Address: Information Technology Institute, Jilin Agricultural University, 2888 XinCheng Street, ChangChun, 130118, P. R. China, Tel: +86-431-84532775, Fax: +86-431-84542775, Email: yuhelong@yahoo.com.cn*

Abstract: Bayesian network is a powerful tool to represent and deal with uncertain knowledge. This paper mainly introduces some technologies and methods of modeling Bayesian network, which are used in the building Maize Diseases Diagnosis system. In the construction of Bayesian network, noisy-or model and transformation from certainty factor to probability are used. Then maize disease diagnosis system based on BN is built by Netica (a BN software package). The practice proves that BN is an effective tool for maize disease diagnosis.

Keywords: Bayesian Network; maize; disease; diagnosis

1. INTRODUCTION

There exists a lot of uncertainty phenomenon and problem. Uncertainty in agriculture is more extensive and complex. So, in order to create an effective intelligent system, uncertain knowledge must be dealt with.

From representation of uncertainty knowledge, there are two methods of dealing with uncertainty. One is rule-based method, and the other is model-based method. The advantage of rule-based method is that its computation is convenient, and its disadvantage is that its syntax is not systemic. The advantage and disadvantage of model-based method are contrary to the rule-based method.

Chen, G. and Yu, H., 2008, in IFIP International Federation for Information Processing, Volume 259; Computer and Computing Technologies in Agriculture, Vol. 2; Daoliang Li; (Boston: Springer), pp. 917–924.

From measurement of uncertainty, the methods of dealing with uncertainty are fuzzy theory and probability theory. Fuzzy theory mainly deals with vagueness, and probability theory mainly deals with randomness.

Therefore, Bayesian network in the article is a model-based probability method.

For the uncertainty in agriculture, there are some good model and application, but mainly rule-based. This method is adapted to knowledge represented by rule. However, uncertainty in agriculture is various. It is not adequate for rule to represent this uncertainty.

In order to solve it, Bayesian network is introduced. It widens knowledge representation that increases the reliability of expert system.

Bayesian network is a combination of probability theory and graph theory. Study of Bayesian network originates from the 1980's. Since 1990's, its study and application have stirred great concern. Compared with rule based method, the syntax of Bayesian network is clearer, which can reason in dual direction and can be constructed and debugged rapidly. The disadvantage of Bayesian network is that the computation complexity is high.

This paper mainly introduces the application of Bayesian network in maize disease diagnosis system.

As far as the computing complexity is concerned, in the Bayesian network construction, noisy or technology are adapted to simplify network structure and condition probability table.

On running the system, we find that the result is conformed to domain expert. It proves that it is effective to use Bayesian network to represent and deal with uncertain knowledge in agriculture.

2. BAYESIAN NETWORK

2.1 Bayesian Network Syntax

BN=(Structure, CPT)

(1) Structure contain nodes and arcs

Nodes: random variable.

- Nodes can be continuous or discrete.
- Nodes can have two states or more.
- Nodes can be deterministic or nondeterministic.

Arcs: relationships between nodes.

- Arcs represent causal relationships of nodes.
- Arc between x and y represents that x has direct causal influence only.

(2) CPT: Condition Probability Table

- Each node has condition probability which is stored in a table (CPT).

- Value in table is $P(X_i|parents(X_i))$, $parents(X_i)$ is the set of parent nodes of X_i.
- Root node is particular, as it has no parent node and has only prior probability: $parents(X_i) = \Phi$, so $P(X_i|parents(X_i)) = P(X_i)$.

2.2 BN Semantics

- Local semantic: represent conditional independence in the net
- Global semantic: represent global probability distribution

$$P(X_1...X_n) = \prod_{i=1}^{n} P(X_i \mid Parents(X_i))$$

We can conclude that Bayesian Network is combination of network structure and CPT, or global probability distribution is combination of conditional independence and local probability.

3. BN BUILDING

Before being deduced, Bayesian network must be constructed. As we know, Bayesian network has two parts: structure and CPT, so the process of constructing Bayesian network is to construct structure and CPT (David J. Spiegelhalter, 1993).

There are three methods to construct Bayesian network: manual construction, machine learning and combination of them. This article mainly introduces manual method, which constructs Bayesian network by domain expert elicitation (E. Charles, J. Kahn, etc, 1997).

3.1 Elication of BN Structure

In this process, variables and relationships between them should be determined.

First, select variable set. It is important to limit the number of variables. So, it is necessary to choose important variables which are

- Query variables: or object variables, they are outputs of net and what we want to know.
- Evidence variables: or observation variables, they are inputs of net and used to reason states of query variables.
- Context variables: or middle variables, they are used to connect query variables and evidence variables.
- Controllable variables: or adjustable variables, they are used to control and adjust net.

If objects are mutex, they can be states of a variable, else being various variables. Arc $cause_i$->effect represent $cause_i$ is one of cause for effect. There are two cases:

- Multi-causes, one effect.
- One cause, multi-effects.

According to the formula $P(X_1...X_n) = P(X_n|X_{n-1},...,X_i)P(X_{n-1} \mid X_{n-2},...,X_1)...P(X_2 \mid X_1)P(X_1) = \prod_{i=1}^{n}(X_i \mid Parents(X_i))$, we can find that the right sequence of adding nodes is:

- Add root nodes.
- Add nodes that be influenced directly by root nodes.
- Repeat the above two steps until leaf nodes are added.

3.2 Elicitation of Condition Probability Table

There are three kinds of probability, namely objective probability, frequent probability and subjective probability, which originate from data, domain experts and literature.

In this process, the state of each variable and qualitative probability should be determined. This can be obtained by domain expert and literature.

In order to decrease the size of network, state number should be limited. In the meanwhile, states should be mutex.

Generally, probability given by domain expert is qualitative, so the transformation from qualitative probability to quantitative probability is necessary (Table 1).

Table 1. The transformation from qualitative probability to quantitative probability

Qualitative Probability	Quantitative Probability
Always	0.99
Generally	0.85
Ofen	0.78
Usually	0.73
Not ofen	0.50
Sometime	0.20
Occasionally	0.15
Usually not	0.10
Seldom	0.30

3.3 Two Methods used in Building Maize Disease Diagnosis System

3.3.1 'Noisy-or' Technology

This model has three assumptions:

- Parents and child are Boolean variables.
- Inhibition of one parent is independent of the inhibitions of any other parents.
- All possible causes are listed. In practice this constraint is not an issue because a leak node can be added (a leak node is an additional parent of a Noisy-or node).

Now, we can have a definition of noisy-or:

- A child node is false only if its true parents are inhibited.
- The probability of such inhibition is the product of the inhibition probabilities for each parent.
- So the probability that the child node is true is 1 minus the product of the inhibition probabilities for the true parents.

For Fig. 1, we can get this formula:

$$P(F \mid H_1, H_2, \ldots H_n) = 1 - \prod_{i=1}^{n} (1 - p_i)$$

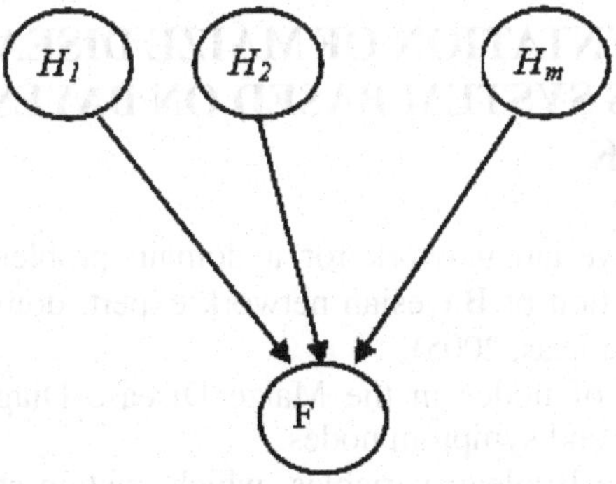

Fig. 1. Noisy-or model

Generally, for node having k parent nodes, if use 'Noisy-or', it needs O (k) parameters, if not, it needs $O(2^k)$ parameters. Obviously, BN is simplified.

3.3.2 Transformation from Certainty Factor to Probability of BN

In the maize disease diagnosis, the knowledge given by domain expert is rule-based, and measurement for the belief is certainty factor, that is:

IF *A* THEN *B* *CF(B|A)*

Definition of certainty factor (CF):

$$CF(B\mid A)=\begin{cases}\dfrac{P(B\mid A)-P(B)}{1-P(B)} & \text{if } P(B\mid A)>P(B)\\ 0 & \text{if } P(B\mid A)=P(B)\\ \dfrac{P(B\mid A)-P(B)}{P(B)} & \text{if } P(B\mid A)<p(B)\end{cases}$$

However, in the Bayesian network, uncertainty is measured by probability. So, in order to construct Bayesian network, it needs to transform CF to probability (F. trai. 1996; Kevin B. Korb, Ann E. Nicholson, 2006; Nevin Lianwen Zhang, 1996). From above formula, we can obtain:

$$P(B\mid A)=\begin{cases}CF(B\mid A)(1-P(B))+P(B) & \text{if } CF(B\mid A)\geq 0\\ (CF(B\mid A)+1)P(B) & \text{if } CF(B\mid A)<0\end{cases}.$$

So, in order to get probability, it needs to know *P(B)*, which is prior probability of node *B*.

P(B) can be obtained from domain expert, literature, or assume *P(B)*=0.5.

4. IMPLEMENTATION OF MAIZE DISEASE DIGNOSIS SYSTEM BASED ON BAYESIAN NETWORK

Construcion of a Bayesian network for a domain problem needs communication and cooperation of Bayesian network expert, domain expert and BN software tool (P.J.F. Lucas, 2005).

There are two types of nodes in the Maize Disease Diagnosis System, which are disease nodes and symptom nodes.

The disease nodes are Boolean variables, which contain states: 'happen' and 'unhappen' [P.J.F. Lucas, 2001; Radim Jirousck, 1997], while Symptom nodes may contain multiple states.

This BN is a two-layer network, in which the upper layer is composed of disease nodes and the lower layer is composed of symptom nodes. Obviously, the arc direction is from disease nodes to symptom nodes.

Fig. 2 is a part of BN structure for maize disease diagnosis. In this structure there are four disease nodes and four symptom nodes, which corresponds to four diseases and four symptoms. The four diseases are maize dwarf mosaic, maize sheath blight, maize northern blight and bipolarismaydis.

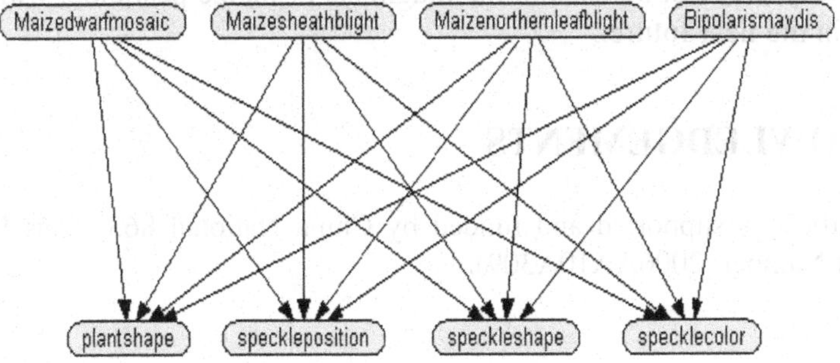

Fig. 2. A part of BN structure

With the 'Noisy-or' technology and probability transforming from CF to probability, node's CPT is achieved. We can find the inference results from Fig. 3, which are the posterior probability of disease when plant shape is normal, speckle position is lamina and speckle shape is others.

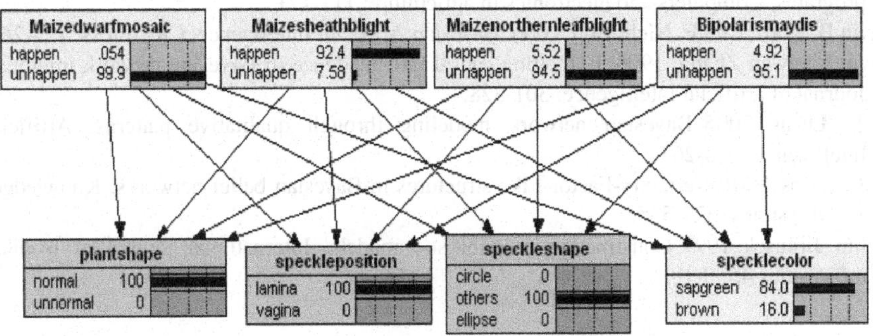

Fig. 3. Inference results of the BN

5. CONCLUSIONS

BN is a strong tool for representing and dealing with uncertain knowledge. There exists a lot of uncertainty knowledge in maize disease diagnosis. So it is natural to use BN to build maize disease diagnosis system.

While building BN, Noisy-or model and transformation from CF to probability are used to decreasing network scale and simplify the network structure.

In running the maize disease diagnosis system, we find that the reasoning result is conformed with the solution given by domain expert, as proves that it is effective to use Bayesian network to represent and deal with uncertain knowledge in disease diagnosis.

Obviously, BN can be used in the diagnosis of maize pesticide, which will be done in the near future.

ACKNOWLEDGEMENTS

This article is supported and funded by China National 863 Plans Projects (Contract Number: 2006AA10A309).

REFERENCES

David J. Spiegelhalter. 1993. Bayesian Analysis in Expert Systems, Statistical Science, Volume 8, Issue 3: 219-247.

E. Charles, J. Kahn, etc. 1997. Construction of a Bayesian network for mammographic diagnosis of breast cancer, Comut. Biol. Med: 19-29.

F. trai. 1996. A bayesian network for predicting yield response of winter wheat to fungicide programs, Computers and electronics in agriculture: 111-121.

Kevin B. Korb, Ann E. Nicholson. 2006. Bayesian Artificial Intelligence, CRC Press: 225-260.

Nevin Lianwen Zhang, 1996. Exploiting causal independence in Bayesian network inference, Journal of artificial intelligence: 301-328.

P.J.F. Lucas, 2005. Bayesian network modeling through qualitative patterns. Artificial Intelligence: 233-263.

P.J.F. Lucas. 2001. Certainty-Factor-Like structures in Bayesian belief networks, Knowledge-based systems: 327-335.

Radim Jirousck. 1997. Constructing probabilistic models, International journal of medical informatics 45: 9-18.

STUDY ON AUTHORIZATION MANAGEMENT MODEL OF ONE-STOP OFFICE SYSTEM

Jianwei Yue [1,2], Hongchun Cai [1,*], Luyao Chen [1], Wei Zhuang [1]
[1] *College of Resources Science and Technology, Beijing Normal University, Beijing, China, 00875*
[2] *Academy of Disaster Reduction and Emergency Management Ministry of Civil Affairs & Ministry of Education, Beijing Normal University, Beijing, China, 100875*
* *Corresponding author, Address: College of Resources Science and Technology, Beijing Normal University, No. 19 XinJie KouWai Street, Beijing, 100875 P. R. China, Tel: +86-10-58807713, Fax: +86-10-58806173, Email: Caihc@lgy.cn*

Abstract: An authorization management model named five-layer-two-section is proposed through studying on problems of authorization management model and role network model. The model inherits the thought of role network model, and at the same time it expands the role network model by putting website into management model, so as to more accurate authority management. And it puts user and post into one system, and role, activity, website into another. The model, combining rough regulation and meticulous adjustment, integrity and flexibility, could greatly decrease the complexity of the authorization management and the probability of making mistake.

Keywords: authorization management, one-stop office system, RBAC

1. INTRODUCTION

Information technology is a key force for social development and it is very necessary for modern society to enhance informatization, a key factor to promote economic development and new countryside construction. Large scale application of information technology is becoming an important technological method to enhance the modernization of countryside and to develop its economy. At present, office systems have been built in governmental management department of all levels. But these systems are mutually independent, which results to some inconvenience. Thus, one of the urgent problems is to integrate these separated systems to set up a one-stop

Yue, J., Cai, H., Chen, L. and Zhuang, W., 2008, in IFIP International Federation for Information Processing, Volume 259; Computer and Computing Technologies in Agriculture, Vol. 2; Daoliang Li; (Boston: Springer), pp. 925–931.

office system which could enable unified access to different systems and make system maintenance more efficiently. Because of the complicated relations among office systems which hold different functions, there are numerous problems when the one-stop system is being built by integrating several systems together. One of these problems is authority management (Ye Xin et al. 2006).

Fig. 1(a) shows common authorization management model, which can be accepted as Role-based Access Control (RBAC) developed by Sandhu (Yan han et al. 2006, Li Huaiming et al. 2006, Han Shengju et al. 2006). This model separates users and specific transaction process, and combines the two with role. In this model, the change of user's roles can meet the requirement of user authority change, which appears efficient and simple. However, this efficiency and complexity are just limited to the number of user's roles. If a user has many roles, the operation of change user's authority must be very complicated. This would especially be true in the following example. There are several systems in a certain department, and the user's authority would be changed a lot when an employee was promoted to a manager. Accordingly, his authority may be confined in one system before, but after promoting his authority would expand to all the systems that the manager can log in. This kind of user authority change in the system is convoluted and complicated.

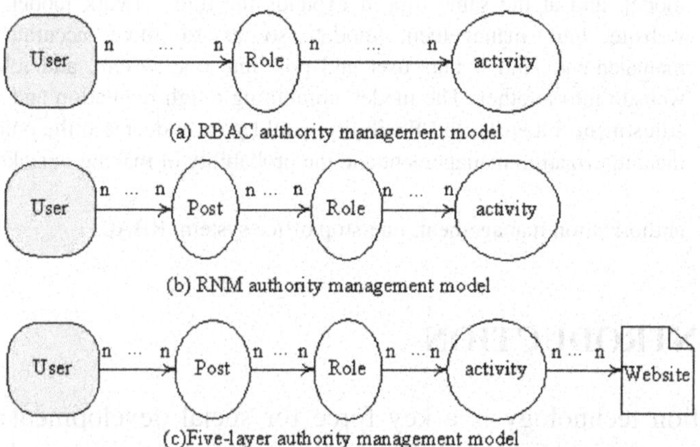

Fig. 1. Three kinds of authority management model

As for the problems existed in the model, Yu and Wang give a role network model (RNM) based on system theory and role theory (Yu Miao et al. 2003) (Fig. 1 (b)). Li etc. develop SODB-RNM based on the RNM mode (Li Huaiming et al. 2006). RNM model and SODB-RNM model extract the role deeply, and separate the role into two parts which named role and post. The authority change in the system would be complex when the user's post changes. Post, a role group, can solve the problem. SODB-RNM model is separated in several layers and applied to the system development and application, but its complexity makes the developing management be uneasy. Besides, RBAC, RNM and SODB-RNM is only limited to level of "activity" when they are applied to authority distribution management.

For solving the problems discussed above, this article put forward an authority management model whose systematic structure has five layers and two sections.

2. FIVE LAYERS AND TWO SECTIONS AUTHORIZATION MANAGEMENT

Inheriting the thought of RNM, the five layer and two sections model separates the role in RBAC model into post and role, in order to solve the authority management problems which could take place when the users' headship change. And it also expands the RNM and puts the website into the model for more accurate authority management. Based on the model, it is studied that the position of model's content in the system development. Personnel, post, role, activity and website (table) are combined to two systems to achieve unified and flexible authority management.

2.1 Five layers authorization management

In the RNM, the information and the access control function are based on the post and role: a user could have several posts and a post could have several roles; a role could handle several actions, an action could be handled by several roles. RNM could solve the big adjustment of authority, but both RNM and RBAC could not solve the problem that tables change in an action. In this article a five layer model is proposed to solve the problems that tables have changed in an action, for example, some tables do not need in an action, or adding a search of some tables, or adding a new table. This model also expands the RNM and introduces the website. As showing in Fig. 1(c), in five layers model, the action authority could be divided more specifically, and an action could handle several table (website), and a table also could be handled by different actions.

Fig. 2 shows kinds of situation when the authority of activity changes (the website of action handling changes). It can be seen that the model could adjust the system conveniently and promptly, so that it responds the service change in time. Without this model, the activities and tables would be coded stubbornly in the system; the system would be recompiled when tables change. The recompiling would not only increase the sustaining cost but also decrease the expansibility of the system.

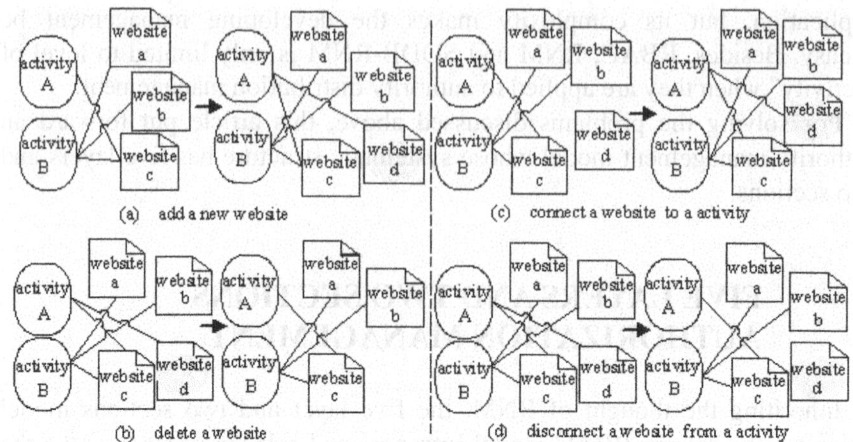

Fig. 2. The change of the activity authority

2.2 Two sections authorization management

Five Layers Authorization Management model can solve the problem of authority setup when user authority changed. However, it is difficult to decide the depth of common authority management, since every subsystem has its own authority managements. How to deal with the distribution of the authority management and how to harmonize the common user management subsystem is becoming a big problem. So a two sections authority management is developed, as Fig. 3 shows.

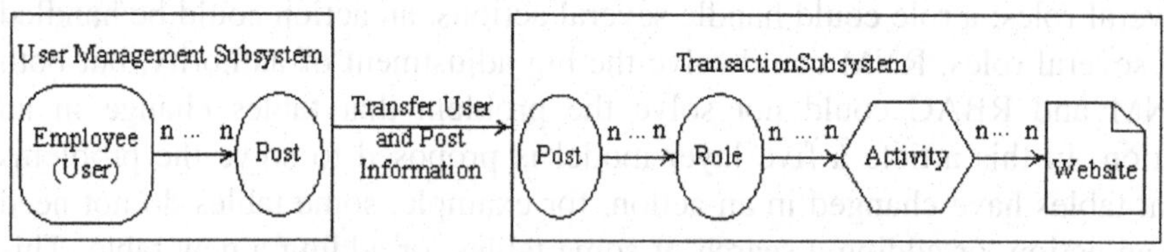

Fig. 3. Five layers and two sections authorization management model

In this "Two Sections Authorization Management", personnel and position management are extracted and are managed in subsystem of user management. The relations of position, role, website and activity are managed in specific operational system. In this way, the five layers model is separated into two sections:

The first section (rough adjustment): indicates administrators, positions and the relations between personnel and positions in subsystem of user management. Fig. 4(a) shows that the method is to read the information of user and post in the user management subsystem, instead of managing it. This method can solve the problem of internal position change; the change would have effect on all user authorities in all transaction subsystems, rather than on specific transactions. That's why it is called rough adjustment;

The second section (specific adjustment): post, roles, activities and the relations between them are managed in the detail operational system, as shown in Fig. 4(b). When the position authority changes or when the operation change results to role managing action change, what has to be changed are the relationsof positions and roles or relations of roles and activities, rather than subsystem of user management and other operational system. So this action cannot cause any influence to other operational systems. All these adjustments are related to authority position, so it can be called "specific adjustment".

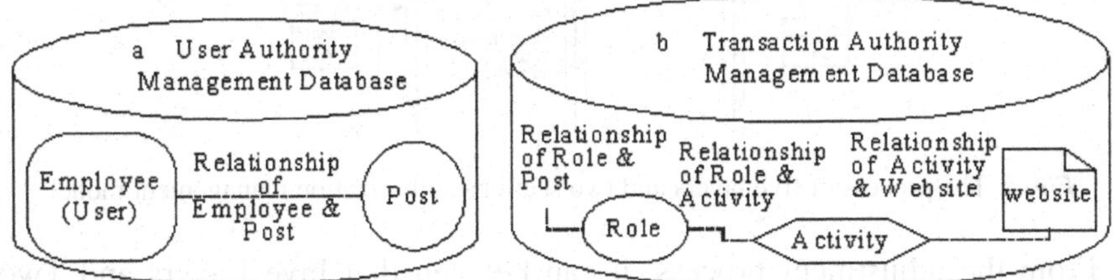

Fig. 4. Storage structure of user management subsystem & other subsystems authority management

2.3 The process of five layers and two sections authorization management

As the Fig. 5 shows, assuming the authority exchanging between user A and B, user B does not hold the role b and B is endued with a new position A which has role c in sub-system A, and he could log in the website 2. Take this situation for example to illustrate the process of Five Layers and Two Sections Authorization Management:

(1) Delete the relation between user B and post B, save the change in the user management database;

(2) Set up the relation between personnel B and post A, save the change in user management database;

(3) Read the information of personnel and post of application system A in user management system, then delete the relation between post A and role b, save the change in the authority management database of the application system;

(4) Set up the relation between role c and action c in the application system A, save the change in the authority management database of the application system;

(5) Set up the relation between the action c and website 2, save the change in the authority management database of the application system.

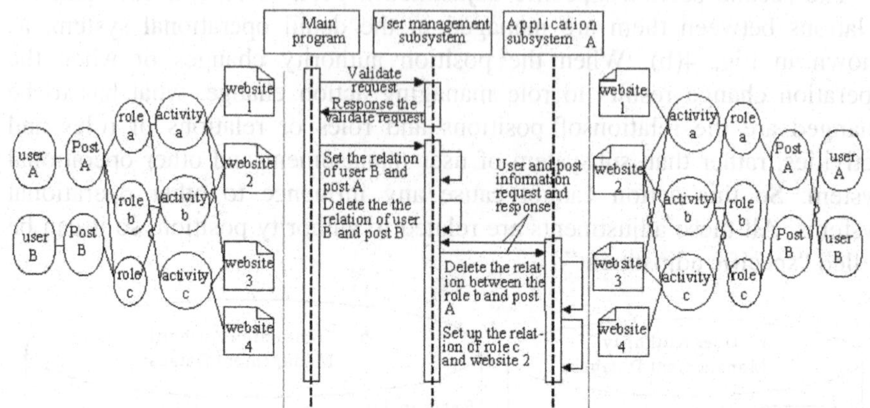

Fig. 5. The process of five layers and two sections authorization management model

From the adjustment process, it can be seen that Five Layers and Two Sections Authorization Management can consider not only the authority adjustment when user's positions have been little changed, but also the specific adjustment when user's authorities have been changed a lot. Therefore, the model is highly flexible so that it meets the requirement of authority management of one-stop office system.

3. APPLICATION

Five Layers and Two Sections Authorization Management model has been successfully applied to many constructions of Electronic Government's System, for example, the Land and Resources Information Management System (LRIMS) of Liuzhou city. The LRIMS contains 6 subsystems, including subsystem of cadastre management, land use and law supervision and so on. The whole system has adopted one-stop operation mode with receiving all kinds of business applications and distributing them to responsible department in the system. Because different operators use different subsystems and has different roles, the role management seems to be complicated. As the periodical internal positions adjustment of the Land Resource Bureau, there is a big change in the authority management mode base on roles. But if adopt the five layer and two section model, the workload will be decrease greatly. With the business development, when some tables or websites are changed, the advantage of flexibility and expansibility of the model could be seen more clearly.

4. CONCLUSION

With the development of informatization, Electronic Government's System must be widely used in government offices. If the application complexity can be reduced, the one-stop office system must be widely used. One-stop office system combines several transaction subsystems together, which leads to the complexity of role management. Five Layers and Two Sections Authorization Management model separates authority management into five layers, and put rough adjustment and specific adjustment respectively in subsystem of user management and subsystem of specific transaction process. In this way, the problem of authority management complexity can be solved and workload can be reduced.

ACKNOWLEDGEMENTS

This study has been funded by China National 863 Plans Projects (Contract Number: 2006AA120102). This study has been applied to some systems programming, such as Liuzhou Land Resource Management Information System, Tangshan Digital Land Information System, and so on. Thank Liuzhou Land Resource Bureau, Tangshan Land Resource Bureau and other organizations for their data and support.

REFERENCES

Han Shengju, Ye Xin, Wang Yanzhang, Li Huaiming. 2006, Research on extended role network models of e-government application integration, Journal of Dalian University of Technology, Vol. 46, No. 2: 286-291.

Li Huaiming, Ye Xin, Wang Yanzhang. 2006, Organization and Empowerment Management of the Complex Government Affair Information System, Systems Engineering, Vol. 24, No. 4: 44-48.

Yan han, Zhang hong, Xu manwu. 2006, Organization and Empowerment Management of the Complex Government Affair Information System, System Engineering, 24, (4), 44-48.

Ye Xin, Li Huaiming, Wang Yanzhang. 2006, Research and Design of One-stop Administrative Permit System, Application Research of Computers, 2006, No. 4: 200-203.

Yu Miao, Wang Yanzhang. 2003, Research on an E-Gov Affairs System Architecture Based on Role Network Model & Its Realization. Computer Engineering and Applications. Vol. 39, No. 12: 31-35.

RESEARCH ON AGRICULTURAL E-COMMERCE PUBLIC TRADE PLATFORM SYSTEM

Qifeng Yang[1,*], Bin Feng[1], Ping Song[1]

[1] *Economics College of Wuhan University of Technology Wuhan, Hubei, 430070 P. R. China*

[*] *Corresponding author, Address: Economics College of Wuhan University of Technology Wuhan, 122, Luoshi Road, Hongshan Disct, Wuhan, 430070, P. R. China, Tel: +86-27-87383196, Fax: +86-27-87651809, Email: yangqifengwhut@163.com*

Abstract: As we know, the decentralized management of agricultural production is inconsistent with the big market and circulation in our country. Facing this situation, we suggest strong enterprise construct third party agricultural public trade platform under the guide of the government in a way of market-based operation. The drive to build such platform from government, enterprises, peasant and agricultural organization was analyzed. The framework, function module and logic structure of the platform system were designed. In order to realize the sustainable development of the platform, we suggest the government should strengthen the function of the role in the platform construction and operation. Besides, we should improve the functional innovation, technological innovation and management innovation of the platform continuously to give full play to its economic and social benefit, and led the three sides that the government, peasant and enterprise to win together.

Keywords: Agriculture, e-commerce, trade platform, system structure

1. INTRODUCTION

China is a large agricultural country, and the agricultural resource in china is abundant, and its distribution is extensive. The agricultural production value accounts for a great proportion of the GDP of China. With the development of the market economy, the problems in the agricultural production come out conspicuously day-by-day. The circulation of

Yang, Q., Feng, B. and Song, P., 2008, in IFIP International Federation for Information Processing, Volume 259; Computer and Computing Technologies in Agriculture, Vol. 2; Daoliang Li; (Boston: Springer), pp. 933–942.

agriculture hinders the development of the countryside and becomes one of the most important factors, which influence the income increasing of peasant and the stability of the countryside. The information technology, network technology and the e-commerce commercial mode, inject life and vigor into agriculture of our country. By integrating each resource of the traditional agriculture, utilizing the advanced, convenient technology to construct agricultural information platform, the online agricultural commerce can be realized. These are useful to improve the management state and strengthen the competitiveness of the agricultural production, and will exert a far-reaching influence on the reform and development of the rural social economy in our country.

Nowadays, the agricultural production in our country mainly is decentralized management mode in which the manpower, financial resources and technologies of each managing enterprises are weak. (Xia et al., 2003) However, it is unrealistic to develop their own e-commerce platform independently, because it will add their operation cost and cause resource wasting. In this situation, the way of constructing a third party agricultural e-commerce platform is effective. Firstly, most peasants do not have high literacy, and it's difficult for them to grasp the use of PC terminal station and Internet's function. Secondly, the economic condition is limited, and it's difficult for them to buy and use the modern network device. Thirdly, the construction and maintenance of the information system will be very difficult. In this case, the construction of the third party e-commerce public trade platform is an effective way. Under the support of social forces and government, the third party agricultural e-commerce platform regards district as centre, and provides e-commerce services for agricultural production of this area. (Yi, 2006) Meanwhile, it radiates to peripheral area, provinces and cities, and can offer e-commerce support to peasant in this area effectively. On the basis of agricultural information website, the platform can supply with agricultural production trade service, including agricultural demand, logistics and price information.

2. DRIVE OF THE AGRICULTURAL E-COMMERCE PUBLIC TRADE PLATFORM CONSTRUCTION

The industrial informational development in some districts lags behind relatively because of low developed economy. So the large agricultural products are produced blindly. This led to social resource to waste seriously. The agricultural e-commerce platform uses the information technology to provide the information promulgation and collection such as supply, demand and price information. Beside, it utilizes the network and agricultural production base and logistics system to make the trade and payment convenient, safe and fast. The advantages of the platform are obvious. It not only cuts down the trade cost by reducing the intermediate link, makes the lagging agricultural economy realize great-leap-forward development, but also led the traditional agriculture break through the space-time limitation; effectively solve the problem that the agricultural production and market information are unsymmetrical. (Yi et al., 2007) Besides, it can improve the organization degree of agricultural production, the agricultural value chain of our country and competitiveness; elude the risk of product price fluctuation. In addition, the need of the government, enterprise, peasant and agricultural organization are the drive of the construction of agricultural e-commerce public trade platform.

2.1 The government needs the social benefit of agricultural e-commerce platform

This platform has typical external benefit, and can help the amalgamation of the decentralized management of agricultural production with the big market and circulation. It also can improve the agricultural market circulation system so as to use the informationization to promote the development of the traditional agriculture. Meanwhile, the construction of the agricultural e-commerce public trade platform helps the government to carry on the macro adjustments and controls of production means. By the mean of constructing information platform between the regulation and control department of the government and the department who provides the production mean, the manager can know the distribution situation of the production mean. Moreover, it makes the operation, regulation and control more transparent, and realizes the facilitation of information technology to national economy.

2.2 The agricultural e-commerce public trade platform improves the peasants' income and living standard

As to peasants and agricultural organization, the agricultural public trade platform is undoubtedly a tool to obtain market information and increase probability of selling the products. By the platform, peasants can get information about price, weather, agricultural machinery and market, etc. So it's possible for them to know which kind of crops is suitable for them to plant. This reduces the blindness of agricultural plant. In addition, they can buy seed, chemical fertilizer and agricultural machinery through Internet, and sell their products at the same time. Regarding agricultural public trade platform as carriers, we can concentrate decentralized peasants, and set up an online agricultural countryside commune. (Yan et al., 2005) Under the organization of powerful agricultural intermediary made by the Internet, the peasant produce according to the order, and do the global agricultural futures business online for 24 hours. This provides new developing opportunity for peasants and agricultural organization, and improve their income and living standard greatly too.

2.3 The agricultural e-commerce public trade platform increases the profits of the enterprises

Because of the support of policy and peasant, enterprises including the user enterprises and platform operating main body enterprise can benefit from the platform. The user enterprises including agricultural producing enterprise, marketing enterprise and processing enterprise can share information together, and cut down the stock cost, improve their ability to adapt to the risk and the perceptibility of opportunity by the platform. It is useful for them to optimize the resource distribution, promote marketing and reduce cost so as to win together. For example, the production of the agricultural producing enterprises can be supplied to agricultural processing enterprises as raw materials, or to be sold wholesale to the agricultural marketing enterprises. At the same time, they can know the demand of the processing enterprises and marketing enterprises to the production in time. So it's possible for them to control production rationally, adjust the stock, reduce cost and increase benefit. After the processing enterprises reprocess the products from producing enterprises, they can sell the new product to the marketing enterprises. Thus the processing enterprises can know the demand of the marketing enterprises and make rational producing plan. Likewise, the marketing enterprises can obtain production from upriver supplier timely, know the market-supplying situation, and make rational price strategy. Similarly, because of the support of the government, with the increasing of the platform users, the platform operating main body enterprise, namely the enterprise that constructs and operates the platform, can regain the cost and make profits gradually while perfecting the function and maintaining the operation of the platform.

3. SYSTEM DESIGN OF THE AGRICULTURAL E-COMMERCE PUBLIC TRADE PLATFORM

3.1 System framework design

The agricultural e-commerce platform public trade platform adopts the multi-layer system structure. System integrates information and trade service of B2B and B2C mode, and its core is the database server. It provides information data service for the authorized users by application server and authentication system. The whole framework of the system is shown below (fig. 1).

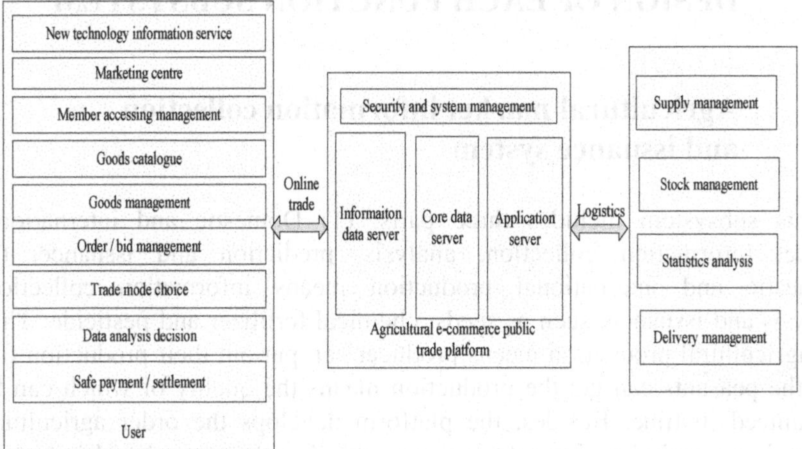

Fig. 1. System framework of the agricultural e-commerce public trade platform

3.2 System function design

(1) Online trade. The online trade is the core function of the system. We need to realize trade negotiation, negotiation management, and contract management. The system combines many ways to realize convenient and fast business negotiation such as E-mail, audio and video communication.

(2) Goods management. System provides function for managing the goods circulation and intelligent goods information inquiry. Based on database of agricultural machinery and byproducts processing equip, system carry on the data mining, analysis, realize the inquiry and statistics function, and predict the market trend.

(3) Payment system. On the basis of safe and reliable certification system, the platform provides many functions to create convenient commerce environment such as account management, payment settlement, client report form management and so on.

(4) Logistics. The logistics system of the platform connects the logistics centre of the user enterprises, the third party logistics company, and the post net logistics.

(5) Member management. According to users of different layers, platform provides service of different layers, such as credit evaluation.

(6) Information and data service. Platform provide functions for practical and new technology intercourse, enterprise and production information

issuance, technology and news dynamic issuance, policy and regulation inquiry, goods information issuance and so on.

4. DESIGN OF EACH FUNCTION SUBSYSTEM

4.1 Agricultural market information collection and issuance system

This subsystem includes three parts: (1) Domestic and international market information collection, analysis, prediction and issuance. (2) Domestic and international production means information collection, analysis and issuance, such as seeds, chemical fertilizer and pesticide. Thus the agricultural production means producer can put out their production fast and the peasants can get the production means the quality of which can be guaranteed in time. Besides, the platform develops the order agriculture, which is becoming an important way to develop the agricultural industrial management in our country. The order agriculture can meet the need of enterprises and peasants, create a new way to earn foreign exchange, increase peasants' income, and expedite the pace of agricultural informational development.

4.2 Market trade digital management system

The platform utilizes the information technology to realize integrative management of person, money and goods of this public trade market. Based on this, the digital management system improves the work quality and efficiency of the market, and its standardized management. The main function of this subsystem include management of store, stall, charge, industrial and commercial license, tax registration, certificates, warehouse, personnel, attendance, price issuance and so on.

4.3 Agricultural trade intelligent communication system

The platform provides safe and accurate information issuance about supply and demand information of both sides of the trade. Besides, we will study the intelligent automatic matching of the information issuance and its issuance standards. Thus, the problem that supply and demand information of both sides is blocked can be solved.

4.4 Visual e-commerce trade system

The platform will develop visual e-commerce trade system that suits products of some kinds of agriculture such as planting and feeding. The user can issue their product information to the trade system by PC, phone or text message. The agricultural production trade information intelligent communication engine then match the information. If there is a match, the engine will inform the both supplier and the buyer. The users can also browse or search for the production information themselves.

4.5 Agricultural e-auction system

This part is to study the feasible auction way, and to keep the e-auction information is dependable, undeniable and identifiable. Based on this, the platform develops the agricultural e-auction system that stress on demand and actual effect.

4.6 System logical structure design

The logical structure of the system is shown in fig. 2, which is made up of local program server, data baking up server and work terminal.

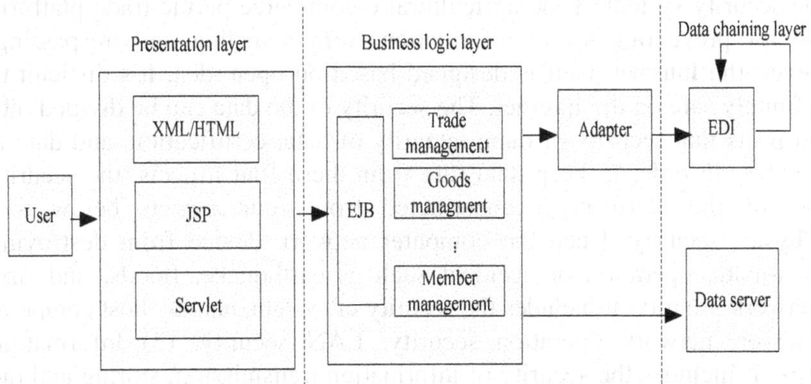

Fig. 2. Logical structure of the agricultural e-commerce public trade platform

4.7 Network system

The basis of the agricultural e-commerce platform is a perfect, safe and reliable network which can be divided into three parts: (1) External network connection groupware, including WAN connection equipment such as router, dial-up accessing server, issuance server such as DNS, mail repeater; (2) Internal network connection groupware such as internal server group.

(3) Network security device and software, including fire wall system, network security scanning software, network real-time monitoring software, Web monitoring and resume system, etc.

4.8 Software system

The basic system structure of the software system of the platform is shown in fig. 3.

Agricultural e-commerce public trade platform is a complicated system, whose software development and data amount are very large. Besides, it will use a lot of complicated technology. So we had better provide a middleware to support it. This middleware should support large-capacity system, complicated trade, various front-end, multi-database, complicated communication way and business logic, and be able to integrate other application system. The platform uses some technologies such as the Web and Java, and all kinds of application server to realize the business management of the platform. Besides, there is an interface to the other information system of the enterprises.

4.9 Security system

The security system of the agricultural e-commerce public trade platform is use for preventing secret information form revealing and trespassing. However, the Internet itself is designed based on open idea. It's difficult to be definitely safe on the Internet. The security of the data can be divided into four aspects that security of data; integrity of data, certification, and data is undeniable. In order to keep data safe from these four aspects, the security system of the platform is constructed from four aspects below too. (1) Physics security. Keep the computer network device from destroying of man-made operation or accident such as earthquake, floods, and fire. (2) Network security. It includes the security of system, namely host computer and server, network operation security, LAN security. (3) Information security. It includes the security of information transmission, storing and the audit of information content. The main technologies are encrypt, digital signing, e-certificate, e-envelope and double signing etc. (4) Trade security. It includes the identity validation, trade affirmation and information transmission security. The platform will provide reliable security services such as differentiation service; access controlling service, confidentiality service and so on.

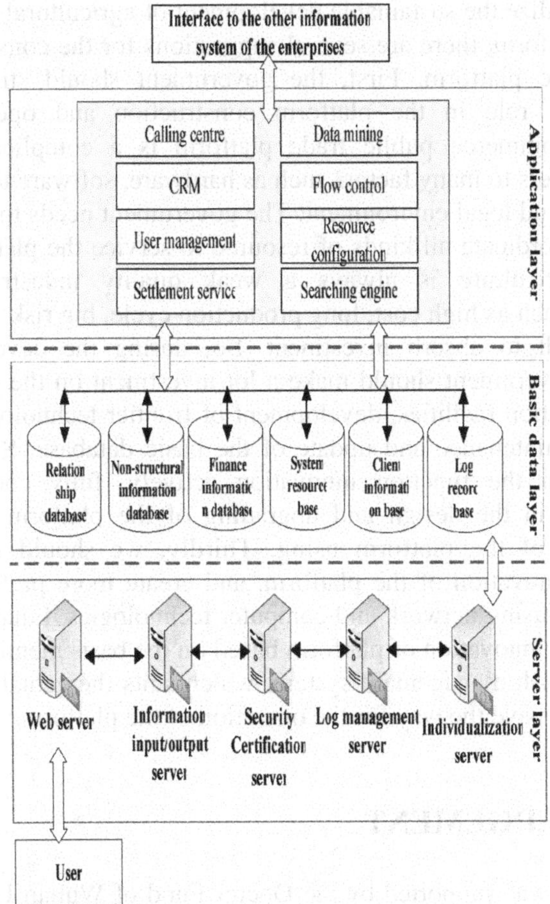

Fig. 3. Software system structure of the agricultural e-commerce public trade platform

5. CONCLUSIONS AND EXPECTATIONS

The mode that certain strong enterprise constructs the agricultural e-commerce public trade platform under the lead and support of the government is an ideal mode to develop agricultural e-commerce. On one aspect, this platform has a typical social benefit, and will improve the development level of the whole agricultural trade so as to improve income and living standard of the peasants. On the other hand, it helps the amalgamation of the decentralized management of agricultural production with the big market and circulation, and perfect the agricultural market circulation system. Moreover, the platform help agricultural product of our country to consolidate the domestic market, and go out to the world and capture the world market.

In order to realize the sustainable development of agricultural e-commerce public trade platform, there are several suggestions for the construction and operation of the platform. First, the government should strengthen the function of the role in the platform construction and operation. The agricultural e-commerce public trade platform is a complicated system engineer that refers to many factors such as hardware, software technologies, finance system and legal environment. The government needs to be active to organize and coordinate all kinds of resource to service the platform better. Meanwhile, agriculture is always a weak quality industry with the characteristics such as high cost, long production cycle, big risk, low benefit, and it's difficult to absorb investment. So, during the development of platform, the government should make a lot investment on the construction of basic information facilities, development of frontier technologies and the construction, maintenance and update of the basic database. Secondly, we should carry on the function innovation actively, fully consider user' education level in the design and upgrading of the platform and reduce the complexity of the platform using. Thirdly, we should develop the technological innovation of the platform, and create more prefect function for the platform-using network and computer technologies. Finally, improve the management innovation of platform based on the basis mentioned above, and try to establish management system, which suits the actual situation in this area so as to pave the way for the operation of the platform.

ACKNOWLEDGEMENT

This research was supported by the Doctor Fund of Wuhan University of Technology under Grant 471-38650316.

REFERENCES

Xia Wen Hui, Logistics operation mode of agricultural production on e-commerce platform, Rural Economy, No. 7, 2003

Yan Chuliang, Tian Zhaofeng, E-commerce platform construction of agricultural machinery and byproduct, Transaction of The Chinese Society of Agricultural machinery, No. 1, 2005

Yi Famin, EC Platform and Electronic Integration of Agricultural production Supply Chain, Finance and Trade Research, No. 6, 2006

Yi Faming, Xia Jiong, Research on agricultural Supply Chain integration based on e-commerce platform, On Economic Problems, No. 1, 2007-7-14

AUTONOMOUS NAVIGATION SYSTEM BASED ON GPS

Zhaoxiang Liu, Gang Liu *

Key Laboratory of Modern Precision Agriculture System Integration Research, China Agricultural University, Beijing, China, 100083

* *Corresponding author, Address: P.O. Box 125, China Agricultural University, Qinghua Donglu 17, Haidian District, Beijing, 100083, P. R. China, Tel: +86-10-62736741, Fax: +86-10-62736746, Email: pac@cau.edu.cn*

Abstract: An autonomous navigation system based on GPS was developed. The system was composed of the under-controlling part and the decision-making part; the two parts communicated with each other via wireless data transmission modules. The under-controlling part included the ARM7 microprocessor, the wireless data transmission module, the GPS receiver and the mobile quadricycle. The decision-making part included the laptop and the wireless data transmission module, the path planning algorithm based on Visual C++ 6.0 and MapObjects. The autonomous navigation and long-distance control of the quadricycle were realized. To evaluate the performance of the system, some tests were performed.

Keywords: GPS, autonomous navigation, wireless data transmission, MapObjects

1. INTRODUCTION

The development emphasis of agricultural engineering is artificial intelligence control and agricultural mechanical automatic system in the 21th century. The application of agricultural intelligent technology will promote automatization of the agricultural mechanism and the automatization of the agricultural mechanism will play an important role in the digital agricultural production system in the future (Jiang Chen, 2005).

With the development of agricultural productivity, the agricultural tractor becomes more and more good-sized, and the farmers more and more rely on

Liu, Z. and Liu, G., 2008, in IFIP International Federation for Information Processing, Volume 259; Computer and Computing Technologies in Agriculture, Vol. 2; Daoliang Li; (Boston: Springer), pp. 943–950.

several high-power tractors in the agricultural production. So it is necessary to maximize the efficiency of the tractors. The autonomous navigation technology of agricultural vehicle can realize high-efficiency fieldwork without interference of human beings for a long time, and improve the accuracy of operation at the same time. What is more, it can also reduce the cost of production, improve the quality of farm produce and relieve the working load of drivers.

As a positioning sensor, GPS has been widely used in autonomous navigation system of vehicle for which to supply high-accuracy positioning information. The application of autonomous navigation technology based on GPS in agricultural vehicle can realize the consistency of operation, improve the performance of the agricultural mechanism, realize working at night, prolong the working time, relieve the working intensity of drivers and make the drivers spend more attention on the status of farming apparatus which can improve the working quality (Defeng Kong, 2007). So it is significant to study on the autonomous navigation technology of agricultural vehicle.

The research target of the autonomous navigation system based on GPS is to make the vehicle have the ability of autonomous positioning and path planning, to realize autonomous driving. In this study, a mobile quadricycle was used as the test vehicle.

2. DESIGN OF SYSTEM

Autonomous navigation system based on GPS was consisted of under-controlling part and decision-making part (Figure 1).

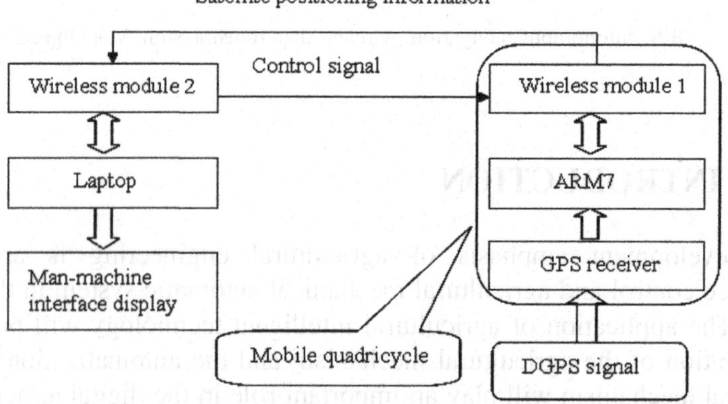

Figure 1. General structure of the system

The under-controlling part included a wireless data transmission module, an ARM7 microprocessor, a GPS receiver and a mobile quadricycle. The decision-making part was composed of a wireless data transmission module and a laptop. The two parts communicated with each other via the wireless data transmission modules. The under-controlling part received the GPS signal via GPS receiver, and transmitted the GPS data to the decision-making part via the wireless data transmission module 1. The decision-making part received the GPS information via wireless data transmission module 2. After processing and analyzing in the laptop, the decision-making part transmitted the control information to the under-controlling part, which controlled the mobile quadricycle to run, turn left or right and stop. And the position of the quadricycle was displayed by the man-machine interface at real-time.

3. IMPLEMENTATION OF SYSTEM

3.1 The under-controlling part

Figure 2 showed the structure of the under-controlling part.

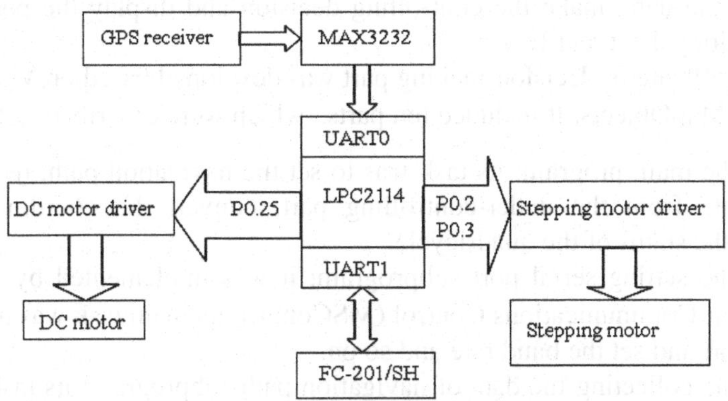

Figure 2. Structure of the under-controlling part

The under-controlling part was made up of ARM7 microprocessor (LPC 2114), wireless data transmission module (FC-201/SH), GPS receiver (AgGPS 132) and mobile quadricycle. The LPC 2114 was used as control unit, its UART0 which connected with the GPS receiver by the level translator of MAX3232 chip was used to receive the GPS data, and its UART1 which connected with the FC-201/SH directly was used to transmit the GPS data to the decision-making part and receive the control information from the decision-making part. The mobile quadricycle assembled a DC motor, which was used to drive the quadricycle, a DC motor driver, two steeping motors, which were used to control the swerve of the quadricycle and two steeping motor drivers. The I/O port P0.25 of LPC 2114 connected with the DC motor driver, I/O port P0.2 and P0.3 connected with the stepping motor drivers.

The software of under-controlling part included the main program initiating the system and waiting interrupt, the interrupt 0 subprogram receiving and transmitting the GPS data, and the interrupt 1 subprogram receiving control information and controlling the action of quadricycle.

3.2 The decision-making part

Figure 3 showed the structure of the decision-making part.

Figure 3. Structure of the decision-making part

The decision-making part included laptop and wireless data transmission module (FC-201/SH). They connected with each other via the level translator of MAX3232 chip. The FC-201/SH was used to receive the GPS data and transmit the control information, the laptop was used to analyze and process the data, make the controlling decision and display the position of the quadricycle at real-time.

The software of decision-making part was developed based on Visual C++ 6.0 and MapObjects. It included ten parts, which were described as follows:

(1) The main program: its task was to set the navigation path, receive the GPS data from the under-controlling part, convert the coordinates and control the status of the quadricycle.

(2) The setting serial port subprogram: it was implemented by adopting Microsoft Communications Control (MSComm), its main task was to choose serial port and set the baud rate and so on.

(3) The collecting the data of navigation path subprogram: its task was to receive the GGA sentences of NMEA-0183 format from a moving GPS receiver along the navigation path, and use the CStdioFile class of Microsoft to save the data in a text file.

(4) The setting navigation path subprogram: the CFile class and CArchive class were adopted to obtain the GGA sentence from the text file, then convert it into Gauss Plane Coordinates, and then save the coordinates in the CArray class, finally, display it in form of point on the map.

(5) The receiving GPS data subprogram: it obtained a GGA sentence by judging two marking bits of '$' and '*' in the GGA sentence via serial port.

(6) The extracting longitude and latitude from the GGA sentence and converting coordinates subprogram.

(7) The displaying position subprogram: its main task was to project the point onto the map according to the coordinates realizing displaying the

position of the quadricycle. It was realized by adopting dynamic tracking layer of MapObjects.

(8) The controlling status with straight navigation path subprogram: its main task was to make a controlling decision by comprehending the position information of quadricycle.

(9) The path transformation of complicated path subprogram: its task was to transform the complicated path into many short straight paths, which were used to match the complicated path.

(10) The manual control subprogram: in order to avoid damaging the quadricycle when autonomous navigation was out of control, manual control program was designed to control the quadricycle to run, turn left, turn right and stop by pressing the control button manually.

3.3 Navigation algorithm

Considering the accuracy of the GPS receiver was not very high and the speed of the quadricycle was slow in actual application, the following algorithm was chosen (Figure 4).

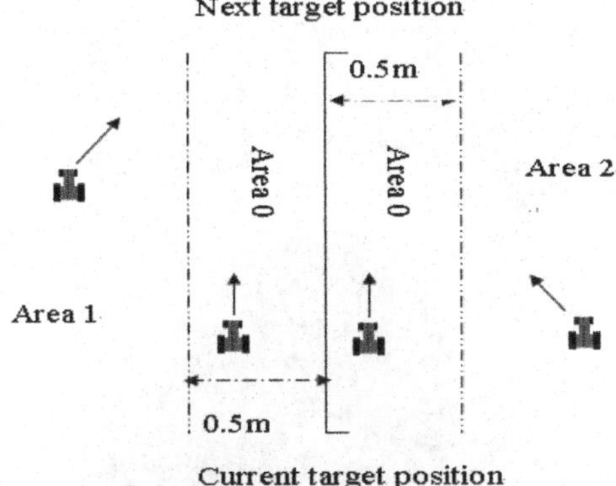

Figure 4. Sketch map of the navigation algorithm

The width of the area 0 was set as 1 m, if the quadricycle was in the area 0, it was instructed to continue running, if in the area 1, it was instructed to turn right, if in the area 2, it was instructed to turn left. This kind of algorithm had higher efficiency because the frequent turning was not needed. The algorithm was described as follows:

It was supposed that the real-time position of the quadricycle was (X_0, Y_0), the current target position was (X_1, Y_1), the next target position was (X_2, Y_2) and the equation of the straight path, which was formed by the current target position and the next target position, was formulated as:

$$AX+BY+C=0 \tag{1}$$

The vertical distance from the real-time position to the straight path formed by the current target position and the next target position was used to determine whether the quadricycle needed turning, the orientation relationship between the real-time position and the straight path was used to determine the direction of the turning.

4. PERFORMANCE TEST

4.1 Experiment equipment

The experiment equipment for performance test included a mobile quadricycle, a DGPS (AgGPS 132), which included a base station and a mobile station, an IBM T43 laptop (1.86GHz CPU, 512M RAM) and a software developed based on Visual C++ 6.0 and MapObjects. Figure 5 showed the experiment equipment.

Figure 5. The experiment equipment

4.2 Result and discussion

Tests were performed on the top of the Computer Network Center building in China Agriculture University. Before the tests, real-time data was collected by walking along a straight path (20 m) and a rectangular path (20 m × 11 m), and then saved as desired paths. The tracking accuracy was measured using the lateral deviation of the quadricycle center of gravity (CG) from a reference line, namely the desired path (Q. Zhang, 2004).

In the navigation test with straight path, the quadricycle was placed with a 1.5-meter deviation from the path to the left at the beginning of the path, and

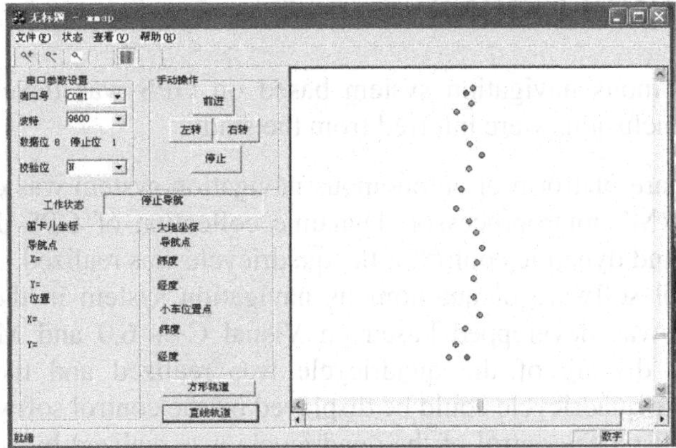

Figure 6. Navigation result with straight path

the speed of the quadricycle was 1.1 m/s. Figure 6 showed the navigation result with straight path.

As shown in Figure 6, two red points formed the straight path, and the green points represented the real-time positions of the quadricycle, the maximum lateral deviation was less than 0.71 m.

In the navigation test with rectangular path, the quadricycle was also placed with a 1.5-meter deviation from the path to the left at the beginning of the path, the speed of the quadricycle near the four corners was about 0.8 m/s, and the rest time was 1.1 m/s. Figure 7 showed the navigation result with straight path.

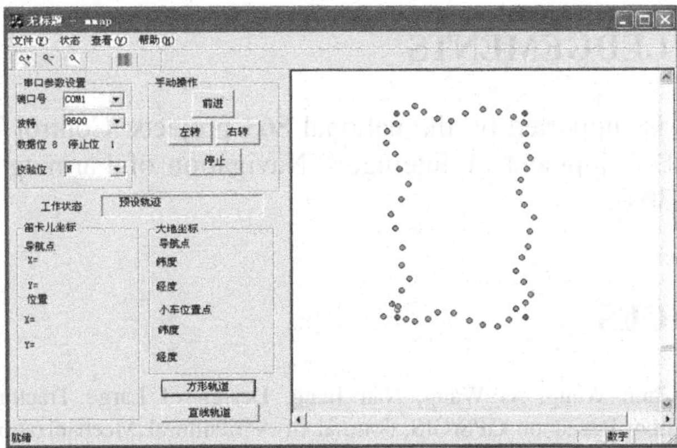

Figure 7. Navigation result with rectangular path

As shown in Figure 7, four red points formed the rectangular path, and the green points represented the real-time positions of the quadricycle. The density of green points at the corners was higher than others, because the speed at the corners was slower than others, the maximum lateral deviation was less than 0.82 m at the corners and others were less than 0.74 m.

5. CONCLUSIONS

An autonomous navigation system based on GPS was developed. The following conclusions were inferred from the study:

(1) Hardware platform of autonomous navigation system was developped based on ARM7 microprocessor. Dynamic collection of GPS data on the quadricycle and dynamic control of the quadricycle was realized.

(2) Control software of autonomous navigation system in the decision-making part was developped based on Visual C++ 6.0 and MapObjects. Autonomous driving of the quadricycle was realized and the real-time position of the quadricycle could be displayed by the control software.

(3) Long-distance control of the quadricycle was realized by adopting the wireless data transmission modules.

(4) Tests were performed on the top of the Computer Network Center building in China Agriculture University. When tracking the straight path, the maximum lateral deviation was less than 0.71 m, and when tracking the rectangular path, the maximum lateral deviation was less than 0.82 m at the corners and others were less than 0.74 m. According to the results, the navigation algorithm could fulfill the task effectively.

(5) In order to improve the accuracy and efficiency of the autonomous navigation system, the following research should be emphasized on increasing the navigation sensor, adopting multi-sensor fusion technology and improving the navigation algorithm.

ACKNOWLEDGEMENTS

This paper is supported by the national 863 projects: Control Technique and Product Development of Intelligent Navigation of Farming Machines (2006AA10A304).

REFERENCES

Defeng Kong, Chun Wang, Xi Wang, Nan Jiang, Design of Large Tractor Automatic Guidance System Based on GPS/GIS, Journal of Agricultural Mechanization Research, No. 3, 2007, pp. 54-55 (in Chinese).

Jiang Chen, Shuaibing Shi, Juncai Hou, Jie Xu, The Development of Digital Agriculture and Chinese Agriculture Machinery, Journal of Agricultural Mechanization Research, No. 4, 2005, pp. 21-23 (in Chinese).

Q. Zhang, H. Qiu, A Dynamic Path Search Algorithm for Tractor Automatic Navigation, American Society of Agricultural Engineers, Vol. 47, No. 2, 2004, pp. 639-646.

DEVELOPMENT OF INTELLIGENT EQUIPMENTS FOR PRECISION AGRICULTURE

Xiaochao Zhang*, Xiaoan Hu, Wenhua Mao

Institute of Machine and Electron Technology, Chinese Academy of Agricultural Mechanization Sciences, Beijing, China, 100083

* *Corresponding author, Address: P.O. Box 25, Institute of Machine and Electron Technology, Chinese Academy of Agricultural Mechanization Sciences, No. 1, Bei Shatan, Deshengmen wai, Beijing, 100083, P. R. China, Tel: +86-10-64882584, Fax: +86-10-64882652, Email: zxc@caams.org.cn*

Abstract: This paper briefly described the general production of intelligent agricultural machine in precision agriculture. It summarized the basic principle and the application in precision agricultural demonstration field of those agricultural machines, which mainly included wheat variable ferti-seeder, yield distribution information acquiring system, variable controlled large irrigation system moving in synchronous, intelligent spraying herbicide machine, ultra-low altitude remote system, fast-analysis system of food quality and on-board computer, GPS, GIS specially designed for precision agriculture.

Keywords: precision agriculture, variable ferti-seeder, yield distribution, variable controlled, intelligent spraying

1. INTRODUCTION

Precision agriculture means using the soil character of every field and the growth character of some one crop to decide the investment of agricultural resources, such as the most seeds, fertilizer and irrigation, and than acquiring the maximum benefit both in economics and environments. The key character of precision agriculture lies in correct prescription, veracious position and accurate agricultural resource inputs.

Intelligent agricultural machine plays a very important role in the research of precision agriculture technology. According to the conception of systematization, intelligent agricultural machine is composed of agricultural machine

Zhang, X., Hu, X. and Mao, W., 2008, in IFIP International Federation for Information Processing, Volume 259; Computer and Computing Technologies in Agriculture, Vol. 2; Daoliang Li; (Boston: Springer), pp. 951–958.

with electronics, intelligent control and information technology, to accomplish the function that normal agricultural machine cannot do.

2. ON-BOARD COMPUTER AND NAVIGATION AND POSITION SYSTEM BASED ON GPS FOR PRECISION AGRICULTURE

Navigation system which was based on the GPS position technology navigates agricultural machine in field according to the scheduled course and navigation indication (Liu, 2003). This system can offer scheduled course for operators of farming plane and tractor, and show the error range of yawing. If the DGPS system was composed of a low-cost OEM border and Pseudo-Range Difference method, the position precision, error range of it could be 2 meters (RMS). If the Carrier Phase Real-time Kinematics Differential Method (RTK) was used, the result of it could be 2 centimeters. The cost of GPS navigation and position will be lower in districts, where have the difference station or can offer network difference data.

In figure 1, a precision agricultural on-board computer together with GPS receiver, which can be used precision agriculture.

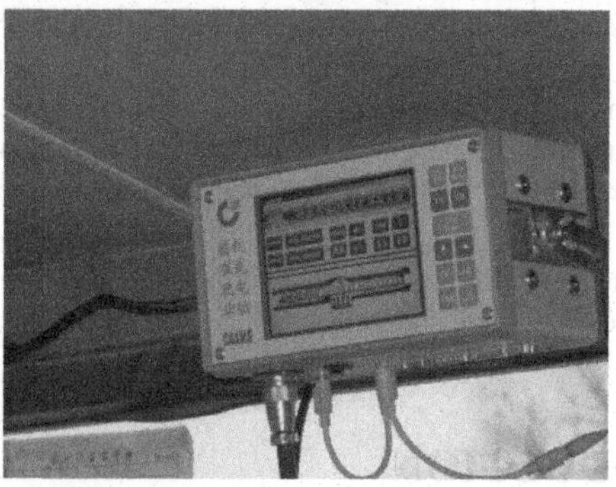

Figure 1. An automatic navigation system based on GPS

3. AUTOMATIC VARIABLE FERTI-SEEDER

The equipment which can adjust fertilizer quantity dynamically was developed for the large scale variable ferti-seeder. It can real-time adjust the amount of fertilizer according to the given prescription map and control instruction. That could improve the reliability of dynamic work and the equipment level of information.

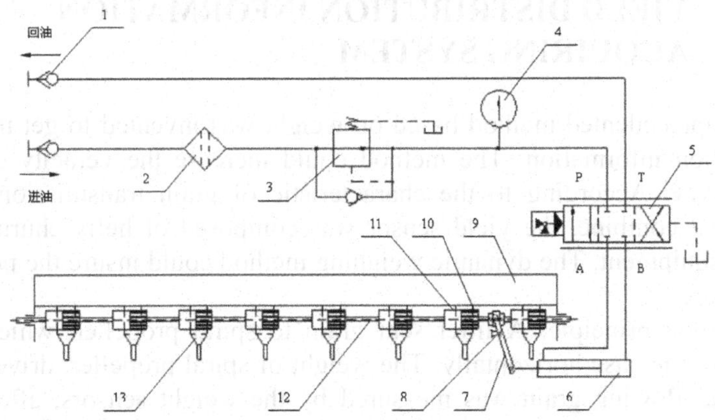

Figure 2. Wheat variable ferti-seeder

1-hydraulic pressure pipeline's tie-in of input and output, 2-filter, 3-one-way relief pressure valve, 4-manometers, 5-electro-hydraulic servo valve, 6-two acting oil cylinder, 7-fertilizer quantity setting lever, 8- pivot, 9-fertilizer trunk, 10-fertilizer ejector wheel, 11-block wheel, 12-fertilizer ejector axis, 13-fertilizer ejector box

In figure 2, the control equipment operated by electro-hydraulic system was geared into the tractor's hydraulic pressure system. When the variable ferti-seeder worked in the field, the control system could real-time adjust the quantity of fertilizer according to the prescription. When the quantity of fertilizer was needed to increase, the electro-hydraulic servo valve pushed the pressure oil from the port A to the left of two acting oil cylinder, then pressed the piston pole to move rightwards. With that, the fertilizer quantity setting lever turned around the pivot counter-clockwise. another end was transversely linked with the fertilizer ejector axis. At the same time, the fertilizer ejector axis moved leftwards that opposite to the fertilizer trunk. The fertilizer ejector wheel moved leftwards, too. Therefore, the amount of fertilizer was increased. If the quantity of fertilizer was needed to decrease, the electro-hydraulic servo valve pushed the pressure oil from the port B to the right of two acting oil cylinder. Then the fertilizer quantity setting lever turned around the pivot clockwise. At last the work length of the fertilizer ejector wheel was decreased. As a result, the fertilizer quantity was reduced. If the quantity of fertilizer was needed to be a constant, the electro-hydraulic servo valve controlled the pressure balance of port A and port B for the piston of two acting oil cylinder could not move. The amount of fertilizer would be invariable, accordingly.

The controls parameters included that the fertilization adjustable range (75 – 375kg/ha), precision (±5% FS), fertilization system response interval (< 30s), seed quantity adjustable range (90 – 225kg/ha), seed system response interval (< 42s).

4. YIELD DISTRIBUTION INFORMATION ACQUIRING SYSTEM

An unprecedented method based on weight was invented to get the yield distribution information. The method could increase the veracity of grain flux survey. According to the characteristic of grain transmission in the traditional combine, the yield sensor was composed of helix churn-dasher weight equipment. The dynamic weighing method could insure the precision of measure.

Operation principle: A lifter sent grain to spiral propeller, which drive grain into the case horizontally. The weight of spiral propeller, drive equipment and flowing grain was measured by the weight sensors, after being amplified by a high-accuracy amplifier, transformed to digital format, then sent into on-board computer. After modified by moisture information, the weight information could be turned into yield information by integral calculation. Finally, the yield map would be mapped according to the yield data, together with GPS positioning system. To improve the precision of measurement, signal processing technology was used, including the wavelet signal filtering, multi-dimension vibration signal compensating, GPS signal post-processing, etc.

Experimentation in figure 3 tell us, weight based yield obtained system achieved an precision of ±3.5%.

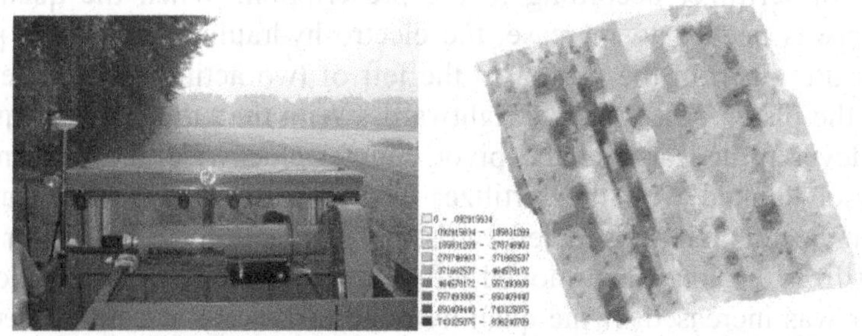

Figure 3. Food yield distributing information acquired system

5. VARIABLE CONTROLLED SYSTEM FOR LARGE IRRIGATOR IN LINEAR MOVE

The GPS OEM board based high precision navigation and positioning technology, together with the collection and analysis of soil moisture, prescription map outputted by the Decision Support System, and remote survey and control system of large-scale irrigator, realized the goal of variable

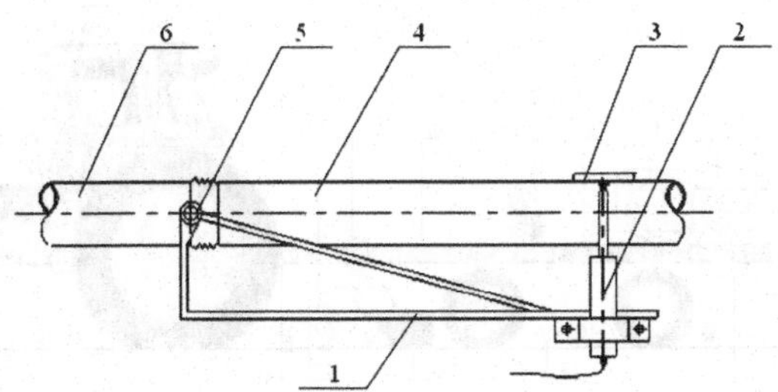

Figure 4. A control construction of large irrigation system

controlled irrigation in precision agriculture. Peasant could control large-scale irrigation machine in office to make it automatically work in field without manned. And the irrigation quantity could be automatically adjusted according to the given or anticipant data in the specified area of field.

The method of return-to-zero feedback of every variety angle included two aspects (Zhang et al., 2004). One was angle sensed in synchronous, another was feedback controlled. The method of angle sensed in synchronous was shown in figure 4.

For every section of a large irrigator could move in synchronous, every angle between two sections was measured. If it was not zero, control system of on-board computer would send adjust signals to every frequency conversion governor respectively. That would adjust the speed of ground wheels to return every angle to zero. As the response time is quite short, and every angle is quite small, the angles between every section could be kept under 0.05°.

The variable control of large-scale irrigation system also included: remote sensing and remote control of irrigation machine, rotate speed of every ground wheel of each section adjusted forward or backward to change the stance of irrigation machine, digitized accurate control technology used to control irrigation quantity in the field according to the given prescription map, automatic navigation control and fuzzy decouple control of every speed of rotating ground wheel, etc.

6. INTELLIGENT SPRAYING MECHANISM BASED ON AUTOMATIC WEED DETECTION

An automatic detection method of weed image, a control method of integration of mechanics and electrics, and a weeding equipment of intelligent and precision spraying herbicide were developed. When the spraying mechanism was worked in the field, it could real-time open the relevant nozzles and variably quantity of herbicide according to the species and distribution of

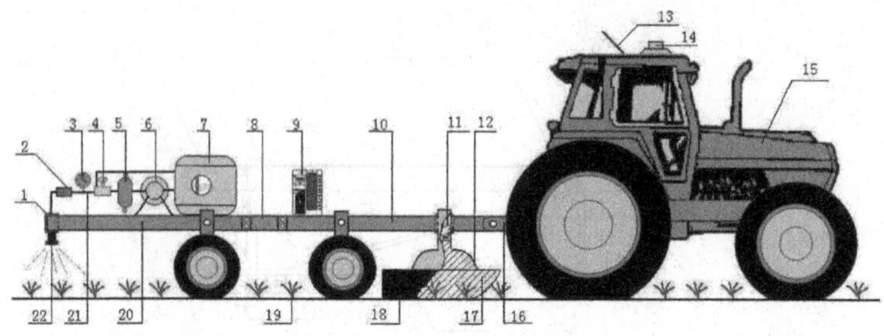

Figure 5. An intelligent weed detection and spraying herbicide machine
1-electromagnetic valves, 2-flow control valves, 3-pressure meter, 4-pressure regulator valve, 5-filter, 6-liquid pump, 7-herbicide trunk, 8-linking frame, 9-computer, 10-frame of image capturing, 11-cameras, 12-close box, 13-antenna of DGPS, 14-GPS, 15-tractor, 16-traction frame, 17-lamp lighting, 18-rubber shield, 19-green plant, 20-spraying frame, 21-pipeline, 22-nozzles

weed infestation. The aim of it was saving herbicide and reducing environmental pollution. The spraying system based on automatic detection of weed images was composed of image detection equipment and spraying equipment (Figure 5).

The image detection equipment was included the frame of image capturing, cameras, box closed in, lamp lighting, rubber shield, and so on. The multi-cameras were transverse mounted on the frame of image capturing. The spraying equipment was made up of electromagnetic valves, flow control valves, pressure meter, pressure regulator valve, filter, liquid pump, herbicide trunk, spraying frame, pipeline and nozzles, etc. The nozzles were corresponded to the electromagnetic valves. A pairs of them were a suite connecting with a flow control valve. The suites were matching with the breadth of captured images. The whole machine was formed by the image detection and the spraying equipment which was linked with the traction frame. The computer processed the control of weed detection and spraying automatically.

The method of automatic detection of crop and weed was done as follow steps:

1) Segmentation of green plant and soils

The extra-green method was adopted to segment crop and weed from soil background. The color index of extra-green was computed as the formula: Extra-green=2G-R-B

2) Detection of between-row weed based on position feature

The pixel histogram of green plant was used to determine the center line and the width of crop row, and then distinguish the between-row weed.

3) Discrimination of intra-row weed based on texture feature

The block of texture was selected on the basis of the center line of crop row. The co-occurrence matrixes of the H color space were computed. Based on it, five texture parameters were extracted. Then, the K-means clustering method was used to recognized weed within crop rows.

The method of crop and weed segmentation line used the H and S color space and the distance clustering method. The nonlinear model of segmentation crop and weed was built by the statistic data. The correct classification rate of weed cluster was achieved 90%, and the mistake classification rate of crop was lowed to 5%.

7. ULTRA-LOW ALTITUDE REMOTE SYSTEM FOR CAPTURING FIELD INFORMATION

The study of ultra-low altitude remote system provided a good hardware platform for field information gathering in precision agriculture (Figure 6). Its establishment prepared for observation of plant diseases and insect pests in field, on measurement of the crop growing and field data, etc.

Figure 6. An ultra-low altitude remote system

The technical indicators of the ultra-low altitude remote system were controllable flight height (2 – 200 m), maximum speed (120Km/h) and load ability (>4kg). The helicopter loaded 5.7kg could work in order. The test result showed that the system could be dynamic flight indeed the condition of two meters or under.

For capturing field information, the ultra-low altitude flight system mainly mounted visible light detectors in the visible light. Hot infrared, near infrared and Doppler's detectors could also be loaded for special survey. The automatic detectors is equipped with wireless data communication system of high speed (1Mbps), allocating large capacity data memory (>40G), and data automatic analysis and process system for remote information. The regional GIS and expert system of insect pest and weeds were set up. Weeds and insect distribution message were display with the graphical method.

8. CONCLUSION

The novel research on precision agricultural equipment has been accomplished in the near years. The industrialization research on precision agricultural technical equipment and products will be developed to turn the precision agricultural technology into production.

REFERENCES

Liu Jiyu. Principle and method on GPS satellite navigation and position. Beijing: Science Press, 2003:229-261.

Zhang Xiaochao, Wang Yiming, Wang Youxiang. Application and study on GPS technology for irrigation system. Transactions of the Chinese society for agricultural machinery, 2004, 35(6):102-105.

WEED DETECTION BASED ON THE OPTIMIZED SEGMENTATION LINE OF CROP AND WEED

Wenhua Mao, Xiaoan Hu, Xiaochao Zhang*

Institute of Mechatronis Technology and Application, Chinese Academy of Agricultural Mechanization Sciences, Beijing, China, 100083

** Corresponding author, Address: P.O. Box 121, Institute of Mechatronis Technology and Application, Chinese Academy of Agricultural Mechanization Sciences, No. 1, Bei Shatan, Dewai, Beijing, 100083, P. R. China, Tel: +86-10-64882584, Fax: +86-10-64882652, Email: zxc@caams.org.cn*

Abstract: Weed detection is a key problem of spot spraying that could reduce the herbicide usage. Spectral information of plants is very useful to detect weeds in real-time for the fast response time. However, the cost of an imaging spectrograph-based weed detection system is too high. Therefore, the main objective of this study was to explore a method to classify crop and weed plants using the spectral information in the visible light captured by a CCD camera. One approach to weed classification was to directly use of G and R component of RGB color space. Another was to utilize the spectral information among the green band that hue was regarded as wavelength, and saturation was represented as reflectance. The result of statistic analysis showed that both of them using the G-R and H-S optimized segmentation line of crop and weeds could be used to detect weed (lixweed tansymnustard) from wheat fields. Moreover, the method of using the H-S optimized model could avoid the affect of lighting.

Keywords: weed detection, color image, image process, spectrum information

1. INTRODUCTION

Research (Thompson et al., 1991; Yao et al., 1999) has shown that weeds are highly aggregated and that if herbicides were only applying over the weed-infected areas rather than uniform application, a reduction herbicide

Mao, W., Hu, X. and Zhang, X., 2008, in IFIP International Federation for Information Processing, Volume 259; Computer and Computing Technologies in Agriculture, Vol. 2; Daoliang Li; (Boston: Springer), pp. 959–967.

usage would occur. This reduction has economical advantage and environmental benefit.

A weed-sensor and spray-control system (Tian, 2002) would facilitate spatially variable herbicide application. However, how to sensor the information of weed infestation in fields is a primary challenge for spot spraying. There are many method of weed detection, such as manual surveying (Stafford et al., 1996), remote sensing (Brown et al., 1995) and near ground-based approaches (Lee et al., 1999). The machine-vision based detection system is mainly used, because it has high spatial resolution sensors and may utilize spectral, spatial or texture information of plants.

Until now, shape feature of plant leaves has been successfully used to classify individual plants, or some plants with little occlusion leaves (Lee et al., 1999; Pérez et al., 2000). Weed detection used texture information of plant canopy can achieve a high rate of correction classification (Tang et al., 1999; Burk et al., 2000), but most recognition algorithms are time-consuming for performed a lot of calculation and compared with co-occurrence matrices. Spectral approach imposes fewer restrictions on the response of time than other approaches (Robert et al., 2002), which greatly increases the potential for field implementation in a real-time spot spraying.

Borregaard et al. (2000) collected spectral data using line imaging spectroscopy in the 660 to 1060nm spectral range. Features selected for the discriminant function classifiers were 694, 970, 856, 686, 726, 897 and 978nm. The correct classification was 90%. A study using an imaging spectrograph was conducted by Vrindts and De Baerdemaeker (2000), over the spectral range of 485 to 815nm. Sugarbeet classifiers used up to 11 wavelength features, and corn up to 9 features. Feyaerts and van Gool (2001) developed an online weed sensor based on an imaging spectrograph. The spectral range of the system could be adjusted to cover interesting wavelengths (441, 446, 459, 883, 924nm). Under field conditions, 86% of the vegetation samples were classified correctly.

However, the cost of an imaging spectrograph based weed detection system is too high, which limits its' application in the field of the Chinese farm. Therefore, the main objective of this study was to explore a method to classify crop and weed plants using the spectral information in the visible light captured by a CCD camera.

2. SPECTRAL PROPERTIES OF GREEN PLANT

The general reflectance properties of green vegetation have been well established (Fig. 1). It could be seen from the figure measured by Vrindts et al. (2000) that the mean reflectance of some green plants is different in the

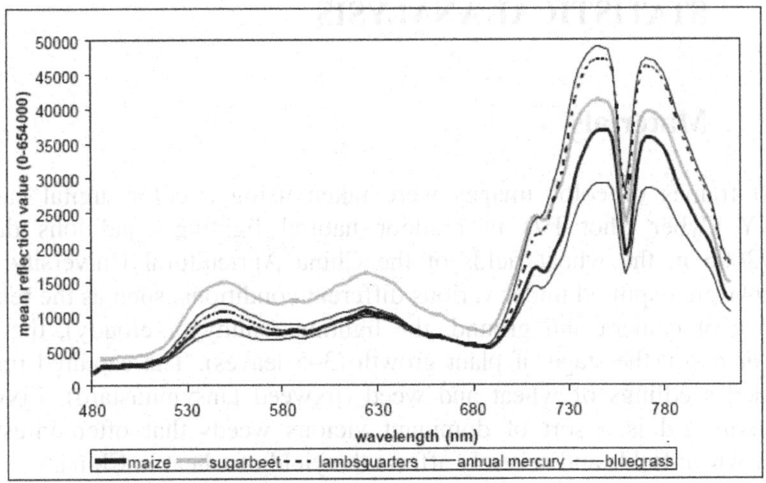

Fig. 1. Average spectra for maize, sugar beet and 3 weed species

green light (approximately 500–550nm). Moreover, there also has a difference in the red light (610–650nm).

Therefore, weed could be detected according to the difference of the mean reflectance among the visible light band. Color Images captured by a CCD camera can combine up to three bands (red, green and blue). Thus, one approach to weed classification was to directly use of G and R component of RGB color space. Another was to utilize the spectral information among the green band.

Color can be described by its wavelength. Hue is largely dependent on the dominant wavelength of light that is emitted or reflected from an object. Thus, a change in wavelength of visible light is manifested by a change in hue. Moreover, saturation depends upon the relative dominance of pure hue in a color sample. Hence, hue can be regarded as wavelength, and saturation can be represented as reflectance. The formula that converts from RGB to HS is described as follows:

$$\begin{cases} H = \cos^{-1}\left[\dfrac{\frac{1}{2}[(R-G)+(R-B)]}{\sqrt{(R-G)^2+(R-B)(G-B)}}\right] \\ S = 1-\dfrac{3}{R+G+B}[\min(R,G,B)] \end{cases} \tag{1}$$

If B is greater than G, then $H = 360° - H$.

3. STATISTICAL ANALYSIS

3.1 Materials

500 frames of color images were taken using a color digital camera (SONY Cyber Shot P7) in outdoor natural lighting conditions during 2005/2006 in the wheat fields of the China Agricultural University. The images were captured under various different conditions, such as the relative distance of camera and ground, the lighting (sunny or cloudy), the time (a.m. or p.m.), the stage of plant growth (3–5 leaves). The captured images included seedlings of wheat and weed (lixweed tansymnustard). Lixweed tansymnustard is a sort of dominant vicious weeds that often infests in winter wheat fields and severely affects the yield of wheat in China.

3.2 Methods

In order to find the useful color information of wheat and weed (flixweed tansymnustard), 500 frames collected color images were used for the statistical analysis as the follow steps:

Step 1. Preprocess
The original image included wheat plants, weed (flixweed tansymnustard) plants, soil and residue. The extra-green color index (Exg = 2G-R-B) was used to segment plants (keeping initial color) and background (transforming black (0, 0, 0)). Then, wheat plants and weed plants were manually extracted from the plant images (Fig. 2).

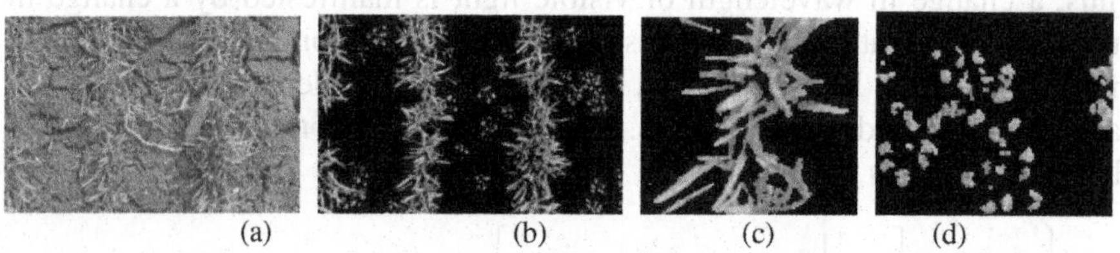

(a) (b) (c) (d)

Fig. 2. The procedure of preprocess
(a) Original image; (b) Segmented plant image; (c) Extracted wheat image; (d) Extracted weed image

Step 2. Data Distribution
The G-R and H-S values of wheat and weed (flixweed tansymnustard) pixels were collected from the extracted plant images to build a G-R dataset and an H-S dataset, respectively. The software SPSS 10.0 for windows was used to statistically analyze two datasets of plant pixels.

The simple scatter-plot was drawn to describe the original distribution of G-R or H-S. The scatter-plot graphs of two datasets were shown as Fig. 3. It could be seen that the distribution of G-R dataset was nearly linear, and that of H-S dataset was curve.

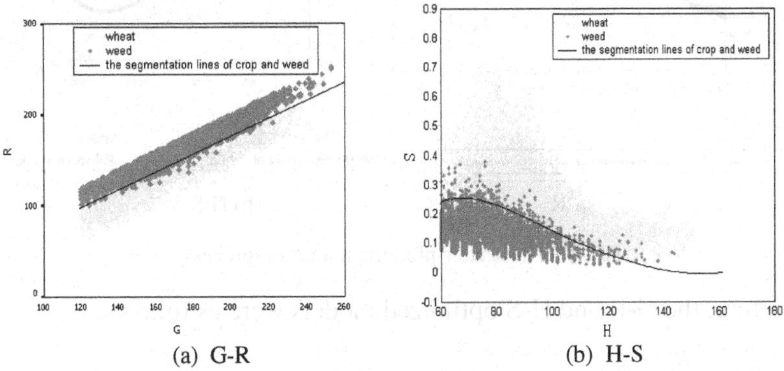

(a) G-R (b) H-S

Fig. 3. The scatter-plot graph

Step 3. Discriminant Analysis

The discrimination procedure was tested for the possibility of weed detection. The G-R and H-S data set were used to construct a discriminant function. The discriminant function, calculated with the enter independents together method is determined by a measure of generalized squared distance, based on the within-group covariance, yielding a quadratic discriminant rule. The built canonical discriminant fuctions were:

$$D=0.087R-0.075G-0.188 \quad (2)$$

$$D=0.059H+5.799S-6.758 \quad (3)$$

Step 4. Regression Analysis

To compress and centre the data, the value of the R dependent variable was replaced by its mean for each of the G independent variable. The H-S data were similarly processed. The regressive functions were shown in Table 1, which were acquired by the linear and nonlinear regression procedure, respectively. The result of regression was shown in Fig. 4.

Table 1. The regression functions

	Regressive function	R2
G-R_crop	Rc=0.957G-22.078	0.992
G-R_weed	Rw=0.994G-8.129	0.999
H-S_crop	Sc=2E-06H3-0.0006H2+0.0619H-1.7474	0.871
H-S_weed	Sw=7E-07H3-0.0002H2+0.0181H-0.3581	0.824

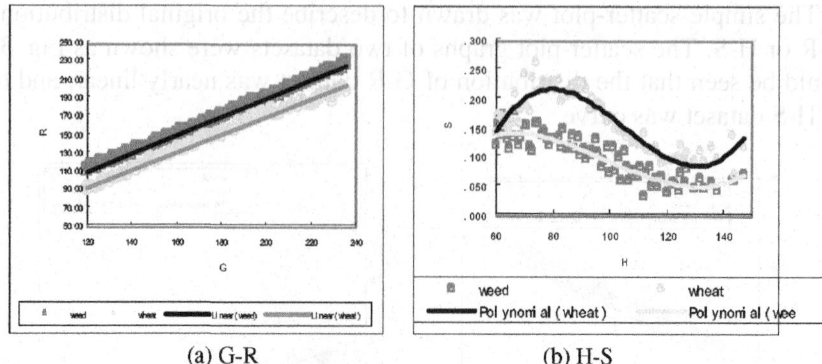

(a) G-R (b) H-S

Fig. 4. The scatter-plot graph after preprocess

Therefore, the G-R and H-S optimized models were as follows:

$$R_d = R_c + k\frac{R_w - R_c}{2} \tag{4}$$

$$S_d = S_w + k\frac{S_c - S_w}{2} \tag{5}$$

4. ALGORITHM DEVELOPMENT AND EVALUATION

The algorithms based on the G-R and H-S discriminant and optimized models were developed using the software MS visual C++ 6.0 to segment wheat field color images into wheat and weed (flixweed tansymnustard). To decrease the effect of noise, the median filter of plant pixels was done in a 3×3 window. The pseudo codes of algorithms were as follows:

The first approach used the G and R component of RGB color space. For each plant pixel point (R_i,G_i,B_i) computed as the followed formula:

$$R_d = R_c + k\frac{R_w - R_c}{2} \tag{6}$$

or

D=0.087Ri-0.075Gi -0.188

If ($R_i > R_d$ or D>0) then the plant pixel point was weed; else the plant pixel point was wheat.

The second method utilized the spectral information among the green band. For each plant pixel point (R_i,G_i,B_i) computed H_i and S_i:

$$S_d = S_w + k\frac{S_c - S_w}{2} \tag{7}$$

or

D=0.059H+5.799S-6.758

If (S_i > S_d or D>0) then the plant pixel point was wheat; else the plant pixel point was weed.

The detection results of using the discriminant function and optimized function were shown in Fig. 5. In the Fig. 5, the detected wheat pixels were green, and the classified weed pixels were red.

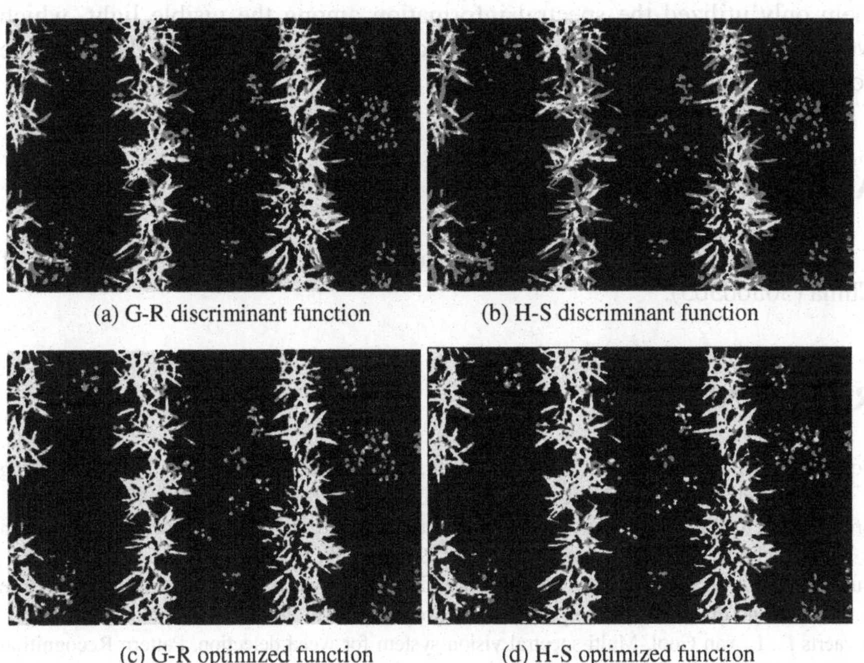

(a) G-R discriminant function (b) H-S discriminant function

(c) G-R optimized function (d) H-S optimized function

Fig. 5. The classification results

5. CONCLUSIONS

For the 500 frames color images captured under the wheat fields, the means of correct classification using the optimized models were shown in Table 2.

It could be seen form the Table 2 that the optimized function greatly improved on the wheat correct classification, and basically held on the weed correct classification.

Table 2. The classification results of using the optimized models

	Weed correct classification	Wheat correct classification
G-R dicriminant model	92.5%	75.3%
H-S dicriminant model	90.6%	56.7%
G-R optimized model	91.4%	87.6%
H-S optimized model	90.1%	88.2%

Therefore, both of approaches of using the G-R and H-S optimized model could be used to detect weed from wheat field. Moreover, the method of using the H-S optimized model could avoid of the affect of lighting. All of them only utilized the spectral information among the visible light, which would increase the feasibility of application under fields and decrease the cost of production.

ACKNOWLEDGEMENTS

This study has been funded by the National Natural Science Foundation of China (30500305).

REFERENCES

Borregaard T., H. Nielsen, L. Norgaard, et al., Crop-weed discrimination by line imaging spectroscopy, Journal of Agricultural Engineering Research, 2000, 75:389-400

Brown R. B., J.-P.G.A. Steckler, Prescription maps for spatially variable herbicide application in no-till corn, Transactions of the ASAE, 1995, Vol. 38, No. 6 :1659-1666

Burks T.F., S.A. Shearer, F.A. Payne, Classification of weed species using color texture features and discriminant analysis, Transactions of ASAE, 2000, Vol. 43, No. 2:441-448

Feyaerts F., L. van Gool, Multi-spectral vision system for weed detection, Pattern Recognition Letters, 2001, 22:667-674

Lee W.S., D.C. Slaughter, D.K. Giles, Robotic Weed control system for tomatoes, Precision Agriculture, 1999, 1:95-113

Perez A.J., F. Lopez, J.V. Benlloch, et al., Color and shape analysis techniques for weed detection in cereal fields, Computers and electronics in agriculture, 2000, 25:197-212

Robert J. Baron, Trever G. Crowe, Thomas M. Wolf, Dual camera measurement of crop canopy using reflectance, AIC 2002 Meeting, CSAE/SCGR Program Saskatoon, July 14-17, 2002

Stafford J.V., J.M. Le Bars, B. Ambler, A hand-held data logger with integral GPS for producing weed maps by field walking, Computers and Electronics in Agriculture, 1996, 14:235-247

Tang L., L.F. Tian, B.L. Steward, et al., Texture-based weed classification using Gabor wavelets and neural network for real-time selective herbicide applications, ASAE, 1999, Paper No. 993036

Thompson J.F., J.V. Stafford, P.C.H. Miller, Potential for Automatic Weed Detection and Selective Herbicide Application, Crop Protection, 1991, 10:254-259

Tian L., Development of a sensor-based precision herbicide application system. Computers and electronics in agriculture, 2002, 36:133-149

Vrindts E., J. De Baerdemaeker, Using spectral information for weed detection in field circumstances, Presented at AgEng 2000 Warwick, 2000, EurAgEng Paper No. 00-PA-010

Vrindts Els, Automatic recognition of weeds with optical techniques as basis for site-specific spraying, [Dissertations DE Agricultura]. Katholieke Universititeit Leuven, 2000

Yao H., L.F. Tian, L. Tang, et al., Smart sprayer performance simulation. ASAE, 1999, Paper No. 99-1103

APPLE MATURITY DISCRIMINATION AND POSITION

He Bei, Liu Gang [*]
Key Laboratory of Modern Precision Agriculture System Integration Research, China Agricultural University, Beijing, China, 100083
** Corresponding author, Address: P.O. Box 125, China Agricultural University, Qinghua Donglu 17, Haidian District, Beijing, 100083, P. R. China, Tel: +86-10-62736741, Fax: +86-10-62736746, Email: pac@cau.edu.cn*

Abstract: The most crucial problem of apple harvesting robot is to get the exact location and the maturity degree of the target. A system of apple maturity discrimination and positioning based on the multi-spectral technology and laser triangulation ranging principle is researched. Signal transmission, signal reception, signal processing, the CPU processing and several PC components are involved in the system. The distance between the apple and the device was calculated according to principle of triangulation. Ripe red apple, unripe green apple, stems and leaves have different reflectivity to red and infrared beam. The different ratio of red and infrared signal can be used to identify whether an apple is ripe. The threshold used to judge whether an apple is ripe is an experiential interval, which is acquired by several experiments.

Keywords: multi-spectral vision technology, triangle ranging, maturity of discrimination, positioning, nonlinear least square method

1. INTRODUCTION

Fruit harvesting is the most time-consuming and laborious part in fruit production chain. Because of its seasonal characteristic, labor intensity and high cost, ensuring harvesting fruit timely, lowering harvesting costs is an important way to increase agricultural income. Due to the complexity of harvesting operations, harvesting is still in a very low degree of automation. At present, domestic fruit harvesting operations is basically done by labor,

Bei, H. and Gang, L., 2008, in IFIP International Federation for Information Processing, Volume 259; Computer and Computing Technologies in Agriculture, Vol. 2; Daoliang Li; (Boston: Springer), pp. 969–976.

and its harvest cost accounts for about 50% – 70% of the total cost. And the harvest time is concentrated. As an important agricultural robot types, harvesting robot is used to reduce the intensity of labor and production costs, improve labor productivity and product quality and ensure harvesting fruit timely, so it has great potential (Van Henten, 2003).

2. SYSTEM DESIGN AND ANALYSIS

2.1 Functional module structure

The system of apple maturity discrimination and positioning based on the multi-spectral technology and laser triangulation ranging principle included signal transmission, signal reception, signal processing, CPU and PC, as shown in the Fig. 1.

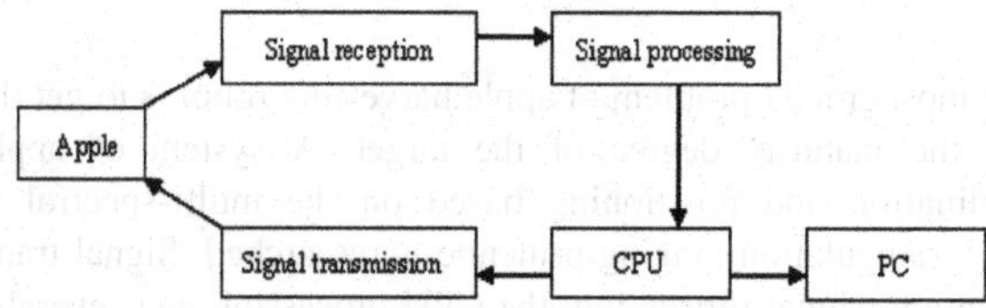

Fig. 1. General structure of the system

2.2 System principle

The system principle is shown in the Fig. 2.

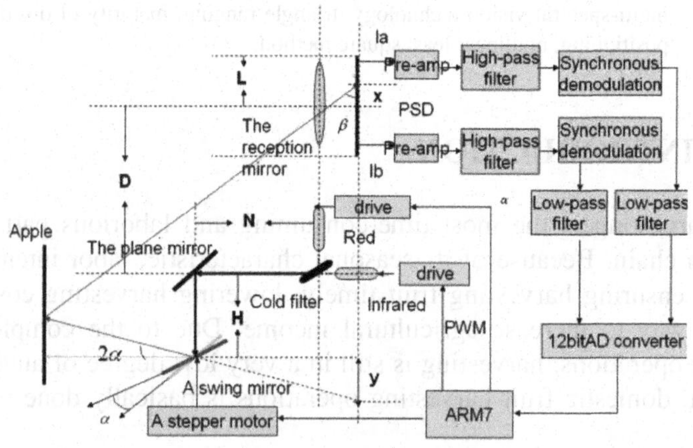

Fig. 2. System principle

ARM7 outputted two PWM pulses, which respectively modulate two laser beams including a red laser beam (685nm) and an infrared laser beam (830nm) (Liu Weiming, 2004). The two modulated beams go through a transmission lens and unite into one beam, then reflect by a fixed plane mirror. The united beam exposes to a rotatable swing lens, which is controlled by a stepper motor, and forms the linear scanning beam. Some beams reflected by apple go through the reception lens, and then focalize and image on the sensitive surface of semiconductor components PSD, which is a position sensitive sensors. PSD output two current associated with the photoelectric position. After the current-voltage conversion, amplification, filtering and demodulation etc., voltage signal was collected by ARM7 through 12 bit A/D converters. ARM7 processed the acquired data, based on the triangulation ranging principle and nonlinear least squares principle. The coordinates of the center of the apple was calculated, and then the maturity degree of the apple was judged, according to the acquired ratio of red signal and infrared signal. At last, ARM7 would transmit the processing results to the PC.

3. SYSTEM HARDWARE DESIGN

3.1 Signal transmission and reception

It was necessary to measure the reflectance characteristics of the apple for designing the system. This characteristic was measured by using a spectrum reflectance analyzer. The measurement result is shown in the Fig. 3.

This shows that the light beam with wavelength 685nm was well reflected by the red apple fruit and was reflected not so much by the other part such as the leaf, green apple, and stem that contain much chlorophyll. The light beam with wavelength 830nm was well reflected by all parts of the apple

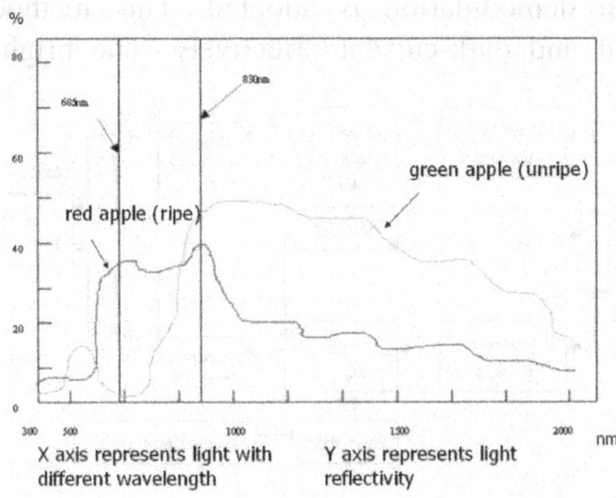

Fig. 3. Reflectance characteristic of apple

plant. Based on this characteristic, the light transmission had two laser triode, one transmit red beam, which wavelength was 685nm; the other transmit infrared beams, which wavelength was 830nm. The intensity was 10mW.

PWM was used to control the laser transmitter to emit red and infrared beams. Two laser diodes were fixed perpendicularly. Two beams were concentrated into one optic axis by a cold filter. As shown in the Fig. 4.

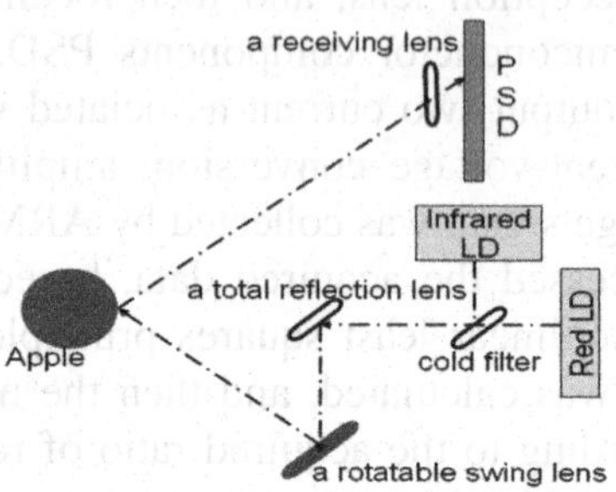

Fig. 4. Block diagram of signal transmission and reception

3.2 Signal processing

The object of this system was to detect the micron dimension motion. So the selection of device was very important (Pei Xiandeng, 2004).

AC modulation frequency method was used to separate background light from dark current. Choosing this method is for the reason that the disturbance of background light is always natural light and artificial light source, whose brightness changes slowly. The PSD response is direct current and low-frequency signal. If do high-frequency modulation to the object, the PSD response is high-frequency pulse signal. So high-pass filter can separate the signal and disturbance. To ensure the completion of position calculation and generate strong light intensity on PSD, square wave modulation and synchro position demodulation is adopted. This method can eliminate background light and dark-current effectively. The brightness of outside

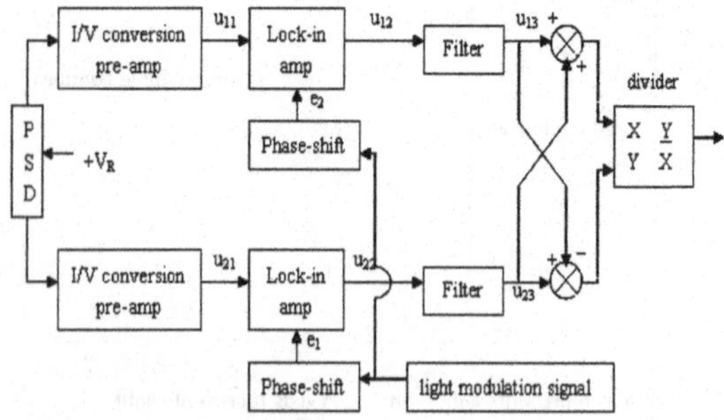

Fig. 5. Block diagram of modulation-demodulation

light is almost unchangeable in this system. It is suitable to use modulation method to eliminate background light and dark-current. So this system takes the method as the signal processing part. As shown in the Fig. 5.

4. SYSTEM SOFTWARE DESIGN

4.1 Main function

The software system mainly had the following functions:

(1) Control the rotation and operation of the stepper motor
(2) Drive the laser diode fire red and infrared beams
(3) Collect AD signals
(4) Process data and transmit the distance and the maturity value to the PC

4.2 Main program

The main program is shown in the Fig. 6.

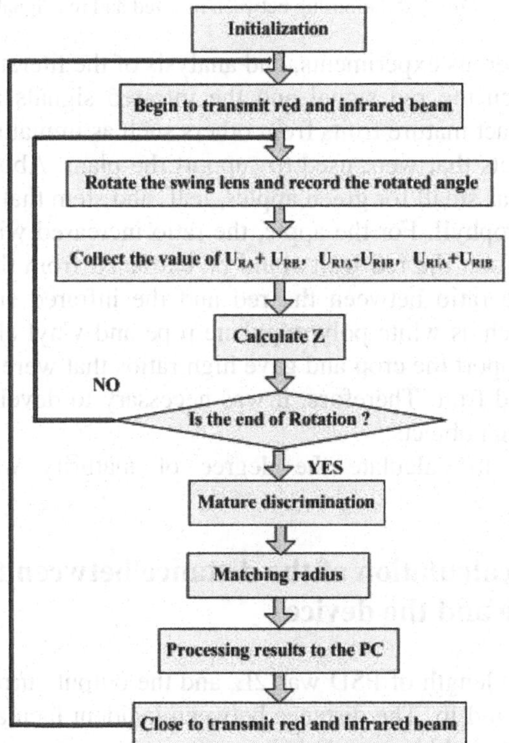

Fig. 6. Software flowchart

4.3 Maturity model

The signal outputted by the PSD A, B pole went through synchronous detection, low pass filter and become red beam signal URA and URB and infrared signal URIA and URIB. As shown in the Fig. 7.

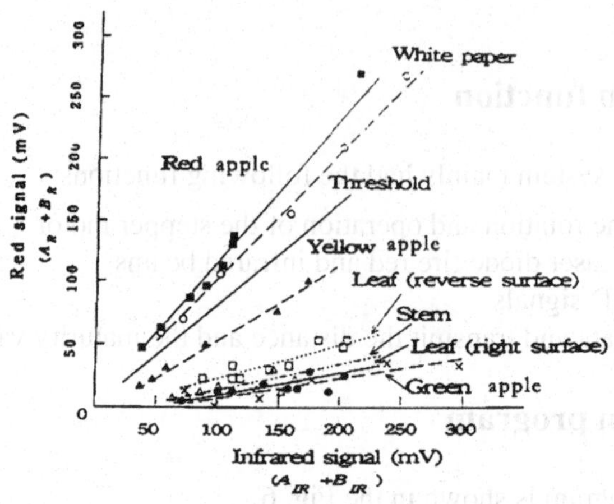

Fig. 7. Relationship between infrared and red signal

Through numerous experiments, and analysis of the literature found, using the ratio between the red signal and the infrared signals as the threshold value could extract mature fruits from others such as immature fruits, leaves, stems, and objects that were used to support the plant. Above figure shows that the ratios was small for green apples, leaf, and stem that contains a large amount of chlorophyll. For the apple, the ratio increased with the degree of maturity. Therefore, the red fruit could be extracted from the other parts of the plant by the ratio between the red and the infrared signals. However other objects such as white polypropylene rope and vinyl chloride pipe that were used to support the crop and gave high ratios that were nearly the same as that of the red fruit. Therefore, it was necessary to develop an algorithm for extracting such objects.

The formula to calculate the degree of maturity was URA+URB/ URIA+URIB.

4.4 The calculation of the distance between the target apple and the device

Supposing the length of PSD was 2L, and the output current of two poles of PSD were Ia and Ib. The distance between incident facular and the center of PSD was X. It could be concluded:

$$X = \frac{(I_b - I_a)\times L}{(I_b + I_a)}$$

As the figure 9 shows, the focus of transmission lens was C, the distance between the transmission lens and total reflection lens was N, the distance between the total reflection lens and the swing lens was H, the distance between transmission lens and reception lens was D. When the swing lens rotated α degree, the beam scanned the region of apple was 2α, the angle between the beam and the PSD was β. The distance y was as the following Fig. 8 shows.

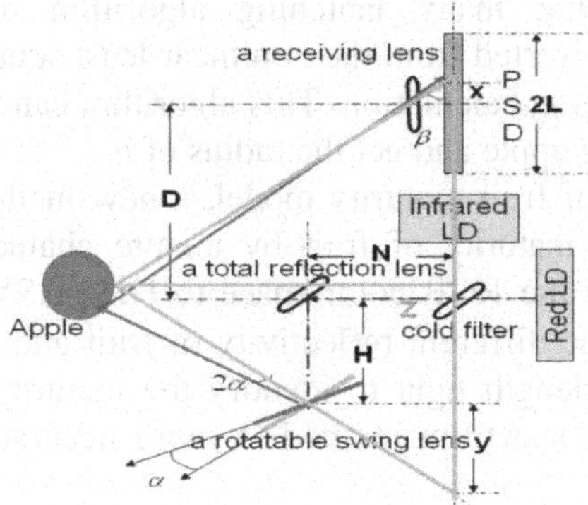

Fig. 8. Diagram of distance calculation

As the figure shows,

$$tg\beta = C/X$$

$$tg2\alpha = y/N \Rightarrow y = tg2\alpha\times N$$

$$\Rightarrow \frac{y+H+D+X}{Sin(180-\beta-90+2\alpha)} = \frac{tg2\alpha\cdot N+H+D+X}{Sin(90-\beta+2\alpha)} = \frac{tg2\alpha\cdot N+H+D+X}{Cos(2\alpha-\beta)}$$

So the formula of calculation of distance between the apple and the device was:

$$Z = Cos2\alpha\cdot Sin\beta\cdot\frac{tg2\alpha\cdot N+H+D+\dfrac{(I_b - I_a)\times L}{(I_b + I_a)}}{Cos(2\alpha-\beta)}$$

The actual distance between the apple and the sensor was:

$$Z_S = Cos2\alpha\cdot Sin\beta\cdot\frac{tg2\alpha\cdot N+H+D+\dfrac{(I_b - I_a)\times L}{(I_b + I_a)}}{Cos(2\alpha-\beta)}+R$$

where, R represents the radius of the apple, so the apple was almost round. According to round matching method, nonlinear least squares principle was used to calculate. Least squares principle was a mathematics optimization method. It found the best matching of an array by the least square error.

5. CONCLUSIONS

The following conclusions are inferred from the study:

(1) A ranging method based on laser triangulation principle, which used to calculate the distance between the target apple and the device is designed. It uses specific wavelength light to line scanning the vision field, receive the reflected light by PSD photoelectric sensor, calculate the characteristic of receiving light, and get the distance between the target apple and the device at last.

(2) After analyzing many matching algorithm, linear least squares principle, which converted from the nonlinear least squares principle is used to implement matching calculation. This algorithm can give a best matching of the contour of the apple and get the radius of it.

(3) In the study of fruit maturity model, a new method is proposed. It is not identifying the maturity of fruit by texture characteristic, shape characteristic or turning the RGB color space to HIS, HSV, YIQ, color space. It is according to the different reflectivity of fruit and leaves of the crop to some specific wavelength light to identify the maturity of fruit. Compared with other method, spectrum method is more accurate and reliable.

ACKNOWLEDGEMENTS

This paper is supported by the national 863 projects: Key Technique of Fruit Harvesting Robot on Machine Vision (2006AA10Z255).

REFERENCES

Ahmad, I.S. & J.F. Reid. 1996. Evaluation of Color Representations for Maize Images. Journal of Agricultural Engineering Research, 63(3):185-195.

D.M. Bulanon, T. Kataoka, S. Zhang. 2001. Optimal Thresholding for the Automatic Recognition of Apple Fruits. Transactions of the ASAE.

Liu Weiming, Mao Hanping, Wang Xinzhong. Application of Multi-spectrum Reflectance Vision Technique for Harvesting Robot [J]. Agricultural Equipment & Technology, 2004, 30(6):15-18.

Pei Xiandeng, Luo Chun, HuangHa. Research of high precision position measurement system and its design based on PSD. Huazhong Univ. of Sci. & Tech. (Nature Science Edition, 2004, 32(2):7-9.

Shigehiko Hayashi, Katsunobu Ganno, Yukitsugu Ishii, et al. Robotic harvesting system for eggplants [J]. JARQ, 2002, 36(3):163-168.

Van Henten E.J., Van Tuijl B.A.J., Hemming J. Field test of an autonomous cucumber picking robot [J]. Biosystems Engineering, 2003, 86(3):305-307.

Wang Xiaodong, Zhao Jie, Wu Wei. Measuring and Control of Laser Range Sensor Using PSD. Journal of Transducer Technology, 1995(5):37-38.

BIOINFORMATICS AND ITS APPLICATIONS IN AGRICULTURE

Jian Xue [1], Shoujing Zhao [1,*], Yanlong Liang [1], Chunxi Hou [1], Jianhua Wang [1]

[1] *College of Biological and Agricultural Engineering, Jilin University, Changchun, China, 130022*

[*] *Corresponding author, Address: College of Biological and Agricultural Engineering, Jiling University, 5988 Renmin Street, Changchun, Jilin, 130022, P. R. China, Tel: +86-431-85095253, Fax: +86-431-85095253, Email: swgc@jlu.edu.cn*

Abstract: The field of bioinformatics emerged as a tool to facilitate biological discoveries more than 10 years ago. With the development of Human Genome Project (HGP), the data of biology increased fabulously and marvelously. The ability to capture, manage, process, analyze and interpret data became more important than ever. Bioinformatics and computers can help scientists to solve it. Here are introduced roles of bioinformatics, meanwhile Web tools and resources of bioinformatics were reviewed. And its applications in agriculture were also discussed.

Keywords: bioinformatics, WorldWideWeb, agriculture

1. INTRODUCTION

Bioinformatics which is coming with HGP brings together the fields of life science, computer science and statistics and strives to understand medical and biological systems by the creative application of statistics and computer analysis.

Xue, J., Zhao, S., Liang, Y., Hou, C. and Wang J., 2008, in IFIP International Federation for Information Processing, Volume 259; Computer and Computing Technologies in Agriculture, Vol. 2; Daoliang Li; (Boston: Springer), pp. 977–982.

Bioinformatics is the use of computer technology to help scientists keep track of the genetic information they find. Using computers, researchers can gather, store, analyze and compare biological data with great speed and accuracy.

Imagine studying gene structures without the help of a computer. It would take many years to compare the 15,000 genes of Arabidopsis to the genes of a similar plant. And keeping track of the 100,000 genes of a human being would be inconceivable. With computers, the process of comparison is automated. By storing information as it is discovered, computers ease the immense job of genome mapping. But computers can analyze as well as store information. They can be used to construct models that reduce the need for experimentation.

In this way, biotechnology has become more efficient. Scientists are able to use fairly reliable computer-assisted predictions of test results on genetic modifications. This complements the time-consuming process involved in growing out every modified plant in the laboratory or greenhouse to test for the desired modification.

2. ROLES OF BIOINFORMATICS

Bioinformatics today has entered every major discipline in biology. In genomics, Bioinformatics has aided in genome sequencing, and has shown its success in locating the genes, in phylogenetic comparison and in the detection of transcription factor binding sites of the genes (Liu et al., 1995; Thijs G. et al., 2002), just to name a few. Microarray technology has opened the world of transcript me in front 'biologists (Spellman et al., 1998; Eisen et al., 1998). Bioinformatics provides analytical tools for microarray data. These tools range from image the processing techniques that read out the data, to the visualization tools that provide a first-sight hint to the biologists; from preprocessing techniques (Durbin et al., 2002) that remove the systematic noise in the data to the clustering methods (Eisen et al., 1998; Sheng et al., 2003) that reveal genes that behave similarly under different experimental conditions. In proteomics, bioinformatics helps in the study of protein structures and the discovery of sequence sites where protein-protein interactions take place. To help understanding biology at the system level, bioinformatics begins to show promise in unraveling genetic networks (Segal et al., 2003). Finally, in the study of metabolome, bioinformatics is used to study the dynamics in a cell, and thus to simulate the cellular interactions.

3. WEB TOOLS AND RESOURCES OF BIOINFORMATICS

The WorldWideWeb provides a mechanism for unprecedented information sharing among researchers. Today, scientists can easily post their research findings on the Web or compare their discoveries with previous results, often spurring innovation and further discovery. The value of accessing data from other institutions and the relative ease of disseminating this data has increased the opportunity for multi-institution collaborations, which produce dramatically larger data sets than were previously available and require advanced data management techniques for full utilization.

As a side effect of these types of collaborations, some tools become de facto standards in the communities as they are shared among a large number of institutions. For instance, consider the BLAST (Altschul et al., 1990) family of applications, which allow biologists to find homologs of an input sequence in DNA and protein sequence libraries. BLAST is an example application that has been enhanced as a Web source, which provides dynamic access to large data sets. Many genomics laboratories provide a Web-based BLAST interface (http://blast.wustl.edu/) to their sequence databases that allow scientists to easily identify homologs of an input sequence of interest. This capability enhances the genomics research environment by allowing scientists to compare new sequences with every known sequence and to have their work validated by other members of the community. The addition of new sequences at an increasingly frequent rate (NIAS DNA Bank, http://www.ncbi.nlm.nih.gov/Genbank/genbankstats.htm) further increases the value of this capability.

There are a number of common bioinformatics analyses one can perform at other sites, such as European Bioinformatics Institute (EBI), BioWeb Pasteur and Canadian Bioinformatics Resource, including BLAST and sequence analyses, primer tools and phylogenetics tree construction. EMBOSS sequence analysis package and SRS bio-database access are among the widely useful web tools available at these and other resource sites.

There are numerous web lists of bioinformatics resources, with many aimed at the biologist looking for software. Some of these, such as Bioinformatics.net, include discussion forums on the use of biology software. These are useful for biologists, as well as bioinformatics engineers looking for tools related to their work, or to be used at service centers. Many of these share a similar organization by functional categories, with many of the same links. It is useful to compare these for their different editorial perspectives, e.g., genomics/molecular biology or proteomics/biochemistry, as well as effort to update and remove obsolete links. General resources such as Google, Amazon's Alexa and Open Directory Project at Mozilla.org include biology and bioinformatics categories in their directories. These directories are populated by robots or from submissions; they tend to lack the comprehensiveness of biologist-maintained lists.

Bioinformatics.ca provides a curated list of links that are well organized in categories, with main sections that include human genome and model organisms, sequences, gene expression, education and computer-related resources. Most or all of these include useful editorial comments on the content and value of the linked resources, making this list especially useful in learning about resources.

The Genome Web at MRC, UK, offers a similar very useful catalogue of links with editorial abstracts. An interesting function at Bioinformatics.ca is provided by an XML standard for web news called RSS, for sharing bioinformatics links. This allows customers and other web sites to have computable access to this catalogue. For instance, you can use an RSS program to notify you of additions and changes to this catalogue.

The Bionetwork project at Pasteur Institute provides an example of resource lists that are searchable by several bioinformatics criteria: Biological Domain, e.g., sequence analysis or structural biology, Resource type, e.g., database or online analysis tools, and Organism. This biology-focused search engine proves especially useful in finding that tool or resource most relevant to one's research. This project also has implemented link maintenance by using semi-automatic scanning of internet news and resources (robot-like) to update the catalogue. A similar project is BioHunt, which uses internet robot technology to search and update molecular biology resources.

BioHunt maintains current entries (it shows update times of this review month for several searches), making it especially useful to find new or updated tools that one has heard of, but lacks crated cataloguing of these to make it easy to find by subject matter.

Bioinformatics.net is a catalogue of online biology resources, specializing in bioinformatics tools. Its focus is towards he needs of molecular biologists and life science professionals, more than for bioinformaticians, and includes discussion and help forums on the use of software and bioscience topics. Jonathan Rees, who developed this resource, also curates biology lists in the Open Directory project. This service is supported in part y advertising, as are others reviewed here, one of the limited options available o maintain such services.

Bioinformatik.de offers a similar directory style collection of crated bioinformatics and biology resource links. The CMS molecular biology resource is an extensive catalogue of biology resources, including software tools. The Southwest Biotechnology Center also maintains a useful catalogue covering a broad range of biology resources.

Bioinformatics.org and SourceForge.net are resources that support software developers and bioinformatics engineers, but are also useful to biologists looking for tools. Open-source software development in bioinformatics and other fields is being invigorated through agencies such as these. The number of active, widely used and valuable bioinformatics projects at these services is growing, including Generic Model Organism Database, Gene Ontology, GeneX Gene Expression Database and Staden Package for sequence analysis. These agencies allow for software archiving, but the primary attractions to software developers are infrastructure and tools that enable collaborative software development. A historical archive or catalogue service of bioinformatics software is limited, and maintenance of software releases is left to developers using this service.

4. APPLICATIONS OF BIOINFORMATICS IN AGRICULTURE

Plant life plays important and diverse roles in our society, our economy, and our global environment. Especially crop is the most important plants to us. Feeding the increasing world population is a challenge for modern plant biotechnology. Crop yields have increased during the last century and will continue to improve as agronomy re-assorting the enhanced breeding and develop new biotechnological-engineered strategies. The onset of genomics is providing massive information to improve crop phenotypes. The accumulation of sequence data allows detailed genome analysis by using-friendly database access and information retrieval. Genetic and molecular genome co linearity allows efficient transfer of data revealing extensive conservation of genome organization between species. The goals of genome research are the identification of the sequenced genes and the deduction of their functions by metabolic analysis and reverses genetic screens of gene knockouts. Over 20% of the predicted genes occur as cluster of related genes generating a considerable proportion of gene families. Multiple alignment provides a method to estimate the number of genes in gene families allowing the identification of previously undescribed genes. This information enables new strategies to study gene expression patterns in plants. Available information from news technologies, as the database stored DNA microarray expression data, will help plant biology functional genomics. Expressed sequence tags (ESTs) also give the opportunity to perform "digital northern" comparison of gene expression levels providing initial clues toward unknown regulatory phenomena. Crop plant networks collections of databases and bioinformatics resources for crop plants genomics have been built to harness the extensive work in genome mapping. This resource facilitates the identification of ergonomically important genes, by comparative analysis between crop plants and model species, allowing the genetic engineering of crop plants selected by the quality of the resulting products. Bioinformatics resources have evolved beyond expectation, developing new nutritional genomics biotechnology tools to genetically modify and improve food supply, for an ever-increasing world population. So bioinformatics can now be leveraged to accelerate the translation of basic discovery to agriculture. The predictive manipulation of plant growth will affect agriculture at a time when food security, diminution of lands available for agricultural use, stewardship of the environment, and climate change are all issues of growing public concern.

REFERENCES

Altschul, S.F., Gish, W., Miller, W., Meyers, E.W., Lipman, D.J. 1990. Basic local alignment search tool, Mol. Biol., 215: 403.

Durbin, B.P., Hardin, J.S., Hawkins, D.M., Rocke, D.M. 2002. A variance-stabilizing transformation for gene-expression microarray data, Bioinformatics, 18(Suppl. 1): s105.

Eisen, M.B., Spellman, P.T., Brown, P.O., Botstein, D. 1998. Cluster analysis and display of genome-wide expression paterns, Proc. Natl. Acad. Sci. (USA), 95: 14 863.

Liu, J.S., Neuwald, A.F., Lawrence, C.E. 1995. Bayesian models for multiple local sequence alignment and Gibbs sampling strategies, J. Amer. Stat., 90: 1156.

Segal, E., Shapira, M., Regev, A., Pe'er, D., Botstein, D., Koller, D., Friedman, N. 2003. Module networks: identifying regulatory modules and their condition-specific regulators from gene expression data. Nat. Gen., 34(2): 166.

Sheng, Q., Moreau, Y., De Moor, B. 2003. Biclustering microarray data by Gibbs sampling, Bioinformatics, 19(Suppl. 2): ii 196.

Spellman, P.T., Sherlock, G., Zhang, M.Q., Iyer, V.R., Anders, K., Eisen, M.B., Brown, P.O., Botstein, D., Futcher, B. 1998. Comprehesive identification of cell cycle-regulated genes of the yeast saccaromyces cerevisiae by microarray hybridization, Molecular Biology of the Cell, 9: 3 273.

Thijs, G., Marchal, K., Lescot, M., Rombauts, S., De Moor, B., Rouze, P., Moreau, Y. 2002. A Gibbs Sampling method to detect over-represented motifs in upstream regions of coexpressed genes, Journal of Computational Biology, 9(2): 447.

OPTIMIZED SELECTION OF WETLAND WATER QUALITY MONITORING POINTS BASED ON INFORMATION ENTROPY AND FUZZY SIMILARITY

Xinjian Xiang

College of Automation & Electrical Engineering, Zhejiang University of Science and Technology, Hangzhou, 310023, P. R. China, Tel: +86-571-85070268, Email: hzxxj@sina.com

Abstract: Known as the kidney of earth, wetland has significant ecological functions such as freshwater conservation, poison elimination, carbon storage, water quality purification, flood storage and drought control, climate regulation and remaining biodiversity etc. So protecting wetland is protecting ourselves. Water environment quality best reflects the ecological environment condition of wetland. According to multi-index and Spatial and Temporal variation of wetland water pollution, combining optimized selection requirements of wetland water quality monitoring, fuzzy similarity is propose. Through constructing multi-index monitoring data samples Decision-making Matrix, fuzzy similarity matrix between sample data and their mean values is established. According to the index value variation, the index weights are calculated based on information entropy theory. With the index weight and sample fuzzy similarity matrix, comprehensive fuzzy similarity of each monitoring point is calculated. Finally, according to comprehensive fuzzy similarity, each monitoring point is reasonably clustered, then representative points is selected from each category, so distribution optimization could be realized. Practical running proves that this scheme is simple and feasible, and extensionally applied to optimize other environmental monitoring points.

Keywords: fuzzy similarity method, information entropy, wetland water quality monitoring, optimized selection

Xiang, X., 2008, in IFIP International Federation for Information Processing, Volume 259; Computer and Computing Technologies in Agriculture, Vol. 2; Daoliang Li; (Boston: Springer), pp. 983–991.

1. INTRODUCTION

Among three biggest ecosystems, known as the Kidney of Earth, wetland has significant ecological functions such as freshwater conservation, poison elimination, carbon storage, water quality purification, flood storage and drought control, climate regulation and remaining biodiversity etc. It plays an important role in remaining ecological balance and prosperous economy. Wetland ecological environment mainly includes water, air, acoustic and soil environment. And water quality best reflects wetland ecological environment condition (Liang et al., 2002). Protecting wetland, and establishing water environment real-time monitoring system is an urgent task.

Compared with sea, rivers and lakes, wetland water environment has its specialties (Jiang, 2007): entire water environment is divided into widely distributed small independent water areas with irregular shapes, different area. Area and depth of water area is easily influenced by season, climate, human interruption and other factors. As the obvious difference, it is necessary to monitor the water environment of each small water area in wetland with multi-points real-time monitoring. In the initial period of wetland water quality monitoring, generally there are a lot of points distributed to fully master the water quality of entire area. After a certain period of monitoring, abundant data are accumulated. To reduce Human resource, material resource and financial resource and other waste, optimized distribution is necessary. Optimized goal of distribution is to search minimum monitoring points and still objectively reflect area environment quality. Currently, there are many optimized methods applied to environment monitoring points, such as Fuzzy Clustering, Matter element and Osculation Value methods. Though with different characteristics, these methods all have complicated programs. Through constructing multi-index monitoring data samples Decision-making Matrix, fuzzy similarity matrix between sample data and their mean values is established. According to the index value variation, the index weights are calculated based on information entropy theory. With the index weight and sample fuzzy similarity matrix, comprehensive fuzzy similarity of each monitoring point is calculated. Finally, according to comprehensive fuzzy similarity, each monitoring point is reasonably clustered, then representative points is selected from each category, so distribution optimization could be realized. Compared with other methods, fuzzy similarity has clear conception, simple calculation, direct image and single conclusion. Practical running proves that after analysis, the results of optimized points are corrects.

2. FUZZY SIMILARITY MONITORING POINTS OPTIMIZATION THEORY

Wetland water quality environment changes with space and time (Xu et al., 2002). In addition, multi-index brings lots of difficulties to monitoring point's optimization. Single index parameter comprehensively reflects water quality. As a result, how to transform multi-index monitoring parameters into single index parameter and then cluster monitoring points is the basic access to optimize monitoring points and the starting point of proposing and applying the concept of fuzzy similarity.

2.1 Sample Decision Making Matrix Establishment

m monitoring points ($\theta_1, \theta_2, \ldots\ldots \theta_m$) n evaluation index ($A_1, A_2 \ldots\ldots A_n$), initial matrix $\mathbf{R_0}$ is constituted:

$$\begin{array}{cc} & \begin{array}{cccc} A_1 & A_2 & \cdots & A_n \end{array} \\ \mathbf{R0} = \begin{array}{c} \theta_1 \\ \theta_2 \\ \vdots \\ \theta_m \end{array} & \begin{bmatrix} c_{11} & c_{12} & \cdots & c_{1n} \\ c_{21} & c_{22} & \cdots & c_{2n} \\ \vdots & \vdots & \vdots & \vdots \\ c_{m1} & c_{m2} & \cdots & c_{mn} \end{bmatrix} = (\mathbf{c_{ij}})_{m\times n} \end{array} \tag{1}$$

C_{ij} is the *i*th monitoring point θ_i's *jth* quantized value of evaluation index, $i \in (1, 2 \ldots \ldots m), j \in (1, 2 \ldots \ldots n)$.

$$r_{ij} = \frac{C_{ij}}{\sqrt{\sum_{i=1}^{m} C_{ij}^2}} \tag{2}$$

Normalize Matrix (1) establish sample decision making matrix $\boldsymbol{R}$:

$$\begin{array}{cc} & \begin{array}{cccc} A_1 & A_2 & \cdots & A_n \end{array} \\ \mathbf{R} = \begin{array}{c} \theta_1 \\ \theta_2 \\ \vdots \\ \theta_m \end{array} & \begin{bmatrix} r_{11} & r_{12} & \cdots & r_{1n} \\ r_{21} & r_{22} & \cdots & r_{2n} \\ \vdots & \vdots & \vdots & \vdots \\ r_{m1} & r_{m2} & \cdots & r_{mn} \end{bmatrix} = (\mathbf{r_{ij}})_{m\times n} \end{array} \tag{3}$$

$\theta_i = (r_{i1}, r_{i2}, \ldots\ldots r_{im})$ $(i = 1, 2 \ldots\ldots m)$ is the sample point to be optimized.

2.2 Fuzzy Similarity Theory

2.2.1 Fuzzy process of measurement value and mean value

Influenced by various factors, there is error between water quality index sensors' measurement value and object real value. Real value is near all the effective values. Supposing measurement value error is random error, measuring distribution is normal distribution decided by mean value and variance. Fuzzy membership function takes Gauss. To satisfy engineering application needs, triangle membership function is adopted here. The center of triangle is sensor's normalized measurement value; the width is four times of measurement data standard various. For sensor i, suppose measuring mean value after L times measurements value of real value A of certain monitoring point is x_i, measuring variance is σ_i and measuring value fuzzy value can be expressed as follows:

$$\tilde{A}_i = (a_{i1}, a_{i2}, a_{i3}) = (x_i - 2\sigma_i, x_i, x_i + 2\sigma_i)$$

Mean value (normalized) fuzzy processing is similar with that of measuring value. Only mean value x_0, mean variance σ_0 are as follows (m is monitoring points number):

$$x_0 = \frac{1}{m}\sum_{i=1}^{n} x_i$$

$$\sigma_0{}^2 = \frac{1}{m-1}\sum_{i=1}^{n}(x_i - x_0)^2 \tag{4}$$

S_o fuzzy value of mean value is:

$$\tilde{A}_0 = (a_{01}, a_{02}, a_{03})\ x_0 - 2\sigma_0, x_0, x_0 + 2\sigma_0)$$

2.2.2 Definition and calculation of fuzzy similarity

Suppose $\tilde{A}_i$ and $\tilde{A}_j$ are two fuzzy values, defining $S = S(\tilde{A}_i, \tilde{A}_j)$, if S satisfies:

(1) $0 \le S \le 1$.

(2) for $\tilde{A}_i = \tilde{A}_j$, S=1.

(3) $S(\tilde{A}_i, \tilde{A}_j) = S(\tilde{A}_j, \tilde{A}_i)$.

(4) If and only if $\tilde{A}_i \cap \tilde{A}_j = \Phi$, $S(\tilde{A}_i, \tilde{A}_j)=0$.

(5) if $\tilde{A}_i \subset \tilde{A}_j \subset \tilde{A}_s$, $S(\tilde{A}_i, \tilde{A}_j) > S(\tilde{A}_i, \tilde{A}_s)$.

Then call S as $\tilde{A}_i$, $\tilde{A}_j$ similarity, that is close extension of $\tilde{A}_i$ and $\tilde{A}_j$.

There are many methods for calculating similarity between fuzzy values (Xiang, 2004). To realize reliability and convenience, define similarity calculation method based on distance measurement:

Suppose $\tilde{A}_i = (a_{i1}, a_{i2}, a_{i3})$ and $\tilde{A}_j = (a_{j1}, a_{j2}, a_{j3})$ are two triangle fuzzy value, then the similarity is:

$$S(\tilde{A}_i, \tilde{A}_j) = \frac{1}{1+d(\tilde{A}_i, \tilde{A}_j)} \tag{5}$$

In the equation:

$$d(\tilde{A}_i, \tilde{A}_j) = \frac{a_{i1} + 4a_{i2} + a_{i3} - a_{j1} - 4a_{j2} - a_{j3}}{6}$$

The bigger $S(\tilde{A}_i, \tilde{A}_j)$ is, the closer $\tilde{A}_i$ and $\tilde{A}_j$. $S(\tilde{A}_i, \tilde{A}_j) = 1$, means $\tilde{A}_i$ and $\tilde{A}_j$ are completely the same, $S(\tilde{A}_i, \tilde{A}_j) = 0$, means $\tilde{A}_i$ and $\tilde{A}_j$ totally different. To convenience describe, define $\tilde{A}_{ij}$ as the normalized fuzzy measurement value of j, index of point i, $\tilde{A}_{0j}$ 为 is fuzzy value of normalized measurement mean value, so $S_{ij} = S(\tilde{A}_{ij}, \tilde{A}_{0j})$ is called as fuzzy similarity of index j of monitoring point i.

3. OPTIMIZED SELECTION OF WETLAND WATER QUALITY MONITORING POINTS BASED ON INFORMATION ENTROPY AND FUZZY SIMILARITY METHOD

Information entropy and fuzzy similarity optimization principle is a level entropy multi-object comprehensive evaluation method. Based on information entropy and fuzzy similarity, basic process of optimizing water quality monitoring points is: construct multi-index parameter sample matrix of monitoring points, normalize the matrix; establish sample data and the fuzzy similarity matrix between ample data and their mean value, with information entropy technology, each index weight is confirmed. According to this, comprehensive similarity of monitoring and mean value points is calculated. And sequencing cluster the monitoring points by similarity, finally realize optimized distribution according to clustering results.

3.1 Index Weight Confirmation Principles and Methods Based Information Entropy

In multi-objects decision making, the bigger the difference of certain index value, the more important it is in comprehensive evaluation. If index values of certain index are all the same, the index is useless in comprehensive evaluation. In information theory, entropy means information quantity got from a group of unknown objects. Information entropy is the measurement of disorder extension of a system (Sun et al., 2000). The bigger the variance extension of a certain index, the smaller the information entropy is, that is the function is much stronger in comprehensive evaluation. So the index weight is bigger; in the contrary, the weight should be smaller. According to the variance of each index, with information entropy, weight of each index can be calculated, specific method as follows:

(1) Calculate output entropy of index j:

$$E_j = -(\ln m)^{-1} \sum_{i=1}^{m} p_{ij} \ln p_{ij} \tag{6}$$

Of which: $p_{ij} = r_{ij} \Big/ \sum_{i=1}^{m} r_{ij}$

If $p_{ij}=0$, then define $p_{ij} ln p_{ij}=0$.

(2) Calculate variety of index j

$$D_j = 1 - E_j \tag{7}$$

(3) Calculate objective weight of index j

$$w_j = D_j / \sum_{j=1}^{n} D_j \tag{8}$$

3.2 Fuzzy Similarity Optimized Selection of Wetland Water Quality Monitoring Points

As shown above, in the foundation of getting the fuzzy similarity of the index j of point i, and calculating relative weight of each index, comprehensive fuzzy similarity of point i can be calculated with the following equation:

$$S_i = \sum_{j=1}^{n} w_j s_{ij} \tag{9}$$

According to above thoughts, the optimized selection of wetland water quality monitoring distribution based on information entropy and fuzzy similarity is as follows:

1) With equations (1), (2), (3), n index evaluation data of m monitoring points are normalized and sample matrix ***R*** is established.

2) With equation (4), (5), the index j of point i fuzzy similarity is calculated. Fuzzy similarity matrix of sample data and their mean value is established.

3) With equation (6), (7), (8), relative weight in index j in clustering data is confirmed.

4) With equation (9), the comprehensive fuzzy similarity of point i is calculated finally.

5) According to the comprehensive fuzzy similarity, points are scientifically clustered. And representative points are selected from each category to realize distribution optimization.

4. EXAMPLE ANALYSIS

With information entropy and fuzzy similarity mentioned above, provided by a certain environment monitoring station in a city's wetland in 2003, data of 7 monitoring points have been optimized. Table 1 proves that, there are 4 pollution factors, respectively as COD, BOD_5, NH_3-N and TP.

Table 1. The water quality monitoring data

Moni-points	COD	BOD_5	NH_3-N	TP
1	14	3.1	1.56	0.138
2	24	2.8	2.06	0.134
3	18	4.2	0.31	0.037
4	33	2.9	0.99	0.099
5	34	5.9	1.85	0.385
6	19	4.8	2.60	0.486
7	17	5.1	1.08	0.265

According to methods mentioned above, we first establish sample matrix $(C_{ij})_{7\times4}$ of 7 monitoring points data, then normalize the matrix to gain dimensionless matrix $(r_{ij})_{7\times4}$. Applying equation (4) and (5), fuzzy similarity S_{ij} is calculated, shown as table 2.

Table 2. The similarity degree between the monitoring points and the standard-point: Ci-m

S_{ij}	1	2	3	4
1	0.3799	0.2481	0.4713	0.1889
2	0.2849	0.3721	0.0709	0.0522
3	0.5223	0.2569	0.2265	0.1396
4	0.5382	0.5228	0.4233	0.5429
5	0.3007	0.4253	0.5949	0.6853
6	0.2691	0.4519	0.2471	0.3737
7	0.3595	0.3645	0.3415	0.3111

According to (6), (7), (8) relative weight of index j in data clustering is confirmed, shown as table 3.

Table 3. The relative weight of index j

Index	1	2	3	4
w_j	0.188	0.167	0.297	0.348

From equation (9) fuzzy similarity of point i is finally calculated, shown as table 4.

Table 4. The similarity degree between the monitoring points and the average values

Points	1	2	3	4	5	6	7
Si	1.09	1.03	1.24	1.10	0.86	0.85	1.02

Seen from table 4: we can cluster points 1 and 4 as the same category; 2 and 7, 5 and 6. According to clustering results, best monitoring points can be selected: points 3, 1, 7 and 5, of which point 5 is the representative point of most serious pollution, and 7 secondary, 1 common, and point 3 is the list.

5. CONCLUSIONS

Monitoring points optimization based on information entropy and fuzzy similarity is a multi-objects decision-making method. It can cluster and sort. Applying level analysis, entropy technology and fuzzy similarity principles, optimized decision-making mode is established, which effectively provides scientific decision-making basis for setting wetland water quality monitoring points.

Practical examples prove that applying this method to optimize distribution of wetland water quality monitoring points, optimized results are acceptable. Information entropy and fuzzy similarity have clear calculation significance and concept, flexible and convenient, which can be extended to optimized distribution in other environments. Meanwhile, seen from the concept of fuzzy similarity concept, it fully mines the abundant information in the sample data and could be called as a good data processing method.

REFERENCES

Jiang Peng, Survey on key technology of WSN-based wetland water quality remote real time monitoring system [J]. Chinese journal of sensors and actuators No. 1, 2007, pp. 83-86.

Liang weizhen, ye jinglun. Water quality evaluation based on optimal fuzzy clustering [J]. The Administration and Technique of Environmental Monitoring No. 6, 2002, pp. 6-7.

Sun Shimin, Shi Haixing, Evaluation of Milking Machine Based on Entropy Technology and Idea 1 Point Principle [J]. Transactions of the Chinese Society for Agricultural Machinery, No. 5, 2007, pp. 82-87.

XiangXinjian, A method to sensor data fusion based on fuzzy and statistics integration Chinese journal of sensors and actuators, No. 2, 2004, pp. 197-199.

Xu lizhong, Zhang jiangshang. Optimization of atmospheric environmental monitoring sites with modified intimate value method [J]. Environmental Engineering No. 18, 2002, pp. 50-53.

DESIGN OF PORTABLE INSTRUMENT FOR MEASURING AGRICULTURE FIELD SIZE BASED ON GPS

Xinjian Xiang, Chunting Yang
College of Automation & Electrical Engineering, Zhejiang University of Science and Technology, Hangzhou, 310023, P. R. China, Tel: +86-571-85070268, Email: hzxxj@sina.com

Abstract: Global Positioning System (GPS) plays an important role in precision agriculture. Applying positioning function of GPS, this paper researches on portable instrument for measuring field size. Instrument consists of low-cost GPS module, data acquisition board in microcontroller, large capacity EEPROM storage, RS232C communication interface, keyboard and LCD display. GPS position information acquisition technology, method for calculating polygonal fields geometric size, software and hardware design principles of instrument, GPS measuring error analysis and accuracy control of instrument are provided. Experiment results prove that measuring instrument can fast measure geometric size of fields in any shape, relative measuring error is less than 2.5%, and the bigger the size of field is, the higher the measuring accuracy is.

Keywords: GPS, field geometric size, measuring instrument, precision agriculture

1. INTRODUCTION

Field geometric size is one of the most basic data. Field geometric size such as distance, perimeter and area directly decides the input amount of seeds, fertilizer, pesticides and other production materials. It is also the main basis for calculating working time and charging when agriculture machine are doing field working. Traditionally, geometric size of fields are most measured by tape, simple and practical. However, it is only suitable for small

Xiang, X. and Yang, C., 2008, in IFIP International Federation for Information Processing, Volume 259; Computer and Computing Technologies in Agriculture, Vol. 2; Daoliang Li; (Boston: Springer), pp. 993–1000.

size field, for which with large are, it cost lots of time and effort, and easily causes mistakes. Also, it is difficult to measure the perimeter and area of irregular fields. As the core technology of supporting precision agriculture (Zhang et al., 2002), GPS global satellites positioning system can provide real-time navigation information such as longitude, latitude and elevation etc. With GPS positioning function, each point's coordinate in field can be calculated, which makes it possible to calculate fields distance, perimeter and area with mathematic methods. Adopting low-cost GPS module, this paper develops portable instrument for measuring field geometric size, realizing outdoor fast GPS positioning measurement for regular or irregular fields, green land, forest, water area and tidal flat in any shape. For seeds, fertilizer and pesticides input, total amount can be calculated by inputting single operation amount. For machine operation, operation single price input, and instrument directly displays the charge. Large capacity EEPROM storage, RS232C communication interface, graphic LCD can input measuring data into computer and display outline of measured field and manage field data base. Instrument has functions of storing, displaying and printing outline graphic. Additionally there are two big functions: for conveniently using the instrument, there is additional manual measuring for geometric size (for relatively regular district or inconvenient passing area, such as building, bridge, big tree and other obstacles, manual measurement can be taken. Only apex data are needed, not necessarily to walk around); and data filling up (when measuring area, if end and starting points not meet, instrument will automatically filling data with the straight line between them and calculate the area).

2. GPS POSITIONING INFORMATION ACQUISITION

By GPS receiver, GPS positioning information is acquired. Commonly used GPS receiver can be divided into measuring, navigation, handwriting SMS, OEM board and many kinds. There is big difference in functions and prices among them. For example, differential GPS receiver Ag GPS132 produced by American company Trimble has positioning accuracy with 10-2 meter level. It is widely used in precision agriculture. However, because of its high price, large volume and high power consumption, it is not suitable for portable instrument. This research takes GPS 25LP OEM board made by GARMIN Company to position. GARMIN Company is global GPS OEM board supplier. 25LP GPS OEM board is single point mode product, whose main function is: parallel 12 channels receiving; recapture time<2s, hot starting time is 15s, cold starting is 45s, automatically searching time is 90s; positioning accuracy is 15m, in differential (DGPS) situation, positioning accuracy <5m; external antenna is provide to help receive information; volume is small and power consumption low; 5V power source supplier.

GPS OEM board's input and output sentence accords serial communication protocol. Default communication baud rate is 4800bps, data structure is 8 data bits, 1 initial bit and 1 stop bit, no even-odd check bit, initialization of output data format is NMEA-0183.NMEA-0183 is the standard format for marine electronic device designed by American Ocean Electronic Association. As the format is ASCII code string, direct and easily processed, it can be directly judged and separated in many advanced languages to acquire data users need. Through input sentences, users initialize GPS OEM board, set data format, communication baud rate, required output categories; output sentence is to output various data GPS information to users.

In fields geometric size positioning measurement, $ GPGGA sentence format recommended is taken. $ GPGGA sentence contain commonly used positioning information as longitude, latitude, speed, date, time, receiver condition. This sentence consist of 75 characters, each data segment has a fixed position. For example, position of longitude data is bit 19 to bit 27 in the string, position of latitude data is bit 31 to 40. Reading relative district data, positioning information is acquired. Then longitude and latitude strings are transformed into numbers, further the distance, perimeter and area are calculated.

GPGGA sentence is the standard data format output by GPS OEM of GARMIN, starting with character "$", ending with carriage return character, data segments and checksum in the middle and comma separating data segments. For instance, if real-time receiving sentence "$GPGGA": $GPGGA,114641,3002.3232,N,12206.1157,E,1,03,12.9,53.2,M,11.6,M, *4A. This GPS positioning data information sentence means: UTC Time is 11h46m41s, locates at north longitude 30° 2.3232', east latitude 122° 6.1157', common GPS positioning method, receiving 3 satellites, horizontal accuracy is 12.9m, the height of antenna from the sea level is 53.2m, location is 11.6m higher than horizon, checksum is 4AH. As GPS OEM sends more than1 sentence, so to completely receive this "$GPGGA" sentence, it is necessary to judge the head of the sentence, that is 7 characters of "$GPGGA". When these 7 characters are completely received, the data can be ensured useful.

3. GEOMETRIC SIZE CALCULATION METHOD

Walking around the field with GPS receiver, a group of positioning data consisting of many points can be measured. According to the order of time they are: (X1, Y1), (X2, Y2)... (Xn ,Yn), of which X is longitude, Y latitude. Connecting these points, then polygon with n apexes is constructed, as shown in Figure 1.

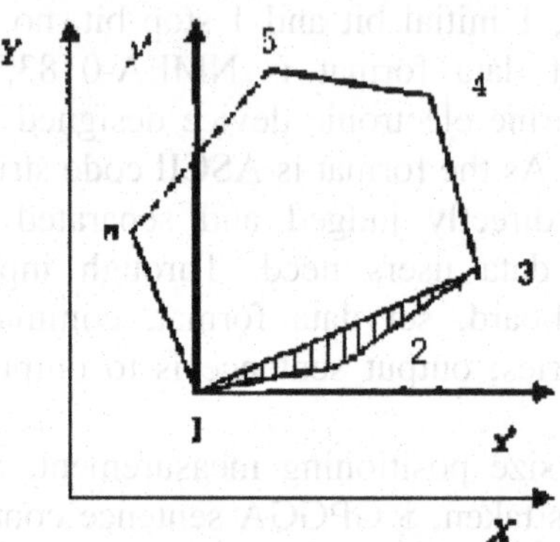

Figure 1. Area calculation for polygon field

As earth is ellipsoid, to accurately calculate distance or area, generally they are transformed into plane coordinate by projecting. In our country, for big size map, Gauss-Krüger projecting is used (Zhu et al., 2000). However, projecting calculation is very complicated and hardly being realized in microcontroller. To simplify calculation, we regard earth as pellet. As the field distance is short, generally less than 1km, and area small, so the error brought by simplification can be neglected. Taking earth radius as 6371116m, then longitude and latitude can be transformed into plane coordinates:

$$x = 2\pi RX/360, y = 2\pi RY \cos Y/360$$

In above equation: R is the earth radius, X longitude/m, Y latitude/m. As x,y values are big, if used to calculate area, big error will be caused. So the first point is set as the original point of coordinate, new coordinate is made by relative displacement, shown as Figure 1:

$$x_i' = x_i - x_1$$

$$y_i' = y_i - y_1$$

Then distance between two adjacent points is:

$$L_i = \sqrt{(x'_{i+1} - x'_i)^2 + (y'_{i+1} - y'_i)^2}$$

Perimeter of polygon is:

$$G = \sum_{i=1}^{n-1} L_i$$

For polygons, triangle method is used to calculate area. When walking around the field once with GPS receiver, for example the triangle consists of apexes 1, 2, 3 (the shadow in diagram 1), the area is:

$$S_1 = \sqrt{p(p-a)(p-b)(p-c)}$$

In the above equation:

$$p = (a+b+c)/2,\ a = \sqrt{x_2'^2 + y_2'^2},\ b = \sqrt{x_3'^2 + y_3'^2},\ c = L_2$$

Then the area of field made of n points is:

$$S = \sum_{i=1}^{n-2} S_i$$

4. MEASURING INSTRUMENT DESIGN

GPS positioning information collecting and displaying system consists of microcontroller 89C51, LCD displaying module MGLS12864 and GPS module GARMIN GPS25LP shown as Figure 2. It can display Beijing Time, longitude & latitude and ground horizontal height and other real time information. As the main controller, microcontroller controls GPS OEM board to receive, read and transfer data. GPS OEM board receives the signals sent from satellites; output them in NMEA 0183 format. After RS223Cand COMS/TTL electrical level transformation, output serial data are sent to the serial port of microcontroller. After processing, information to be displayed is selected by keyboard, finally sent to LCD MGLS12864, Liquid Crystal Display will update the content to be displayed in time.

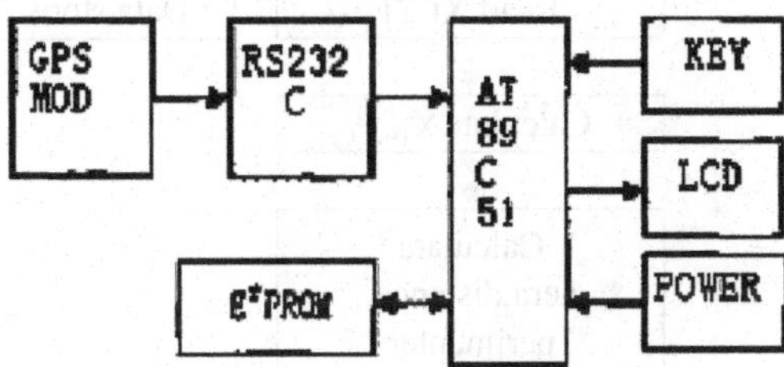

Figure 2. Block diagram of instrument

GPS module is applied to receive GPS satellite signals, output positioning information in NMEA format. System takes 25LP GPS OEM board made by

GARMIN to realize positioning. This module integrates GPS receiver, interior antenna, electrify restoration circuit, signal interface circuits etc, 3.0-5.5V wide range supplying electricity, 12 channel 5m positioning accuracy. Microcontroller is the core of instrument. System adopts AT89C51 microcontroller as system processor, with high cost-performance, realizing data receiving, processing, displaying, storing and printing etc.

System working principles as follows: GPS module receives GPS satellite signals, outputs positioning information through RS-232 serial ports; communication interface transforms RS-232 serial signals into COMS/TTL electrical level and input them into the serial ports of microcontroller. Microcontroller processes positioning information under the control of software, receives orders from keyboard and display the geometric size calculation results on LCD or store them into serial E2PROM storage.

C51 language is used to write microcontroller software, and in the environment of Keil C51 compiling and debugging is carried out. Software mainly includes data receiving, calculating, keyboard scanning, printing module and etc. Microcontroller receives positioning information by serial port interruption method. Serial parameters are set: 4800bps, 8 data bits, 1 stop bit, no check bit. Serial ports interruption is core software, flowchart is shown as Fig. 3.

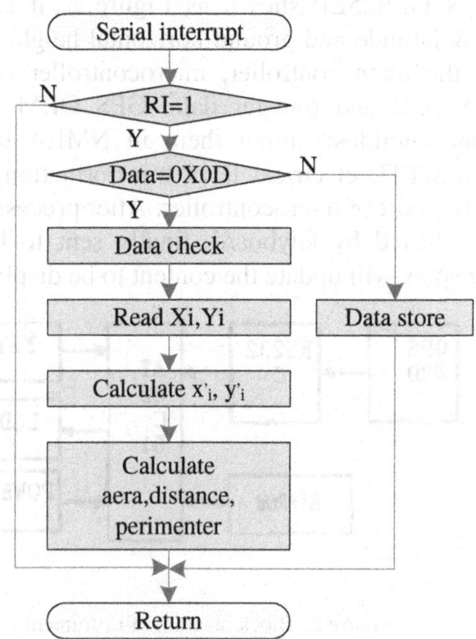

Figure 3. Flowchart of serial interrupt

5. EXPERIMENT AND ERROR ANALYSIS

Experiment is carried out in the standard 400m playground in school. According to first running track parameters, we can calculate that the area is 10695m2. Walking around the first track when measuring, walking with normal speed in straightway and lower speed in curve way. Table 1 provides statistical data of 10 times measurements. LCD displays 3 standard playground outlines in different locations after 3 GPS positioning measurements. Though there is obvious deviation, they all keep good standard shape of playground, with similar area. It proves that low-cost GPS positioning has drift error relating to time, the accuracy is not good enough. However, in a short period, drift is not big, so the acquired track is basically similar, and the results are basically the same.

Table1. Statistics of experimental data

Standard playground	
Area/m2	10695
Times	10
Mean value/m2	10567
Mean relative error (%)	1.2

Seen from outline, when moving fast, positioning points distribute sparsely, some key points are possibly missed. These missed points have big influence for calculating area. So in the curve parts, speed should be lowered down. With high dense points to measure, the key points can be ensured not missed. Measuring straight line has no big influence on area, so the speed can be high. Measuring statistics in Table 1 proves that relative error of measuring playground area is 1.2%, which means GPS measuring instrument has good accuracy. Playground includes semicircle area prove that the instrument can well measure non-rectangle field area.

Hand-held GPS positioning error is mainly of system error and coarse error. Coarse error mainly comes from observation conditions, such as poor satellite visibility conditions; and system error mainly comes from GPS positioning satellites and atmospheric delay of positioning signal. Due to random error has characteristics as measuring times approaches infinity and expect value approaches 0, so with mean value revised method to recognize and modify observation error, 2.5m hang-held GPS positioning accuracy and area measuring error less than 2% is eventually acquired.

6. CONCLUSION

In precision agriculture management, it is very necessary to fast real-time measure the geometric size of the field. Traditional measuring technology can hardly fast measure geometric size of irregular field and satisfy the requirements of precision agriculture. This design complete software and hardware development of portable instrument for measuring field geometric size. After practical testing and debugging, system runs normally. The development of this instrument provides a new device for fast measuring geometric size of fields. The main characteristic of the instrument is fast measuring geometric size of fields in any shape, with high measuring accuracy, providing accurate basis for managing production and agriculture machinery operation in precision agriculture.

REFERENCES

Zhang Nai-gian, Wang Mao-hua, Wang Ning. Precision agriculture-a worldwide overview [J] Computer sand Electronics in Agriculture, No. 3, 2002, pp. 113-132.

Zhu Yun-giang, Gong Hui-Ii, XU Hui-ping. Map projection transformation of GIS [J]. Journal of Capital Normal University, No. 22(3), 2001, pp. 88-94.

A SOLUTION OF RURAL INFORMATION NETWORK ACCESSING

Lu Yang*, Jun Xiao

College of Information and Electrical Engineering, China Agricultural University, Beijing, China, 100083

* *Corresponding author, Address: P.O. Box 142, China Agricultural University, 17 Tsinghua East Road, Beijing, 100083, P. R. China, Tel: +86-10-62342590, Email: yanglumail@263.net*

Abstract: Along with the development and permeation of Internet, it has become an important commercial platform to spur economic development. Aimed at practical problems such as laggard network infrastructure, low economic supportability and inadequate literacy in country area, China has brought up rural information utilities like agricultural information machine, which can provide country families with basic agricultural information and meet their demand because of its rather cheap hardware cost and simple operation. However, the Internet resource available for agricultural information machine is limited, and it will fail to satisfy the country users as their need for Internet information keeps enhancing. This paper is to present an appropriate scheme of information network accessing to solve the deficiency of agricultural information utilities, and a pilot study on the rural information prolongation with a simulation experiment designed to validate.

Keywords: agricultural information machine, set-top box, information network

1. INTRODUCTION

Internet has become more and more important. For the development of agricultural in particular, the Internet can provide information promptly, and has the potential to promote China's agricultural development in an obvious way. However, some practical factors are restraining agriculture employees from obtaining abundant information on the Internet. The living condition in countryside keeps improving, though, practical difficulties such as laggard communication infrastructure, farmers' inadequate literacy, low family

Yang, L. and Xiao, J., 2008, in IFIP International Federation for Information Processing, Volume 259; Computer and Computing Technologies in Agriculture, Vol. 2; Daoliang Li; (Boston: Springer), pp. 1001–1010.

income and economic supportability are ubiquitous in countryside. Therefore a network accessing solution suitable for nowadays rural actuality is a crying need.

2. INTELLIGENT AGRICULTURAL INFORMATION MACHINE

Aimed at practical problems such as laggard network infrastructure and inadequate literacy in country area, China has brought up several rural information utilities, which have already been put into use in some districts.

The Ministry of Science and Technology of the People's Republic of China has cooperated with Beijing Guangcai Agricultural Information Network Technology Co., Ltd. to develop the System of China's Agricultural Information Accessing into Countryside, which has been successively demonstrated as part of the Technology Information Accessible to Every Country Project in eleven provinces and fourteen districts. It has created good economic benefit for local farmers, helping them to head towards a better-off life (He Jiankun et al., 2003).

Heilongjiang GM Telecom Paging Co., developed GM agricultural information machine to change the situation that some rural districts in the province lack of technology information and difficulties arisen in breeding and planting cannot be solved in time.

In June 2007, 17 sets of agricultural information machine arrived at 17 administrative villages in Machangying town of Pinggu district, Beijing, opening the door of convenience for all the farmers to detailed information on agricultural information, prices of agricultural products, medical treatment and health care, whether forecast, social news and the like.

Among all the agricultural information utilities, the intelligent agricultural information machine developed by Beijing Zhongxunxiongfeng Technology Co., Ltd. of Chinese Academy of Science is relatively developed and representative. Here is the brief introduction below.

2.1 Primary applications

Agricultural information machine is connected to a certain server then presents information from the server on a TV set, and users operate with a remote controller to obtain information.

For the time being, agricultural information machine has several main functions as below:

1. Information downloading. Servers provide various kinds of information to rural families such as information on market, supply & demand and the like. Users download through agricultural information machine the

information renewed on servers everyday to the local storage so that they can take a browse at any time.

2. Information release. Users can also upload their information to servers. Released information is mainly about supply and demand. After servers' simple process, the information will be released on the servers for other users to download.

3. Communication. Service providers will give each user an account similar to an E-mail box, each corresponding to some information storage space. Users can send information to each other through accounts to communicate.

4. Surfing. The information machine makes simple WAP site accessing available.

2.2 Strongpoint

Agricultural information machine takes China's urban actuality into consideration, thus becomes acceptable to rural users. Its main strongpoint is:

1. Low hardware cost. Of all the system, the only thing that rural families need to buy is agricultural information machine since TV set is made use of as the vision device. Compared with PC, the cost has been lowered greatly to be affordable to rural families.

2. Convenient operation. Using remote controller as the controlling device saves a lot of time of study. With simple introduction of the manual, users can easily operate following the direction of interfaces or icons.

3. Easy maintenance. Agricultural information machine is designed with a simple structure to ensure a simple installation. All users need to do is to connect wires and setup user account, password and gateway under the direction of the manual.

2.3 Shortcoming

Agricultural information machine has its own shortcomings, among which the most extrusive one is the shortage of information. The current information machine only has access to appointed servers and WAP sites. Servers provide rather fixed services and only WAP sites are up to users' choice. Since WAP sites are primarily for mobile users, its information content is quite small compared with abundant information on Web sites.

The development of countryside will lead to rural families' unsatisfactory with information supplied by the information machine, and it will become a necessity to make the widest range of Web information accessible to rural families.

3. A RURAL INFORMATION NETWORK ACCESSING SOLUTION DESIGN

3.1 Problems

In practical use, problems below become obstacles to Web site visiting.

1. Bandwidth of rural network

The infrastructure of Chinese rural network is relatively laggard, and the main way to visit Internet is depended on dial-up technology through low bandwidth. Normally, the size of a Web page is above 100kb, and the demand for bandwidth becomes larger with multimedia information like pictures and sound added.

2. Information presentation and operation

First of all, most Web sites are designed for computer users with the resolution of 1024*768 as the displaying area, which is not possible on a TV set. If the visual effect of normal browser is directly presented on a TV set, only a small part of the page can be seen. Secondly, normal browser presents in a way suitable for mouse and keyboard to operate, but rural users mainly use remote controllers and it's hard to operate by a few buttons.

3.2 Solution

1. Web information selection

To solve the problem of low bandwidth in countryside, it's not practical to widen bandwidth. Under the condition of low bandwidth, the only solution is to work on information quantity transmitted, that is, to cut down the information quantity with the integrality of information well reserved.

Let's start with an analysis of the structure of a typical Web site. Take Sina.com.cn as an example, its Web site structure is demonstrated in Fig. 1. The home page is divided into two parts: one contains navigation of all the channels of the Web site such as finance and entertainment, the other contains important news of every channel. The entertainment channel constitutes two parts as well: classification of its blocks and important news of them. All the news in blocks can be seen with a click on the block.

News included in the upper level is in the corresponding block as well, and a great amount of redundancy of information accumulates in this way. Thus, the integrality of Web site information can be ensured by reserving news texts in blocks.

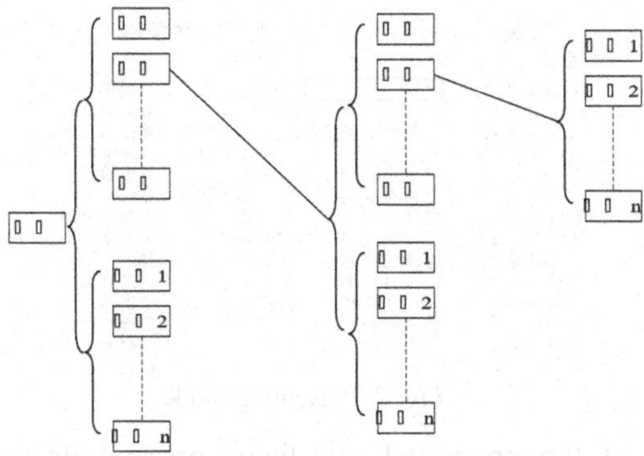

Fig. 1. The structure of web site

In the next place, a lot of multimedia information and advertisement are included in Web pages, which is unnecessary for rural users.

Therefore we can go through an appropriate selection of web information. For information on the home page of a Web site, we can just keep down navigations of its channels. As for the home page of a channel, only classifications of its blocks are going to stay. When it comes to blocks, we are going to pick up all the links to the texts. If users click to read news, only the text part of it is to be presented, eliminating other unnecessary information. Sina again, Table 1 is a comparison between before and after selection.

Table 1. Comparison result

Page arrangement	Before selection	After selection
Home page	200kb±	<10kb
Home page of channels	200kb±	<10kb
Blocks of channels	100kb±	20kb±
News text	100kb±	<10kb

After selection process, the information quantity has been cut down by large, lowering enormously the demand for network bandwidth. Meanwhile, the integrality of information on Web sites is preserved to satisfy rural users.

2. A design of suitable presenting mode

The presenting mode of a common browser is not suitable for a TV set, neither for operating with a remote controller. So it's important to design a suitable presenting mode that applies to TV's low resolution and easy to operate with a remote controller.

User's operation is an interaction between people and machine, and a good interaction is the key of the product's success. In accordance with principles of user-oriented and consistency (Huo Youren et al., 2006), the design of presenting mode is as Fig. 2 shows, which is easy to learn and cognize and can be accepted and mastered very soon by users.

发改委称将提高天然气价格与国际接轨

主页/财经/国内

美国成品油...	发改委称将...	上海房价真...	中国第一天...
个税自行申...	万亿美元储...	280万存款?..	石油业兼职...
三大电力巨...	盖茨电力之...	近500亿资?..	建设银行原...
建设部官员...	新黄埔置业...	石油业兼职...	2006胡润百...
身价41亿 ...	双色球5000...	2006胡润百...	美国成品油...
周正毅隐姓...	胞弟开口谈...	登榜消息刺...	7.8940 人...
胡润百富榜...	观澜湖2006...	连跌四日 ...	发改委称将...
上海房价真...	2006胡润百...	发改委定调...	美国成品油...
中国房地产...	喜来健赞助...	发改委: 天...	280万存款...
怀念计划经...	个税自行申...	请发改委告...	新浪财经调...
上海房价真...	油价电价水...	裘祖贻三封...	中国第一天...
樊纲: 人民...	栋梁新材造...	外资银行全...	第六届中国...
古铁雷斯第...	国家发改委...	中国外汇储...	个税自行申...

第1页/共4页

Fig. 2. Presenting mode

The first line of the presented interface corresponds to the part below framed in black, representing the link user choose. It is in larger font to emphasize the importance and make operation easier. The second line is used to indicate where users are, avoiding users getting lost in the Web. The following part is most important in the interface, which concludes the information of the Web site that users need. To enhance the efficiency of user's operation, links are arranged in matrix, and in this way the content of one page is enlarged. The last line represents the quantity of information in the page to show to users.

By this means, users can browse and choose information through a simple click on the buttons of the remote controller. On the whole, the presenting mode is simple, clear, focused and pleasant to use, providing a good solution to the problems caused by the difference of presenting device and operating tools.

3.3 Scheme design

Combined with the strongpoint of agricultural information machine, the design provides a solution of network accessing; actualizing country family visiting Web sites, and the structure of it is showed in Fig. 3. The system constitutes four parts: user, TV set, STB (set-top box) and information server.

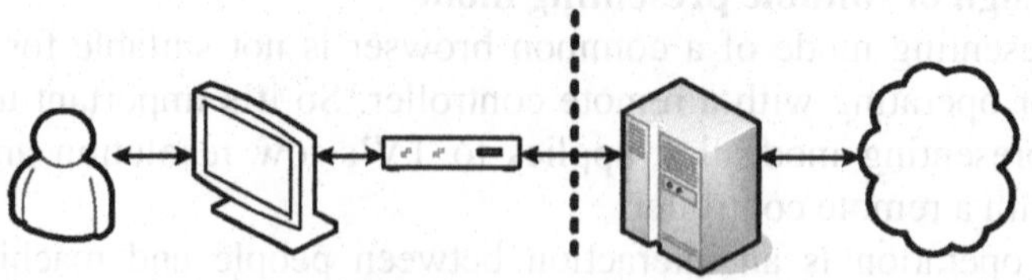

Fig. 3. The structure of system solution

3.3.1 Functions of STB

In a country family there is a STB (Lu Feng, 2006) which is connected to the Internet through low bandwidth Dial-up. Users get a link to information server by designating the gateway address to STB. The interaction between users and information servers is in the charge of STB.

First of all, STB receives instructions from users. If it is asking information server to visit Web site, STB will transmit the request to the information server. If it is asking STB to process cached information, STB will follow the instruction and submit corresponding cached information, for example, the choice of links.

In the next place, STB is responsible for presenting information to users. It shows the information returned from the information server on a TV in the presenting mode designed in section 3.2 for users to browse.

3.3.2 Functions of information server

Information server manages STB in its range. As the processing ability of servers varied, a server could be used for one county, or for one town. Information servers can be connected to Internet by fiber or other means to have a wide Bandwidth.

In the architecture of the Internet, an information server plays the role of an explorer, obtaining information from Web servers through HTTP protocol. Meanwhile, it acts as a server when communicating with STB, through HTTP protocol as well.

Firstly, an information server functionates as an explorer to visit Web sites. It receives the request of visiting from STB in its range, then transits the request to HTTP request to send to relevant Web sites, and at last obtains the information returned from Web sites.

Secondly, an information server has a strong ability of information processing. For information already in local-storage, it can reply the request of STB directly. At the same time, it manages the redundancy of information, and sets up a sententious information image for users. In process of information server's design and implementation, we need focus on resolutions of problems like information classification and management, redundancy elimination, maintenance of information validity and information search. Solutions of those problems will be described in other papers.

4. ILLUSTRATION OF EXPERIMENT RESULT

4.1 Simulation scheme

Based on the above scheme, we have designed a simulative experiment, the structure of it showed in the Fig. 4. The structure contains two computers, computer 1 simulates information server; computer 2 simulates STB and TV, with monitor screen representing the screen of TV and buttons on keyboard representing those on remote controller.

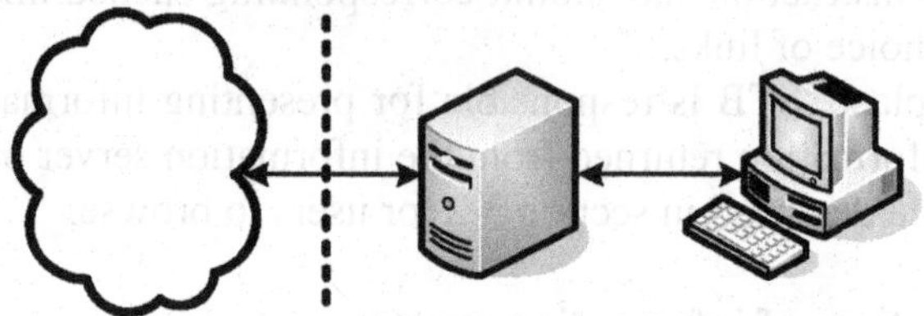

Fig. 4. The structure of simulative experiment

In the simulation, computer 2 connects with computer 1, sending the visiting request to computer 1, and then presents the information returned from computer 1 in the designed mode demonstrated in section 3.2.

Computer 1 connects to the Internet, and after receiving the request from computer 2, it transmits it to HTTP request to send to corresponding Web sites. Having obtained information from a Web site, computer 1 processes the information in the way designed in section 3.2, and sends the processed information back to computer 2 as the final step.

The overall simulation contains the important parts of the solution and has implemented the key functions, thus we can say it is conform to the requirement of an actual simulation.

4.2 Illustration of result

The result of experiment has achieved the purpose of design, the demonstration of result showed below. Let's visit www.163.com for illustration.

User sends the request to server, which gets the information on homepage, after selection, what shows up is like in Fig. 5, user selects sports channel from the news channel of the Web site.

体育

主页

VIP邮箱	新闻	体育	NBA
娱乐	明星	财经	科技
汽车	手机	女人	游戏
军事	探索	图片	博客
奥运	英超	电影	商业
证券	基金	车型	数码
时尚	情爱	招聘	历史
播吧	杂志	论坛	NULL
NULL	NULL	NULL	NULL
NULL	NULL	NULL	NULL
NULL	NULL	NULL	NULL
NULL	NULL	NULL	NULL
NULL	NULL	NULL	NULL

第1页/共1页

Fig. 5. Experiment illustration 1

User clicks to enter the chosen channel, what presents itself after selection of the server is showed in Fig. 6 classifications of blocks in the sports channel.

CBA

主页/体育

滚动	评论	专题	08奥运
相册	旅游	拍卖	姚明
CBA	男篮	国际	意甲
海外	彩票	西甲	德甲
欧冠	比分	国内	中超
国足	中甲	女足	综合
网球	F1	刘翔	乒羽
NULL	NULL	NULL	NULL
NULL	NULL	NULL	NULL
NULL	NULL	NULL	NULL
NULL	NULL	NULL	NULL
NULL	NULL	NULL	NULL
NULL	NULL	NULL	NULL

第1页/共1页

Fig. 6. Experiment illustration 2

User chooses a block, and then clicks to enter. All the news links that user wants are presented as Fig. 7 shows.

姚明：我愿牺牲一切只为北京奥运会奖牌

主页/体育/CBA

揭秘巴特尔...	意外负伤中...	北京做好巴...	成功卫冕全...
广东青年队...	江苏南钢俱...	王磊正式加...	悉尼国王大...
卫冕全国青...	广东青年队...	朱芳雨：球...	宏远推出易...
走进杜锋疗...	朱芳雨肯定...	陈江华终于...	父亲节男篮...
代代长盛生...	王磊正式加...	优先保证国...	男篮励志教...
异乡生活失...	大郅篮球基...	亚篮网介绍...	江苏南钢俱...
南京晨报：...	父亲节男篮...	江苏南钢洋...	夏季联赛并...
难忘阿联精...	阿联赛后拒...	易建联两双...	意外负伤中...
男篮比赛孙...	大郅阿联难...	姚明：我愿...	姚明：为了...
承诺回国家...	错过训练营...	WCBA全明星...	WCBA球员家...
陈楠最后一...	女篮三杰WN...	2007年WCBA...	WCBA明星赛...
WCBA全明星...	隋菲菲苗立...	陈楠35分徒...	王磊正式加...
悉尼国王大...	东莞今战悉...	东莞新世纪...	国王看上东...

第1页/共3页

Fig. 7. Experiment illustration 3

Fig. 8 shows what happens after user clicks to read a chosen piece of news.

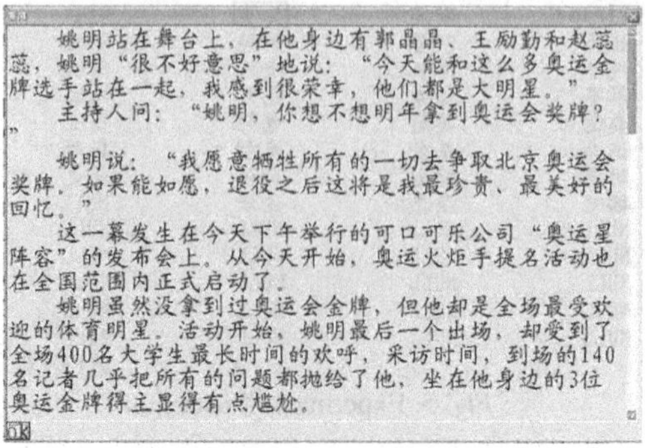

Fig. 8. Experiment illustration 4

User finishes with text reading thanks to buttons of up roll and down roll. By far, a successful browse of web information has been completed.

5. CONCLUSION

A key issue of rural economic development is how to advance informatization in countryside, which has a serious problem that information provided by agricultural information machine is too limited to meet the demand of country's families. In order to extend county's information network and solve the shortage of information provided by agricultural information machine, this paper analyses the status quo in countryside and designs an applicable solution for information network accessing validated by experiments. We hope that it will become a reference for agricultural informatization. The design and implementation of information server is very important for this solution, and we will discuss key techniques in later paper.

REFERENCES

He Jiankun, Ji Gang 2003, Record of Technology Information Comes to Villages Project Helping towards Better-off. http://www.stdaily.com/gb/misc/2003-11/14/content_171007.htm.

Huo Youren, Xie Zhibin 2006, Exploring Designing Method for Human-machine Interface of Information Products. Journal of Ningbo University (science and technology edition), 1:107-109 (in Chinese).

Lu Feng 2006, Functions and Development of IP Set-top Box. Computer Knowledge and Technology, 9:21-24 (in Chinese).

DECISION SUPPORT SYSTEM OF VARIABLE RATE IRRIGATION BASED ON MATHEMATICAL MODEL AND GIS

Jianjun Zhou [1], Gang Liu [1*], Su Li [2], Xiu Wang [3], Man Zhang [1]
[1] *Key Laboratory of Modern Precision Agriculture System Integration Research, China Agricultural University, Beijing, 100083*
[2] *Computer College, Beijing Technology and Business University, Beijing, 100037*
[3] *National Engineering Research Center for Information Technology in Agriculture, Beijing, 100089, China*
[*] *Corresponding author, Address: P.O. Box 125, China Agricultural University, 17 Tsinghua East Road, Beijing, 100083 P. R. China, Tel: +86-10-62736741, Fax: +86-10-82377326, Email: pac@cau.edu.cn, zhoujianjun@cau.edu.cn*

Abstract: Owing to the difference of soil water in the farm, a system of variable rate irrigation was developed with Visual C++ and MapObjects that can save water as well as can improve economy benefit. Soil water content can be forecasted by using the forecast model of soil water, according to the real soil water content of different sampling sites in this system. Irrigation or not can be decided by comparing the current soil content with the light drought and serious drought index of the crop. The water amount of irrigation can be decided by the analysis of economy benefit according to the model of water consumption-yield, thus the irrigation prescription map can be created. This system can also query the information of fields, manage and analysis the data of soil. The test was performed in national precision agriculture demo farm.

Keywords: variable rate irrigation; geographic information system; model; prescription

1. INTRODUCTION

China is one of the countries that are short of water resource. Farmland irrigation consumes most of water resource, so how to use water reasonably is an insistent demand in agriculture production. (Wang haijiang et al., 2001)

Zhou, J., Liu, G., Li, S., Wang, X. and Zhang, M., 2008, in IFIP International Federation for Information Processing, Volume 259; Computer and Computing Technologies in Agriculture, Vol. 2; Daoliang Li; (Boston: Springer), pp. 1011–1019.

Spray irrigation is a kind of advanced irrigation technology for saving water and has been used in many places. Generally spray irrigation can save between 30% and 50% of water. Parallel moving machine of spray irrigation can move in the fields independently, and it can save water and work force (King B A et al., 1999). Currently Geographic information System (GIS), Global Positioning System (GPS), Expert System (ES) have been quickly developed, which make variable rate irrigation possible. Spray irrigation machine can irrigate automatically according to prescription map. The decision support system was developed for variable rate irrigation, which can irrigate according to real water content in every manipulation cell and the forecast model of soil water. Irrigation or not can be decided by comparing with the index of water requirements. The amount of irrigation can be determined by the analysis of economy benefit. Thus water can be saved as well as better benefit can be obtained with less water consuming. This system can provide the prescription map for irrigation.

The first aim is to build a model for variable rate irrigation. The second is making a map of irrigation prescription. The third is to develop a decision support system for variable rate irrigation of winter wheat and implement of variable rate irrigation.

2. SYSTEM DESIGN

2.1 Software environment

Visual C++, MapObjects and MS SQL Sever2000 were adopted in this system. Visual C++ has many characters as follows. Firstly, the character of object oriented programming make the system can support the modularized design, according to the inner character and function. Secondly, VC++ has high speed of implementing and uses less memory. MapObjects OCX is the earliest component offered by ESRI Company, which is the biggest software supplier of GIS. It provides the common function of GIS by the least interface, even the dynamic characteristic of GPS. MapObjects has the reasonable construction and can be understood easily (Zhu yan et al., 2003).

2.2 Function analysis

The decision support system of variable rate irrigation is a geographic information system based on the module, which can import a few data formats such as text, table, etc. In this system, the property data can be queried the spatial data with each other. It also can offer many GIS function involving zooming in the map, moving the map or eagle eye etc. Irrigation prescription can be created according to the soil water data and the climate parameter. This system can be directly applied in the region of Beijing. For the expandability of the system, the module for expansion is designed. After the consumer imports the relative data of other region, this system can be immediately applied in that region.

3. SYSTEM REALIZATION

3.1 Interpolation of the sampling sites

The Kriging interpolation and the inverse distance weighted interpolation are accomplished. Through the interpolation, the distributing map of water content in a field can be immediately gained.

3.2 Database design

Three greatest databases such as the spatial database, the property database and basal database are designed. The spatial database includes the background map of national precision agriculture demo farm and the sampling sites' distributing map. The property database includes the soil water content in the sampling sites, which is acquired for the decision. The foundation database uses MS SQL Server2000 to establish relational database, among which there are four tables. They are the climate parameter table, farm crop parameter table, the price of water and wheat price table and the table of water index needed by wheat. There are seven fields in climate parameter table such as region, date, the sun radiation (Ra), the biggest possible sunshine hour (N), experience constant a, experience constant b and wind velocity correction coefficient (C). These seven fields are required for calculation the potential evapotranspiration (EToi). There are four fields in farm crop parameter table such as region, the period of crop growth, farm crop coefficient and yield reflection coefficient, which can be used to compute the amount of practical evapotranspiration. There are four fields in the price of water and wheat price table such as region, the price of wheat, the price of water and machine depreciation expenses, which can be used for the analysis of economy benefit. In the table of the index of water needed by wheat, there are five fields such as the region, the type of soil, the growth stage of the crop, the light drought index and the serious drought index, which can be used to decide to irrigate or not. The data in the later three tables can be extended by users, thus the extensibility of the system is increased. SQL Server is safe and data can not be changed without permission, which increases the security of the system.

4. GENERATION OF PRESCRIPTION MAP

The main function of the system is to generate the prescription map of variable rate irrigation. The prescription is made according to the soil water forecast model, the model of water consumption-yield, the optimal decision model and basal data.

4.1 Foundation data acquirement

The soil water content in 40 cm depth, field moisture capacity of the plot (Fc) has been measured. GPS was used to give the longitude and latitude of every sampling site, which can be transferred into layers of GIS. Through the interpolation, the surface data can be obtained. The experiment data can be got from national precision agriculture demo farm, where have taken 100 sampling sites in the winter wheat plot. Soil water content is recorded once every 15 days.

4.2 Forecast module of soil water content

4.2.1 Potential evapotranspiration model of farmland

The potential evapotranspiration amount of farmland is computed using Penman formula recommended by FAO1979. (Hu jichao, 2002)

$$ET_{0i} = \left\{ \frac{5.08\times10^7\times10^{(8.5(T-273)/T)}}{P\times T^2}\left[0.75\,Ra\left(a+b\frac{n}{N}\right)-2*10^{-9}\right.\right.$$

$$\left.\times T^4\left(0.56-0.079\times\sqrt{6.1\times10^{(8.5(T-273)/T)}\times r}\right)\left(0.10+0.90\frac{n}{N}\right)\right]+0.26 \quad (1)$$

$$\left.\times 6.1\times10^{(8.5(T-273)/T)}(1-r)(1.0+C\cdot U)\right\}/\left(\frac{5.08\times10^7\times10^{(8.5(T-273)/T)}}{P\times T^2}+1.00\right)$$

where: EToi is the potential evapotranspiration amount, P is the air pressure in the field plot (hPa), T is the average temperature of a day (absolute zero), Ra is the sun radiation, n is an actual sunshine hours, N is the hour of biggest and possible sunshine, a and b are experience constant, r is the relative humidity of the atmosphere, C is the revised coefficient of wind velocity and U is the wind velocity above the ground 2 meters (m/s). In these parameters, except the T, P, U, r, n need user to input, others can be obtained from database.

4.2.2 Practical evapotranspiration model of soil

The practical evapotranspiration amount (ETai) can be computed using formula (2) (Gong yuanshi, et al., 1998) . Crop coefficient (k_c) can be got from database.

$$ETai = EToi * k_c \qquad (2)$$

4.2.3 Water equilibrium model in farmland

The basis of soil water forecasting is an equilibrium equation of water in soil:

$$W_{T+1} = W_T + Pj + G - ETai \qquad (3)$$

where, WT+1 is soil water content end period (mm), WT is the soil water content start period, Pj is the effective rainfall inside the time (mm) and G is the quantities of the groundwater replenishment inside the time (mm). The amounts of groundwater replenishment can be zero in Beijing region. Formula (3) can reckon soil water content when decision-making from the soil water content that were measured.

4.3 The irrigation amount model

The soil water content in every cell of spray irrigation by decision-making can be obtained from above, and formula (4) or formula (5) can compute the irrigation amount in every irrigation cell.

$$H1 = Hg * (Fc - \theta) \quad \text{(mm) sufficient irrigation} \qquad (4)$$

$$H2 = Hg * (k * Fc - \theta) \quad \text{(mm) insufficient irrigation} \qquad (5)$$

where, Fc is field moisture capacity (volumetric soil water content), θ is the volumetric soil water content when the crop is short of water (coming from the model), k is the lower limit of irrigation under the condition of insufficient irrigation (the percent to the field moisture capacity, commonly it use 80%), Hg is the depth of irrigation management (mm), usually 600mm (Annandale J G et al., 1999).

4.4 Water consumption-yield model of winter wheat

The practical irrigation amount can be determined by the economy benefit brought by the different irrigation amount. The yield increment aroused by different irrigation amounts can be got from formula (6).

$$\Delta Y = Ki*(-0.00008H^3 + 0.0095H^2 + 0.6355H - 1.5217)*(-3.5714W + 2.5357) \quad (6)$$

where: ΔY is the increment of yield (kg/666.7 m2), H is the quantities of irrigation (mm), W is the relative humidity of the soil (%), Ki is the response coefficient of yield (kg/mm 666.7m2). The value of Ki can be got from the farm crop parameter table.

4.5 The optimal decision model

Owing to the amount of the spraying water is coincident in the perpendicular direction of the spray machine moving, whereas the irrigation amount computed may be different in each grid in this direction, so the analysis of economy benefit is needed to decide a uniform amount in this direction. The function for the economy benefit is inducted.

$$B = C_1*\Delta Y - C_2*H - C_3*S \quad (7)$$

Its restrictive condition is $H_{ij} \leq (1000*Fc\text{-}W_{T+1})$. The depth of the layer of the soil is 400 mm in this research. H_{ij} is the amount of irrigation (mm/666.7 m2). Bij is the economy benefit received after irrigation (yuan/666.7 m2), ΔY is the potential increase of yield from the irrigation (kg/666.7 m2), Si is a variable that show whether or not irrigation. Irrigation occur when Si is equal to 1, and not occur when Si is equal to 0. C1 is the price of the wheat (yuan/kg), C2 is the price of water (yuan/mm 666.7 m2), C3 is the expenditure of labor and machine depreciation after proceeding irrigation once in the unit area (yuan/666.7 m2). The restriction condition controls the irrigation amount, is used for fear occurring leakage. When decision-making, firstly Si is supposed equal to 1, namely proceeding irrigating. The economy benefit (Bij) of each grid in vertical direction can be computed from the amount of irrigation in each grid. When Bij is less than or equal to 0, irrigation isn't done. When Bij is more than 0, take the biggest value of Bij under different irrigation amount circumstance, then the opposite Hij is the irrigation amounts of all the grids in the upright direction against the machine moving.

4.6 Flow of prescription map generation

In the framework, Wp is the light drought index, and Wd is the serious drought index. They can be got by experiment. The flow chart of generating prescription map is Fig. 1.

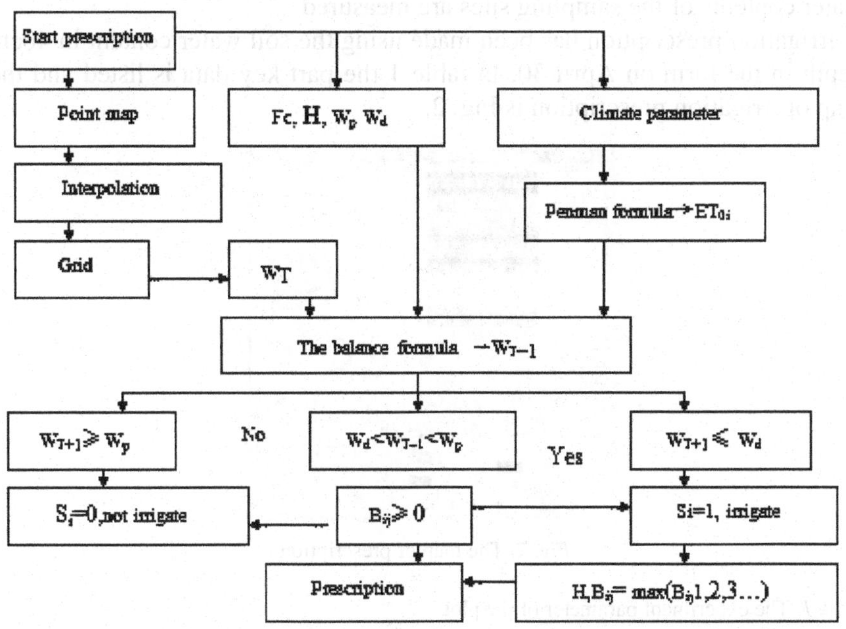

Fig. 1. Flow chart of making prescription map

4.7 Extension of module

After the user inputs the right password, the modules of this system can be extended and the domain of system can be spread. Users can input the parameters such as the crop coefficient of other regions, the yield reflecting coefficient and the field moisture capacity etc., thus this system can be applied in other regions to generate prescription of irrigation.

5. APPLICATION

An Embedded computer was equipped in the irrigation machine. The variable rate irrigation system run in the computer and the map of irrigation prescription can be import into the system. The GPS fixed in the top of machine was connected with the computer through the COM1 port. The system can receive the GPS signal, thus the location of the machine can be obtained in real time. The location of the machine matched with the prescription map. Then the system knew the count of the irrigation in real time and control flow capacity of valve to realize variable rate irrigation (Xu Di; Li Yi-nong, 2007).

This decision support system has been put into use in national precision agriculture demo farm, where the soil type is clay loam. On April 17, the soil water contents of the sampling sites are measured.

Irrigation prescription has been made using the soil water content in 40cm depth in the farm on April 30. In table 1 the part key data is listed and the map of irrigation prescription is Fig. 2.

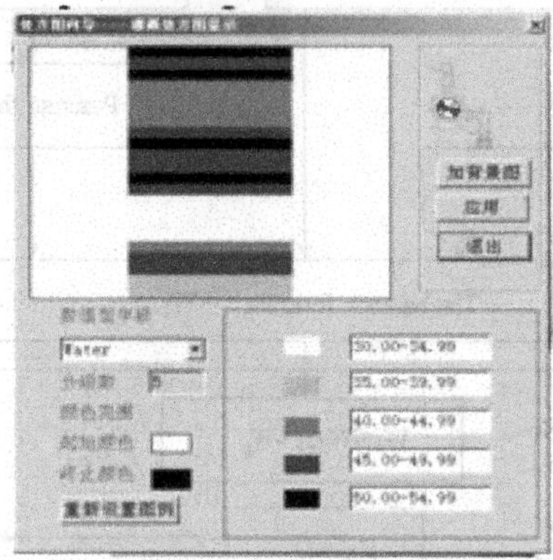

Fig. 2. The map of prescription

Table1. The experiment parameter of the plot

Parameter	Value
ET_{0i}	5.1
Kc	0.8
ET_{ai}	4.08
The effective rainfall (mm)	2.2
The days of interval	13
Wp (mm)	287
Wd (mm)	192.7
Fc (%)	41%
Total evaporation	53.04

6. CONCLUSIONS

Soil, atmosphere and crop are sufficiently taken into account in the irrigation model. The decision support system for variable irrigation can forecast soil water content by using the forecast mathematical model of soil water content which is measured according to the different (sampling) sites using GIS. Whether or not irrigation can be decided by comparing the current soil water content with the light drought and serious drought indexes of the crops. The amount of irrigation can be determined by water equilibrium model in farmland and water consumption-yield model of winter wheat.

The irrigation prescription map obtained can be used for parallel moving spray irrigation. This decision support system of variable rate irrigation was developed which have GIS plain function. Further more two kinds of interpolation method were achieved, through which soil water distributing map can be created.

Variable rate irrigation can save water in a distant degree, which is important to the development of china's agriculture. Variable rate irrigation machine is diversiform, so irrigation prescription map for other type of irrigation machine can be researched later.

REFERENCES

Annandale J G, Jovanovich N Z, Benadè N, et al. 1999. Modeling the long-term effect of irrigation with gypsiferous water on soil and water resources. Agriculture, Ecosystems and Environment, (76):109-119.

Gong yuanshi, Li zizhong, Li chunyou. 1998. Measurement of crop water requirements and crop coefficients using time domain reflectometry. Journal of China Agricultural University, 3(5):61-67.

Hu jichao, 2002. The research on the moisture connection of crop and moisture manage system. Nanjing Agriculture University Transaction, 9.

King B A, McCann I R, Eberlein C V, 1999. Computer control system for spatially varied water and chemical application studies with continuous-move irrigation systems. Computers and Electronics in Agriculture, 24:177-194.

Wang haijiang, Wang bo, 2001. Reasonable crop irrigation system based on GIS. Bulletin of science and technology, 17 (2):36-33.

Xu Di, Li Yi-nong, 2007. Review on advancements of study on precision surface irrigation system. Journal of Hydraulic Engineering, 38(5)529-534.

Zhu yan, Cao weixing, Wang shaohua etc. 2003. Application of soft component technology to design of intelligent decision-making system for crop management, Transaction of the CSAE, 19(1):132-137.

ELIMINATING CROP SHADOWS IN VIDEO SEQUENCES BY PROBABLE LEARNING PIXEL CLASSIFICATION

Tanghai Liu[1,*], Xiaoping Cheng[1]

[1] *Faculty of Computer and Information Science, Southwest University, Chongqing, China, 400715*

[*] *Corresponding author, Address: Faculty of Computer and Information Science, Southwest University, 1 Tiansheng Road, Beibei, Chongqing, 400715, P. R. China, Tel: +86-13647662870, Email: liutangh@yahoo.com.cn*

Abstract: Shadows have been one of the most serious problems for vegetation segmetation, espescially under conditions of natural random airflow and human or vehicle disturbance. A video sequence processing method has developed in this paper to identify and eliminate crop shadows. The method comprises pixel models and algorithms explained in a probable learning framework. Expectation maximization (EM) for mixture models is established and an incremental EM method is proposed. This method performs a probable reasoning unsupervised classification of pixels for real-time implementation. The results show that the method is quite robust and can successfully remove shadows under natural lighting conditions.

Keywords: probable learning, shadows, vegetation segmentation, video processing

1. INTRODUCTION

In earlier work of remote sensing technology as part of the agriculture related, it was shown that images acquired from satellites or aircrafts can provide effective help in agricultural related information such as woods, cropland, soil, and plant density.

Liu, T. and Cheng, X., 2008, in IFIP International Federation for Information Processing, Volume 259; Computer and Computing Technologies in Agriculture, Vol. 2; Daoliang Li; (Boston: Springer), pp. 1021–1028.

Recently, precision agriculture (PA) is popular as a management strategy that uses information technologies to bring data from multiple sources to bear on decisions associated with agricultural production (NRC, 1997; Robert, 2002). In addition, PA needs low altitude images (Vioix et al., 2002), even images very close to examine crops (Shrestha et al., 2003), which have been acquired for analysis.

Advances in computer technology and digital video technology had increasingly opened up using video cameras in real-time field application for the economical acquisition of images. (Easton and Easton, 1996) developed a mechanical sensing system for counting young corn plants, which was mounted on a one-wheeled, human-powered cart. (Shrestha et al., 2003) developed a machine vision-based corn plant population sensing system to measure early growth stage corn population. Video was acquired from a vehicle-mounted digital video camera under different daylight conditions. (Steward et al., 1999) mounted a video camera at a height of 3.35 m on a custom-made camera boom and took video streams at a slow forward travel speed to keep the full horizontal resolution available in the video signal.

The next step after image acquisition was segmentation of vegetation from background. Different methods were available for separating vegetation from non-vegetation regions. (Pérez et al., 2000) used a normalized difference index (NDI) along with morphological operations for plant segmentation. (Tang et al., 2003) used Gabor wavelets and an artificial neural network for classification of broadleaf and grass weeds. (Steward et al., 2004) developed a method called reduced-dimension clustering (RDC) for vegetation segmentation. Besides, segmentation of monochrome field scene images were typically accomplished by thresholding the intensity histograms, which typically had bimodal distributions of pixel gray levels (Meyer et al., 1998; Andreasen et al.,1997).

Various vegetation segmentation researches, however, had little to do with shadows. Some researchers tried to avoid shadows by taking top down video (Steward et al., 2004; Steward et al., 1999; Shrestha et al., 2003; Vioix et al., 2002).

We learn that shadows are significant consideration in vegetation segmentation. Since sunshine cast is not always perpendicular, vegetation must have naturally shadows. Shadows pose serious difficulty in correct segmentation (see Figure 1). Eliminating shadows can improve qualities of segmentation. Also, with the help of shadows elimination approach, the field observation time can be prolonged. Although sunshine casts have usually angles, we need not await special sunshine.

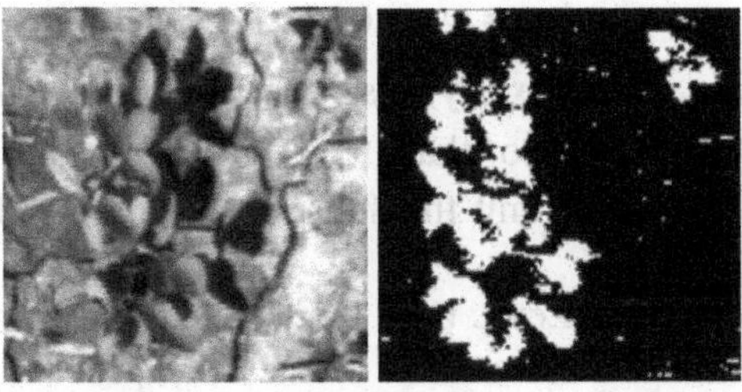

Figure 1. The left image has significant shadow components; obviously, the right image was segmented imprecisely because of existence of shadows. (Steward et al., 1999)

Furthermore, there are other natural conditions. For example, in the field, wind and breeze are common. At those cases when video acquisition processing needs human or vehicle participation, which should unavoidably disturb airflow, crop will swing. Then shadows are also not still.

Most researchers utilized video cameras for just acquiring still image frames, whose practices didn't exploit video camera capabilities fully. A video sequence provides various possibilities for improving field image processing. In fact, multiple temporal consecutive frames of images produced by video cameras can be used in removing noise, subtracting images from the background and so on. In this paper, we propose a video-processing-based approach to remove crop shadows. By using a probable reasoning method, our approach is quite robust and can successfully remove shadows under conditions of natural random airflow and human or vehicle disturbance.

"Background subtraction" is an old technique for finding moving objects in a video sequence. It succeeds not only in detecting moving object, but also their shadows. Each pixel must be classified before being used to update the background model. We show how this can be done properly, using a probabilistic classifier and a stable updating algorithm.

2. METHOD

2.1 Pixel model

Consider a single pixel and the distribution of its values over time. Some of the time it will be in its "normal" background state. Some of the time it may be in the shadow of swinging plants, and some of the time it may be part of a plant. Thus, we can think of the distribution of values $i_{x,y}$ of a pixel (x, y) as the weighted sum of three distributions $c_{x,y}$ (crop), $f_{x,y}$ (field), and $s_{x,y}$ (shadow):

$$i_{x,y} = w_{x,y} \cdot (c_{x,y}, f_{x,y}, s_{x,y})$$

The model for pixel (x, y) is parameterized by the parameters $\Theta = \{\omega_l, \mu_l, \Sigma_l : l \in \{c, f, s\}\}$ so that $w_{x,y} = (\omega_c, \omega_f, \omega_s)$, $f_{x,y} \sim N(\mu_f, \Sigma_f)$, and so on. For clarity, we omit the subscript x, y from the names of these parameters.

Let i be a pixel value. Let L be a random variable denoting the label of the pixel in this image. Our model defines the probability that $L = l$ and $I(x, y, t) = i$ to be:

$$P(L = l, I(x, y, t) = i|\Theta) = \omega_l \bullet (2\pi)^{-\frac{2}{d}} |\Sigma|^{-\frac{1}{2}} \exp\{-\tfrac{1}{2}(i - \mu_l)^T \Sigma_l^{-1} (i - \mu_l)\}.$$

Where $I(x, y, t)$ is the instantaneous pixel value for the (x, y) pixel at time t.

Given these probabilities, we can classify the pixel value. Namely, we choose the class l with highest posterior probability $P(L = l \mid I(x, y, t))$.

2.2 Algorithms

2.2.1 EM for mixture models

Suppose we observe a sequence of pictures 1,…, T, and that $I(x, y, t)$ is the value of pixel (x, y) in the t-th image. We want to learn the parameters of the distribution $c_{x,y}$, $f_{x,y}$, and $s_{x,y}$, as well as the relative weights $w_{x,y}$. Then, we define the likelihood of a set of parameters Θ to be the probability of the data given Θ: $\prod_{t=1}^{T} P(L = l_t, I(x, y, t) \mid \Theta)$. We want to choose the parameters that maximize the likelihood. The sufficient statistics for this mixture estimation is N_l, M_l, and Z_l, where

N_l is the number of images for which $L_t = l$;

M_l is the sum of the input vectors, $\sum_{t=1,\ldots,T, L_t = l} I(x, y, t)$; and

Z_l is given by $\frac{1}{T} \sum_{t=1,\ldots,T, L_t = l} I(x, y, t) \cdot I(x, y, t)^T$, the sum of the outer products of the input vectors with themselves.

From these sufficient statistics, we can compute:

$$\omega_l = \frac{N_l}{\sum_l N_l} \tag{2}$$

$$\mu_l = \frac{M_l}{N_l} \tag{3}$$

$$\Sigma_l = \frac{1}{N_l} Z_l - \mu_l^T \mu_l \tag{4}$$

The likelihood with respect to the observable data is $L(\Theta) = \prod_{t=1}^{T} P(I(x,y,t) \mid \Theta)$. The EM algorithm explores a sequence of parameter setting, where each setting is found by using the previous one to classify the data. Formally, we compute the expected value of the sufficient statistics as follows:

$$E[N_l \mid \Theta^k] = \sum_{n=1}^{T} P(L_t = l \mid I(x,y,t), \Theta^k)$$

$$E[S_l \mid \Theta^k] = \sum_{n=1}^{T} P(L_t = l \mid I(x,y,t), \Theta^k) I(x,y,t)$$

$$E[Z_l \mid \Theta^k] = \sum_{n=1}^{T} P(L_t = l \mid I(x,y,t), \Theta^k) I(x,y,t) I(x,y,t)^T$$

Then defining Θ^{k+1} by using Equation 2-4 with the expected sufficient statistics.

2.2.2 Incremental EM

The standard EM procedure we just reviewed requires us to store the values of pixel (x, y) for all the images we have observed. This is clearly impractical for our application. We now describe an incremental variant of EM that does not require storing the data. This procedure was introduced by (Nowlan, 1991), and is best understood in terms of the results of (Neal and Hinton, 1993).

Neal and Hinton show that we can think of the EM process as continually adjusting the sufficient statistics. In this view, on each iteration when we process an instance, we remove its previous contribution to the sum and replace it with a new one. Thus, for example, when we update N_l, we remove $P(L_t = l \mid I(x,y,t), \Theta^{k'})$ and $P(L_t = l \mid I(x,y,t), \Theta^k)$, where $\Theta^{k'}$ are the parameter setting we used to compute the previous estimated statistics from $I(x,y,t)$, and Θ^k are the current parameter settings. Neal and Hinton show that after each instance is processed, the likelihood of the data increases. Whenever we observe a new instance, we add its contribution to the sufficient statistics. Thus, in the long run, this process converges to a local maximum with high probability.

The resulting procedure for each pixel (x, y) has the following structure:
Initialize parameters Θ.

$t \leftarrow 0$

for $l \in \{c, f, s\}$

$N_l \leftarrow k\omega_l$

$S_l \leftarrow k\omega_l \cdot \mu_l$

$Z_l \leftarrow k\omega_l \Sigma_l + \mu_l \cdot \mu_l^T$

Loop the following

$t \leftarrow t+1$

for $l \in \{c, f, s\}$

$N_l \leftarrow N_l + P(L_t = l \mid I(x,y,t), \Theta)$

$S_l \leftarrow S_l + P(L_t = l \mid I(x,y,t), \Theta) I(x,y,t)$

$Z_l \leftarrow Z_l + P(L_t = l \mid I(x,y,t), \Theta) I(x,y,t) I(x,y,t)^T$

Compute Θ from $\{N_l, S_l, Z_l\}$.

The initialization step of the procedure sets the statistics to be the expected statistics for the initial choice of Θ. Then, in each iteration we add the expected statistics for the new instance to the accumulated statistics.

3. EXPERIMENTS AND CONCLUSIONS

For testing the performance of the method for eliminating shadows, we take several continuous video streams of maize crops. Figure 2 shows some typical frames.

In order to distinguish shadows from vegetation, pixel models and incremental EM are determined. The related results are shown in Figure 3.

It can be seen from following examples; our real-time approach has quite good shadows elimination capability, which paves the way for further processing in various field applications. Also, our method has consistently high accuracy in different video streams, which means robustness desired in natural condition.

However, our approach cannot deal with situation that shadows, which vegetation cast, cover vegetation. Our algorithm could be further integrated to eliminate shadows for robustness enhancement and possible accuracy improvement. We are quite sure our shadows identification method is a promising pre-processing technique for various agricultural applications.

Figure 2. Frames from video sequences in which crop cast shadows

Figure 3. Shadows casting on soil are eliminated via our approach

ACKNOWLEGEMENTS

This research is supported by Southwest University Scientific Developing Fund: SWUF 2006002.

REFERENCES

Andreasen, C., M. Rudemo, and S. Sevestre, Assessment of weed density at an early stage by use of image processing, *Weed Research* 37, 1997: 5–18.

Easton, D. T., and Easton, D. J, Device to measure and provide data for plant population and spacing variability, U. S. Patent No. 5, 568, 405, 1996.

Jean-Baptiste Vioix, Jean-Paul Douzals, Frédéric Truchetet, Louis Assémat, and Jean-Philippe Guillemin, Spatial and spectral methods for weed detection and Localization, *EURASIP Journal on Applied Signal Processing* 2002, 7: 697–685.

Meyer, G. E., T. Mehta, M. F. Kocher, D. A. Mortensen, and A. Samal, Textural imaging and discriminant analysis for distinguishing weeds for spot spraying, *Trans. ASAE* 1998, 41 (4): 1189–1197.

National Research Council, *Geospatial and Information Technologies in Crop Management*, Washington, D. C. National Academy Press, 1997.

Neal, R. M., and G. E. Hinton, A new view of the EM algorithm that justifies incremental and other variants, Unpublished manuscript, 1993.

Nowlan, S. J., *Soft Competitive Adaptation: Neural Network Learning Algorithms based on Fitting Statistical Mixtures*, PhD thesis, School of Computer Science, Carnegie Mellon University, 1991.

Pérez, A. J., F. López, J. V. Benlloch, and S. Christensen, Colour and shape analysis techniques for weed detection in cereal fields, *Computers and Electronics in Agriculture* 2000, 25 (3): 197–212.

Robert, P. C, Precision agriculture: a challenge for crop nutrition management, *Plant and Soil* 2002, 247 (1): 143–149.

Shrestha, D. S., and B. L. Strward, Automatic corn plant population measurement using machine vision, *Transactions of the ASAE* 2003, 46 (2): 559–565.

Steward, B. L., and L. F. Tian, Machine-vision weed density estimation for real-time outdoor lighting conditions, *Transactions of the ASAE*, 1999, 42 (6): 1897–1909.

Steward, B. L., L. F. Tian, D. Nettleton, and L. Tang, Reducee-dimension clustering for vegetation segmentation, *Transactions of the ASAE*, 2004, 47 (6): 609–616.

Tang, L., L. Tian, and B. L. Steward, Classification of broadleaf and grass weeds using gabor wavelets and an artificial neural network, *Transactions of the ASEA*, 2003, 46 (4): 1247–1254.

A WEB-BASED KNOWLEDGE AIDED TUTORING SYSTEM FOR VEGETABLE SUPPLY CHAIN

Hui Li[1,2], Zetian Fu[1,2,*], Yan Li[3], Jian Zhang[1,2]

[1] *College of Engineering, China Agricultural University, Beijing, China, 100083*

[2] *Key Laboratory of Modern Precision Agriculture System Integration, Ministry of Education, Beijing, China, 100083*

[3] *China Agricultural University, Beijing, China, 100083*

[*] *Corresponding author, Address: P.O. Box 121, China Agricultural University, 17 Tsinghua East Road, Beijing, 100083, P. R. China, Tel: +86-10-62736717, Fax: +86-10-62736717, Email: fzt@cau.edu.cn*

Abstract: With the recent rapid progress of computer technology, researchers have attempted to adopt artificial intelligence and computer networks to develop computer-aided tutoring systems. Meanwhile, some have also attempted to develop more effective programs to test and enhance the learning performance of students. This study surveyed the current theoretical basis and design principle and proposed a web-based knowledge aided tutoring system for vegetable supply chain. The main idea of this paper was focused on the interactive design and feedback control. The proposed concept was illustrated by the functions of chat system, discussion section, questionnaires and online learning record, etc.

Keywords: Computer aided tutoring, Vegetable supply chain, Interactive design, Feedback control

1. INTRODUCTION

With accelerated growth of computer and communication technologies, researchers have attempted to adopt computer network technology for research on education. Notable examples include the development of computer-aided tutoring and testing systems (Hopper, 1992). In 1989, Johnson et al. proposed a software design and development research program called Microcomputer Intelligence for Technical Training (MITT).

Li, H., Fu, Z., Li, Y. and Zhang, J., 2008, in IFIP International Federation for Information Processing, Volume 259; Computer and Computing Technologies in Agriculture, Vol. 2; Daoliang Li; (Boston: Springer), pp. 1029–1036.

Vasandani et al. developed an intelligent tutoring system that helps to organize system knowledge and operational information to enhance operator performance (Vasandani et al., 1995). Meanwhile, Gonzalez and Ingraham designed an intelligent tutoring system, which was capable of automatically determining exercise progression and remediation during a training session according to past student performance (Gonzalez et al., 1994). Hwang also proposed an intelligent tutoring environment which can detect the on-line behaviors of students (Hwang, 1998). In 2000, Ozdemir and Alpaslan presented an intelligent agent for guiding students through on-line course material. This agent could help students to study course concepts by providing navigational support according to their knowledge level (Ozdemir et al., 2000). Hwang proposed a testing and diagnostic system (Hsu et al., 1997; Hwang, 1999), which involved some sub-tasks, such as the investigation of network-based tutoring systems, network learning environments, knowledge-based systems for tutoring process control and interaction pattern analysis.

Clearly, the development of tutoring systems and learning environments has become an important issue in both computer science and education. The rapid development of computer networks is allowing access to information and communication free of spatial and temporal constraints (Hwang, 2003). Network communications allow problems to be solved through on-line discussion, and thus the implementation of computerized testing and practice systems has become an important issue.

2. METHODOLOGY

2.1 Theoretical Basis

The web-based computer aided tutoring system is mainly based on psychology and systems science.

The theory of cultural and historical development of psychologist Л.С.Выготский. According to his points, the thinking and intelligence were developed in the activities, subjected to internal digestion of interactions. The learners could turn the ability obtained during cooperative and interactive activities into their independent development gains. He proposed the concept of nearest developing section, which meant the difference between actual level and potential level. Full development depended on abundant social interactions. Achievements made by the tutors' instructions and partner' cooperation were far more than that of independent learning. And participation in collaborations and thinking could promote the individuals' cognitive development (Zhai et al., 2005; Cen et al., 2005).

Constructivism learning theory. In this theory, knowledge was obtained by the learners through interactions and cooperation under certain situations, using necessary learning materials by way of meaning constructivism. Situation, cooperation, conversation and meaning construction were the four key factors. Accordingly, the web-based computer aided tutoring system provided an integrated leaning environment, abundant interactive opportunities, covering learners, learning resources, man-machine interaction on the learning platform as well as the interaction between learners themselves. They were involved in the cooperation, conversation and learning through the resource system and network support platform and thus the meaning construction was realized (Zhai et al., 2006; Shao 2004).

Systems science – feedback control theory. The web-based computer aided tutoring system was a complicated dynamic system with multi-layers, various types and factors. Evaluation on its effects was a sophisticated systems engineering. It had the function of information feedback, systematically collecting feedback information, processing them to make judgments on teaching effects, and further regulating the teaching methods and scheduling so as to perform the internet learning activities more effectively. Therefore, the web-based computer aided tutoring was a way of macro control over the network tutoring activity system via information transfer, exchange and feedback. Only the control could guarantee stable development of the system with relative balance. So the designs including online learning record of learners and observation of learning attitude, etc. could be introduced into the web-based computer aided tutoring system, so as to know about the learning status, progression and effects of the learners in the internet learning environment. Accordingly the instructors could further improve the teaching methods and strategies according to the feedback.

2.2 Design Principle

In the design of web-based computer aided tutoring system, the following principles and requirements should be satisfied. (1) Centering around the learners. The network tutoring had to submit to and serve the learning activities of learners and promote their learning by enlightening their independent thinking and estimation. (2) The demands of communication between learners should be fully considered so as to make them have consciousness of cooperation and interaction, and encourage them being active in the tutoring activities. (3) The technical advantages should be used as much as possible to remedy the deficiency of human resources and realized the optimized integration of both. (4) The interaction should be timely, controllable, deep and wide. (5) The feedback should be emphasized. If without feedback design, the web-based computer aided tutoring system could not be a complete one. A scientific tutoring system pays much attention to not only how the instructors teach, but also how the learners study. If the learners make achievements, they should be given timely praise and if they encounter difficulties, they should be given timely encourage and help. To track the status of learners and know about their learning attitude and tendency are favorable to stimulate their enthusiasm, make the tutors obtain effective feedback information and thus properly regulate the teaching procedure.

3. CASE STUDY

The web-based knowledge aided tutoring system research for vegetable supply chain was taken as an example in this paper, which was part of the Network-based China Vegetable Supply Chain Management Technology Promotion Project of European Union. The proposed method was illustrated by the example with the analysis of system characteristics from the interactive design and feedback control as well as the function implement.

3.1 Interactive Design

To accord with the features of network teaching, firstly, the design should have excellent interactive ability, in which the learners can do self-adjustment according to the progression and selection of teaching materials and learning approaches. And the teachers can make corresponding feedback on the learners' activities and regulate the teaching strategy. Secondly, the design has excellent cooperative ability. The cooperation is advantageous to cultivate the team work spirits of learners and improve cognitive ability. And the network is an ideal site for cooperative learning, in which there are communication platforms such as forums, chat system, BBS, etc. The predominant characteristic of the network is its convenience of intercommunication, free of spatial and temporal constraints. The web-based knowledge aided tutoring system for vegetable supply chain integrated the two merits and realized the cooperative learning centering on interaction, which reflected the 'individual to individual' interaction in the network tutoring system. Therefore, it could provide learners with tools of cooperative learning and network space. Its functions mainly included:

The excellent interaction was realized by multi-channels as follows: a) BBS (Bulletin Board System) and online chat system. These functions could lead the learners to open discussion and enlighten them to put forward some questions with emanative thinking, so as to improve their learning and creative ability. b) Interactive reply and query system via E-MAIL. c) Cooperative learning group via forum. Exchange and discussion between learners happened within group. In conclusion, the learners could use these network communication platforms to participate in various conversation, negotiation and discussion and thus cultivate their independent thinking, difference thinking, innovative ability and team work spirits (Fig. 1). By these means, in the web-based knowledge aided tutoring system for vegetable supply chain, information communication was realized between learners and teachers and among learners themselves.

Cooperation. In the cooperative learning, the learners participated by way of groups to achieve mutual goals with certain inspiring system, and cooperated to make the maximum benefits. During this course, interaction was the basic unit of stimulating and producing mutual knowledge and conversation was the basis for cooperation. The web-based knowledge aided tutoring system for vegetable supply chain could set up learning groups by various interactive multi-channels and realize group discussion on the internet without getting together in realities.

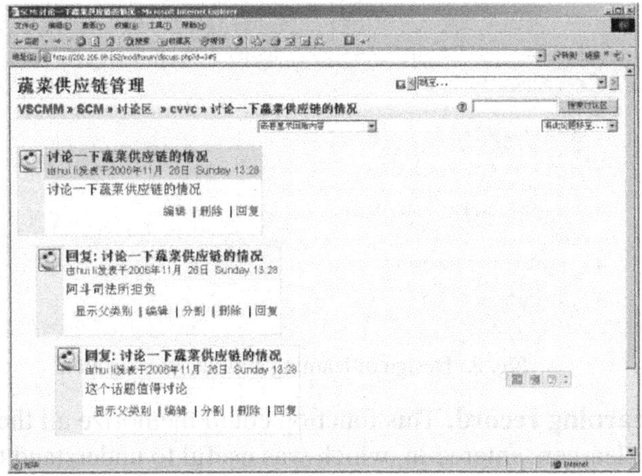

Fig. 1. Asynchronous communication: discussion section

3.2 Feedback Control

A good computer aided tutoring system should emphasize the feedback and control in order to guarantee the effective teaching and learning. The learners could do self-adjustment and select teaching materials and learning approaches according to their progression. And the teachers could have timely feedback on the learners' activities and attitude and adjust teaching strategy. In the web-based knowledge aided tutoring system for vegetable supply chain, to facilitate the feedback control, the design of learning attitude investigation as well as online learning record were concerned in this study.

Design of learning attitude investigation. The design of the system emphasized on the feedback and control of learning effects, in which the learning attitude investigation of the learners was introduced into the module design. It was an effective and direct way to evaluate on online learning environment based on constructivism theory. The module of questionnaire provided instructions for investigation, which was testified valid in online learning situations. The tutors could make use of these questions to get information from the learners, supervise and find out heir learning attitude and tendency, etc. so as to improve tutoring strategy and accord with the learners' cognitive characteristics and current knowledge background. (Fig. 2)

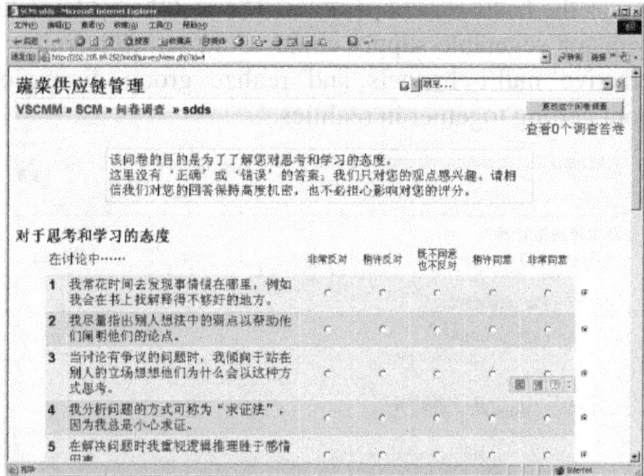

Fig. 2. Design of learning attitude investigation

Online learning record. This function could memorize all the courses the last time the learners entered in, which was useful to understand their various learning status. For instance, it could help to find out the participation enthusiasm and degrees of the learners in the online network learning, and help the learners to get to know the activities and progression of others, so as to inspire them and promote the cooperative atmosphere. It also could let the tutors to know the status, progression and positivity of each learner and accordingly give proper individual instructions and make teaching strategy regulation. This system could fully represent the idea of centering on the learners. (Fig. 3)

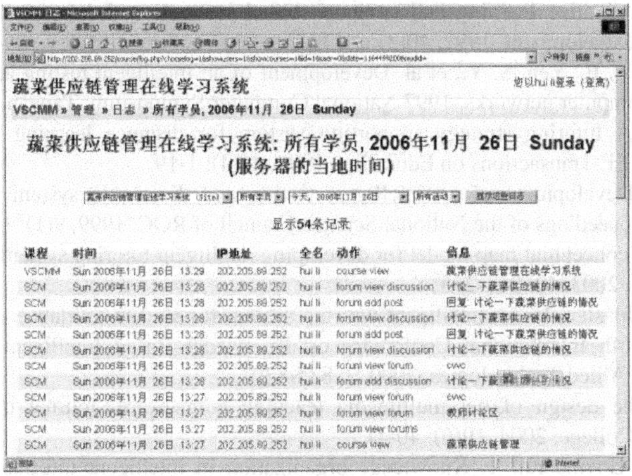

Fig. 3. Online learning record of the learners

4. CONCLUSIONS AND FUTURE WORKS

This study proposed a web-based knowledge aided tutoring system for vegetable supply chain. Interaction has been compare to 'conversation' and feedback control compared to 'diagnosis' in the web-based tutoring aided system. Therefore, interactive design and feedback control are of consequence in the system. The usability and practicability of the web-based tutoring aided system have been improved in this study to guarantee the realization of tutoring and learning. Further research will be focused on more effective interaction development.

ACKNOWLEDGEMENTS

We acknowledge financial support from EU Aided Asia IT & C (CN/ASIA-IT&C/005(89099)), and the Ministry of Science & Technology of China (Integration and application of technology on rural compositive information services).

REFERENCES

Cen, J. J., Wang, L., Lin, Q. S. Research on Interaction of Online Courseware. Journal of Henan Mechanical and Electrical Engineering College, 2005, 13(5): 113-115.

Gonzalez, A. J., Ingraham, L. R. Automated exercise progression in simulation-based training, IEEE Transactions on Systems, Man and Cybernetics, 1994, 24(6): 863-874.

Hopper, S. Cooperative learning and computer-based instruction. Educational Technology Research & Development, 1992, 40(3): 21-38.

Hsu, C. S., Tu, S. F., Yeh, S. Y., et al. Development of an intelligent testing and evaluation system on computer networks. 1997 National Computer Symposium, Taiwan, 1997.

Hwang, G. J. A tutoring strategy supporting system for distance learning on computer networks. IEEE Transactions on Education, 1998, 41(4): 1-19.

Hwang, G. J. Development of an intelligent testing and diagnostic system on computer networks. Proceedings of the National Science Council of ROC, 1999, 9(1): 1-9.

Hwang, G. J. A conceptual map model for developing intelligent tutoring systems. Computers & Education, 2003, 40: 217-235.

Ozdemir, B., Alpaslan, F. An intelligent tutoring system for student guidance in Web-based courses. Fourth International Conference on Knowledge-Based Intelligent Engineering Systems and Allied Technologies, 2000, 2: 835-839.

Shao, Z. R. The design of net multimedia teaching courseware. Journal of Chongqing Polytechnic College, 2004, 19(6): 40-41.

Vasandani, V., Govindaraj, T. Knowledge organization in intelligent tutoring systems for diagnostic problem solving in complex dynamic domains. IEEE Transactions on Systems, Man and Cybernetics, 1995, 25(7): 1076-1096.

Zhai, R., Feng, S. Z. A study of the model of web-based cooperative learning. Journal of Beijing University of Posts Telecommunications, 2005, 7(1): 39-43.

Zhai, R., Feng, S. Z. A study of interaction design based on the instructional system of multimedia networks. Journal of Beijing University of Posts Telecommunications, 2006, 8(3): 71-75.

THE DESIGN OF WIRELESS SENSOR NETWORKS NODE FOR MEASURING THE GREENHOUSE'S ENVIRONMENT PARAMETERS

Cheng Wang, Chunjiang Zhao*, Xiaojun Qiao, Xin Zhang, Yunhe Zhang

National Engineering Research Center for Information Technology in Agriculture Shuguang Huayuan Middle Road 11#, Beijing China, 100097

** Corresponding author, Address: P.O. Box 121, EU-China Center for Information & Communication Technologies, China Agricultural University, 17 Tsinghua East Road, Beijing, 100097, P. R. China, Tel: +86-10-51503411, Fax: +86-10-51503705, Email: Zhaocj@nercita.org.cn*

Abstract: Wireless sensor networks are integrated of sensor techniques, embedded system techniques, distributed computation techniques and wireless communication techniques. It's a hot research area in the world. It will be widely used in agriculture. This paper introduces the hardware and software design of wireless sensor networks node for the measurement of greenhouse environment parameters. These nodes can form networks themselves, measure the environment parameters precisely, transfer data safely, and solve the difficulty of wiring in greenhouse, making the earlier research of wireless sensor networks used in greenhouse

Keywords: wireless sensor networks, node, greenhouse, environment parameters, msp430

1. INTRODUCTION

As a new cropper plant technology, greenhouse is not limited by many factors such as plant zone, nature environment, climate and so on. It is important for agricultural production. Greenhouse environment parameters' measurement is a key tache of greenhouse's automation and efficiency. In

Wang, C., Zhao, C., Qiao, X., Zhang, X. and Zhang, Y., 2008, in IFIP International Federation for Information Processing, Volume 259; Computer and Computing Technologies in Agriculture, Vol. 2; Daoliang Li; (Boston: Springer), pp. 1037–1045.

traditional sensor data collection system, a lot of cable has been used, it brings many difficulties into system installment and greenhouse production.

Because of a new research area and broad application prospect, wireless sensor networks have been paid a lot of attention by many scholars. It is essential for agricultural workers bring wireless sensor network into agriculture production.

We have designed and made the wireless sensor networks node for measuring temperature, humidity, dew point, using its general analog or digital port, it can also measure soil moisture, CO_2 concentration, nourishing cream's PH, EC and cropper's physiological parameters and other environment parameters in greenhouse.

The paper will take the node used in greenhouse for example to state hardware and software design.

2. THE HARDWARE OF THE WIRELESS SENSOR NETWORKS NODE

The hardware of the wireless sensor networks node (Fig. 1)

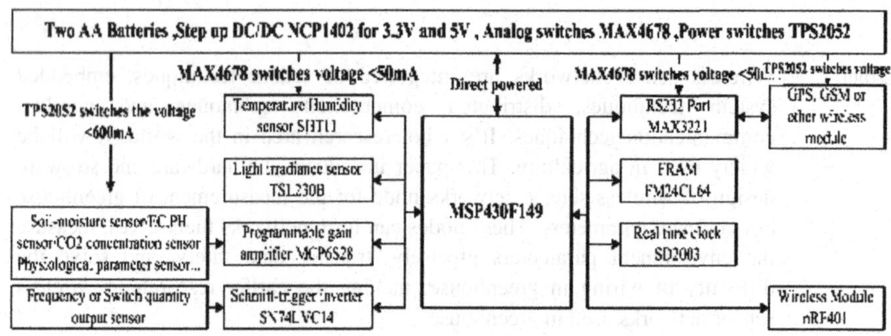

Fig. 1. Block diagram of the system hardware

2.1 Performance of TI's MSP430F149

The ultra low-power micro-controller 1MSP430F149 has 60KB flash ROM and 2KB RAM. It has powerful processing ability (RSIC architecture, 125ns instruction cycle time when using 8M crystal) and rich internal peripheral (12bit A/D converter, on-chip comparator, two USART/SPI ports, 48 I/O pins (Hu Dake, 2002) Ultra low-power is MSP430's most marked feature. It has low supply voltage range: 1.8-3.6V, five power-saving modes. When powered by 2.2V, 1 MHz, the consume current is 280uA. As it is powered by battery, node's energy is limited. This kind of micro-controller is the best choice for wireless sensor networks node, (Anton Muehlhofer, 2007, LutzBierl, 2002, E. Welsh, 2003).

2.2 Sensors

In this system the basic detected parameters are air temperature, air humidity, dew point, light irradiance.

Temperature, humidity and dew point are measured by SHT11 which is produced by Sensirion Company Switzerland. The SHT11 digital humidity and temperature sensor is fully calibrated, it offers long term stability and ease of use at very low cost. The digital CMOSens Technology integrates two sensors and readout circuitry on one single chip. Its measurement range is 0-100% RH, accuracy is ±3%RH. Temperature range is -40°C – +123.8°C, and its accuracy is ± 0.4°C (Wang Hanzhi, 2004). Base on the data of temperature and humidity, we can work out the dew point precisely (accuracy is ±1°C).

Light irradiance is detected by TSL230B. This IC combines a configurable silicon photodiode and a current-to-frequency converter on single monolithic CMOS integrated circuits. The output can be either a pulse train or a square wave with frequency directly proportional to light intensity. The sensitivity of the device is selectable in three ranges, providing two decades of adjustment. The full-scale output frequency can be scaled by one of four preset values.

By timer's capture function, MSP430F149 can work out the value of frequency. Base on the output frequency and irradiance chart which is offered by the chip producer (Zou Wei, Kang Longyun, 2004), we can get the light irradiance (unit: uW/cm2).

2.3 General Analog and Digital Input Port

For other parameters' detection, we designed the analog port to solve the standard 0-5V, 1-10V, 4-20mA, 0-10mA signals and even smaller signals which are exported by other sensors. It can be connected with eight kinds of different sensors. Standard current signals are transferred into voltage (0-2.5V) signal by resistor, and standard voltage signal can be divided into the range of 0V-2.5V.

MCP6S28, the single-ended, rail-to-rail, programmable gain amplifiers (PGA), can be configured for 8 gains from +1 – -32V/V and the input multiplexer can select one up to eight channels through an SPI port. [4] When does not want to amplify the signal, the PGA's gain is +1 as a follower to match the resistor between A/D convert and sensor.

The signal through the PGA and connects to MSP430F149's 12bit A/D converter.

Digital input port using six-Schmitt-trigger inverter SN74LVC14 to receive the data signal and filter the wave such as frequency, switch quantity, making the signal easy to receive by the micro-controller.

Users can choose different sensors to measure the different environment parameters in greenhouse.

Wireless Transfer Module:

Very low power consumption UHF transceiver chip nRF401 is designed to operate in the 433MHz ISM (Industrial, Scientific and medical) frequency band. It features Frequency Shift Keying (FSK) modulation and de-modulation capability. It's transmit power can be adjusted to a maximum of 10dBm. This system uses the differential antenna to increase communication length. This IC is connected to one of MSP430F149's USART port. In application the bit rate of this wireless transfer module is 20kbit/s, and communication length is over 300 meters (Fig. 2).

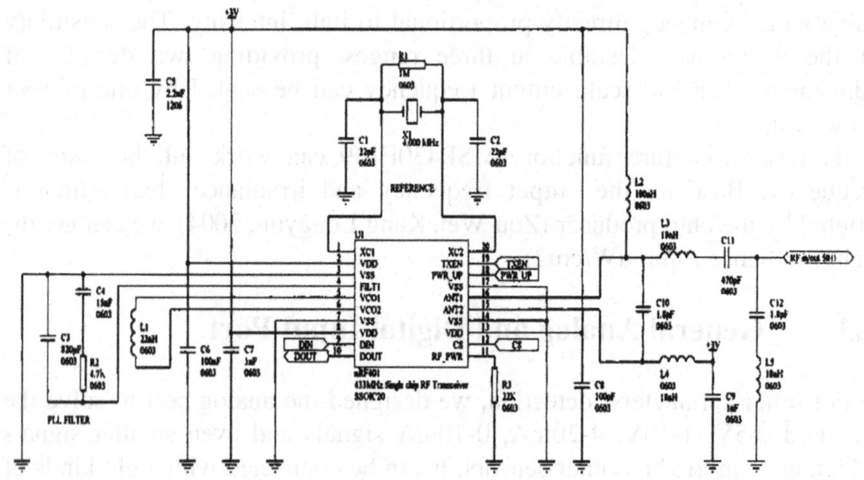

Fig. 2. nRF401 circuit diagram

2.4 Power Design and Management

This system is powered by two AA batteries. Energy consumption is very important for system design. All systems take high efficiency, step-up DC/DC IC NCP1402 to booster the batteries' voltage up to 3.3V. The MAX4678 quad analog switches has 2 ohm on-resistance. Each switch handles Rail-to-Rail analog signals and the current through it is 50mA. So MAX4678 has been used for switching the power of sensors, PGA and so on.

TPS2052 is also power switch for larger current.

When system is not collecting data, the MSP430 should turnoff the power of each sensor and switch into the sleep model to decreasing the power consumption. Main ICs' working current is show in Table 1.

Table 1. Working current of main ICs

ICs		Active Model	Sleep Model
MSP430F149		0.5mA	2.0uA
nRF401	Receive	0.28mA	8.0uA
	Transfer	8.0mA	—
SHT11		0.3mA	—
FM24CL64		0.2mA	—
SD2003		0.2mA	0.25uA
TSL230B		1.2mA	—
Other Sensors		60mA (Max)	—
Other consumption		0.5mA	20uA

In case of each node works an hour on average, it consumes 5mA (exclude other sensors). The system can work for over one year, so it is fit for greenhouse environment.

3. SOFTWARE DESIGN

3.1 System Software

It is composed of the main program, power management subroutine, different sensors measurement subroutine, nrf401 Rx/Tx subroutine.

The main program scheme (Fig. 3). The node is working in sleep model; other subroutine is called by interruption. In this operation, the system and networks' life can be prolonged.

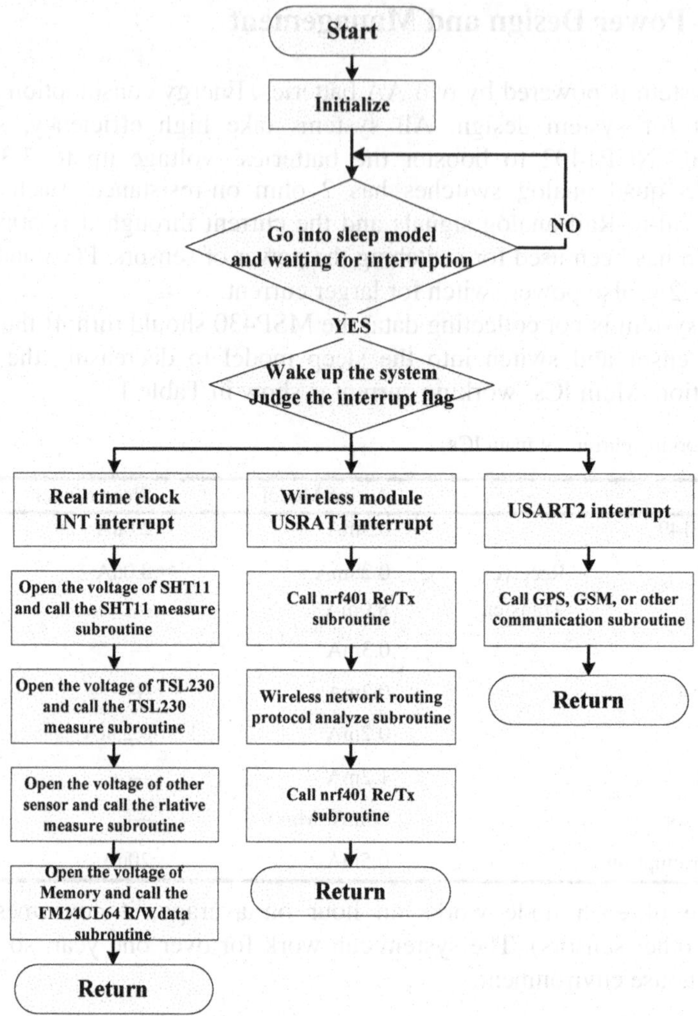

Fig. 3. Main program scheme

3.2 Communication Protocol

The key of the wireless data transmit subroutine. It is similar to the TCP/IP used in Internet. This protocol can achieve mesh, star or point to point networks (Akyildiz I.F., 2002).

Because of the limit of length, this paper only introduces simplify protocol. Its data packet format is designed as Table 2 shows.

Table 2. Wireless transmission data packet format

Lead1	Lead2	Mode	Length	MiddleID	LocalID	Destination	Data	Checksum
0xFF	0xAA	1Byte	1Byte	1Byte	1Byte	1Byte	nByte	2Byte

In the test, 0Xff followed by 0xAA can restrain noise effectively. Mode is defined as data or instruction and master-slave transmit or retransmit mode. Length is the sum of MiddleID, LocalID, Destination and Data. Destination is the master node's address. LocalID is the local node's address. MiddleID is used when node is in retransmit mode. All this protocol keeps the data transfer safely. The exact network and communication protocol (Fig. 4) (Sohrabi K., 2000, L. Eschenauer, 2002, Warneke B., 2001).

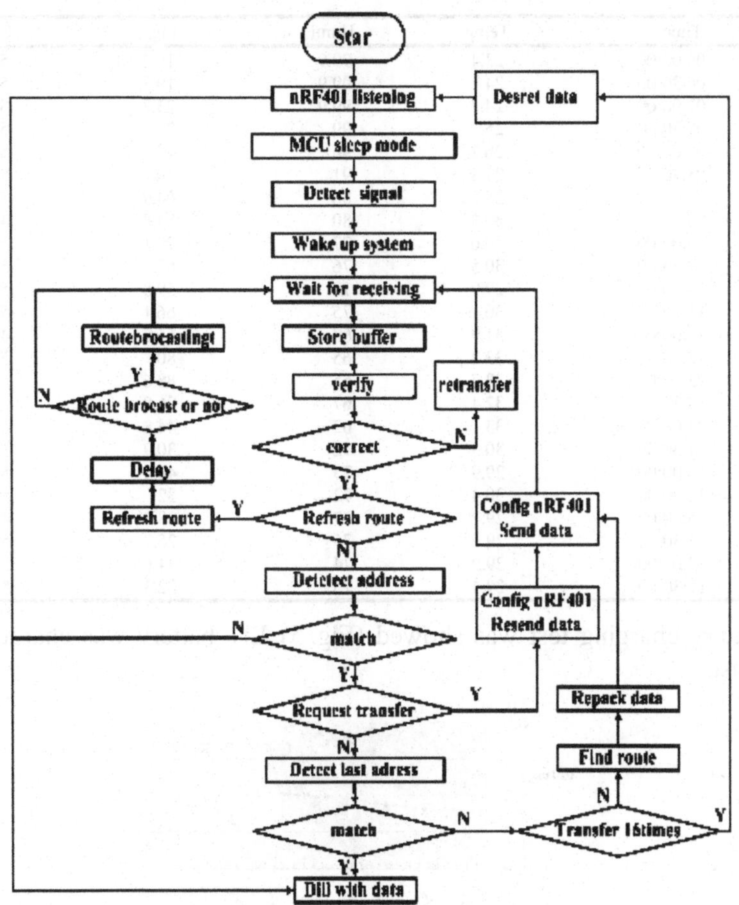

Fig. 4. Communication protocol

4. TEST AND RESULT

In order to test the performance of nodes and networks and examine the ratio of the error code in transmitting, the test was conducted under the condition that a PC was used as the master and four nodes formed wireless sensor networks. Different parameters were collected and transmitted to the PC. There were no data lost or wrongly transmitted during the test. In addition, the distance of each node should be less than 300 meters, when in retransmit pattern the networks' area was not limited. Table 3 show the greenhouse's data collected by PC.

Table 3. Data collected in July 7 2007

Time	Temp	Humi	Light	CO2
06:00:00	23.4	-99.9	10.2	570
06:30:00	24.1	-99.9	19.3	586
07:00:00	24.7	-99.9	23.2	570
07:30:00	25.3	99	26	578
08:00:00	26.2	96	42.5	566
08:30:00	27.8	91	50	542
09:00:00	28.5	86	64.9	506
09:30:00	30.8	80	84.8	498
10:00:00	30.6	77	79.1	510
10:30:00	30.5	76	63.1	510
11:00:00	29.9	79	59.8	554
11:30:00	30.1	75	66.1	546
12:00:00	31.9	69	65.7	490
12:30:00	33.2	65	80	526
13:00:00	33.7	65	88.6	538
13:30:00	32.4	67	96.8	510
14:00:00	33.1	62	44.8	486
14:30:00	30.3	72	30.9	522
15:00:00	29.9	73	40.2	522
15:30:00	30.4	72	32	510
16:00:00	29.7	77	28.9	494
16:30:00	29.6	75	25.7	502
17:00:00	29.9	74	34.6	514
17:30:00	29.4	75	22.3	530

Battery charging test was showed (Fig. 5), Li+ battery was charged safe and fast.

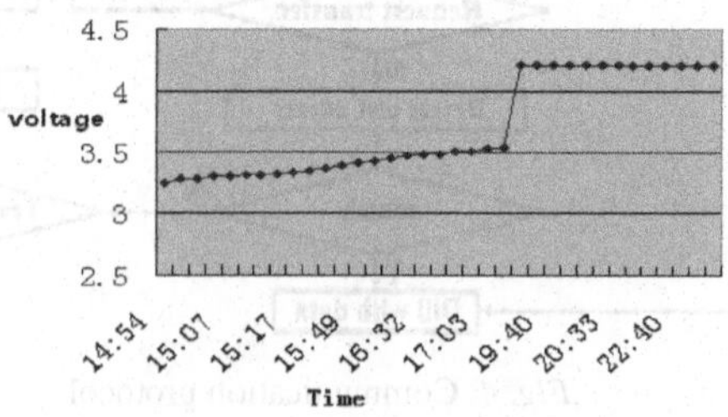

Fig. 5. Battery charge

5. CONCLUSION

The design of wireless sensor networks node caters to the tendency of sensor's development: wireless and network. Bring the study of wireless sensor networks node into the measurement of greenhouse environment parameters, have solved a lot of difficulties in greenhouse. This node can not only be used in agriculture, but also in other fields, such as industry, academe, civil domain and so on (M. Srivastava, 2001).

ACKNOWLEDGEMENTS

This research is supported by the National High Technology Research and Development Program of China (863 Program, Grant 2006AA10A311 and Grant 2006AA10Z253) and Beijing Science & Technology Program (Z0006321001391).

REFERENCES

Hu Dake. MSP430 series low power 16 bit micro-controller. Beijing University of Aeronautics and Astronautics Press, 2002 (3):5-10

Wang Hanzhi, Niu Zhenquan. Digital Humidity & Temperature Sensor SHT11 and its Applications Based on the Technology of CMOSens®. Sensor World. 2004 (9):35-38

Zou Wei, Kang Longyun. The Device based on MSP430 Microcontroller for Wind Speed and Light Irradiance Testing Synchronously. Information of Microcomputer, 2003 (19)11:56-58

Xian Lijuan, Li Shengyu, Yang Shizhong. The application of programmable gain amplifier MCP6S2X in signal collection. International Electronic Elements, 2004 (2):71-73

Akyildiz I. F., Su Weilian, Cayirci E.A survey on sensor networks. IEEE Communications Magazine, 2002 (8):102-114

Sohrabi K., Gao J., Ailawadhi V., et al. Protocol for self-organization of a wireless sensor network. IEEE Personal Communication, 2000 7(5):16-27

L. Eschenauer and V.D. Gligor. A key-management scheme for distributed sensor networks, In Proceedings of the 9th ACM Conference on Computer and Communication Security, 2002 (22):41-47

Warneke B., Last M., Liebowitz B., et al. Smart dust: communicating with a cubicmilli-meter computer. IEEE Computer, 2001 34(1):44-51

Anton Muehlhofer Controlling the DCO Frequency of the MSP430x11x. Texas Instrument Deutschland Application Report SLAA074 2007

LutzBierl. Interfacing the 3V MSP430 to 5V Circuits. Application Report SLAA148 2002.10

M. Srivastava, R. Muntz, and M. Potkonjak, Smart Kindergarten: Sensor-based Wireless Networks for Smart Developmental Problem-solving Environments, Intl. Conf. on Mobile Computing and Networking 2001, Rome, Italy, July 2001, pp. 132-138

Rice's GNOMES: E. Welsh, W. Fish, P. Frantz, GNOMES: A Testbed for Low-Power Heterogeneous Wireless Sensor Networks, IEEE International Symposium on Circuits and Systems (ISCAS), Bangkok, Thailand, 2003 (5): 15-20

RESEARCH OF GREENHOUSE EFFICIENT AUTOMATIC IRRIGATION SYSTEM BASED ON EVAPOTRANSPIRATION

Chunjiang Zhao, Yunhe Zhang*, Cheng Wang, Xiaojun Qiao, Ruirui Hao, Yueying Yang

National Engineering Research Center for Information Technology in Agriculture, Beijing, P. R. China, 100097

* *Corresponding author, Address: Shuguang Huayuan Middle Road 11#, Beijing, P. R. China, 100097, Tel: +86-10-51503409, Fax: +86-10-51503449, Email: zhangyh@nercita.org.cn*

Abstract: Traditional water-saving irrigation management turns on water pump in a settled interval decided by grower's experience. The method just saves water on the back of losing yield. In order to result the problem, a kind of greenhouse efficient irrigation management system based on Evapotranspiration has been developed. The system includes greenhouse environmental monitoring subsystem, irrigation analysis subsystem, control methods, can instruct users to irrigation greenhouse through quantitative calculating greenhouse crop water requirement. The system has been examined in greenhouses of ShengFangYuan Test Station in Beijing, China; the result shows that the system can instruct greenhouse irrigation efficiently, increase yield 18%, and save water 28%.

Keywords: greenhouse, automatic irrigation, crop water requirement, irrigation practice

1. INTRODUCTION

Greenhouse agriculture has become the main mode of production facilities and because of the character of high input and high output has been applied all over the world, but the large water rate of greenhouse makes its development limitedly (Baille etc. 1994).

Zhao, C., Zhang, Y., Wang, C., Qiao, X., Hao, R. and Yang, Y., 2008, in IFIP International Federation for Information Processing, Volume 259; Computer and Computing Technologies in Agriculture, Vol. 2; Daoliang Li; (Boston: Springer), pp. 1047–1054.

Now, the strategic position of water resources is more and more important. Development of efficient water-saving irrigation is the main method to mitigate the shortage of water resources. So there is great significance to research and develop efficient greenhouse irrigation management systems.

2. PROBLEM FORMULATION

Traditional water-saving irrigation management turns on water pump in a settled interval decided by grower's experience. Mean of Efficient water-saving irrigation, is not the irrigation of traditional sense, which losing yield to save water, but the water-saving irrigation which improve the efficiency of water use. In order to result these problems, a Greenhouse Efficient Automatic Irrigation System Based on Evapotranspiration has been developed in this paper. It relies on modern science and technology to use water resources efficiently, in order to achieve high quality and crop yield with the littlest water source.

3. PROBLEM SOLUTION

Greenhouse Efficient Automatic Irrigation System Based on Evapotranspiration can collect greenhouse environmental parameters such as air temperature, air humidity and net radiation real time, and save the data in its mass data storage according to the interval set by user. The system then calculates the irrigation amount and time based on the data above by formula, and tell user in voice.

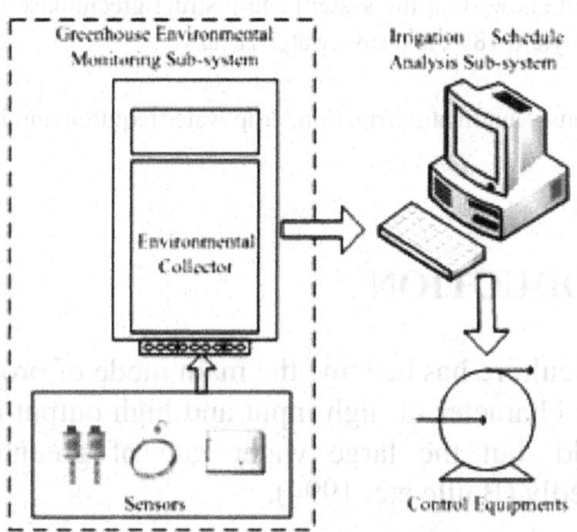

Figure 1. System structure

The system can communicate with PC in several kinds of way such as USB, Ethernet and RS232/485. There are three parts in the system, including greenhouse environmental monitoring sub-system, irrigation schedule analysis sub-system and all the irrigation control equipments in greenhouse. The real-time data collected in the greenhouse environmental monitoring sub-system, while the irrigation schedule analysis sub-system can calculate the irrigation amount and time based the data above.

3.1 Greenhouse Environmental Monitoring Sub-system

Main function of the Greenhouse Environmental Monitoring Sub-System is to collect and save data of greenhouse environmental parameters real time. In order to calculate the irrigation amount in different time based on the formula introduced in part of The Irrigation Schedule Analysis Sub-system, we must know the maximum and minimum daily temperature, net radiation of crop canopy and the relative humidity of air. So the Greenhouse Environmental Monitoring Sub-System consists of Sensors and Environmental Collector.

Air temperature and humidity sensors made by National Engineering Research Center for Information Technology in Agriculture are used in the Greenhouse Environmental Monitoring Sub-system. TDD-1 net radiation meter is used to measure crop canopy net radiation. Measuring accuracy of the air temperature sensor is ±0.2%, and measuring range is from -20°C to 70°C; Measuring accuracy of the air humidity sensor is ±2%, and measuring range is from 0 to 100%; Shortwave radiation measuring range of The TDD-1 net radiation meter is 0.27 – 3μm, earth radiation measuring range of The TDD-1 net radiation meter is 3 – 50μm.

In order to collect and save environmental parameters above, an Environmental Collector is developed. The core part of the Environmental Collector is single chip; control program is inserted in it. For the advantage of lower power consumption and mass storage, MSP430F149 is used in this system; system clock module is DS1302; voice module consists of AT89C4051 and PM50100 to tell the information and alarm; storage module consists of FM24CL64 and AT45DB08 to save mass data; 128*64 LCD display is used to show data; lithium charge managing chip TC4055; linearity power NCP500SN33 supply power to voice system; power chip TPS7333 supply power for whole system; the first serial interface of MSP430F149 communicates with voice system, the second serial interface of MSP430F149 and CP2101 are transform into USB interface, SPI interface is expanded into 3 serial interfaces by GM8141, connected with

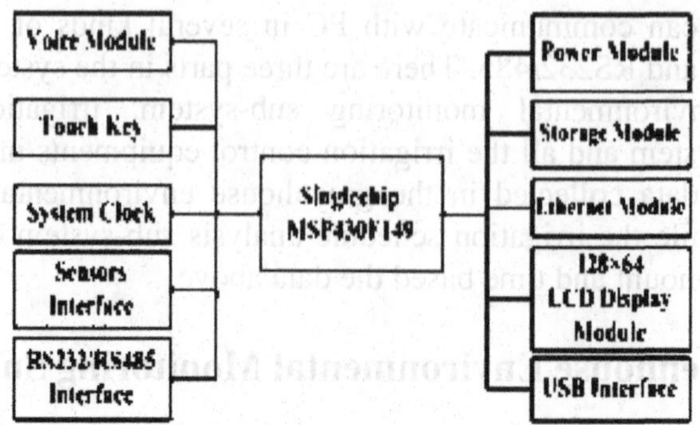

Figure 2. Structure of environmental collector

Ethernet module DNE-18, RS232 CMOS chip MAX3221 and RS485 CMOS chip SN65HVD3082, these consist of multiple serial communication module.

3.2 Irrigation Schedule Analysis Sub-system

The irrigation analysis is the core part of Greenhouse Efficient Automatic Irrigation System Based on Evapotranspiration. Through analysis the environmental data collected by monitoring sub-system, we can compute the less water volume to supply for crops' growth. Then carry on the instruction irrigation under the guidance of the crops water demand in the greenhouse.

The Computation of crops water demand based on the computational method of "Kc•ET0", where Kc is crop coefficient of certain crop, and ET0 is reference crop evapotranspiration. The product KC•ET0 is the crops evapotranspiration under the proper condition (in non-water-stress).

The crop coefficient Kc reflects the water demand difference between crop and the reference crop. One coefficients can be used to describe the infection of evaporation and transpiration, which also can be described by two coefficients separately, namely so-called single crop coefficient and double crop coefficient. When crops are small or planted quite sparsely, after rainfall or irrigation, soil evaporation plays the main role which can be accounted for a large proportion. Especially on a condition of soil surface frequently moist, the computed result of double crop coefficient method is closer to actual value. Therefore, the double crop coefficient method is used to calculate Kc in this paper.

FAO in 1998 published the "Computation Guide of Crops evapotranspiration volume and the Crops Water requirement (the FAO Irrigation And Draining water Handbook - 56)", in which Penman - M.Tess formula recommended to calculate ET_0 as the standard computational method. It has made extremely detailed stipulation to the formula exploitation conditions. And it has been approved an effective method (Allen,1998; Nandagiri etc. 2005; Temesgen etc. 2005; Utset etc. 2004).

FAO's recommendation Penman - M.Tess formula in fact is the composition of the radiation and the aerodynamics. And its combination equation form is as follows:

$$ET_0 = \frac{0.408\Delta(R_n - G) + \gamma\frac{900}{T+273}u_2(e_s - e_a)}{\Delta + \gamma(1+0.34u_2)} \tag{1}$$

In the formula:

ET0: represents crop evapotranspiration, defined as: "transpiration and evaporation of the assumption crop similar to green grasses that in full water supply and exuberant growth condition, and they are 0.12m high; the fixed leaf resistance is 70 s/m, reflectivity is 0.23, broad surface, highly even";

Δ : represents the slope of saturation vapor pressure - temperature curves;

R_n : represents the acceptable net radiation of reference crop canopy surface;

G : represents soil heat flux;

γ : represents the thermometer constant;

T : represents the mean temperature;

U_2: references the wind's velocity Of 2m height;

e_s : represents the stream pressure of saturation vapor;

e_a : represents: the stream pressure of actual vapor.

The formula can be used to calculate the evapotranspiration of reference crop monthly, every 10 days, daily, even hour (Li Yulin etc. 2002). Because the formula calculates the amount of evapotranspiration to guide the greenhouse crop irrigation scheduling, it is calculated using daily (Hussein M. 2002).

An analysis of each parameter in the formula (1):

$$\Delta = \frac{4098e_s}{(T+237.3)^2} \tag{2}$$

$$e_s = \frac{e^0(T_{max}) + e^0(T_{min})}{2} \tag{3}$$

$$e^0(T) = 0.611\exp(\frac{17.27T}{T+237.3}) \tag{4}$$

$$e_a = \frac{e^0(T_{min})RH_{max}/100 + e^0(T_{max})RH_{min}/100}{2} \tag{5}$$

$$G=0 \tag{6}$$

Note: In the daily calculation of reference crop evapotranspiration, it can be completely ignored (D. Itenfisu etc. 2003).

$$\gamma = 0.0016\frac{p}{\lambda} \tag{7}$$

$$p = 1010 - 0.115H + (0.00175H)^2 \tag{8}$$

Note: Beijing is located north latitude 39°48' east longitude 116°28', therefore elevation elevation H is 31.3m.

$$\lambda = 2.501 - 0.0023617 \tag{9}$$

Note: T is the average temperature.

U_2: surveys by the wind speed sensor.

R_n: Uses the Measure of net radiation.

Put all parameters' values into the formula to calculate KC • ET_0, obtain the daily evapotranspiration of greenhouse crops.

In order to satisfy the crops high production demand, the water stored in the water accepting layer of crop's Root should be maintained at the appropriate scope in random time interval. It means that average soil moisture content of moist soil layer is usually not less than the minimum content allowed by crops and not more than the maximum content allowed by crops. The minimum moisture content and the maximum moisture content allowed by different crops, and the initial moisture content are known, The system according to calculations of the crop evapotranspiration rate calculate when the soil moisture content turn to the minimum moisture content allowed, and calculate the irrigation volume from the minimum moisture content allowed to the maximum moisture content allowed by crops. The system turns on irrigation equipment in the suitable date.

4. TESTING AND DISCUSSION

The system has been tested in six greenhouses with the same condition of ShengFangYuan in Beijing, China. Number the six greenhouses as 1, 2, 3, 4, 5, 6. The same amounts of cucumber seeding have been planted, greenhouse of NO.1, NO.2 and NO.3 irrigated with the traditional experience methods, greenhouse NO.4, NO.5 and NO.6 irrigated under the instruction of the Efficient Automatic Irrigation system. The result of testing shows that the system can instruct greenhouse irrigation efficiently, average increase yield 18%, and average save water 28%.

Table 1. Comparison of the crop yield and water consumption between greenhouses irrigated with the traditional experience methods and greenhouses irrigated under the instruction of the Efficient Automatic Irrigation system

	NO.	Crop Yield (Kg/ha)	Water Consumption (m^3/ha)
Traditional Irrigation method	1	75300	3090
	2	76500	3180
	3	80000	3300
Irrigate by the Efficient Automatic Irrigation System	4	90000	2000
	5	93700	2310
	6	91800	2540

5. CONCLUSIONS

With the further development of world's economy and society, the strategic position of water resources is more and more important Greenhouse is a kind of spending mass water plant, so development of efficient water-saving irrigation is significant. Traditional irrigation method saves water but losing yield, in order to result these problems, a Greenhouse Efficient Automatic Irrigation System Based on Evapotranspiration has been developed in this paper. The system improves the efficiency of water use to save water, relying on modern science and technology to achieve high quality and crop yield with the littlest water source.

The system has been examined in greenhouses of ShengFangYuan Test Station in Beijing, China; the result shows that the system can instruct greenhouse irrigation efficiently, increase yield 18%, and save water 28%.

ACKNOWLEDGEMENTS

This research is supported by the National High Technology Research and Development Program of China (863 Program, Grant 2006AA10Z202 and Grant 2006AA10Z253).

REFERENCES

Allen R.G., Pereira L.S., Raes D. and Smith M. Crop Evapotranspiration-Guidelines for Computing Crop Water Requirement [M], Rome, Food and Agriculture Organization of the United Nations, 1998.

Baille A. (1994). Irrigation management strategy of greenhouse crops in Mediterranean Countries. Acta Horticulturae, 361: 105-22

Baille M., Baille A. and Delmoa D. (1994a). Microclimate and transpiration of greenhouse rose crops. Agric. For. Meteorology. 71:83-97.

D. Itenfisu, R. Elliott, R.G. Allen et al. Comparison of reference evaportranspiration calculations as part of the ASCE standardization effort [J]. Journal of Irrigation and Drainage Engineering, 2003, 129(6):440-448.

Li Yulin, Cui J ianhuan, Zhang Tonghui. Comparative study on calculation method of reference evapotranspiration [J]. Journal of Desert Research, 2002, 22(4).

Nandagiri, Lakshman; Kovoor, Gicy M. Sensitivity of the food and agriculture organization Penman-Monteith evapotranspiration estimates to alternative procedures for estimation of parameters [J]. Journal of Irrigation and Drainage Engineering, 2005, 131(3):238-248.

Temesgen, Bekele; Eching, Simon; Davidoff, Baryohay et al. Comparison of some reference evapotranspiration equations for California [J]. Journal of Irrigation and Drainage Engineering, 2005, 131(1):73-84.

Utset, Angel, Farre, Imma, Martinez-Cob, Antonio et al. Comparing Penman-Monteith and Priestley-Taylor approaches as reference-evapotranspiration inputs for modeling maize water-use under Mediterranean conditions a [J]. Agricultural Water Management, 2004, 66(3):205-219.

A MEASURING INSTRUMENT FOR MULTIPOINT SOIL TEMPERATURE UNDERGROUND

Cheng Wang, Chunjiang Zhao*, Xiaojun Qiao, Zhilong Xu

National Engineering Research Center for Information Technology in Agriculture, Beijing, P. R. China, 100097

* *Corresponding author, Address: Shuguang Huayuan Middle Road 11#, Beijing, 100097, P. R. China, Tel: +86-10-51503411, Fax: +86-10-51503449, Email: zhaocj@nercita.org.cn*

Abstract: A new measuring instrument for 10 points soil temperatures in 0–50 centimeters depth underground was designed. System was based on Silicon Laboratories' MCU C8051F310, single chip digital temperature sensor DS18B20, and other peripheral circuits. It was simultaneously able to measure, memory and display, and also convey data to computer via a standard RS232 interface.

Keywords: Multi-point Soil Temperature; Portable; DS18B20; C8051F310

1. INTRODUCTION

The temperature of soil is a vital environmental factor, which directly influences the activity of microorganisms and the decomposition of organic substances. It can affect roots absorbing water and mineral elements. It also plays an important role in the growth rate and range of roots. Statistically, roots of most plants are within 50 centimeters underground, so it becomes very significant to measure the soil temperature of different depth in this level.

The Soil Temperature Measuring Instruments used nowadays mainly fall into three types, the first type is the measure temperature by making use of the relationship between the soil temperature and the temperature-sensitive resistor. Before using this sort of instruments, the system parameters need to

Wang, C., Zhao, C., Qiao, X. and Xu, Z., 2008, in IFIP International Federation for Information Processing, Volume 259; Computer and Computing Technologies in Agriculture, Vol. 2; Daoliang Li; (Boston: Springer), pp. 1055–1061.

be adjusted; it is inconvenient to repair when the system runs into trouble. The second type is non-contact Soil Temperature Measuring Instrument which use infrared ray to measure temperature, this sort of instruments is quite expensive. The third type is instrument measure temperature by making use of digital thermometer, at the present time, this sort of instruments can only measure one point of soil temperature, and the data can not be stored or transmitted.

In all, the products mentioned above can hardly become popular for they are either costly too expensive or functionally too simple. So a new kind of cheaper and more advanced instrument is required to be invented.

2. MATIERIALS AND METHOD

This system applies the high quality Single Chip C8051F310 (Li Gang et al. 2002; He Limin, 2000) as the core controller, it mainly includes some functional blocks such as Data Collection Block, Display and Storage Block, Real Clock Block, Serial Communication Block, Keying Control Block and Power Source Block. Fig. 1 shows what the system consists and how it functions.

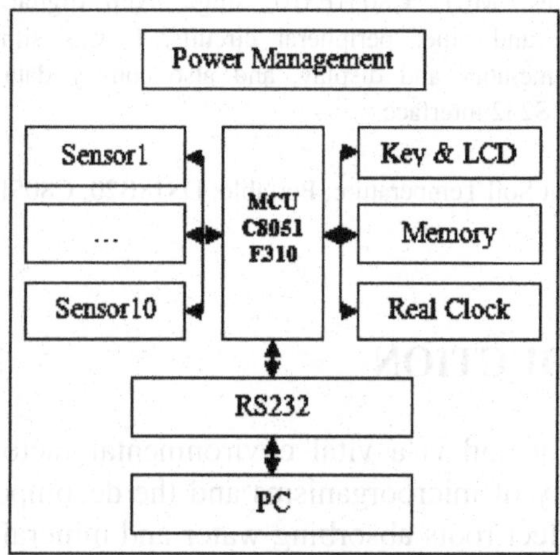

Fig. 1. System block diagram

The system can measure soil temperature of ten points in different depth, it can display and store both the data of temperature and the time, at which the data is collected, after that, it can transmit the data to the computer through serial communication port. The user can set system parameter or operate the system by pressing keys. By experiment, this cost-effective and portable instrument works stably and operates well.

2.1 Hardware design

In the hardware design, the system utilizes parts including MCU C8051F310, Digital Thermometer DS18B20s, power charge Chip ISL6292, voltage management Chip NCP500 and real-clock Chip DS1302, combining with corresponding peripheral circuits, and these parts make the main structure of the system. Some of the main parts and its peripheral circuits will be introduced as follows.

2.1.1 High quality MCU C8051F310

C8051F310 device is fully integrated mixed-signal system-on-a-chip MCU, whose microcontroller is compatible with 8051 instruction set. C8051F310 mainly composes of microcontroller core CIP-51, analog peripherals, digital I/Os and the power unit. The CIP-51 core employs a pipelined architecture that greatly increases its instruction throughput, with a maximum system clock at 25MHZ, it has a peak throughout of 25MIPS. The CIP-51 core offers all the peripherals included with a standard 8052, which is familiar to Chinese technologists. The Digital Crossbar allows mapping of internal digital system resources to Port I/O pins; C8051F310 device includes a total of 29 I/O pins.

2.1.2 The DS18B20 digital thermometer

The DS18B20 Digital Thermometer (He Xicai. 2001; Chen Liangguang 2001). measures temperatures from -55°C to +125°C. DS18B20 includes three pins, respectively are data I/O Pin DQ, power supply Pin VDD and the GND Pin. Fig. 2 shows the DS18B20 application chart.

Because each DS18B20 contains a unique silicon serial number, multiple DS18B20s can exist on the same 1-Wire bus. This allows for placing temperature sensors in many different places and provides convenience for the hardware design in this system.

This system utilizes ten DS18B20s to measure temperature of ten points soil in different depth within 50 centimeters. The first DS18B20 and the

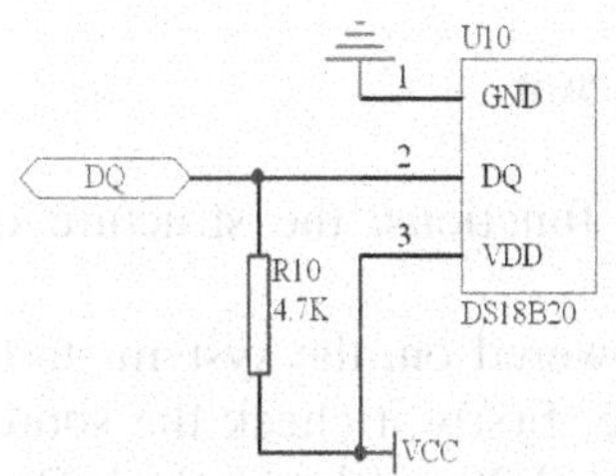

Fig. 2. Application chart of DS18B20

second share a 1-Wire bus, the third and the fourth share one, the other six DS18B20s communicate through their own 1-Wire bus respectively. The system applies external power source, as there are too many DS18B20s. The data collected by DS18B20s is filtered by a 0.1uF capacitor, and then lead to the I/O port of MCU by interface circuits.

2.1.3 Power source and recharge circuits

The whole system is contained in a sealed box; it employs a rechargeable lithium battery so as not to open the box frequently.

The ISL6292 is an integrated single-cell Li-ion or Li- polymer Battery Charger, which is capable of operating with an input voltage as low as 2.4V. The ISL6292 can be used as a traditional linear charger.

For the output voltage of lithium battery ranges from 2.8V to 4.2V, and the system works at a constant voltage of 3V, so it utilizes a voltage variation chip NCP500 to supply the system with a stable voltage. Fig. 3 shows the application chart of NCP500, the Pin VIN connects to the output of lithium battery; Pin AIN0 outputs to an analog pin of C8051F310, the digital value of lithium voltage can be acquired after a A/D conversion.

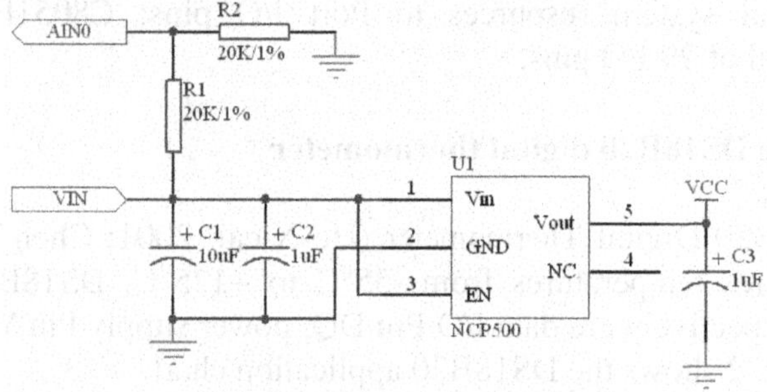

Fig. 3. Application chart of NCP500

2.2 Software design

The MCU program written in C Language was assembled and debugged in Keil C Assembler (Ma Zhongmei et al. 1998).

2.2.1 The main program

In order to realize its functions, the structure of the main program is designed as follow.

When the system is powered on, the system starts initialization, then the program runs into a circle, firstly it check the source voltage, secondly the MCU reads data from DS18B20s and real clock DS1302, thirdly the data is displayed and stored, after that, the main program check whether the interruption flag is set or not, if it equals one, the program runs into the part of interruption program, if not, the main program runs into another circle. Fig. 4 shows the flow chart of the main program.

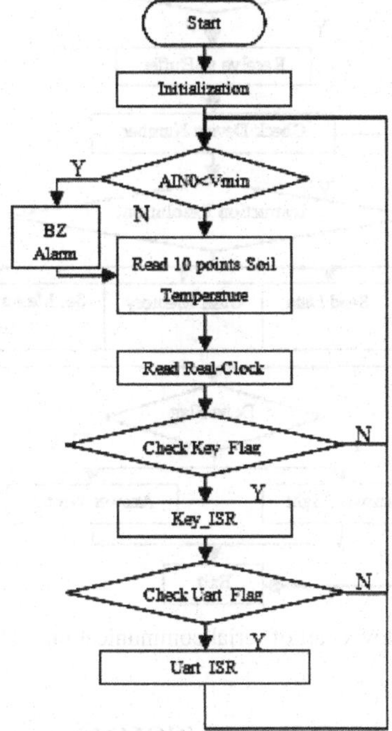

Fig. 4. Flow chart of main programme

2.2.2 The interruption sub-programs

The interruption sub-programs are at the end of the main program, when the main program discovers the interruption flag true, it will runs into the interruption sub-programs.

The Key-Interruption helps to set the system parameters by pressing keys on the box to intrigue the interruption program. The structure of Key-Interruption Sub-Program is somehow similar to the Serial Communication Interruption Sub-Program, so only the later is described in detail.

The instrument and the computer are connected via RS-232 Serial Communication Port. According to the communication protocol, every instrument is assigned a unique machine number. When the instrument receives instructions from the computer, the program first compare its own machine number with the machine number sampled from the instructions, if the machine number is matched, the instrument then respond to the instructions. Fig. 5 shows the flow chart of Serial Communication Interruption Program.

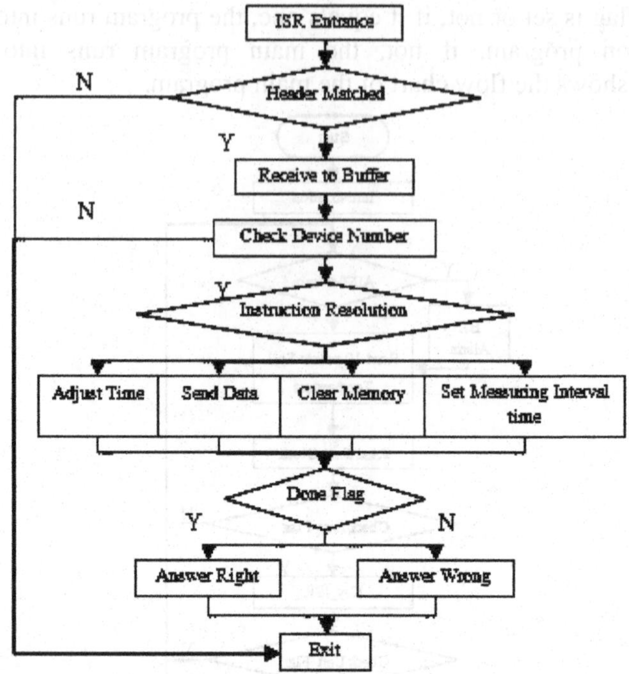

Fig. 5. Flow chart of serial communication interruption

3. RESULTS AND DISCUSSION

Combining with functions such as display, storage and transmission, this system is mainly used in measuring soil temperature. User can connect this instrument to computer via RS-232 Serial Port, through which the data of temperature and time can be transmitted to the PC. It makes a lot of improvements as well as makes full use of the previous products. First, this instrument is capable of obtaining temperature of multipoint soil, so the temperature data is adequate. Second, this system employs rechargeable lithium battery, which largely prolongs the span of the instrument. Furthermore, in the process of hardware and software design, low power consumption was always taken into consideration. In a word, this instrument is cost-effective, portable and precise.

ACKNOWLEDGMENTS

This work is funded by the project for Beijing Science and Technology Plan (Contract Number: Z0006321001391).

REFERENCES

Chen Liangguang. 2001. Principle and Application of Digital Thermometer [M]. Beijing: CIM Press.

He Limin. 2000. An Advanced Course on MCU [M]. Beijing: BUAA Press.

He Xicai. 2001. Sensor and Its Application Circuits [M]. Beijing: PHEI Press.

Li Gang, Lin Ling. 2002. High Quality 8051 Compatible MCU C8051Fxxx [M]. Beijing: BUAA Press.

Ma Zhongmei, Jie Shunxin. 1998. Design of C Language Program on MCU [M]. Beijing: BUAA Press.

THE REALIZATION NETWORK CORRESPONDENCE WITH WINSOCK CONTROL COMPONENT OF VB

Chunjiang Zhao [1], Cheng Wang [1], Jun Yang [1,2,*], Xiaojun Qiao [1]

[1] *National Engineering Research Center for Information Technology in Agriculture, china, 100097*

[2] *China Agricultural University, Beijing, China, 100083*

[*] *Corresponding author, Address: Shuguang Huayuan Middle Road 11#, National Engineering Research Center for Information Technology in Agricultur, Beijing, China, 100097, Tel: +86-10-51503409, Fax: +86-10-51503449, Email: yang1jun2@sohu.com*

Abstract: This paper mainly expounds that using the WINSOCK controller which is provided by VB6.0 to program the procedures in the coop intelligent control system, it introduces the character of the TCP protocol and the UDP protocol, and according to the demands of the system environment to choose the suitable protocol. This system uses the WINSOCK controller and the UDP protocol for network communications, and also it has the communication function of network data between all the acquisition controllers and PC.

Keywords: Network Communications, Environment control, Winsock control component

1. INTRODUCTION

With the emergence of large-scale, intensive rearing methods, information technology has been widely applied to the aquaculture industry. Foreign environmental control technology for the earlier study, it began in the 1970s, using a combination of analog instrumentation to collect site information to instruct, record and control. In the end of 1980s, distributed control system developed. Now the research and development is multi-factor integrated control system of computer data acquisition and control system. On the domestic, coop environmental monitoring system mostly used the traditional on-site control of RS485, CAN and so on[1-2]. we sleeted a technology

Zhao, C., Wang, C., Yang, J. and Qiao, X., 2008, in IFIP International Federation for Information Processing, Volume 259; Computer and Computing Technologies in Agriculture, Vol. 2; Daoliang Li; (Boston: Springer), pp. 1063–1071.

program with universal, scalability and upgrading in hardware and software systems of environment intelligent control system, and it adapt to the domestic situation. All of these make the research and development of coop environmental monitoring technology to have good expansion and upgrading foreign coop environmental monitoring system exists the shortcomings of expensive, incompatible with the status of domestic. To change this situation, According to the coop environment and the system's needs, We embed UDP network protocol in our systems, through the network to make a real-time monitoring to coop, greenhouses and other agricultural installations and environment. By the way, we construct the network data transmission system of coop intelligent monitoring system.

2. WINSOCK CONTROL NETWORK PROGRAMMING

In the coop intelligent monitoring system, including acquisition module, transmission module, and analysis modules of environment parameters. In this paper, we introduct the network transmission of environment parameters based on the UDP protocol. In our reasearch of network applications using VB, Winsock Control was used to achieve function of system it is visible to users. It provide a convenient way to visit TCP and UDP network services. Microsoft Access, Visual Basic, Visual C + + or Visual FoxPro developers all can use it. Winsock Control prepares the client and the server application procedures. Without understanding the details of TCP or calling Junior Winsock API. By setting the control attributes and calling its methods can easily connect to a remote computer, and can also achieve a two-way exchange of data[3].

2.1 Method of Winsock control

Listen: Server to be used for creating socket and it is set to interception pattern, waiting for called by customer.

Connect: Clients send request to the server.

Accept: Server aware of the connection request of Client, Using this method to agree to connect.

SendData: Data will be sent to the other part.

GetData: Receiving the data come from the other side of the information and storing the data.

Close: Close the using link.

2.2 Event of Winsock control

ConnectingRequest: The incident occurred when Server "listen" to the connect request from client, at this time Server should express receiving or not.

DataArrival: The incident occurred if the other sides send data by the method SendData. At this point these data was deal with, For example, saved it to the database or displayed it.

2.3 Property of Winsock control

LocalPort: For the sever, Setting up a local port, approach "listen" listen to news from local port.

RemotePort: Customers set up a the remote port to visit, namely the server localport value.

RemoteHost: Clients designated computer to connect, namely IP address. Like "192.168.4.144", we also can use the computer name "admin".

State: The states of Winsock Control, namely the linking of two state machines.

Protocol: Returning or setting up protocol of Winsock Control used (TCP or UDP).

3. NETWORK COMMUNICATION PROTOCOL

In Internet, the main protocol is the TCP/IP protocol,there has two types of protocol to achieve information transmission in TCP/IP transport layer, They were the TCP and UDP[4]. Characteristics of TCP and UDP were introduced as following.

3.1 TCP Basis

TCP is a connection-oriented protocol, the protocol that need to establish a reliable link between the two points, Which includes a special mechanism to ensure the transfer, When the receiver has received reports from the sender information and automatically send to information confirmed to sender. The sender continue to send other information after receiving confirmation of information, Otherwise the sender will have to wait until the information of confirmed received. TCP used for non-real-time data business normally, it can be compared with the telephone system. At the beginning of data transmission, users must establish a connection.

Data Transfer Protocol allows to creat and maintenance connecting with the remote computer, Two computers can be connected with each other for data transmission. If client applications was created, we must know the server computer name or IP address, We must also know the "bugging" of the port, then called connect Method. If the server application procedures was created, we should set up a "listen" ports, and called listen Methods. The client computer will need to connect when the ConnectionRequest of incident occurred to complete the connection, called the incident ConnectionRequest of Accept Method. After establishing a connection, any one computer can be send and receive data, in order to send information, called the SendData of method. When receiving data, DataArrival incident occurred. Called the incident DataArrival GetData method to get data.

3.2 UDP Basis

UDP is a news-oriented protocol, it does not need to establish communication link and not provide data transmission mechanism of the guarantee. If the data was lost from the sender to the receiver of the transfer process, the protocol itself does not make any detection or prompt. So UDP is known as unreliable transmission protocol. Between the two computers similar to the transmission of mail transmission, message is sent from a computer to another computer, but the two is no clear link. In addition, each transmission of the largest volume of data depended on the specific network.

UDP is a connectionless protocol, UDP is different operation of TCP. It is not establish a link between computers. In addition, the UDP applications procedure can be both client and server. To transmit data, we must first set up a client computer's LocalPort attributes. Then, the server computer can be set up RemoteHose as customers computer Internet address. RemotePort attribute was set to the same as client computer with LocalPort port, called SendData Method to send information. Client computer have access to the sended information using GetData method of DataArrival incident.

4. PROTOCOL CHOICE

In the use of Winsock, the first we must consider the kind of communication protocol. Network for data communications, it needs to mark the mainframe network using address, in order to ensure correct data sent to the mainframe. We have introduced the use of the protocol. Including TCP and UDP protocol, the difference between the two protocols is that their connection status, TCP is a connection-oriented protocol, and UDP is a connectionless protocol[3].

In the system design which protocol should be chosen in the end, protocol was choosed by Establishing Application Programm. Following several aspects of the introduction and analysis of our system suitable agreement.

(1) At the time of sending and receiving data, application procedures, whether they need to be client or server confirmation message? If you need to use TCP, the sending and receiving data before establishing a clear link to ensure reliability

(2) Transmission data whether it is particularly large (like images and audio files)? If yes, after connected, TCP protocol can safeguard connection and ensure the integrity of data, but, this link requires more computing resources, so it is more "expensive" in Cost and efficiency.

(3) Whether this data is sent the interval in a conversation or completed? If Application procedures required to notify a computer in a particular task completed, then UDP is more appropriate, it is suitable for a small amount of data to send or immediate data.

Through the analysis we have come to this conclusion, UDP protocol suitable for the two sides what do not need confirmation communication of information, UDP has small resource consumption, with kept its speed more advantage than TCP. Although TCP protocol was implanted various security functions, TCP will take up a lot of consumption cause speed to be severely affected in actual executive process. UDP excluded the reliability of information transmission mechanism and transferred to the upper application to complete security and scheduling of protocol greatly reducing the execution time, the speed of a guarantee, Although UDP protocol lost some packets in application, the receiver will not have much impact on the results. Comparing with the reliability example, in the practical application, We pay more attention to the actual performance, so in order to obtain better performance, we can sacrifice some reliability.

Therefore, after considering the effect of communication UDP, speed and the requirements of environmental facilities, we adopted UDP protocol between controllers in the environmental monitoring system. Any one collector can send data to any other collector, without getting confirmed by other information, Using broadcasting, a collector broadcast their own data to other collector. In Practical, all collectors collect information outside the station also, namely information of station is shared by all controllers as a reference to the factors.

Comprehensive analysis, the system adopted linking-less of UDP, with support of network communication between PC and controller, the network communication structure shown in the following.

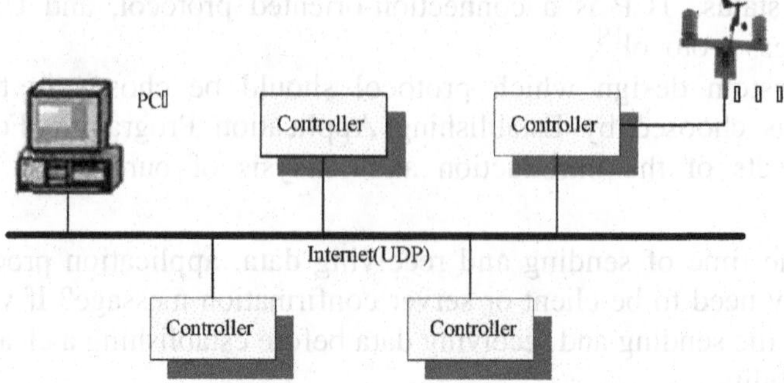

Fig. 1. The structure chart of the network communication

Controller equivalent to a PC in the network in system, System can assign an IP address for each controller. Not only Controller exchanged data with pc, but also realized data sharing between the controllers. The station information as a shared data can publish information in form of broadcasting in network. Other network nodes can be set up monitoring ports in network and monitor different types of data; we can set up port what is corresponding to weather station nodes, in order to receive the data and data sharing.

UDP must know device's IP address and port number in the process of communication. We can read all the information in Initialization procedure. we know to who communication and read out its corresponding IP and Port in the process. In order to transmission of data between multi-machine, we must set up separately LocalPort for each attribute, we set up Remote Host attribute of PC as Controller IP addresses and Remote Port attribute of PC as controller LocalPort attribute value. Then PC called SendData method to realize transmission of data. Controllers use GetData method of DataArrival to obtain information what pc sent to controller. If controller send data to PC, the process is the same as above, some codes are as follows:

```
frmMDIMain.WinsockUDP.RemoteHost=typDevice(iDevice). iRemoteHo
st /* seting up RemoteHost attribute of PC as Controller IP addresses
frmMDIMain.WinsockUDP.LocalPort = 1234 /* Setting up local port
frmMDIMain.WinsockUDP.RemotePort =
typDevice(iDevice).iRemotePort/*setting up RemotePort attribute of PC as
controller LocalPort
frmMDIMain.WinsockUDP.SendData string /*calling SendData method
Private Sub WinsockUDP_DataArrival(ByVal bytesTotal As Long)/*
Calling the incident DataArrival GetData method to access data
Dim rec As String
WinsockUDP.GetData rec, vbString
ReceiveBufferUDP = rec
ReceiveOK = True
End sub
```

5. CONCLUSIONS

The system has realized the sheds environmental parameters monitoring network in Environmental parameters of the transmission using UDP data transmission network. In the coop, the system adopted a network communication Technology between controllers, as well as between the controller and PC, namely UDP protocol. UDP protocol can greatly enhance communication efficiency and lift velocity. UDP protocol can greatly reduce the polling time and reduce the execution time when pc collected

Data form collector. The purpose of systems development makes environmental management of chicken house toward more intelligent and more network in the direction.

ACKNOWLEDGEMENTS

This research is supported by the National High Technology Research and Development Program of China (863 Program, Grant 2006AA10A311 and Grant 2006AA10Z253) and Beijing Science & Technology Program (Z0006321001391).

REFERENCES

ChenWei, Shixi Yang. Visual Basic Programming from basic to practice. Beijing: electronic industrial press, 2005:210-271.

Deze Zhou, Naner Yuan, Yingying. The Design and Application of Computer intelligent Control System. Beijing: Tsinghua University press, 2002:57-60.

Changhua Lu, Wuzi, Lifang Wang etc. Establishment and application of computerized production management system for large scale poultry farm. Journal of Agricultural Engineering, 2003, 19 (6):256-259.

Niuli. Visual Basic6.0 programming. Beijing: electronic industrial press, 2005.

Yadong Luo. Implementation of network real-time communication with Winsock. Chengdu Information Technology Institute, 2004, 19(2):198-201.

Yinong Hu, Shaochun Dou, Lifang Wang etc. Hens of scale farms with Automatic Monitoring System. Jiangsu Agriculture Journal., 2002, 18(3):176-180.

Zhou Juan, Xianghua Chen. With Winsock UDP-based establish applications procedure In VB. Electronic machinery College Journal of Chengdu, 2005, 31(2):17-20.

RESEARCH OF DECISION SUPPORT SYSTEM (DSS) FOR GREENHOUSE BASED ON DATA MINING

Cheng Wang[1], Lili Wang[2,*], Ping Dong[2], Xiaojun Qiao[1]

[1] *National Engineering Research Center for Information Technology in Agriculture, Beijing, China*

[2] *Information engineering institute University of Science and Technology Beijing, Beijing, China*

[*] *Corresponding author, Address: Shu Guang Hua Yuan Middle Road 11#, Haidian district, Beijing, 100097, P. R China, Tel: +86-010-51503409, Fax: +86-010-51503349, Email: lippery@sina.com.cn*

Abstract: Most of expert knowledge in agriculture is descriptive and experiential, so it is difficult to describe in mathematics and build decision support system (DSS) for greenhouse. Therefore, the decision support system (DSS) for greenhouse constructed of data warehouse and date mining technology was introduced in this paper. In the system, data warehouse was founded to memory diversified date, the using of on-line analytical processing and date mining enriches knowledge base with new agriculture information. Implementation of system adopted SQL Server analysis, as a result, tightness coupling of data warehouse, date mining and application, improved efficiency of date mining. Combined data warehouse with on-line analytical processing and date mining to construct a novel DSS.

Keywords: Data warehouse, Date mining, OLAP, Decision support system (DSS)

Wang, C., Wang, L., Dong, P. and Qiao, X., 2008, in IFIP International Federation for Information Processing, Volume 259; Computer and Computing Technologies in Agriculture, Vol. 2; Daoliang Li; (Boston: Springer), pp. 1071–1076.

1. INTRODUCTION

Dates in agriculture have the follow characteristics: large quantity, multi-dimension, no integrity, incertitude and so on. How to effectively find the interrelationship of dates effects agricultural development, economic benefit, social benefit. Using decision support system (DSS) can obtain benefits (Zhao et al., 2006).

Now, expert knowledge base in most DSS is established on a domain, but acquisition domain expert knowledge is a complex process, describe in mathematics is difficult, along with quantity of data augment, the conventional DSS shows its inherent shortage. Development of data warehouse, on-line analytical processing (OLAP) and date mining improve DSS, and have breakthrough (Helen, 2001).

Data warehouse is dates gather aimed at thematic, co positive, steady, and multi-time. It contains basic data, historical date, synthetic data, and source date (Michael, 2001).These dates restructure according to decision convenient for user to extract each wished date and information. On-line analytical processing (OLAP) offers multi-dimension analysis method: slice up, slice block, rotation, etc. it also convenient for user to extract the needed date and information in differ point of view (Dorian, 1999). Based on OLAP and Data warehouse, Date mining can using correlation analysis, similitude search, trend analysis and forecast, pattern analysis to acquire new knowledge, find conclude reasoning and potential mode, help making correct decision. Above three Technologies have internal relation and complementarily, combining them can construct a new DSS (Cody, 2002).

2. THE GENERAL DESIGN

In this system, data warehouse memory dates from idiosyncratic information source, and that, these information source itself is a gigantic data warehouse, OLAP dedicated to date analysis, Date mining apply itself to acquire new knowledge, when design and construct DSS, apply above technology, can apply improve processing capacity.

Architectural structure of DSS as graph 1, data warehouse can realize storage of decision-making date, sampled data from source date, then cleaning, integrating, transferring, offering date view, and then these dates have good quality. OLAP can realize multi-dimension date analysis via building multi-dimension data model in multi-view. date mining automatically dig out the hidden mode and information, forecast the prospective trend, direction OLAP, expert system (ES) can utilize knowledge reasoning qualitative analysis, we call it competitive synthetic

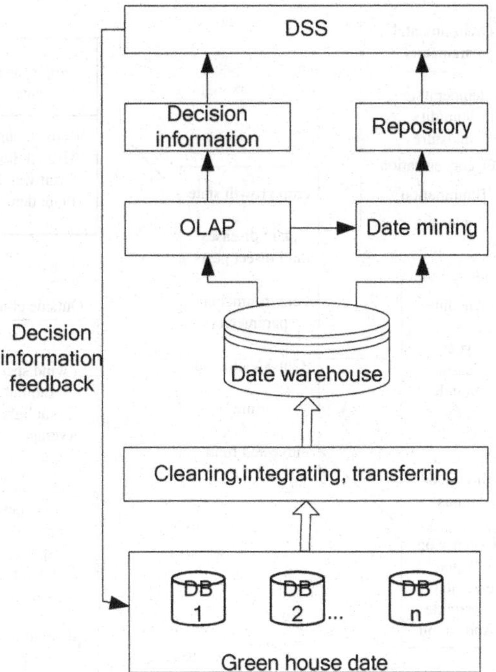

Fig. 1. System whole structure chart

decision support system, complementation, exert each other superiority, realize effective aid decision making (Inmon, 2002).

3. DATA WAREHOUSE MODEL DESIGN

First, Extraction related data environmental monitoring, building data warehouse. Then, On base of date warehouse, using date mining technology, dig out the hidden information, relationship of environmental element and restrict factor, improve utilance of monitoring dates, propose pertinence resolve means, advanced environment administrative department decision-making ability.

According as analysis, determinate issue include time, outside climate, environmental parameters, crop growth state, plant diseases and insect pests, invested funds. In this paper, we choose star model, for this model has the merit: convenience to build model, easy to understand, support multi-demotion date analysis. Structure is Fig. 2.

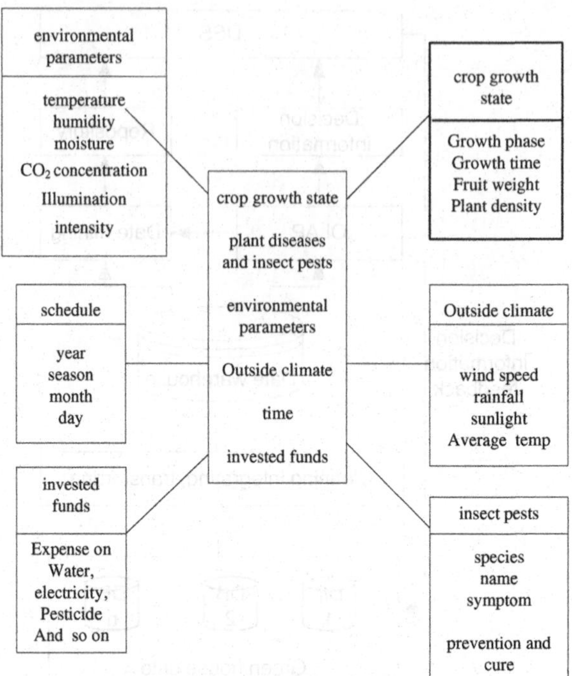

Fig. 2. Star model of data warehouse

4. BUILD GREENHOUSE DSS

4.1 Build OLPA module

To express the data in the data-base which using Service module of Microsoft OLAP Analysis Services, and basing on the fact dimension table. The method is slice up, slice block, rotation, to analyze relate dates in different view. OLAP model can complete analyzing environmental parameters, crop growth state, plant diseases and insect pests, then give out result. The result is easy to understand in mode of report forms, chart etc. PivotTables Service provide interface, adopt XDM sentence to perform multi-dimension date set search.

4.2 Build date mining module

Using multi-dimension date set build in OLAP to build date mining model. API – OLEDB for Data Mining is one of Analysis Services, it is a program interface for date mining application program, by means of API, using different kind of arithmetic to complete mining task.

4.2.1 Trend analysis and forecast module

For greenhouse, under the condition of forecast precision, can using prior three days dates to forecast the fourth day, that is to say, found a four section window. When greenhouse sensor fault or don't send dates, manager can take correspond means to maintain environmental parameters in perfect.

4.2.2 Similitude search module

In system, familiar module in different environmental parameters is similitude, find out comparability of modules. Similar module is temperature and illumination, air humidity and soil moisture and so on.

4.2.3 Associate rule analysis module

In DSS, based on plant diseases and insect pest's experience, using Apriori arithmetic, discover the associate rule of plant diseases and insect pest and environmental parameters.

4.3 Program UI

Visual C++ is a very good program tools, applications have friendly user interface, user can clean, integrity, transfer dates to date warehouse, also can build date warehouse, search date, give out result, etc. Build interested date mining module is allowed.

4.4 System architecture

This DSS consist of client, application server, database server. In client, using Visual C++ programmed applied UI, OLAP and date mining run in application server, bottom date warehouse build in database server. Client is used to report result for user, application serve handing applied logic, acquire date from database server when necessary, and deliver date to client. Database server automatic update maintenance date warehouse.

5. SUMMARIZE

In this paper, we attempt to apply date mining and data warehouse technology in DSS of greenhouse, considering of greenhouse's characteristic, preliminary design the frame of data warehouse, on the base of date warehouse, we perform date mining, in order to offer useful content such as environmental parameters,, it's relation, plant diseases and insect pests for agriculture expert. In future, along with the development of data warehouse, OLAP and Date mining, DSS for greenhouse will have a perfect futurity.

ACKNOWLEDGEMENTS

This research was supported by the National High Technology Research and Development Program of China (863 Program, Grant 2006AA10A311) and National High Technology Research and Development Program of China (863 Program, Grant 2006AA10Z253).

REFERENCES

Helen Hasan, Peter Hyland. Using OLAP and multidimensional data for decision making. IT Professional, Vol. 3, No. 5, 2001, pp. 44–50

Michael Blaha. Data warehouses and decision support systems. M Computer, Vol. 34, No. 12, 2001, pp. 38–39

Dorian Pyle. Data Preparation for Data Mining. Morgan Kaufmann, 1999

Inmon W H. Building the Data Warehouse (Third Edition). John Wiley & Sons, 2002, pp. 20–53

Y W F, Kerulen J T, Krishna V, Spangler W S. The Integration of Business Intelligence and Knowledge Management. IBM Systems Journal, Vol. 41, No. 4, 2002, pp. 697–714

Zhao Yu, Q I Guoqiang. Application of data warehouse in decision support system of rice cultivation management [J]. Journal of Northeast Agricultural University, Vol. 37, No. 4, 2006, pp. 557-562

RESEARCHES OF DIGITAL DESIGN SYSTEM OF RICE CULTIVATION BASED ON WEB AND SIMULATION MODELS

Hongxin Cao [1,*], Zhiqing Jin [1], Yuwang Yang [2], Chunlin Shi [1], Daokuo Ge [1], Xiufang Wei [1]

[1] *Institute of Agricultural Resources and Environment Research/Engineering Research Center for Digital Agriculture, Jiangsu Academy of Agricultural Sciences, Nanjing 210014, Jiangsu, China*

[2] *School of Computer Sciences, Nanjing University of Sciences & Technology, Nanjing 210014, Jiangsu, China*

[*] *Corresponding author, Address: Institute of Agricultural Resources and Environment Research/Engineering Research Center for Digital Agriculture, Jiangsu Academy of Agricultural Sciences, 50 Zhongling Street, Nanjing 210014, Jiangsu, China, Tel: +86-25-84390125, Fax: +86-25-84390248, Email: caohongxin@hotmail.com*

Abstract: In order to integrate web with crop growth simulation and decision-making support system, the field experiment of different basal levels was carried out in experiment area of Jiangsu Academy of Agricultural Sciences in 2005 adopting 4 cultivars such as “Wuyungeng 7”, “Yangdao 6”, “Yueyou 948” and “Nangeng 41”, which mainly were used in collecting cultivar parameters and updating database of them. The database of rice cultivars, soil and weather data were developed using SQL Server 2000. The pages of digital design system of rice cultivation based on web and simulation model (DDSRCBWSM) were designed using Visual Studio.Net, which included register, the main page, cultivar parameter management, site data management, parameter adjustment, decision making for rice cultivation, and so on. The DDSRCBWSM accorded with TCP/IP agreements, which could be installed and run in server (IIS5.0), and be browsed on internet, it inherited mechanism, universal adaptability and utility of rice cultivation simulation-optimization-decision making system (RCSODS), combined web techniques with research of rice growth models, set up web system of RCSODS, made agricultural technicians of main rice production area of china gain pre-sowing optimization cases and rice management suggestion of the current year with dynamic, goal and digital characteristic in accordance with soil, cultivar and weather conditions online, and to fulfill technical direction through many kind manner such as paper, internet, email, television, wall newspaper, and so on, eventually.

Cao, H., Jin, Z., Yang, Y., Shi, C., Ge, D. and Wei, X., 2008, in IFIP International Federation for Information Processing, Volume 259; Computer and Computing Technologies in Agriculture, Vol. 2; Daoliang Li; (Boston: Springer), pp. 1077–1086.

Keywords: Web, Simulation model, Rice cultivation, Decision-making support system, Digital design

1. INTRODUCTION

China is the largest country of rice production and consumption in the world. Rice accounted for more than 50% of commodity food in China, its plant area was about from 28,000,000 to 31,000,000 ha and the total yield was from 120,000,000 to 180,000,000 ton in the normal year, which was 30% of total plant area and 43.6% of total yield of food crops in China, respectively. The matching of both good cultivars and its optimum cultivation techniques was an important scientific guarantee of steady increase for rice production in China; the Rice Cultivation Simulation-Optimization-Decision Making System (RCSODS) (Gao and Jin et al., 1992) provided an available way for the matching. However, RCSODS was still not integrating with web, and its extensive application was limited to some extent.

There were reports inland and overseas for studies on integrating database (DB), agricultural expert system (AES), decision-making support system (DSS), agricultural model, and so on with web. Many researchers integrated fertilization models, groundwater models, soil and water quality models, DSS or DB with web, respectively (Comis, 1999; Shaffer, 2002; McCown, 2002; Winston, 2002; Gunn et al., 2002; Bostick et al., 2004; Miller et al., 2004). Researches of integrating DB, AES, soil eroding models, DSS or GIS with web had also been reported (Shi et al., 1999; Yu, 2000; Gao et al., 2000; Liu et al., 2001; Yang et al., 2002; Liao et al., 2002; Li et al., 2003; Wang et al., 2003; Chu, 2003; Chen et al., 2003; Wang et al., 2004). However, the studies of integrating crop growth models and agricultural DSS with web were not reported in literature inland and overseas.

The objective of this research were integrating DB of rice cultivars, soil and weather data, rice growth models, rice optimization models with web platform, implementing internet running of rice growth models and optimization models based on RCSODS.

2. MATERIALS AND METHODS

2.1 Materials

In order to update cultivar parameter DB, WUYUNGENG 7, YANGDAO 6, YUEYOU948, and NANGENG 41 were adopted as experiment materials.

2.2 Methods

2.2.1 Field experiment

The trail was conducted in 2005 in experiment area of Jiangsu Academy of Agricultural Sciences, Nanjing, China, as split-plot arrangements (main plots were basals including 2 levels (fertilizer and no fertilizer (CK), thereinto, fertilizer included basals of nitrogen rate at 75 kg·ha^{-1}, P_2O_5 at 120 kg·ha^{-1}, and K_2O at 45 kg·ha^{-1}, nitrogen rate during tiller period at 60 kg·ha^{-1}, and nitrogen rate during ear and grain period at 45 kg·ha^{-1}), sub-plots were cultivars with 4 levels) with 8 treatments, 3 replications, and 24 plots, each plot size was (10×3.029)m^2 , and there were 13 rows per plot.

The former crop of the experiment field was fallow, its soil fertility was at middling-crackajack, the soil from the topsoil (0-40cm depth) contained soil organic content at 19.8 g•kg^{-1}, alkali soluble nitrogen at 138.6 mg•kg^{-1}, available phosphorus at 19.9 mg•kg^{-1}, available potassium at 141.9 mg•kg^{-1}, and pH at 6.51. The deep plowing in the land preparation was conducted on May 18, planting date was on May 15, the basals was applied on Jun 15, transplant was conducted on Jun 16, urea 6.5 kg (nitrogen rate 60 kg·ha^{-1}) was applied on Jun 21 and 4.65 kg (nitrogen rate 45 kg·ha^{-1}) on Aug. 17, the other field managements were the same as general rice field.

2.2.2 Data to be collected

They included phenology recordation, tagging and recording leaf age, investigating plant density, yield, and yield components. The shoot and tillering dynamic were investigated 1 time every 5 days, beginning from tillering date, 2 rows 30cm length each plot. LAI, the dry matter, population spectrum feature, plant nitrogen content and phosphorus content also were determined at rice main phenology period such as transplanting date, reviving date, tillering date, elongation date, heading date, anthesis, filling date, and mature date, sampling by the conventional means, thereinto, population spectrum feature was determined using EXOTECH100BX spectrum radiometer. Plant analysis (conventional items) at mature was conducted in laboratory.

3. DESIGN OF STRUCTURE AND FUNCTIONS OF THE SYSTEM

3.1 Design of DB

The DB of rice cultivars, soil, and weather data were developed by SQLServer2000. The rice cultivar DB included model parameters (rice phenology, leaf age, photosynthesis, yielding components, internode numbers, tillering rate, leaf area per plant, optimal season, and temperature index at all heading date), genetic characteristics of cultivars and knowledge DB of controlling pest and weed. The soil parameter DB involved soil types, texture, organic matter, total nitrogen, available phosphorus, available potassium, pH, ratio of volume to weight, saturation water content, field hold water content, and wilting point etc., the weather DB dealt with monthly or daily average temperature, maximum temperature, minimum temperature, sunlight time, precipitation, and raining days etc.

3.2 Design of The System's Functions

The system's functions mainly involved management of cultivar DB and site DB, cultivar parameter adjustment, and rice cultivation decision-making (see Fig. 1). The DB, knowledge DB and model class libraries, and functional modules accorded with TCP/IP agreements, which could be installed and run in server (IIS5.0), and be run on any browser of internet.

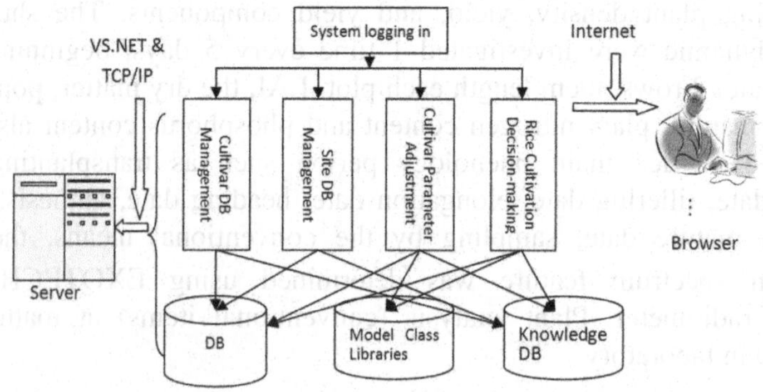

Fig. 1. The function frame chart of the DDSRCBWSM

3.2.1 DB management

In order to fulfill appending, deleting and modifying of data online, the application program of ASP.NET Web in VS.NET needed to be run when calling cultivar and knowledge DB, and site DB of SQL Server, and weather simulator.

3.2.2 Cultivar parameter adjustment

In order to fulfill adjusting of cultivar parameters online, the application program of ASP.NET Web in VS.NET needed to be run when calling cultivar and knowledge DB of SQL Server, and model class libraries (rice growth models and optimum models).

3.2.3 Rice cultivation decision-making

In order to fulfill the normal and the current year decision-making of rice cultivation, show curves of the optimum LAI, shoot and tillering dynamic for rice, and build tables of cultivation case online, the application program of ASP.NET Web in VS.NET needed to be run when calling cultivar and knowledge DB of SQL Server, and model class libraries (rice growth models and optimum models).

3.3 Design of Web Pages of The System

They included enter page (Fig. 2), the main interface page of the system (Fig. 3), the cultivar DB management page (Fig. 4), the site and weather DB management page (Fig. 5), the parameter adjustment page (rice phenology, leaf age, photosynthesis, yielding components, node numbers and tillering rate, leaf area per plant, optimal season, and temperature index at all heading date) (Fig. 6) and the rice cultivation decision-making page (Fig. 7).

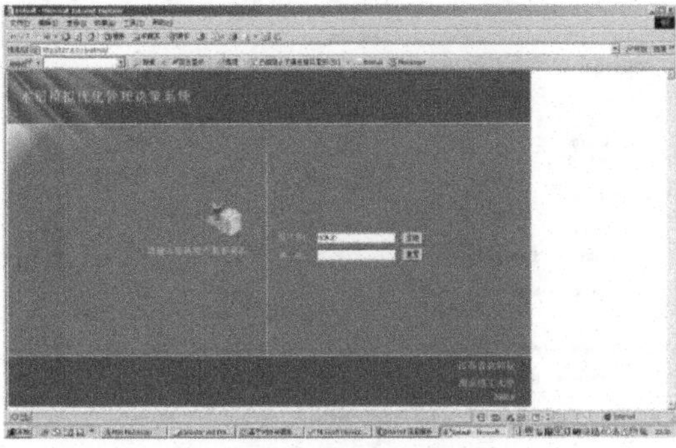

Fig. 2. The enter page

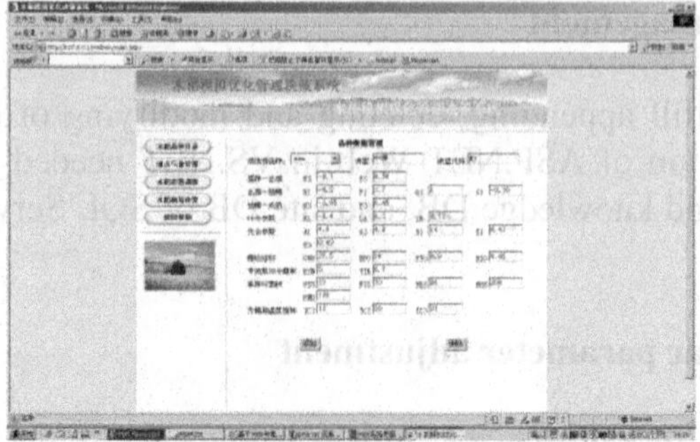

Fig. 3. The main interface page of the system

品种数据管理

请选择品种:	aaa		类型	中粳		类型代码	MJ
播种一出苗	KS	-0.7	PS	0.56			
出苗一抽穗	K1	-4.2	P1	1.7	Q1	1	G1 -0.36
抽穗一成熟	K3	-3.65	P3	0.45			
叶令参数	KL	-0.4	LA	0.3	LB	0.72	
光合参数	A1	4.8	A3	4.8	B1	14	E1 0.43
	E3	0.49					
穗粒结构	GWO	27.5	EPO	24	FTO	0.9	HIO 0.45
节间数与分蘖率	ESN	5	TIR	0.7			
单株叶面积	FS7	23	FTS	70	FES	90	FHS 200
	FMS	100					
齐穗期温度指标	TC1	12	TC2	26	TC3	23	

添加 删除

Fig. 4. The cultivar DB management page

常年气候数据库目录

请选择地点: 福州 纬度: 26.01

	月份	均温	最高温	最低温	日照时数	月雨量	月雨时
编辑	1	11.5	15.8	8.7	96.0	52.1	10.4
编辑	2	11.8	15.9	9.1	73.1	83.8	14.3
编辑	3	14.1	18.6	11.3	81.6	144.2	16.9
编辑	4	18.8	23.5	15.5	109.7	143.4	16.8
编辑	5	23.2	27.5	20.0	123.4	177.5	18.3
编辑	6	26.7	30.8	23.6	139.9	185.1	16.5
编辑	7	29.6	34.3	26.0	222.9	98.4	10.4
编辑	8	29.1	33.8	25.7	191.6	177.0	12.4
编辑	9	26.4	30.6	23.4	141.7	180.1	12.0
编辑	10	22.7	27.0	19.5	141.6	32.2	6.8
编辑	11	18.7	23.0	15.5	116.6	36.5	7.3
编辑	12	13.7	18.3	10.5	117.3	30.4	7.9

添加地点 删除地点 查看每日数据

Fig. 5. The site and weather DB management page

参数调整

调整的品种 汕优63 地点 南京 调整步骤 生育期参数

生育期参数

	播种期		出苗期		抽穗期		成熟期	
	月	日	月	日	月	日	月	日
模拟值	5	15	5	18	8	20	9	30
实际值	5	15	5	18	8	20	9	30

	播种～出苗	出苗～抽穗	抽穗～成熟	播种～成熟
误差（天）	0	0	0	0

生育期模型参数

1. 播种 ～ 出苗 KS= -0.8 PS= 0.4
2. 出苗 ～ 抽穗 K1= -4.38 P1= 1.165 Q1= 0.777 G1= 0
3. 抽穗 ～ 成熟 K3= -3.59 P3= 0.226

调整规则: 误差大于/小于0,则分别减小/增大KS,K1,K3,G1的绝对值

调整后刷新 保存修改后的数据 下一步

Fig. 6. The parameter adjustment page

Fig. 7. The rice cultivation decision-making page

3.4 The Run Environment of The System

It needed to install WIN2000Server or upward operating system, dot NET framework, SQLServer2000 or upward DB system, EXCEL5.0 or upward, and IE6.0 or upward software in server, WIN98 or upward operating system, EXCEL5.0 or upward, and IE6.0 or upward software in client.

3.5 The System Application Cases

The system may establish and down the optimum sheet of pre-planting for rice online by browser through selecting weather data in the normal year in Nanjing, cultivar data of XIANYOU 63, and soil nutrition etc., adjusting cultivar parameter of XIANYOU 63, and conducting the optimum decision-making sheet of pre-plant (Fig. 8), which included the rice growth and development period, the optimum population dynamic, and the optimum fertilizer rate, the assign of nitrogen, and the control of pest and weed, and so on, with strong technique and maneuverability.

4. CONCLUSIONS AND DISCUSSIONS

The DDSRCBWSM implemented integration of the rice growth models, the rice optimum models, and DB with web, developed internet platform of rice simulation-optimization-decision making, and provided basis for application of rice growth models online.

The DDSRCBWSM inherited mechanism, universal adaptability and utility of RCSODS, combined web techniques with research of rice growth models, can establish pre-sowing optimization cases and rice management suggestion of the current year with dynamic, goal and digital characteristic

SHEET OF GOOD CULTIVARS AND ITS OPTIMUM CULTIVATION TECHNIQUES XIANYOU 63 IN NANJING REGION

ILLUMINATING: The case followed the principles of 'high and stable yield, high benefit and agility', its main optimization steps were as follow: under precondition of adopting good cultivars, 1. choosing the optimum season; 2. cultivating strong seedling by suitable density; 3. deciding the optimum density in field; 4. increasing ratio from tiller to panicle or promoting valid tiller and controlling invalid tiller early; 5. covering row space in optimum time to gain the optimum population; 6. gaining big ear or fertilizer in anthesis; 7. increasing grain ratio by delaying leaf function late; 8. fertilizing N, P and K according to soil and yield.

Cultivar:	XIANYOU 63	Rice crop type:	Single season rice	Phenology (M-D):	02-12
Cultivar type:	Hybrid Indica Ting	Seedling and transplanting:	Dryland, transplanting	Transplant (M-D):	06-16
Goal yield (kg·666.7m⁻²):	680	Life-cycle days (D):	152		

Cultivar characteristics and cultivation outlines: Strong seedling and plant, purple sheath, deep leaf colour, middle tillering, more tighten plant type, big ear, high seed rates, good disease resistance, and wide adaptability. Gaining ear number enough, big ear, sowing in the optimum date, wide row, the ratio to basal, tiller and ear fertilizer at 6:4

Actual season					
Seedling age and density	Seedling age (D): 33	Optimum seed rate (kg·666.7m⁻²): 30.2	[illegible]: 4	
Period	From sowing to seedling	From transplant to elongation	From elongation to heading	From heading to mature
Phenophase	Sowing · Seedling · 3-leaf · Draw seedling	Transplant · Re-growth · Valid tiller	Elongation · Flower step · Flag leaf	Heading · Yellow · Mature
Prediction date (M-D)	02-12 · 02-18 · 02-27 · 06-16	06-16 · 06-20 · 07-12	07-24 · 08-02 · 08-10	08-20 · 09-21 · 09-30
Leaf age (number)	1 · 3 · 4	4 · 4 · 12	14 · 16 · 17	
Main aims	Fostering heavy seedling	Valid tiller	Strong plant and big ear	High grains and 1000-grain weight
The optimum shoot number (10⁴·666.7m⁻²)	148.1	4 · 4 · 16.1	28.2	16.1 · 16.1
The optimum LAI		0.1 · 0.1 · 1.5	4.8	7.6 · 4.1
The optimum biomass (kg·666.7m⁻²)	0.1 · 26.6 · 333.3	23.4 · 23.4 · 70.8	186.3	630.3 · 1301.2
Water management	Wet, thin water	Thin water, little water	Thin water, dry and wet turning	Dry and wet turning, no water
Soil nutrition	Fertilizing period	Basal · Tillering · Tillering	Anthesis	Ear or grain
Organic matter (%) 1.5; Total nitrogen (%) 0.1	Nitrogen rate (kg·666.7m⁻²): 13.1	6.6 · 1.5 · 2	2.6	0.7
P_2O_5 (ppm) 15; K_2O (ppm) 138	P_2O_5 rate (kg/666.7m²): 10	10		
pH 6.2	K_2O rate (kg/666.7m²): 8	3.8	1.2	Ear number (10⁴·666.7m⁻²): 16.1; Grains per ear (kernel): 170.5
Nitrogen rate with strong seedling		Decreasing rate (−80~100%)	Decreasing rate (−10~80%)	Ratio of valid grains: 0.82
Nitrogen rate with weak seedling		Increasing rate (+20~40%)	Increasing rate (+20~30%)	Valid grains per ear (kernel): 142.5
Nitrogen rate under high temperature, more light		Decreasing rate (−10~20%)	Increasing rate (+10~20%)	1000-grain weight (g): 28
Nitrogen rate under low temperature, less light		Increasing rate (+10~20%)	Decreasing rate (−10~20%)	Yield in normal year (kg 666.7m⁻²): 608.2
Main disease or insect	Planthopper	Sharp Eyespot	Rice Stripe Disease	Rice Blast Fungi

1. Seed treatment: Control of bakanae disease (Sporta 25% EC, 2000~3000 dilution).
2. [illegible]
3. [illegible]
4. Control of weed in seeding bed: 42% Herbicide milk oil 1 ml plus water 5~8 kg/66.67m², spraying after sowing.

Note	[illegible]
	*Development unit: Jiangsu Academy of Agricultural Sciences, May, 2006

Fig. 8. The pre-plant optimum cases for XIANYOU 63 planted in Nanjing

in accordance with soil, cultivar and weather conditions online, and to fulfill technical direction through many kind manner such as paper, internet, email, television, wall newspaper, and so on.

The DDSRCBWSM also provided internet platform combining with expert knowledge, images, video, and field task record and traceable management etc., can help to direct rice standardization production.

At present, the DDSRCBWSM has not integrated with GIS, however, the realization of region distributing and large-scale direction functions will require supports of GIS, it will need to be strengthened on this aspect.

REFERENCES

Bostick M, Koo J, Walen V K , Jones J, Hoogenboom G. A web-based data exchange system for crop model applications. Agronomy Journal, 2004, 96(3):853-836

Chen S H, Mao D M, Xu Y M. Development of soil nutrition regional management system in Xinjiang Uigur Autonomous WEBGIS-based. Modernization Agriculture, 2003, (11):27-29 (in Chinese)

Chu J X. Design and Development of the Web Based Forest Fire Forecast System. FOREST RESOURCES MANAGEMENT, 2003, (5):58-60 (in Chinese)

Comis D. Model takes the guesswork out of fertilizing, Agricultural Research,1999, 47(10): 15

Gao D M, Yang P H. Implement of web database in internet corn expert system. Journal of Taiyuan University of Sciences & Technology, 2000, 31(5):477-480 (in Chinese)

Gao L Z, Jin Z Q, Huang Y, Li B B. Rice Cultivational Simulation-optimization-decision Making System. Beijing: Chinese Agri Sci & Tech Press, 1992, pp. 21-40 (in Chinese)

Gunn R L, Mohtar R H, Engel B A. World-Wide-WEB–based soil and water quality modeling in undergraduate education, Journal of Natural Resources and Life Sciences Education, 2002, 31:141-147

Li X, Yang B Z, Guo T C, Zhao C J, Chen L P. Researches of field crop management geographical information system based on integration of WebGIS and ES. Journal of North China Agronomy, 2003, 18(2):106-109 (in Chinese)

Liao G P, Li A P, Wu Q Y, Guan C Y. Knowledge Representation of Prescribed Fertilization in Web-Based Expert System of Rapeseed (B. napus) Production. Journal of Hunan Agricultural University (Natural Sciences), 2002, 28(5):378-382 (in Chinese)

Liu H T, Qin Q M. Researches and application of soil eroding model WEBGIS-based. Journal of Water and soil Preserve, 2001, 15(3):52-55 (in Chinese)

McCown R L. Changing systems for supporting farmers' decisions: problems, paradigms, and prospects *Agricultural Systems*, 2002,74(1):179-220

Miller R C, Guertin D P, Heilman P. Information technology in watershed management decision making, Journal of the American Water Resources Association, 2004, 40(2): 347-357

Shaffer M J. Nitrogen modeling for soil management, Journal of Soil and Water Conservation, 2002, 57(6):417-425

Shi J P, Lu R K, Wang D J. Design and establishment of databases for fertilizing decision-making support WEB-based. Soil, 1999, (6):299-316 (in Chinese)

Wang L S, Qi G Q, Fan Y C. Design and development based on net of agricultural decision consulting system. Agricultural Mechanization Research, 2003, (2):136-137 (in Chinese)

Wang S L, Zhang L W, Liu G S, Wang J L. Researches of agricultural weather information service technology online. Chinese Agricultural Meteorology, 2004, 25(1):1-4 (in Chinese)

Winston R B. Ground water modeling software on the Internet, Ground Water, 2002, 40(4):335-336

Yang B Z, Zhao C J, Li A P, Wu Q Y, Sun X, Wu H R. Researches and application of agricultural expert system platform (PAID) based on WEB and components. High Technology Newsletter, 2002, (3):5-9 (in Chinese)

Yu G H. Farm information services system FIIS based on the web. Computer & Agriculture, 2000, (9):13-17 (in Chinese)

FUZZY CONTROL OF THE SPRAYING MEDICINE CONTROL SYSTEM

Yan Shi[1], Chunmei Zhang[2*], Anbo Liang[3], Haibo Yuan[2]

[1] *Architecture and engineering college, Qingdao Agricultural University, Qingdao, China, 266109*

[2] *College of Engine & Electronic Engineering, Qingdao Agricultural University, Qingdao, China., 266109*

[3] *International Intercommunion School, Qingdao Agricultural University, Qingdao, China, 266109*

[*] *Corresponding author, Address: College of Engine & Electronic Engineering, Qingdao Agricultural University, 700 Changcheng Road, Qingdao, 266109, P. R. China, Tel: +13863901209, Email: zcm9092@126.com*

Abstract: The continuous spraying was adopted by most of spraying pesticide systems, which can not accord to the extent and the characteristic change of targets to automatically regulate dose of drugs. This way not only resulted in waste of pesticides, but also caused pollution of environment. The designing studied the variable spraying pesticide system of real-time, which based sensory technology, at the same time, fuzzy control was applied in this system. The paper introduced the composition and control principle of the spraying pesticide system, at the other hand, the linking of Simulink and the fuzzy controller was introduced by this paper. Fuzziness of inputs and the outputs, the determination of subordinate functions, the establishment of fuzzy controlling rules and fuzzy control simulation of the spraying system were designed. Simulating result indicated that transiting time is short and the system is steady.

Keywords: fuzzy control, variable spraying pesticide, Matlab, Simulink

1. INTRODUCTION

Mechanization of Chinese plant protection is relatively laggard in China, the mechanization of plant protection about accurate spraying pesticide is

Shi, Y., Zhang, C., Liang, A. and Yuan, H., 2008, in IFIP International Federation for Information Processing, Volume 259; Computer and Computing Technologies in Agriculture, Vol. 2; Daoliang Li; (Boston: Springer), pp. 1087–1094.

also the same. In recent years, our country intensified research in this area and has made a great deal of progress. For example, Zhao Maocheng and others in Nanjing Forestry University designed the variable spraying pesticide control system based on trees feature which can accurately spray pesticide to targets. He Xiongkui and others in China Agricultural University carried on the design of electrostatic spraying machine to targets about orchards and experiment study.

At present, study of variable spraying pesticide machines mainly has the map-based automatic spraying pesticide system to targets and the automatic spraying pesticide system to targets in real-time sensory technology. Performance of the former is not strong, but, the latter has the advantages in this area (Shi Yan, 2004).

This paper designed a variable spraying pesticide which could conduct decision of spraying pesticide according to weed area, the control system marches variable spraying according to decision of spraying pesticide.

2. MATIERIALS AND METHOD

This part introduced designing process of the control system.

2.1 The total structure of spraying pesticide system

Figure 1 is the total block diagram of the spraying pesticide system.

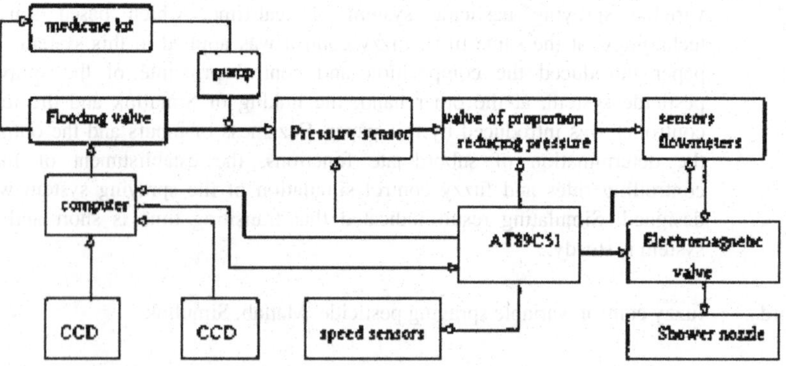

Fig. 1. The total block diagram of the spraying pesticide system

We may know that the spraying pesticide system is composed of medical kit, pump, flow meter, pressure sensor, speed sensors, single-chip, CCD sensor, computer, and shower nozzle from the figure.

2.2 Hardware of spraying pesticide control system

2.2.1 Components of control system

The control system is mainly composed of computer, AT89C51 single-chip, interface circuit, photoelectric isolation driving circuit, and A/D converter. The figure 2 is the diagram of the hardware circuit about the control system.

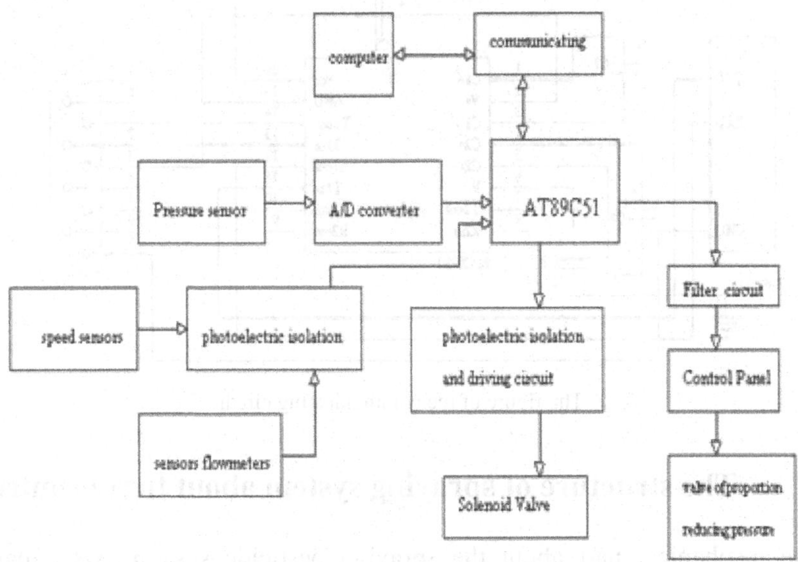

Fig. 2. The diagram of the hardware circuit about the control system

2.2.2 The control principle

The single-chip collects pressure signals, flow signals and speed signals in the spraying pesticide system. After preliminary treating, the data is sent to the computer. The computer gives the single-chip an order after it synthetically handles the signal gained and the result of pattern recognition. The single-chip controls valves switch and switch amounts size in control circuit which controls pressure in pipeline, change of pressure controls flow of variable shower nozzles.

2.2.3 The communicating circuit of the computer and single-chip

Ways of Computer communication have the parallel communication and the string of lines communication. The speed of the parallel communication is much quick, and real time nature is good, but, it is unsuitable to miniaturize a product because it occupies many end lines. Even though the rate of the string of lines communication is much lower, it appears simple, convenient, nimble in the circuit that data handling capacity is not very big (Shi et al. 2002). Therefore, the system adopts the string of lines communication.

Fig. 3 is the figure of the communicating circuit (Li, 2003).

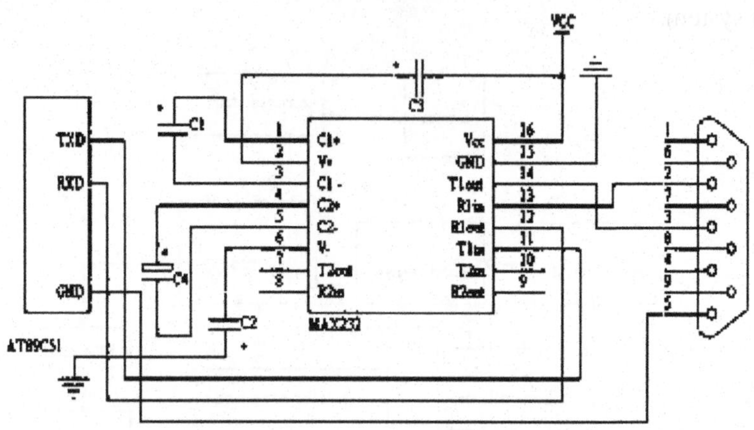

Fig. 3. The figure of the communicating circuit

2.3 The structure of spraying system about fuzzy control

The mechanical part about the spraying pesticide system exists many uncertain factors in the variable spraying pesticide system. It is difficult to set up the accurate mathematic model, therefore, we applied fuzzy control to the control system. Figure 4 is the structure of spraying system about fuzzy control.

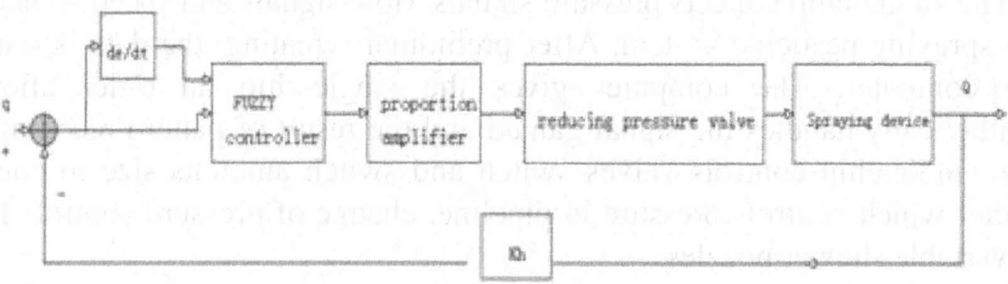

Fig. 4. The structure of spraying system about fuzzy control

2.4 The design of fuzzy controller

2.4.1 Fuzziness of inputs and outputs

The fuzzy controller has two inputs and one output. One between of inputs is the difference of flow rate about the input and actual flow rate, another is variable of e: ec=de/dt. The output is u which is the input voltage of the proportion amplifier. E, EC and U are their fuzzy language variable.

Fuzzy language value:

E={NB, NM, NS, ZO, PS, PM, PB}

EC= {NB, NM, NS, ZO, PS, PM, PB}

U={NB, NM, NS, ZO, PS, PM, PB}

2.4.2 Theory region of fuzzy variables

The fundamental theory region of e is (-e, e). The fundamental theory region of ec is (-ec, ec). The fundamental theory region of u is (-u, u).

Theory region after quantization about e:

{-7, -6, -5, -4, -3, -2, -1, 0, 1, 2, 3, 4, 5, 6, 7}

Theory region after quantization about ec:

{-6, -5, -4, -3, -2, -1, 0, 1, 2, 3, 4, 5, 6}

Theory region after quantization about u:

{-6, -5, -4, -3, -2, -1, 0, 1, 2, 3, 4, 5, 6}

The "a" represents the quantization factor of deviation, "b" represents the quantization factor of deviation change, "c1" represents the proportion factor of control quantity. Therefore,

a=7/e

b=6/ec

c1=6/u

2.4.3 Membership functions of fuzzy variables about inputs and outputs

Fuzzy language value is a fuzzy subset in fact, but language value is described by membership functions ultimately. The membership function of the language value is also known as linguistic semantic rules, it appears with continuous functions sometimes, and it appears with quantification levels of discrete form sometimes. They have their characteristic. For example, the description of continuous functions is accurate; however, the description of quantification levels of discrete form is concise, clear and definite. Common membership functions have the triangle function and the Gauss type, membership functions of E, EC and U are all the triangle function in this paper.

2.4.4 Establishing of fuzzy control rules

There are no ready-made control regulations in objective world, which are collected by designers who go on observing and extracting experiment data according to the structure of the fuzzy controller, part regulations of the paper are listed.

If E is NB and EC is NB, then U is PB
If E is NM and EC is NB, then U is PM
If E is NM and EC is NM, then U is PM
If E is PM and EC is PS, then U is NM

2.4.5 Fuzzy solution

Purpose of fuzzy solution is to obtain real scatter which may reflect control quantity according to the result of fuzzy reasoning. Common methods have three kinds. The dynamic performance of median method is better than the other; therefore, we collected median method.

3. RESULTS AND DISCUSSIONS

MATLAB has a lot of tool case functions to carry on fuzzy analyzing and designing. We can open the artwork interface of Fis editor, conduct fuzziness of inputs and outputs and edit fuzzy rules, when orders about fuzzy are keyed in MATLAB.

Figure 5 is the simulation model of fuzzy control about the spraying pesticide system.

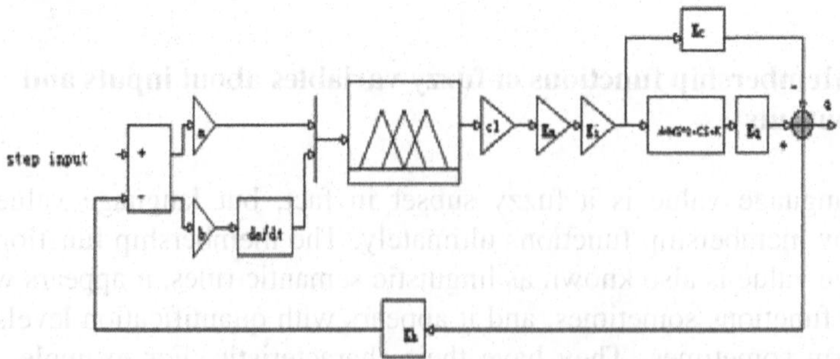

Fig. 5. The simulation model of fuzzy control about the spraying system

Simulink that MATLAB provides is a software package which can be used to build models, simulate and analyze, and which may realize perfect combination with the fuzzy logic tool case.

After opening the simulation model under the environment of simulink, we strike Fuzzy logic controller, and import the file name of the fuzzy controller which has been designed by us. At last, the ok button is keyed. Fuzzy controller and simulink are linked in this way.

After collecting simulation parameters… under menu simulation, a conversation frame will appear. We set type among solver options to Variable-step, algorithm is order45. Simulation result can be gained from the oscilloscope.

Figure 6 is the result of simulation.

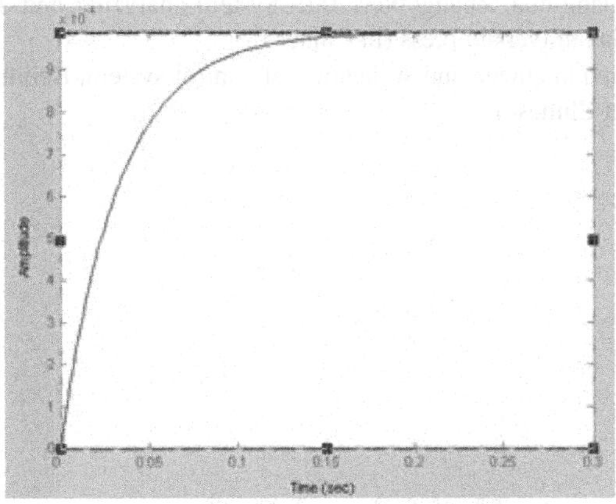

Fig. 6. The result of simulation

We can know transition time is 0.19 second from the figure. Transition time is very short and the system is very stable.

4. CONCLUSIONS AND FUTURE WORKS

From the above analysis, the main outcomes can be outlined as follows:

(1) The spraying pesticide control system can satisfy the timely request and the controllability is much better.

(2) Future work is constructing the platform to do dynamic experiment.

REFERENCES

Shi Yan 2004. Study on the System of Spraying Rate Varied by Pressure of Liquid Chemical Application, Chinese Agriculture University (in Chinese).

Shi Donghai, Hu Xiao, Zhou Xusheng 2002. Data communicating technology of single-chips from elementary course to mastering, xi'an: Xidian university press, 99 (in Chinese).

Li Chaoqing 2003. Data communicating technology about PC and single-chips with DSP, Beijing: Beihang university press, 30 (in Chinese).

Zhang Guoliang, Zeng Jing, Ke Xizheng, etc. 2002. Fuzzy control and application of matlab, xi'an: xi'an jiaotong university press (in Chinese).

Zhu Jing 1995. Theory of fuzzy control and system principle, Beijing: mechanical industry press (in Chinese).

Zhu Jing 2005. Theory of fuzzy control and system principle, Beijing: mechanical industry press, 233–234 (in Chinese).

Lou Shuntian, Hu Chang-hua, Zhang Wei 2001. System analyzing and designing based on matlab, xi'an: xidian university press (in Chinese).

Sun Liang 2004. matlab language and simulation of control system, Beijing: Beijing industry university press (in Chinese).

ANALYSIS AND TESTING OF WEED REAL-TIME IDENTIFICATION BASED ON NEURAL NETWORK

Yan Shi[1], Haibo Yuan[2,*], Anbo Liang[3], Chunmei Zhang[2]

[1] *Architecture and engineering college, Qingdao Agricultural University, Qingdao, China, 266109*

[2] *College of Engine & Electronic Engineering, Qingdao Agricultural University, Qingdao, China, 266109*

[3] *International Intercommunion School, Qingdao Agricultural University, Qingdao, China, 266109*

[*] *Corresponding author, Address: College of Engine & Electronic Engineering, Qingdao Agricultural University, 700 Changcheng Road, Qingdao, 266109, P. R. China, Tel: +15864743768, Email: haibo762@sohu.com*

Abstract: Contrasting the two green strength genes of soil, wheat, corn, and the weed, the paper designed a system to identify the weed from the crop. It used the 2G-R-B and BP neural network, with the help of pixel-position-histogram diagram, to calculate the area and position of weeds. The result showed that it could identify the weed from the field and crop with an accuracy of 93%. The program gave the result that running time of identifying weed in wheat field was 273.31ms. As far as the corn was concerned, the time was 321.94ms, In a word, the system can satisfy the request of real-time.

Keywords: variable spray, weed identify, picture disposal, visual c++

1. INTRODUCTION

Large amounts of pesticides are applied to the field by Chinese farmers each year. By the time of 2006, the Chinese total output of pesticides has reached 2.15 billion pounds. Typically, herbicides are applied with a blanket treatment to a whole field without regard to the spatial variability of the weeds in the field. Automatic target-activated herbicide sprayer can

Shi, Y., Yuan, H., Liang, A. and Zhang, C., 2008, in IFIP International Federation for Information Processing, Volume 259; Computer and Computing Technologies in Agriculture, Vol. 2; Daoliang Li; (Boston: Springer), pp. 1095–1101.

selectively target herbicide according to the allocation and the characteristic change of the target. It can increase the efficiency of chemical usage, and reduce the waste of chemical and environment pollution as well. Weed identification system is one essential part of the automatic sprayer (Thornson J.F., Stafford J.V., 1991). So far much of the machine vision weed sensing research has been done. Burks (2000) used the Co-occurrence Matrix and Color Features to identify 5 weeds from the soil. It used the 11 texture-characteristic parameters of tone and saturation to distinguish weeds from plant and soil, the accurate rate was 93%. However, the division arithmetic of texture needs lots of calculation, which slows down the time of identifying. This method is no good for real-time sensing as well. Plant identification had been accomplished with the use of shape features of plant (Guyer et al., 1993a; Woebbecke et al., 1995; Ji Shouwen and Wang Rongben, 2001). In this way, the system could identify weeds from the corn and soil by the projective area. But it couldn't be applicable to the instance that corn is shorter than weed and its accurate rate was influenced by the light and the condition of the soil. The purpose of the study is to develop a weed-identifying system to detect the crop and gather field information. The system bases on investigating the wheat and corn of the east of Shandong Province to identify weed with the program made by visual c++ (VC++) and artificial neural network (ANN). The result showed that the system could satisfy the request of real-time and identify weeds from the wheat and corn.

2. MATIERIALS AND METHOD

The contrast of the two green strength genes of soil, wheat, corn and the weed in the field is shown in Fig. 1 and Fig. 2. The green genes of plant and the non-plant are almost not overlap. The green genes of plant and the non-plant is distinguished well when the threshold value of 2G-R-B are 10-30, or the value of 2G/(R+B) are 1.0-1.15. There is a range of each one of the two threshold value. When the value slows down, the interference in the picture

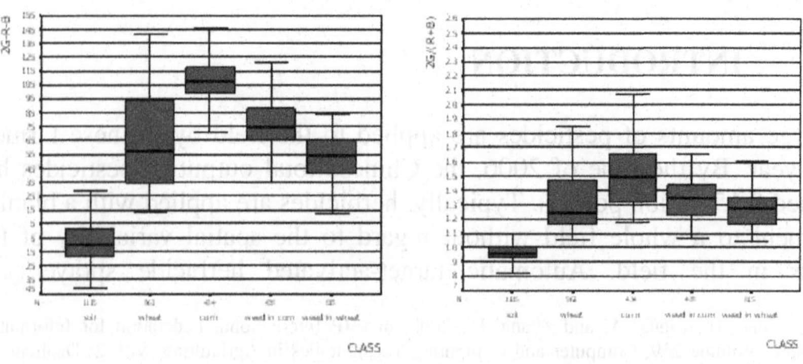

Fig. 1. 2G-R-B *Fig. 2.* 2G/(R+B)

grows up and vice versa. The accuracy of the system can be 95% or more when the threshold value of 2G-R-B is 20 or 2G/(R+B) is 1.05.

In order to reduce the effect of illumination, the color component RGB normalized by brightness can be used to show. That is chroma coordinate, and the coordinate is also called 2g-r-b. Though the characteristic of Excess Green and the 2G/(R+B) factor are affected little by the strength of illumination and illumination angle, they has brought certain miscounting and lead partial characteristics of Excess Green between plants and non-plants to fold because that it has the floating number operation in the computation process. Its deviation absolute value is big in using the characteristic of Excess Green 2G-R-B of the RGB space directly, but its relative deviation is small and the folding doesn't exist between two parts. It is more advantageous that the natural environment influence is infirm. It's more effective to use the characteristic of Excess Green 2G-R-B form than other forms, and in this way it can save the time in weed background segmentation. In this paper, the characteristic of Excess Green 2G-R-B was used as parameter to filter the excess green binary, and the image wasn't dealt with before it is filtered the green binary.

2.1 Segmentation of weed and plant

As illustrated in Fig. 1 and Fig. 2, both 2G-R-B and 2G/(R+B) can't be used to detect crop and weed because they fold seriously to each other. Position distributed characteristic method, shape characteristic method, texture characteristic method and other methods can be used to detect crop. The wheat and other drilling crop are planted by men, so its position distribution is regular and spreads in row; but weeds growth naturally and spread irregularly between crop rows. So in this paper, position characteristic method was used for segmentation of wheat and weed. The corn is dibbling crop, which has leaves from 3 to 5 cotyledons stage, and there is certain distance between plants and row space, so crop leaves are not severely occluded. For the corn, their shape characters were firstly extracted and artificial neural network (ANN) was trained to detect the weed.

2.2 Detection of the row

The central line of the plant is the line formed by the scion during the time of growing. There are many methods to sense the line such as Hough switch and pixel decision histogram. The dealing time of Hough switch is longer than that of pixel decision histogram. Thus in this real-time system, the histogram was adopted. The Fig. 3 and Fig. 4 are the center lines in wheat area disposed by VC++. After detecting lines of the plant, the system used

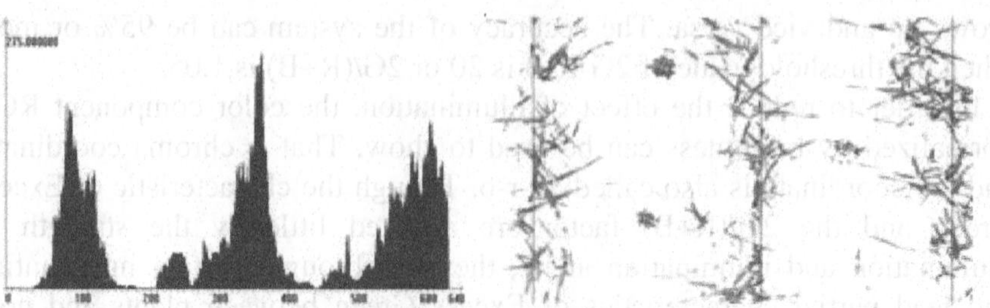

Fig. 3. Chart of plumb projection *Fig. 4.* Center line in wheat area

area planting method to fill the plant area by taking the point in the central line as the seed point. The area connecting with the plant line was considered as plant, and the others were taking for the area of weed.

2.3 Dibbling crop and BP neural network

The framework of neural network is shown in Fig. 5. The network has three layers, and the input layer has 6 nodes. This paper selected 6-20-1 network and tested the neural network by the software of Neural Shell2. The stylebook was suggested as the following: if the S (area) <T1 (threshold value) or the P (perimeter) < T2 (threshold value), the stylebook was regarded as weed, and these stylebooks couldn't be used to train the network. The threshold value is determined by the growing condition of the corn. T1 and T2 were 100 and 50 respectively and the unit was pixel in testing. The result showed that the selected stylebook could improve the training precision and the accuracy of the network.

3. RESULTS AND DISCUSSIONS

3.1 Dibbling crop and BP neural network

The input and output stylebook of the ANN are shown in Table1.

Table 1. Result of input and output

	Input						Output
	Aspect	Roundness	Elongation	PTB	LTP	Compactness	class
1	3.982	2.676	0.597	0.754	0.104	0.667	1
2	1.184	6.464	0.084	1.486	0.130	0.613	0
3	1.140	4.963	0.066	1.127	0.158	0.564	1
4	1.185	4.914	0.084	1.194	0.153	0.629	0

The Table 2 shows the accuracy which the error rate ≤ 0.05 and the error of different network structures. These data were gotten in the condition that the learning rate was 0.5 and momentum gene was 0.5. The number of stylebook was 100.

Table 2. Effect of hidden node (learning rate 0.5 and momentum 0.5)

Number of connotative layer notes	12	16	20	24	26	30
Mean absolute error	17.65	15.01	16.16	17.0	16.78	18.67
STDEV	13.63	12.72	12.05	14.35	14.18	16.24
Min. absolute error	1.48	1.95	1.25	1.35	0.20	0.78
Max. absolute error	69.41	54.34	49.97	70.71	75.06	73.74
Mean related error (%)	3.24	2.76	3.05	3.12	2.94	3.24
STDEV	2.00	1.90	1.80	1.90	1.92	1.89
Min. related error (%)	0.30	0.34	0.21	0.02	0.14	0.16
Max. related error (%)	6.89	7.02	7.51	6.45	7.56	7.31
Percent within 5%	81.82	81.82	87.36	81.82	81.82	81.82

Table 2 shows that the learning rate and accuracy of the stylebook are high in the 6-20-1 network. And the training number is little.

Secondly a test to determine the effect of different learning rate and momentum was performed (Table 3).

Table 3. Effect of identification under different learning rate and momentum

Leaning rate and momentum	0.5, 0.5	0.5, 0.7	0.5, 0.9	0.7, 0.5	0.7, 0.7	0.7, 0.9	0.9, 0.5	0.9, 0.7	0.9, 0.9
Mean absolute error	13.16	13.48	20.27	9.56	5.09	17.03	10.36	15.80	14.70
STDEV	12.05	11.82	15.09	11.99	12.87	13.27	12.01	12.26	13.63
Min. absolute error	1.25	0.65	1.95	1.74	1.53	1.46	0.54	0.93	0.29
Max. absolute error	49.97	48.49	45.40	41.55	38.13	46.97	48.39	54.70	52.59
Mean related error (%)	3.05	3.18	3.85	3.13	3.17	3.27	3.17	3.17	2.70
STDEV	1.80	1.85	3.14	1.72	1.60	2.66	1.91	1.65	2.28
Min. related error (%)	0.21	0.10	0.39	0.30	0.31	0.35	0.09	0.19	0.06
Max. related error (%)	7.51	7.36	14.96	7.21	7.08	9.03	7.35	7.28	8.50
Percent within 5%	87.36	86.36	81.82	89.82	94.91	77.73	88.36	86.36	81.82

Table 3 shows that the accuracy is 94.91% when both the learning rate and momentum are 0.7.

3.2 Design of software

A program was made by VC++. The stylebook was collected in the field of east Shandong Province under different illumination and temperature. We take 321 photos to test. The software runs well and the running time was shown in Table 4.

Table 4. Contrast of consume time

Wheat	Running time (ms)	Corn	Running time (ms)
Pretreatment	61.08	Pretreatment	60.03
Segmentation of plant and the soil	24.14	Segmentation of plant and the soil	26.15
Distill the line	26.39	Unit of areas	65.60
Area planting method	59.07	Distill the characters	98.91
Areas demarcation	41.27	ANN	13.24
Areas calculating	29.90	Areas calculating	26.56
Data save and sending	31.46	Data save and sending	31.45
Total	273.31	Total	321.94

The Table 4 shows that both the dispose time used in corn and wheat are lass than 0.35s. So the system can satisfy the request of real-time.

4. CONCLUSIONS AND FUTURE WORKS

The system runs well in taking the pixel decision histogram to determine the central line of wheat and taking the method of ANN to sense the weed in corn. The software can identify weeds from the plant and soil. It can calculate the area of the weed and satisfy the request of real-time

This paper used the BP neural network and 2G-R-B, with the help of pixel-position-histogram diagram, to calculate the area and position of weed. The result showed that it could identify the weed from the field and crop, and its correct rate was 93% under different illumination. The program gave the result that running time of identifying weed in wheat field was 273.31ms. As far as the corn was concerned, the time was 321.94ms. The test showed that the system could identify weeds from the plant and soil in good condition.

Future work is constructing the platform to do dynamic experiment.

REFERENCES

Blackmore B.S. Developing the principles of precision farming. Proceeding of Agrotech, Barretos. Brazil, 1999, 11(6):15–19

Burks T.F., Shearer S.A., Gates R.S. Back propagation neural network design and evaluation for classifying weed species using color image texture. Transactions of the ASAE, 2000, 43(4):1029–1037

Guyer D.E., G.E. Miles, L.D. Gaultney, and M.M. Schreiber. 1993. Application of machine vision to shape analysis in leaf and plant identification. Transactions of ASAE 36(1):163–171

Ji Shouwen, Wang Rongben, Chen Yajuan. Research on Recognizing Weed from Corn Seedling by Using Computer Image Processing Technology, Transactions of the Chinese Society of Agricultural Engineering, 2001(02) (in Chinese)

Shi Yan, Study on the System of Spraying Rate Varied by Pressure of Liquid Chemical Application, Chinese Agriculture University, 2004 (in Chinese).

Shi Yan, Qi Lijun, Fu Zetian. Model development and simulation of variable rate of pressure spray. Translations of the CSAE, 2004 (in Chinese)

Thornson J.F., Stafford J.V., Potention for automatic weed detection and selective herbicide application. Corp Protection 1991, 10:254–259

APPLICATION OF GENETIC ALGORITHM (GA) TRAINED ARTIFICIAL NEURAL NETWORK TO IDENTIFY TOMATOES WITH PHYSIOLOGICAL DISEASES

Junlong Fang, Changli Zhang, Shuwen Wang
Engineering College Northeast Agricultural University Harbin, Heilongjiang 150030 China
swanhaha@163.com

Abstract: We synthetically applied computer vision, genetic algorithm and artificial neural network technology to automatically identify the tomatoes that had physiological diseases. Firstly, the tomatoes' images were captured through a computer vision system. Then to identify cavernous tomatoes, we analyzed the roundness and detected deformed tomatoes by applying the variation of fruit's diameter. Secondly, we used a Genetic Algorithm (GA) trained artificial neural network. Experiments show that the above methods can accurately identify tomatoes' shapes and meet requests of classification; the accuracy rate for the identification for tomatoes with physiological diseases was up to 100%.

Keywords: tomato with physiological disease; computer vision; artificial neural network; genetic algorithms

1. INTRODUCTION

In China, most traditional fruit quality tests still remain at the preliminary stage in which identification is determined by human sense organs. These subjective evaluations are affected by personal abilities, color resolution, emotions, fatigue and rays of light etc. Therefore these techniques are inefficient and inaccurate, and most of them are observational appreciations whose objectivity and accuracy are inadequate. The result is that China's export fruits are of inferior external appearances, and lack competitive power in international markets. Thus it is rather indispensable to improve the

Fang, J., Zhang, C. and Wang, S., 2008, in IFIP International Federation for Information Processing, Volume 259; Computer and Computing Technologies in Agriculture, Vol. 2; Daoliang Li; (Boston: Springer), pp. 1103–1111.

national test level of fruit quality (YingYi bin, 1999). Contrasted with other test technologies, the advantages of computer vision include high speed, multiple functions, capability to do assignments that man or mechanical grading machines are incompetent to do, and the ability to automatically achieve single time identifications of shape, size, degree of maturation, and surface flaws; all of which greatly help increase grading precision (Panwei, 2000). Early in 1985, Sarkar and Wolfe et al. successfully developed a kind of quality grading equipment for fresh tomatoes, using computer vision with an applicable lighting fixture and direction finding mechanism. Its grading error is only 3.5%, far superior to the grading precision of manual tests, although still low speed (Sarkar N., 1985).

2. CAPTURE OF THE TOMATO'S IMAGE

The tomato's image is captured by a computer vision system, which consists of computers, image acquisition cards, CCD (Charge Coupling Device) cameras, 20-watt annular incandescent lamps, etc. Because of the tomato's extremely-smooth surface, illumination results in serious specular reflection. It was found that if a soft-light sheet was added between a 20-watt annular incandescent lamp and the tomato, the tomato's image would no longer contain specular reflection (Zhang Changli, 2001). Figure 1 shows the tomato's acquired image.

3. NETWORK'S ESTABLISHMENT

3.1 Brief introduction to Genetic Algorithm

Genetic Algorithm (GA) is a kind of search and optimized algorithm that have been produced from simulating biologic heredities and long evolutionary processes of creatures. It stimulates the mechanism of "survival. Competitions; the superior survive while the inferior are eliminated, the fittest survive." The mechanism searches after the optimal subject by means of a successive iterative algorithm. Ever since the late 80s, GA, as a new cross discipline which has drawn people's attention, has already shown its increasing vitality in many fields (Sun YanFeng, 1995).

GA stimulates reproduction, mating, and dissociation in natural selection and natural heredity procedures. Each possible solution to problems is taken as an individual among population, and each individual is coded as character string; each individual is evaluated in response to predefined objective

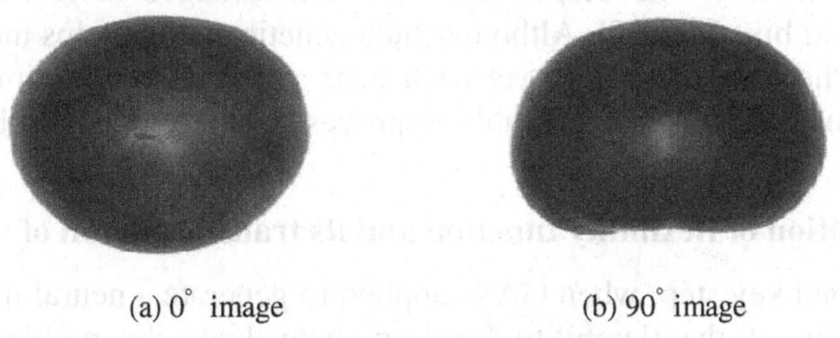

(a) 0° image (b) 90° image

Fig. 1. Tomato's image

functions and a flexibility value given. Three of its elemental operators are selection, crossing, and mutagenesis (Sarkar, 1985).

Its main features are as follows:

(1) GA is to acquire the optimal solution or quasi-optimal ones through a generational search rather than a one-point search; (2) GA is capable of global optimum searching; (3) GA is a parallel process to population change, and provides intrinsic parallelism; (4) The processed object of GA is the individuals whose parameter set is coded rather than the parameters themselves, and this very feature enables GA to be used extensively.

3.2 Model establishment

1) Determination of the genetic encoding scheme

The first step, when GA is applied, is the determination of a genetic encoding scheme, namely to denote each possible point in the problem's search space as a characteristic string of defined length. This is in order to be sure that GA will not only optimize network configuration but, in the meantime, genetic training will proceed on weight values. In this paper, weight values between each layer of the multi-layer feed-forward neural network are simultaneously coded as one chromosome.

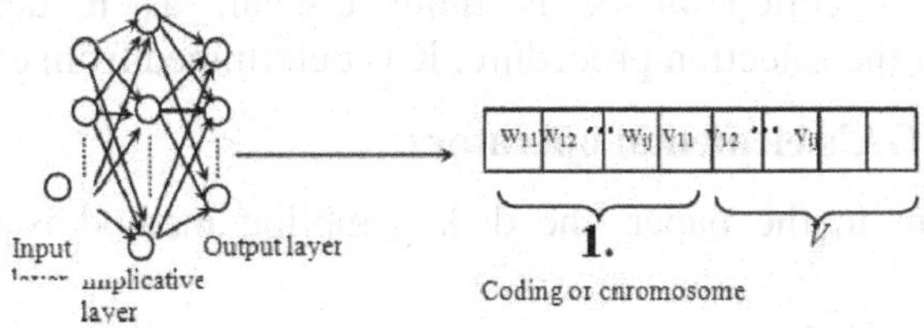

Fig. 2. Neuron network structure and coded chromosome

Structure and weight value of the neuron network are mapped to this very chromosome, one of whose layers is the implicative layer whose nodes are determined in accordance with actual demands. Both W as the weight array from the input layer to the implicative layer and V as that from the implicative layer to the output layer are still arranged in row-to-column sequence and binary-coded. Although such genetic coding helps increase the length of chromosome string, yet each gene value is strictly simplified to either "0" or "1", which remarkably improves the algorithm's global search ability.

2) Definition of flexibility function and its transformation of scale

The second key step, when GA is applied to generate a neural network, is the definition of the flexibility function to evaluate the problem-solving ability of the neural network, which is denoted by a certain specific chromosome string. In this paper, objective function is generated from the cost factor mean-square error (MSE) of the neural network output, and then converted into a function via reciprocal transformation. Its computational formula is as follows:

$$MSE=\frac{1}{mp}\sum_{P=1}^{p}\sum_{j=1}^{m}(d_{pj}-y_{pj})^2 \tag{1}$$

where; m=sum of output nodes; p=sum of trained samples; dpj=expected output of network; ypj=actual output of network.

In GA's later search period, sufficient varieties will probably still exist among population, whereas population's average flexibility value may approximate the optimal variation. If such status isn't changed, then amongst subsequent generations, the individuals that have average value and those optimal ones will approach identical reproduction quantities, so actually here almost no competition exists, which consequently slows down the convergence rapidity.

To avoid the above situation, reproduction quantity in the neural network evolutional process will be adjusted to improve algorithmic performance. Reciprocal scale transformation to the objective function is done as:

$$f\,(MSE) = \frac{K}{MSE} \tag{2}$$

in which the selection of K is quite crucial, as it determines the coerciveness of the selection procedure. K is determined from experiments.

3) Three of GA's elemental operators

(1) Selection. In the paper, the disk gambling method is adopted, and calculated as

$$P_i = \frac{f_i}{\sum_{i=1}^{N} f_i} = \frac{f_i}{f_{sum}} \tag{3}$$

In which; fi=flexibility value of individual i; fsum=total flexibility value of population; Pi=selective probability of individual.

It's obvious that individuals with high flexibility values are more likely to be reproduced (Buckley, 1994).

(1) Crossing. In the paper, one-point crossing is adopted. The specific operation is to randomly set one crossing point among individual strings. When crossing is executed, partial configurations of the very point's anterior and posterior individuals are exchanged, and give birth to two new individuals.

(2) Mutagenesis. As for two-value code strings, mutagenic operation is to reverse the gene values on the gene bed, namely covert 1 to 0 or 0 to 1.

(3) Genetic control parameters

Genetic control parameters in the paper are chosen as follows: N as population scale is 50, Pc as crossing probability is 0.4, Pm as mutagenic probability is 0.003, and network's terminative condition is MSE≤0.05.

4. EXTRACTION OF CHARACTERS

Deformed fruits and cavernous ones result from two major physiological diseases. Cavernous fruit refers to ones with corner angles on their surfaces, and its tangential faces are mostly triangle in shape (Figure 3a). The chief characteristic of deformed fruits is its irregular shapes, such as abnormal, fingerlike projections or papillary (Figure 3b) fruitages, etc. Furthermore, the cavernous fruit can be identified through roundness value of 0° image, and the deformed one identified through variations of the fruit's diameter of 90° images.

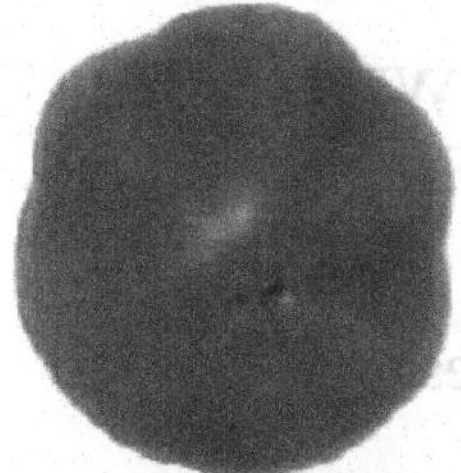

(a) Cavernous tomato

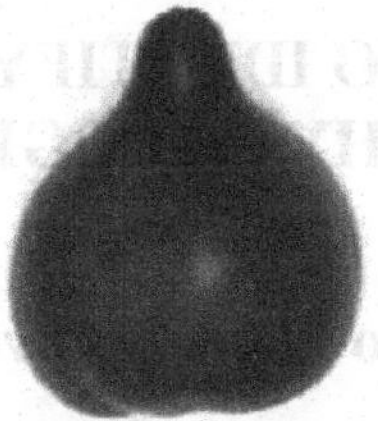

(b) Abnormal tomato

Fig. 3. Tomatoes with physiological diseases

4.1 Roundness character

Roundness, which is derived from the area and the perimeter, measures the shape complexity. Its formula is (Ni B., 1997).

$$e = \frac{4\pi A}{p^2} \tag{4}$$

In which A – the area; P – the perimeter.

It can be seen that, e=1 if round in shape. Since tangential faces of the cavernous fruit are triangular, its roundness is less than 1.

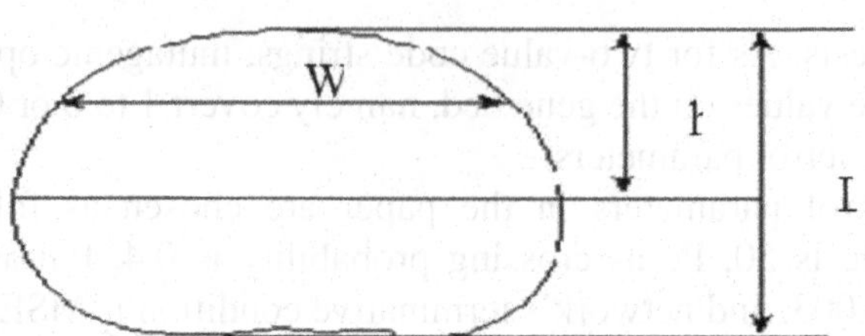

Fig. 4. Characteristic parameters of tomato's shape

4.2 Variations of fruit's diameter

Variations of the fruit's diameter refers to the variations of diameter from tomato's top to its bottom (Liu He, 1996) (Figure 4), which can be defined as the distance Wi between the two intersections derived from a line perpendicular to fruit's axis and tomato's edge. In this way, the tomato is divided perpendicular to fruit's axis into twelve equal parts, then W1–W12 are extracted as characteristic parameters of variations of the fruit's diameter. Due to the unsmooth variations of its diameter, deformed fruits are detected by this means.

5. TO IDENTIFY TOMATOES WITH PHYSIOLOGICAL DISEASES

5.1 To identify cavernous tomatoes

Roundness is calculated via the perimeter P and the area. P is measured in millimeters, and A is measured in square millimeters, while both the length and the area in the plane coordinate system of image are measured in pixels. The quantitative relationship obtained from experiments of the tomato's perimeter in pixels at maximum transverse diameter is shown in Table 1, and the quantitative relationship of the tomato's area at maximum transverse diameter is shown in Table 2.

According to Table 1, the linear regression equation of the tomato's measured perimeter in pixels at maximum transverse diameter on the edge of 0° image is defined as

$$P = 0.2492\,n + 8.4294 \tag{5}$$

In which P=tomato's estimated perimeter at maximum transverse diameter; n=pixels on the edge of 0° image; and their correlation coefficient R=0.9990. The relationship obtained from experiments of the tomato's actual perimeter to estimated perimeter at maximum transverse diameter is shown in Table 3.

Table 1. Quantitative relationship between measured perimeter at maximal diameter and pixel in plane coordinate system of image

Sample	1	2	3	4	5	6	7
Pixels	928	787	667	861	806	937	767
Actual perimeter (mm)	240.19	203.59	175.90	223.10	210.41	241.51	200.23

Table 2. Quantitative relationship of tomato's area at maximum transverse diameter to pixels in plane coordinate system of image

Sample	1	2	3	4	5	6	7
Pixel	61918	44543	32127	53121	46822	62584	42579
Actual area (mm2)	4590.94	3298.43	2462.13	3960.87	3523.02	4641.52	3190.41

Table 3. Relationship of tomato's actual perimeter to that of evaluated perimeter at maximum transverse diameter

Sample	Actual perimeter (mm)	Estimated perimeter (mm)	Absolute Error (mm)	Relative error (%)
1	197.31	199.57	2.26	1.15
2	217.43	217.26	0.17	0.08
3	173.13	173.65	0.52	0.30
4	221.59	220.25	1.34	0.60
5	237.08	234.95	2.13	0.90
6	218.95	217.26	1.69	0.77
7	245.32	244.92	0.40	0.16
8	247.74	244.67	3.07	1.24
9	194.17	193.09	1.08	0.56

Table 4. Relationship of actual area to evaluated area at maximum transverse diameter

Sample	Actual area (mm2)	Estimated area (mm2)	Absolute error (mm2)	Relative error (%)
1	3098.03	3142.41	44.38	1.43
2	3762.08	3764.49	2.41	0.06
3	2385.34	2365.10	20.24	0.85
4	3907.50	3881.47	26.03	0.67
5	4472.82	4335.38	137.44	3.07
6	3814.99	3751.77	63.22	1.66
7	4789.09	4732.63	56.46	1.18
8	4884.31	4775.91	108.40	2.22
9	3000.11	2961.22	38.89	1.30

Seen from Table 3, the maximal error of the test perimeter exists within 3.07 mm, and the estimated perimeter at maximum transverse diameter gained by this means is of higher precision, which meets the demand for identifying tomatoes.

According to Table 2, the linear regression equation of the tomato's estimated area in pixels at maximum transverse diameter inside the edge of 0° image is defined as

$$A = 0.0719n + 130.09 \tag{6}$$

In which A=estimated area at maximum transverse diameter; n=pixels inside the edge of 0° image; and their correlation coefficient R=0.9995. The relationship obtained from experiments of the tomato's actual area to estimated area at maximum transverse diameter is shown in Table 4.

Seen from Table 3, the maximal error of the test area exists within 137.44 mm2, and the estimated area at the maximum transverse diameter gained by this means is of higher precision, meeting the demand for identifying tomatoes.

5.2 To identify deformed tomatoes

Characteristic parameters of variations of fruit's diameter W1–W12, after being pretreated, are taken as inputs of the characteristic identification network. The network outputs are divided into three states: normal, passable and abnormal, which, if the thermometer method is adopted, can be respectively denoted as: (0, 0, 1), (0, 1, 1), (1, 1, 1).

The selection of GA's control parameters are as follows: population scale N is 50, crossing probability Pc is 0.4, mutagenic probability Pm is 0.001, and network's terminative condition is MSE≤0.05.

Through repeated selection, crossing, and mutagenesis, the sum of nodes that was ultimately obtained, in the optimal implicative layer of the characteristic network, was 8. Thus the valid network configuration was 12-8-3, and related connective weight values were stored in weight files. Twenty tomatoes were chosen as trained samples among which both normal fruits and deformed ones each composed half. Another twenty tomatoes were chosen as test samples. Experimental results are shown in Table 5.

Table 5. Identification results of deformed tomatoes

Variations of fruit's diameter / Identification results	Normal	Deformed
Normal fruits	10	0
Deformed fruits	0	10

The experimental results shown above indicate that identification rate of deformed tomatoes derived from GA trained neural network reaches 100%. It's necessary to emphasize that deformed fruits that are collectively selected for training must be sufficiently representative and comprehensive.

6. CONCLUSION

In the paper, roundness character is extracted from the tomato's image to identify cavernous fruit; and the character of fruit shape variations is used to identify deformed fruit. The character of fruit shape variations is analyzed by artificial neural network trained by GA. Research results indicate that through analyses of tomatoes' shapes, fruits with physiological diseases can be absolutely identified.

REFERENCES

Buckley, J.J, Hayashi, Y. Fuzzy Genetic Algorithm and Applications [J]. Fuzzy Sets and Systems 1994 Vol. 61:129–136.

Chen Genshe, Zhen Xinhai. Study and Development of Genetic Algorithm [J]. Information and Control 1994; 23(4):215–21.

Liang Min, Sun Zhongkang. Quick Learning Algorithm and Simulation Study of Multi-layer Feed-forward Neural Network [J]. Systematic E and Electronic Technology 1993 (9):47–57.

Liu He, Wang Maohua. Research on Artificial Neural Network Expert System of Fruit Shape's Identification [J]. Transactions of the Chinese Society of Agricultural Engineering 1996; 12(1):171–6.

Ni, B., Paulsen, M.R., Reid, J.F. Corn Kernel Crown Shape Identification Using Image Processing [J]. Transactions of the ASAE 1997; 40(3):833–838.

Pan Wei. Study on Automatic Identification and Classification of Agricultural Products Using Computer Vision – Automatic Identification and Classification of Tomato [D]. Northeast Agricultural University, 2000.

Sarkar, N., Wolfe, R.R. Computer Vision Based System for Quality Separation of Fresh Market Tomatoes [J]. Transactions of the ASAE 1985; 28(5):1714–8.

Sarkar, N., Wolfe, R.R. Feature Extraction Techniques for Sorting Tomatoes by Computer Vision [J]. Transactions of the ASAE 1985; 28(3):970–4.

Sun Yanfeng, Wang Zhongtuo. Parallel Genetic Algorithm [J]. Systematic Engineering 1995; 13(2):14–6.

Ying Yibin, Jing Hansong, Ma Junfu, Zhao Jun, Jiang Yiyuan. Application of Computer Vision Technology to Test the Size of Virgin Pears and Flaws of Fruit's Surface [J]. Transactions of the Chinese Society of Agricultural Engineering 1999; 15(2):197–200.

Zhang Changli, Fang Junlong, Pan Wei. Research on Genetic Algorithm Trained Multi-layer Feed-forward Neural Network's Automatic Measuring to Tomato Maturity [J]. Transactions of the Chinese Society of Agricultural Engineering 2001; 17(3):153–6.

DEVELOPMENT OF A MODEL-BASED DIGITAL AND VISUAL WHEAT GROWTH SYSTEM

Liang Tang, Hui Liu, Yan Zhu, Weixing Cao*
High-Tech Key Laboratory of Information Agriculture of Jiangsu Province, Nanjing Agricultural University, Nanjing, 210095, China
** Corresponding author: Tel: +86-25-84395845, Fax: 86-25-84396565, Email: Caow@njau.edu.cn*

Abstract: Driven by soil, variety, weather and management databases and integrating process-based growth simulation model, morphological model and visualization model, a model-based digital and visual wheat growth system (MDVWGS) was developed using component-based software and visualization techniques. The system was programmed by the .Net framework with the language of C# and CsGL Library was used for realizing 2D and 3D graphics application and visualization. The implemented system could be used for predicting growth processes and visualizing morphological architecture of wheat plant under various environments, genotypes and management strategies, and has the functions as data management, dynamic simulation, strategy evaluation, real-time prediction, temporal and spatial analysis, visualization output, expert consultation and system help. The MDVWGS should be useful for construction and application of digital farming system and provide a precise and scientific tool for cultivar design, cultural regulation and productivity evaluation under different growing conditions.

Keywords: Wheat; Growth model; Morphological model; Functional-structural plant model; Visualization

1. INTRODUCTION

Crop modeling based on ecophysiological processes of plant growth and development has become an important field of research during the past decades. The main emphasis in crop growth modeling has been put on developing models by description of fundamental biophysical, biochemical

Tang, L., Liu, H., Zhu, Y. and Cao, W., 2008, in IFIP International Federation for Information Processing, Volume 259; Computer and Computing Technologies in Agriculture, Vol. 2; Daoliang Li; (Boston: Springer), pp. 1113–1120.

and physiological processes of growth and yield formation (Bouman et al., 1996; van Ittersum et al., 2003). This type of models is commonly referred to as process-based models, while plant architecture is normally addressed in a simplified manner. In recent years, approaches have been developed to describe the geometric structure of plant. L-system or similar approaches were used to simulate the architecture of plants (de Reffye et al., 1988; Prusinkiewicz and Lindenmayer, 1990), and these models gained in versatility and provided with graphical capability.

A new modeling approach named functional-structural plant model (FSPM) concerned with integration of architecture and resource allocation as aspects of plant function was developed in the mid 1990s (Godin, 2000; Sievanen et al., 2000). For agricultural crops several reports have been published during the last years dealing with FSPM under development (Drouet and Pages, 2003; Fournier et al., 2003; Hanan et al., 2003; Yan et al., 2004). This type of model incorporates the physiological processes into the architecture model, aiming at 3D geometric structure of plant become mechanism and process-based. But most of these models link an architecture model with a canonical mathematical model (Renton et al., 2005), thus these models can simulate only a few ecophysiological processes.

Many crop growth model-based decision support systems have been established, such as APSIM (McCown et al., 1996), DSSAT (Jones et al., 2003) and GMDSSWM (Pan, 2005), which can simulate crop growth and development with different management strategies and making reasonable decisions about crop management based on the results of crop growth simulation. On the other hand, plant architecture software as AMAP (Jaeger M and de Reffye P, 1992; Godin et al., 1997) and vlab (Federl and Prusinkiewicz, 1999) integrating computer graphics aim at plant architecture modeling and visualization. So far both crop simulation and decision support and architecture visualization remain to be incorporated into one system or software for a comprehensive representation of crop growth system.

The present study developed a digital and visual wheat growth system in order to gather different eco-physiological modules and visualization techniques for (i) simulating eco-physiological processes as development, growth, yield and quality and making reasonable decisions about crop management (ii) simulating topological and 3D structure of wheat at organ, individual, population levels; (iii) demonstrating the real-time visualization of wheat growth responses to different growth conditions and varieties.

We hereafter achieved the above objectives by integrating a process-based comprehensive growth simulation model WheatGrow (Yan, 1999; Liu, 2000; Pan, 2005), architecture model WheatArch (Cheng, 2004, Tan,

2006) and visualization model for producing a new functional-structural model WheatFM. This included linking the functional processes of WheatGrow with dynamic structures of WheatArch, as well as integrating component-based software and computer graphic technology for visualization.

2. OVERALL STRUCTURE OF MDVWGS

The MDVWGS could simulate growth, architecture, yield and quality of wheat growing on a uniform area of land under prescribed or simulated management as well as under the changed soil water and nitrogen conditions. It is comprised of database, models, applications and interface (Fig. 1).

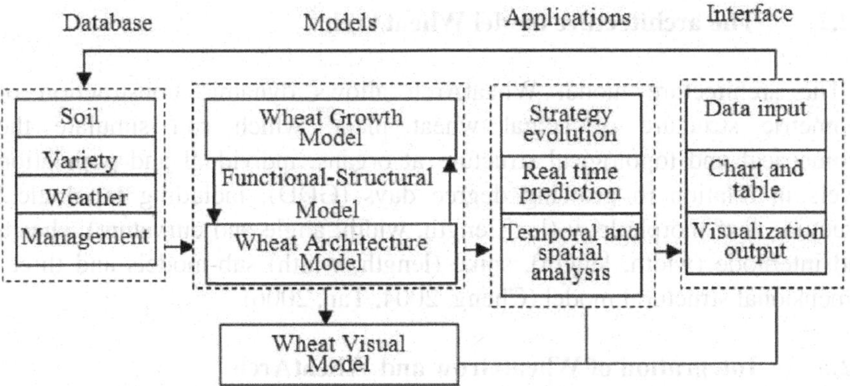

Fig. 1. Structural framework of MDVWGS

2.1 Database

The database includes weather, soil, variety and management data. The weather data has daily records of date, maximum and minimum air temperature, sunshining hours and rainfall. The soil data has typical soil parameters for soil module, e.g. soil water content, ammonium N and nitrate in soil layers, thickness of a layer, actual number of soil layers. The variety data includes genotype-specific parameters, e.g. variety name, phenological parameters, thousand-seed weight and parameters of architecture model. Management data contains some common practices of wheat management.

2.2 Models

The main models include growth simulation model WheatGrow, the architecture formation model WheatArch, visualizaton model, and module for model integration.

2.2.1 Growth simulation model WheatGrow

WheatGrow is a field scale, weather-driven, process-based dynamic simulation model (Yan, 1999; Liu, 2000; Pan, 2005), which operates with a daily time-step, including 6 submodels for simulating phasic and phenological development, morphological and organ formation, photosynthesis and dry matter accumulation, yield and quality formation, soil water relation soil water relation and dynamic nutrient (N, P, K) balance.

2.2.2 The architecture model WheatArch

The architecture model WheatArch allows dynamic construction of geometric structure of actual wheat plant, which can simulate the geometrical and topological structure at organs, individual and population levels in relation to growing degree days (GDD), including topological structure, leaf morphology (leaf length, width, angle and curvature), sheath and internode (width, length), spike (length, width) sub-models and three-dimensional structural model (Cheng, 2004, Tan, 2006).

2.2.3 Integration of WheatGrow and WheatArch

The WheatGrow runs continuously and independently driven by data of weather, variety, soil and management, simulating the functional processes of wheat and providing basic input for WheatArch model. The model WheatGrow was used to calculate: (1) phasic and phenological development to control the runtime of WheatArch from emergence to maturity; (2) characters of population architecture; (3) assimilate accumulation and partitioning, e.g. WheatGrow provides leaf area and leaf weight, while WheatArch simulates individual leaf initiation and growth, and both would be linked to build process-based leaf morphological model (4) water and nutrient factors for quantifying the limiting factors of water and NPK nutrients for WheatArch. These were taken as the inputs of WheatArch.

2.2.4 Visualization model

Visualization model includes geometry, texture, illumination submodels. Geometric submodel includes simulation of several organs, using different geometric shapes (e.g. cylinder) to model leaf, stem and spike; the texture model is constructed by using the photos of organs at different stages with reality as textures and integrated with geometric model; illumination model in OPGL is adopted for simulating the illumination effect in the reality.

2.3 Applications

Different types of applications were accomplished in MDVWGS by using different functions on a daily basis, such as temporal and spatial analysis, strategy evaluation and real time prediction.

2.4 Interface

The interface of MDVWGS included the initial data input, digital and visualization output. The results of simulation strategy in digits can be represented by chart and table, and the results also can be displayed by visualization through the computer graphic technology, including scene controlling as zoom in or zoom out, changing the point of view.

3. FUNCTION DESCRIPTIONS

In MDVWGS, Multiple functions as simulating time-course processes of growth and development, architecture, yield and quality formation, soil water and nutrient dynamics under various environmental conditions, production levels and genetic parameters were developed so that the system could realize the functions of dynamic simulation, strategy evaluation and real time forecasting which have been implemented by Pan (2005) (Fig. 2). Result of these functions can be displayed by digits and visualized as computer images.

4. SYSTEM IMPLEMENTATION

The system was operated under windows 2003 server on PC with 1G of RAM and AMD althon 2500+. The MDVWGS applied Access 2003 to designing database, C# in the framework of .NET to programming the

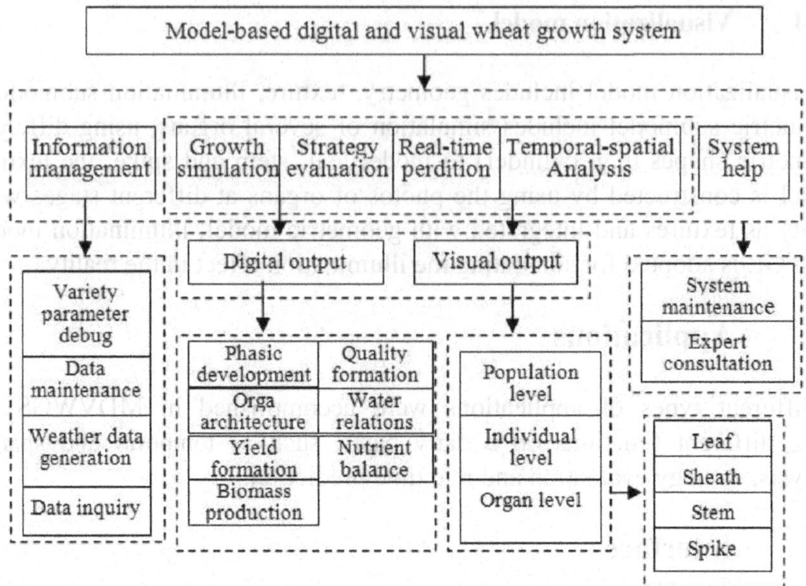

Fig. 2. Functions of MDVWGS

interface and programming growth model components based on the COM standard. Visualization model is designed by CSGL (C sharp Graphics Library) which implements a wrapper for the powerful C-library OpenGL allowing use of any .NET language.

4.1 System display

Part of system application is presented as follows, including the interface of MDVWGS for main menu (Fig. 3 a), digital results displayed by table (Fig. 3b), the visual output of dynamic individual plant growth (Fig. 3c, d).

5. DISCUSSIONS

Based on a process-based growth simulation model WheatGrow, architecture model WheatAM and visualization model, a digital and visual wheat growth system (MDVWGS) was developed using component-based software and visualization techniques, which enable ecophysiologists to investigate the interaction of organs, single plants or plant stands with their biotic and abiotic environment in a unique way. The implemented system integrated prediction function and decision support function, can simulate eco-physiological processes as development, growth, yield and quality, the

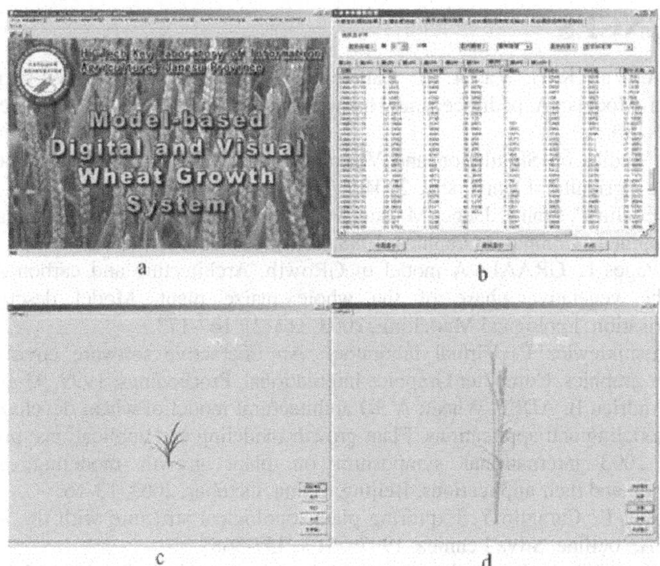

Fig. 3. System exhibition of interface for main menu (a), table output (b), visual output (c, d)

structure of wheat dynamics in mechanism and make decisions for mamagement, demonstrate the visualization of wheat growth under various environments, genotypes and management strategies, which providing a precise and scientific tool for cultivar design, cultural regulation and productivity evaluation under different growing conditions.

Systems like MDVWGS would be put into potential for applications of this technology in research, decision support, education and extension (Room, et al., 1996). In order to make the system more applicable and stability, the next steps will be: (1) to incorporate more physiological processes in the model to enhance the application of the model, e.g. calculation of radiation interception and inter-plant competition which would be feedback to growth model; (2) to extend the variety, soil and weather database, so that the system could be used in further comprehensive conditions; (3) to incorporate the root architecture model and visualization into the MDVWGS; (4) to link the MDVWGS with GIS, RS and management knowledge model for using in precision and digital farm.

ACKNOWLEDGEMENTS

We acknowledge the financial support from the State Hi-tech R&D Plan of China (2006AA10Z219) and the Hi-tech R&D Plan of Jiangsu Province (BG2004320).

REFERENCES

Bouman B A, Van K H, Van L H, Rabbinge R. The'School of de Wit'Crop Growth Simulation Models: A pedigree and Historical Overview. Agricultural Systems, 1996, 52(2): 171-198.

Cheng G Q. Studies on Simulation and Visualization of Morphogenesis in Wheat. Taian: Shangdong Agricultural University, 2004. (in Chinese)

de Reffye P, Edelin C, Fran J, Jaeger M, Puech C. Plant models faithful to botanical structure and development. Computer Graphics, 1988, 22(4): 151-158.

Drouet J L, Pages L. GRAAL: A model of GRowth, Architecture and carbon ALlocation during the vegetative phase of the whole maize plant. Model description and parameterisation. Ecological Modelling, 2003, 165(2): 147-173.

Federl P, Prusinkiewicz P. Virtual laboratory: An interactive software environment for computer graphics. Computer Graphics International, Proceedings, 1999, 93-100.

Fournier C, Andrieu B. ADEL-Wheat: A 3D architectural model of wheat development plant growth modeling and applications. Plant growth modeling and applications: proceedings-PMA03: 2003 international symposium on plant growth modeling, simulation, visualization and their applications, Beijing, China, October, 2003, 13-16.

Godin C, Costes E, Caraglio Y. Exploring plant topological structure with the AMAPmod software: An outline. Silva Fennica, 1997, 31: 3, 357-368.

Godin C. Representing and encoding plant architecture: A review. Ann. For. Sci. 2000, 57: 413-438.

Hanan J S, Hearn A B. Linking physiological and architectural models of cotton. Agricultural Systems, 2003, 75(1): 47-77.

Jaeger M, De R P. Basic concepts of computer simulation of plant growth. Journal of biosciences, 1992, 17(3): 275-291.

Jones J W, Hoogenboom G, H., Porter C, Boote K J, Batchelor W D, Hunt L A, Wilkens P W, Singh U, Gijsman A J, Ritchie J T. The DSSAT cropping system model. European Journal of Agronomy, 2003, 18: 235-265.

Liu T M. Simulation on photosynthetic production and dry matter partitioning in wheat. Nanjing: Nanjing Agricultural University, 2001. (in Chinese)

McCown R L, Hammer G L, Hargreaves J N G, Holzworth D P, Freebairn D M. APSIM: A Novel Software System for Model Development, Model Testing and Simulation in Agricultural Systems Research. Agricultural Systems, 1996, 50: 255-271.

Pan J. Study on wheat growth simulation and decision support system. Nanjing: Nanjing Agricultural University, 2005. (in Chinese)

Prusinkiewicz P, Lindenmayer A. The algorithmic beauty of plants. Springer-Verlag New York, Inc. New York, NY, USA, 1990.

Renton M, Kaitaniemib P, Hanana J. Functional-structural plant modelling using a combination of architectural analysis, L-systems and a canonical model of function, Ecological Modelling, 2005, pp. 277-298.

Room P M, Hanan J S, Prusinkiewicz P. Virtual plants: New perspectives for ecologists, pathologists and agricultural scientists. Trends in Plant Science, 1996, 1(1): 33-38.

Sievanen R, Nikinmaa E, Nygren P, Ozier-lafontaine H, Perttunen J, Hakula H. Components of functional-structural tree models. Ann. For. Sci., 2000, 57: 399-412.

Tan Z H. Studies on simulation model of morphological development in wheat plant. Nanjing, 2006. (in Chinese)

Van Ittersum M, Leffelaar P A, Van K H, Kropff M J, Bastiaans L, Goudriaan J. On approaches and applications of the Wageningen crop models. European Journal of Agronomy, 2003, 18(3): 201-234.

Yan H P, Kang M Z, de Reffye P, Dingkuhn M. A dynamic, architectural plant model simulating resource-dependent growth. Annals of botany, 2004, 93(5): 591-602.

Yan M C. A process-based mechanistic model for phasic and phonological development and organ formation in wheat. Nanjing: Nanjing Agricultural University, 1999. (in Chinese)

DESIGN OF CONTROL SYSTEM OF LASER LEVELING MACHINE BASED ON FUSSY CONTROL THEORY

Yongsheng Si [1], Gang Liu [2,3,*], Jianhan Lin [2,3], Qingfei Lv [4], Feng Juan [1]
[1] *Agriculture University of Hebei, Baoding, China, 071001*
[2] *College of Information and Electrical Engineering, China Agricultural University, Beijing, China, 100083*
[3] *Key Laboratory of Modern Precision Agriculture System Integration Research, Beijing, 100035, China*
[4] *Hangzhou Dianzi University, Hangzhou, 310018, China*
[*] *Corresponding author, Address: P.O. Box 125, China Agricultural University, 17 Tsinghua East Road, Beijing, 100083, P. R. China, Tel: +86-10-62736741, Fax: +86-10-62736746, Email: pac@cau.edu.cn*

Abstract: The control system of laser leveling machine was redesigned and tested to improve the performance and decrease the cost. In this paper, the control system of the machine which contains transmitter, receiver and controller was introduced. The receiver had silicon photoelements as photoelectric sensors. A kind of light filter was designed which was composed of red color optical glass and interference filter. Fussy control theory was applied in the controller design. The experiments conducted in fields showed that the controller and the laser receiver developed matched well. The control system worked well and the leveling accuracy was improved.

Keywords: fussy control, leveling machine, laser-controlled

1. INTRODUCTION

Land leveling, which is an important method to improve water coverage in field, can improve the irrigation efficiency and uniformity (Xu, 1999). Research has shown large decrease in water consumption and large increase in crop yield and quantity due to good field leveling. The sensitivity of the laser sensor system is at least 10 to 50 times more precise than the visual

Si, Y., Liu, G., Lin, J., Lv, Q. and Juan, F., 2008, in IFIP International Federation for Information Processing, Volume 259; Computer and Computing Technologies in Agriculture, Vol. 2; Daoliang Li; (Boston: Springer), pp. 1121–1127.

judgment and manual hydraulic control of an operator on the tractor. Consequently, the land leveling operation is correspondingly more accurate (Walker, 1992). Laser controlled land leveling technology has been already used widely abroad. In China, some farms and departments have imported laser controlled land leveling machine to conduct land leveling experiments since the eighties of the 20th century, while this technology was not popularized for the reason of high price. The water resource is scarce seriously in China, it is necessary to develop low-cost laser controlled leveling system with independent intellectual property right to improve the efficiency of water utilization and raise the crop yield.

The laser leveling system (Fig. 1) is composed of laser transmitter, laser receiver, controller, hydraulic system and bucket. The laser transmitter transmits a laser beam, which is intercepted by the laser receiver mounted on the leveling bucket. The controller mounted on the tractor interprets the signals from the receiver and opens or closes the hydraulic control valve, which will raise or lower the bucket. The controller transmits various control signals to the hydraulic system according to the signals received from the receiver and deals with unusual information. The transmitter, receiver and controller, which constitute the control system of the machine, take large proportion in cost of the system and play a important role in the machine. A domestic product - JP3 Rotation Laser was selected as the transmitter to lower the cost. This paper introduces the development of the receiver, the controller and the experiments analysis of the machine.

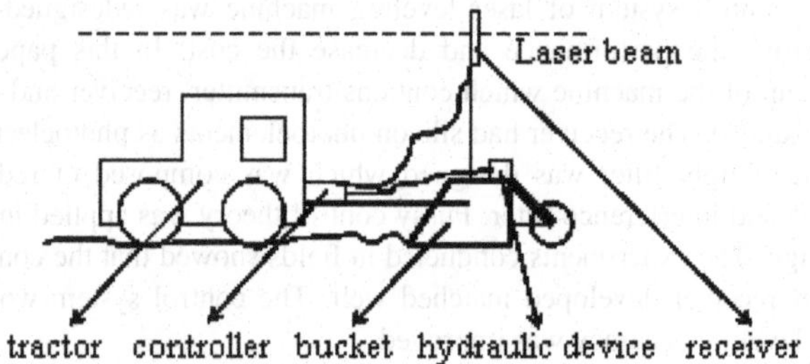

Fig. 1. Laser leveling machine

2. DESIGN OF CONTROL SYSTEM

2.1 Hardware Design of Receiver

The receiver had 8 layers with 32 pieces of silicon photoelements as photoelectric sensors (Lin, 2006). Laser filter was installed on the receiving surface of each photoelement. Each photoelement was designed to work with angle range no less than 90 degrees. In this way the whole working angle of each layer of photoelements comes up to 360 degrees (Fig. 2).

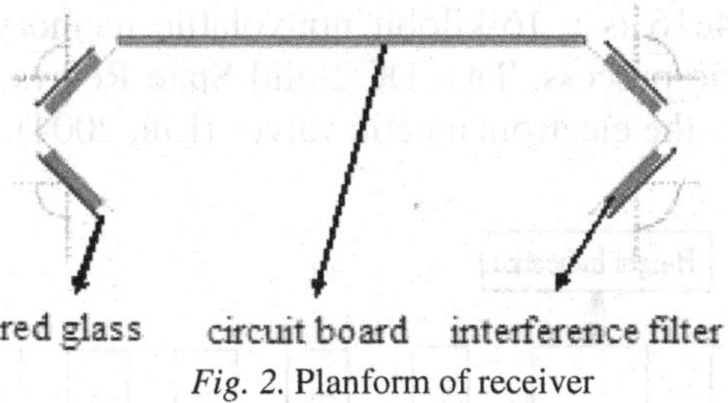

Fig. 2. Planform of receiver

The key problem in receiver design was to avoid or weaken the interference of sunlight. When developing the receiver, 2CR93 silicon photoelements were used to receive the laser signals which indicated the relative heights of the leveling scraper. The price of the interference filter is related with half-bandwidth, transmission rate and depth of the background. In order to lower the cost of the receiver, a kind of light filter, which was composed of red color optical glass and interference filter, was designed to restrict most noisy rays. A signal processing circuit was designed to amplify four channels of low-power analog signals from photoelements and process them into four channels of digital height signals respectively (Fig. 3).

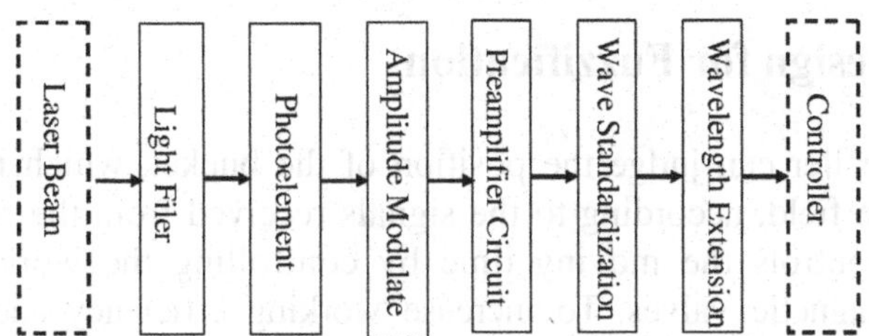

Fig. 3. Block diagram of receiver

2.2 Hardware Design of Controller

The controller raises or lowers the bucket by controlling the hydraulic system according to the signals received from the receiver. The controller is composed of CPU (89S51), pre-processing circuit, watchdog circuit, memory circuit, interlocking circuit, drive circuit and solid state relay (SSR) circuit etc. (Si, 2004). In many cases, the operator need control the system manually. For instance, if the depth of cutting exceeds the depth which can be cut with the power of the tractor and the operator must override the automatic controls in order to keep the equipment operating. Auto/Manual switch circuit was designed for the reason above. There are two output signal channels in the system. At any time, only one channel is positive signal, otherwise the hydraulic system will be destroyed. The controller has memory function to record the topographic data. Fm24C16 was selected as the memory chip. Fm24c16 is a 16-kilobit nonvolatile memory employing an advanced ferroelectric process. Two DC Solid State Relays (SSR) are used in the system to drive the electromagnetic valves (Lin, 2004).

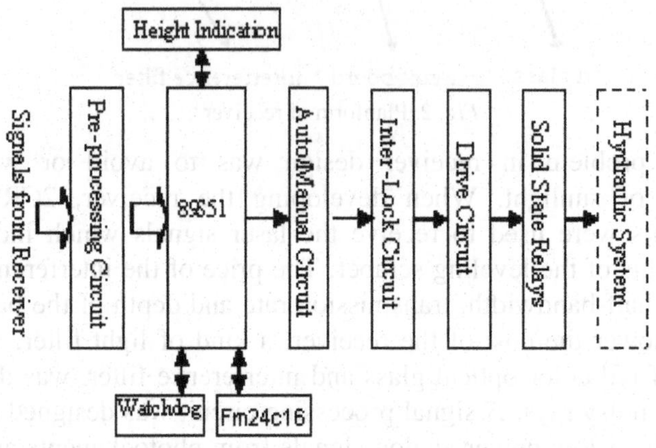

Fig. 4. Block diagram of controller

2.3 Design for Fuzzification

The controller can judge the position of the bucket, which indicates the height of the field, according to the signals received from the receiver. The controller controls the moving time by controlling the working time of the electromagnetic valves. To increase working efficiency and reduce the wearing and tearing of the equipments, the system has Rough/Precision leveling switch functions. The function of fuzzificaton is to translate e (the deviation between the measured height of the field and the expected height of the field) and etc. (the changing rate deviation between the measured height of the field and the expected height of the field) into the language variables E and EC, where the language variables are the fuzzy terminologies, such as “Very High”, “ High”, “Low”, “Very Low”, and so on. Nine fuzzy sets are defined for the language E: positive very big (PVB), positive big (PB), positive middle (PM) and positive small (PS), zero (ZE), negative small (NS), negative middle (NM), negative big (NB) and negative very big (NVB). Two fuzzy sets are defined for the language EC: positive big (PB) and positive small (PS). The language U (compensation quantity of working time of the electromagnetic valves) are divided into nine optimal fuzzy areas: left full open (LFO), left long open (LLO), left middle open (LMO), left short open (LSO), all close (AC), right short open (RSO), right middle open (RMO), right long open (RLO) and right full open (RFO).

2.4 Design of Control Rules

Based on a great deal of experimental data and based on the principle of responding fast and working stable, the optimal working frequency of 8MHz was selected (Lv, 2005). The electromagnetic valves and the hydraulic system work well at this frequency. The system has two working modes: rough leveling and precision leveling. Higher accuracy can be got under the precision leveling mode and under rough leveling mode the system has high working efficiency and low wearing and tearing of the equipments. Under the precision leveling mode, the hydraulic system and the bucket act more frequently than under the rough mode. The system can switch the two modes automatically according to the EC. The fuzzy control rules were summarized and shown in Table 1.

Table 1. Fuzzy control rules

	U	EC	
		PS	PB
E	PVB	LFO	LLO
	PB	LLO	LLO
	PM	LMO	LLO
	PS	LSO	LLO
	ZO	AC	AC
	NS	RSO	RLO
	NM	RMO	RLO
	NB	RLO	RLO
	NVB	RFO	RLO

3. EXPERIMENTS AND RESULTS

Experiments have been conducted in Beijing Shangzhuang Experiment Station on 2 May 2007 to check the efficiency and quality of the new control system. During the experiments we used the bucket and other parts of the leveling machine made by China Agriculture University. There are two parameters to evaluate the quality of field leveling. One is Sd (standard deviation) (Li, 1999):

$$S_d = \sqrt{\sum_{i=1}^{n} (h_i - \overline{h})^2 / (n-1)} \tag{1}$$

Where hi represents the ith height measured, is the mean height and n is the number measured. The other parameter is P (percentage):

$$P = n_c / n \tag{2}$$

Where n_c represents the number of points whose absolute difference, between the measured height and the mean height, is less than a certain value. Before and after the leveling, the experiment field was partitioned into many square grids, whose sides are 10 meters. The field height was measured at every corner point. The experiment data were shown in Table 2 and Table 3.

Table 2. Heights of measured points before leveling (cm)

hxy	x=0	x=1	x=2	x=3
y=0	135.5	146.5	142	142.5
y=1	139	138.5	133	141
y=2	141	140	135	135.5
y=3	140	133	137.5	132
y=4	141.5	138	139	140
y=5	142	141.5	145	135.5
y=6	141.5	141	135	139.5
y=7	141.5	140	140	139
y=8	142	138	138	135.5
y=9	139.5	140	135.5	139
y=10	142	138.5	139	135.5
y=11	142	141.5	143	140
y=12	137.5	143	140	137.5

The Sd (standard deviation) before leveling is 3.03cm and the Sd after leveling is 0.48cm. Before and after the leveling, the percentage of the number of the points whose absolute difference between the measured height and the mean height less than 1.5cm is 38% and 100% respectively. The performance of leveling distribution was very good for the max percentage of the absolute difference of other leveling machines was less than 80% (Dedrick et al., 1982).

Table 3. Heights of measured points after leveling (cm)

hxy	x=0	x=1	x=2	x=3
y=0	175	175	175	176
y=1	175.5	175	174.5	175
y=2	175	175	175.5	175
y=3	175	174.5	175.5	175.5
y=4	176.5	175	175	174.5
y=5	176	175	175	175
y=6	175.5	175	175	176
y=7	175	174.5	175	175.5
y=8	176	175.5	175.5	175
y=9	176	175.5	175.5	175.5
y=10	176.5	176	175.5	175.5
y=11	175.5	176	175	175.5
y=12	175.5	175	175	176

4. CONCLUSION AND FUTURE WORK

The cost of the control system was reduced through the new design of the receiver. The control system matched the other parts of the leveling machine well. High accuracy was got through fuzzy control system during the experiments. The performance of leveling distribution was highly improved especially. Moreover, the application of fuzzy control theory made the system more intelligent and the labor intensity was reduced. In this experiment the field is a regularly sized and shaped field. To test the performance of the control system and to make the control system more stable, more experiments should be conducted in different conditions, such as glutinous field and sandy field, dry field and moist field, field with and without straw and weed and so on, because leveling accuracy is high related with the field condition before leveling. It should be mentioned that the leveling efficiency and the result were high related with the operator and in this experiment the operator was experienced.

ACKNOWLEDGEMENTS

This paper is funded by the national 863 projects: Laser Controlled Land Leveling and Precision Surface Irrigation Equipment. (2006AA100210).

REFERENCES

Dedrick A R, Erie L J, and Clemmens A J, Lever basin irrigation. In Advances in Irrigation. Vol. 2, Hillel D I, ed, Acade mic Press, 1982, New York.

Li Yinong, Xu Di. Application and Evaluation of Laser Controlled Leveling Technology. Transactions of the Chinese Society of Agricultural Engineering 1999 6: pp. 79-83 (in Chinese).

Lin Jianhan, Liu Gang. A lower-cost controlled system for laser leveling, 2004 CIGR International conference Beijing, 2004.10.

Lin Jianhan, Wang Maohua, Liu Gang, Si Yongsheng. Development of a Laser Controlled System for Land Leveling Machinery, CIGR World Congress 2006, 2006.8

Qingfei Lv, Gang Liu, Jianhan Lin. An Innovated Laser Control System for Land Leveling Equipment, The International Symposium on Innovation and Development of Citied Agricultural, 2005.10.

Si yongsheng, Liu Gang, Lian Jianhan. Research on controller of laser leveling system, 2004 CIGR International conference Beijing, 2004.10.

Walker W R. Guidelines for designing and evaluating surface irrigation systems. FAO Irrigation and Drainage Paper No. 45, 1992, Rome.

Xu Di, Li Yinong, Li Fuxiang etc, Study on combination of conventional and Laser-controlled land grading procedures. Journal of Hydraulic Engineering 1999 10:52-56 (in Chinese)

NEAR INFRARED SPECTRUM DETECTION OF SOYBEAN FATTY ACIDS BASED ON GA AND NEURAL NETWORK

Changli Zhang[1], Kezhu Tan[2], Yuhua Chai[2], Junlong Fang[1], Shuqiang Liu[3]

[1] *Engineering College, Northeast Agricultural University, Harbin, China, 150030*
[2] *Cheng Dong College, Northeast Agricultural University, Harbin, China, 150030*
[3] *Hei longjiang Engineering College, Harbin, China, 150050*

Abstract: This paper represented a way to build mathematical model on genetic multilevel forward neural network. Building the relationship between chemistry measurement values and near infrared spectrum datum. The near infrared spectrum data was input in this network, five kinds of content of fatty acids, which measured by chemistry method, were output. Training the weight of multilevel forward neural network by genetic algorithms, building the soybean fatty acids neural network detection model, and exploring the network model which can realize near infrared spectrum detection exactly and efficiently. The authors designed a multilevel forward neural network trained by genetic algorithms. Test showed that relative coefficient in five fatty acids of soybean can be round about 0.9, and can satisfy init detection of soybean breeding.

Keywords: near infrared, multilevel forward neural network, genetic algorithms, soybean, fatty acids

1. INTRODUCTION

With the development of soybean seed breeding and soybean production, breeding experts in breeding has attached more and more importance to the quality of breeding soybean. The quality of oil depends much on the proportion of fatty acid and fatty acid is composed of Kitool acid, Stearic

Zhang, C., Tan, K., Chai, Y., Fang, J. and Liu, S., 2008, in IFIP International Federation for Information Processing, Volume 259; Computer and Computing Technologies in Agriculture, Vol. 2; Daoliang Li; (Boston: Springer), pp. 1129–1136.

acid, Oleic acid, Suboleic acid, Flax acid. While the process of breeding detects soybean fatty acid more frequently and the traditional methods of detection exist some disadvantages, such as time-consuming detection, high cost and a few detecting samples. To these difficulties, near infrared spectrum can solve those disadvantages efficiently.

Methods applying to near infrared spectrum include MLR, SMR, PCR and PLS. Only some spectrum information can be calculated in the methods of MLR and SMR when the samples are analyzed, losing a big surplus of information, the instructed model will generate much adaptability easily. Although PCR eliminate the question of linear relationship, but when decomposing spectrums matrix, inner relationship between the matrix and content in density can't be considered. The model was constructed based on the method that GA and neural network in this paper and forecasting soybean fatty acid by this constructed model.

At present, the method of GA combining neural network has been applied in every field GeHong. etc. Combined neural network fuzzy controller, training method of variation textures with modified GA. This method not only realized self-study of membership grade but also realized regular self organization, it settled a difficult question in fuzzy control, and it improved primary BP arithmetic which had the disadvantages in getting into local part optimality, realized the control goal of object. (LuChun. etc, 2001) raised a new arithmetic: BP-GA which combined GA with neural network. Because there is a strong complementation between GA and BP, GA-BP's convergence probability and convergences pace are good. (Wen Shaochun. etc, 2001) discussed the main appliance of the method which based on GA and neural network and got a lot of experiment data, the experiment indicated that GA has the function with studying net weighting fast. (Luo Jian, Liu, Junxiang, 2001) raised a new operator-BP operator.

To summarize, with the rapid development of computer technology and artificial intelligence, the method that GA combined with neural network developed rapidly. They beat out a new path for agriculture production.

2. MATERIALS AND METHODS

Haizhou opencast colliery which is the largest opencast colliery in Asia is located in the north temperature zone and is characterized by mountainous topography. This natural barrier has a strong influence on the meteorological conditions determining the air pollution situation. This site has been studied for approximately 5 years by researchers at China Agricultural University.

2.1 Instruments and devices

The soybean fatty acid contents detector of the experiment uses which is the perten8620 near-infrared instrument produced by the Sweden Bo Tong company. This instrument has 20 light filters, separately corresponds 20 wavelengths, and has the NIR software procedure.

The instrument to menstruate soybean fatty acid contents by Chemistry method uses GC-9A gas chromatography which produced by DaoJin in Japan.

2.2 Experimental materials

The experiment selects 25 different varieties soybeans which produced by Heilongjiang Province as the examination sample, and uses chemistry method to menstruate 5 varieties fatty acids content, concrete like table 1 show:

Table1. Examples of experiment

Sample name	Kitool acid %	Stearic acid %	Oleic acid %	Suboleic acid %	Flax acid %
DongNong01-1357	10.82	4.116	23.7	50.69	10.27
KenNong19	10.33	3.774	25.81	51.09	8.671
HeiNong41	10.83	4.237	23.78	52.79	7.835
HeFeng42	10.55	3.193	29.48	49.45	7.056
Hei7-50	12.05	3.755	21.82	53.22	8.641
HeiNong44	10.57	4.345	25.61	50.49	8.335
KenFeng9	11.64	4.261	22.06	52.27	9.302
HeFu93155-16	12.29	4.1	21.05	53.95	7.985
SuiNong10	11.38	3.651	18.66	56.95	8.849
HeiHe99	9.768	4.403	24.99	52.43	7.969
HeiNong37	11.94	3.624	17.44	57.33	9.187
HeiHe97-1225	12.18	3.612	23.78	51.27	8.627
BaoFeng9	11.23	4.356	24.56	50.83	8.364
HeFeng41	9.907	4.28	36.62	41.41	7.178
Sui98-579	10.02	3.033	25.57	52.17	8.679
JiYu47	10.6	3.518	23.16	51.45	10.65
DongNong300	10.58	3.519	23.19	51.53	10.66
JiGuang441	11.81	3.599	25.44	50.92	7.671
BeiFeng16	11.66	4.229	21.38	54.06	8.123
NongDa5236	11.98	3.8	25.95	48.33	9.396
SuiHuaWuXing soybean	12.63	3.415	27.02	45.92	10.25
BeiJiang171	12.06	3.971	26.94	47.32	9.141
DongNong1620	10.28	3.95	24.83	52.57	7.789
ShanHe711	11.13	3.692	23.16	53.96	7.368
HeFeng43	11.33	3.867	22.2	5.308	8.142

3. EXPERIMENTAL METHODS

When the near-infrared spectroanalysis technology carries out quantitative analysis. Firstly, choose having representative and can coverage the inspected parameter sample collection, according to the chemistry determining value (as table 1 show) which being allotted for picket age assembles and rectify-forecast assembles. The chemistry determines value coming from corn of Department of Agriculture and products quality monitoring checking testing centre (Harbin) detecting result, the method among them base on the national standard gas appearance chromatography. In putting soybean powder sample inside the near-infrared Perten 8620 standard forms admeasurements instrument (the Sweden Bo tong instrument company produced) receiving sample system, collecting the light absorption value of every sample in each wavelength, and changes into (actual light absorption of log (1/R) amounts), finally stores up the belt ready-made a computer document , forming a spectrum strip.

The spectrum data uses chemistry metrology method to build the quantify mathematic model between determine value and near-infrared chemistry spectrum data with picket age collection, the spectrum data uses intersection to confirm law later with the collection rectifying a forecast, checks what be built a model mass coming the relevance modulus by the fact that forecast standard deviation and the chemistry determining value and the near-infrared spectrum forecasting the value room. If built the model precision reached request, been available, it carries out quantitative analysis on the sample. This method characteristic is to analyze speed quickly, the sample is not to assumes any pretreatment, less dosages, no contaminating.

The main body of a paper is adopt to inherit the quantify mathematic model between determine value and near-infrared multilayer front make a present of neural networks (Zitzler et al., 1999) building-up chemistries spectrum data, the entering data looking on near-infrared spectrum data log (1/R) as a network, takes using chemistry to follow the five kinds fatty acid contents determining as network output, making use of inheritance algorithm to train multilayer front make a present of neural networks right value again, the neural networks building soybean fatty acid is check a model, final, carry out a checkout again with the model that the collection correct rectifying a forecast gets.

4. DATA PROCESSING AND RESULT ANALANCE

4.1 The principle of neural network combined with GA

Artificial neural network has strong ability of non-linear mapping, it uses the method like black box to remember and find the relationship between input and output. But, the designer of neural network still depends on experiential experts. In the actual study, we found that train the network using the traditional BP algorithm and then the actual calculation, the effect is not very satisfactory, there is a major problem as follows: (1) Convergence speed is very slow, sometimes shock; (2) Smallest error is local minimum; (3) Unable to accurately determine the optimal network structure. The appearance of GA brought neural network training with a new look, The objective function is neither to be consecutive, nor differentiable, only that the problem computable, And it always search throughout the entire solution space, easy access to the global optimal solution. Therefore, combine genetic algorithm with BP network, the mixed neural network training is a feasible way.

4.2 The principle of constructing model

The model was constructed based on the method that GA and neural network in this paper and forecasting soybean fatty acid by this constructed model. GA optimizes the weighting of neural network, BP arithmetic realize local part precise search.

Input of the network is X=[x1, x2, …x10] T, output of the network is Y=[y1,y2,…y25] T, Fig. 1 is the NN model of soybean fatty acid:

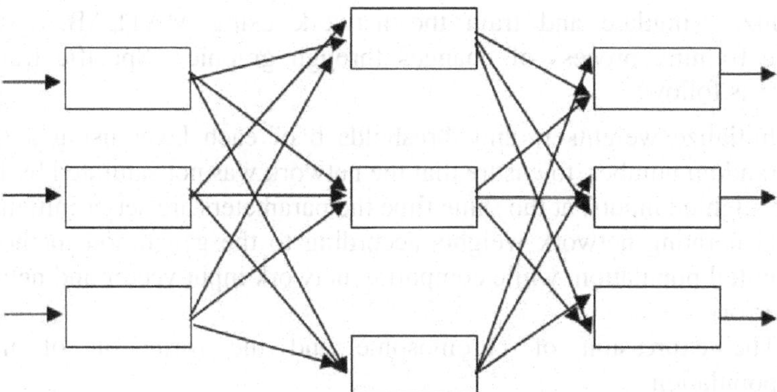

Fig. 1. The NN model of soybean fatty acid

4.3 Calculation process of the model

Network training is a key step in artificial neural network running, BP network training is based on weight modify principles of error gradient reduce, Convergence in local minimum points is its own ineluctable shortcoming. Genetic algorithm is based on the theory of biological evolution mathematical algorithm system, It is an effective global search tool, but not in the precise local search. The combination of artificial neural network and genetic algorithm achieved the complementary of strengths between them. For fixed structure neural network, evolutionary process of connection weights can be divided into the following steps:

(1) Determine weights coding schemes, generate initial population;
(2) Decode each individual in the population, construct the corresponding neural network;
(3) Compute network fitness according to certain performance evaluation criteria;
(4) Determine reproduce probability of each individual According to fitness size, then complete seed selection;
(5) Impose genetic operator on the selected population according to certain probability, in order to get a new generation of population;
(6) Return to step 2, until the performance meets requirements.

4.4 Realization of BP neural network and Genetic Algorithm program

Initialize, simulate and train the network using MATLAB; observe dynamic training process of changes through graphics. Specific training steps are as follow:

(1) Initialize weights w and thresholds b of each layer using a small random number, to ensure that the network was not saturated by large weighted inputs, at the same time the parameters are set or initialized.
(2) Generating network weights according to the generation method of initial population, while computing network input vector and network errors.
(3) The expression of chromosome and the formation of initial population.
(4) Genetic Algorithm operating steps that are selection, crossover and mutation.

Run self designed MATLAB program, Choose 20 samples training the network, Five Outputs are the content of palmitic acid, hard fatty acids, oleic acid, linoleic acid and linolenic acid, Specific steps are shown table 2:

Table 2. The output value of neural network

Sample name	Kitool acid %	Stearic acid %	Oleic acid %	Suboleic acid %	Flax acid %
DongNong01-1357	10.8265	4.1191	23.7055	50.6953	10.2773
KenNong19	10.3310	3.7741	25.8107	51.0906	8.6719
HeiNong41	10.8255	4.2212	23.7773	52.7862	7.8312
HeFeng42	10.5460	3.1964	29.4757	49.4461	7.0509
Hei7-50	12.0401	3.7390	21.8080	53.2071	8.6254
HeiNong44	10.5784	3.3444	25.6183	50.4973	8.3483
KenFeng9	11.6533	4.2640	22.0697	52.2788	9.3123
HeFu93155-16	12.3085	4.1113	21.0668	53.9660	8.0066
SuiNong10	11.3742	3.6677	18.6534	56.9456	8.8419
HeiHe99	9.7861	4.3923	25.0053	52.4430	7.9842
HeiNong37	11.9262	3.6378	17.4288	57.3207	9.1798
HeiHe97-1225	12.1858	3.6072	23.7862	51.2755	8.6316
BaoFeng9	11.2029	4.3638	24.5359	50.8088	8.3384
HeFeng41	9.8998	4.2939	36.6132	41.44049	7.1733
Sui98-579	10.0190	3.0303	25.5684	52.1683	8.6769
JiYu47	10.6032	3.5136	23.1628	51.4514	10.6530
DongNong300	10.5733	3.5205	23.1864	51.5265	10.6549
JiGuang441	11.8237	3.5941	25.4525	50.9310	7.6856
BeiFeng16	11.6543	4.2915	21.3755	54.0553	8.1141
NongDa5236	11.9803	3.7917	25.9495	48.3291	9.3939

5. CONCLUSIONS

Having analyzed characteristics of artificial neural networks and genetic algorithms, we put the combination of them into the modeling of soybean fatty acid content detection. The established model has the following characteristics:

(1) High contrast, BP neural network simulates the way of human thinking, it judges on the basis of intuitionist ratiocination according to the essence of things, rather than some kind of pre-established pattern.

(2) Strong manipulation, providing unknown spectral data samples the trained network will output results of the analyzing, so it is simple and easy.

REFERENCES

Mroczyk W B, Michalski K M. Analyzed elementary compositions of beet tops by NIR [J]. Computers Chem. 1995, 19(3):299-301.

Bochereau L, Beurgine Petal. Classical date analysis and neural network to NIR to predict quality of the apple [J]. Journal of Agricultural Engineering Research, 1992, 51:207-216.

Fogel D B. An introduction to simulated evolutionary optimization [J]. IEEE Trans on Neural Networks, 1994, 5(1):3-14.

Srinivas M, Patnail L M. Adaptive Probabilities of Crossover and Mutations in Gas [J]. IEEE Trans on SMC, 1994, 24(4):656-667.

Zitzler E, Thiele L. Multi-Objective Evolutionary Algorithms: A Comparative Case Study And the Strength Pareto Approach [J]. IEEE Transactions of Evolutionary Computation, 1999, 3(4):257-271.

Williams P C, Norris K, Gehrke C W and Bernstein K. 1983. Comparison of near-infrared methods for measuring protein and moisture in wheat. Cereal Foods Worle. 150:149-152.

Williams P C, Stevenson S, Starkey P M and Hawtin G. 1978. The application of near-infrared reflectance spectroscopy to protein-testing in pulse breeding-programmes. Journal of Science Food Agriculture. 29:285-292.

Morgan J E and Williams P C. 1995. Starch damage in wheat flours: A comparison of enzymic, iodometric and near-infrared reflectance techniques. Cereal Chemistry. 72 (2):209-212.

Murray L and Williams P C. 1987. Chemical principles of near-infrared technology. In Williams P C and Norris K ed. Near-infrared Technology. AACC St. Paul, MN.

Norris K H, Barnes R F, Moore J E and Shenk J S. 1976. Predicting forage quality by infrared reflectance spectroscopy. Journal of Animal Science. 43:889-897.

Norris K H. 1984. Multivariate analysis of raw material. In: Schmilt L W, ed. Chemistry and world food supplies. Manila: Reihold Publisher, 155-164.

SOIL CLASSIFICATION VIA MID-INFRARED SPECTROSCOPY

Raphael Linker
Faculty of Civil and Environmental Engineering, Technion-Israel Institute of Technology, Haifa, 32000, Israel, linkerr@technion.ac.il, *http://www.technion.ac.il/~civil/linker/*

Abstract: The need for rapid and inexpensive techniques for soil characterization has led to the investigation of modern technologies, and in particular those based on reflectance spectroscopy. While near-infrared has been traditionally used, mid-infrared in the 400-4000 cm-1 range is becoming increasingly common due to the specificity of the absorbance bands in this spectral range. The present work discusses two methods based on mid-infrared spectroscopy for soil classification: attenuated total reflectance (ATR) and photoacoustic spectroscopy. The ATR method requires a soil sample close to water saturation, and as a result only the 800-1600 cm-1 interval of the spectrum yields a useful signal. Typical ATR soil spectra consist mostly of several broad bands in the 800-1200 cm-1 region and a calcium carbonate band around 1450 cm-1. By comparison, photoacoustic measurements are conducted with air-dried samples, and the photoacoustic spectra exhibit a larger number of clearly-defined bands. Both methods were tested on data sets containing over 100 samples of various soils commonly used in Israeli agriculture. Data analysis was conducted by wavelet decomposition and neural network classifiers. Very good classification performances were achieved, with correct classification rates of the validation samples typically above 95%.

Keywords: Fourier transforms infrared (FTIR); attenuated total reflectance (ATR); photoacoustic spectroscopy (PAS); wavelets; neural networks

1. INTRODUCTION

Precision farming and similar modern approaches for efficient management of land resources require fast and accurate methods for soil

Linker, R., 2008, in IFIP International Federation for Information Processing, Volume 259; Computer and Computing Technologies in Agriculture, Vol. 2; Daoliang Li; (Boston: Springer), pp. 1137–1146.

characterization. Standard laboratory techniques for soil analysis are labour- and time-consuming, and extensive research has been devoted to the development of new methods for rapid screening of large number of soil samples (Viscarra et al., 1998; McBratney et al., 2006). Among the approaches investigated, spectroscopy, both in the near-infrared (NIR) (Ben-Dor et al., 1995; McCarty et al., 2002; Daniel et al., 2003) and mid-infrared ranges, has yield very promising results (Viscarra et al., 2006). While NIR spectra consist of non-specific overtones that are difficult to interpret, mid-infrared spectra consist of specific bands that can be directly associated with soil constituents. With respect to soil analysis, most mid-infrared studies were conducted in transmittance (Haberhauer et al., 1998; Haberhauer et al., 1999; Gerzabek et al., 2006), diffuse reflectance (DRIFT) (McCarty et al., 2002; Haberhaue et al., 1999; Janik et al., 1998; Nguyen et al., 1991) and attenuated total reflectance (ATR) (Linker et al., 2004; Linker et al., 2005; Linker et al., 2006) modes. Transmittance studies revealed numerous absorbance bands that could be associated with organic as well as inorganic soil components. However, this technique requires the time-consuming preparation of KBr pellets and is not suitable for routine analysis of large amounts of samples. In addition, such measurements involve very small quantities of soil (typically less than 1 mg per sample), and the representativeness of the sample may be questionable. Although some of these limitations are overcome with the DRIFT technique which does not strictly require the preparation of pellets, soil grinding and dilution with KBr usually improves the results significantly. By comparison, the ATR technique requires very minimal sample preparation. However, contrary to transmittance and DRIFT techniques, ATR requires very good contact between the sample and a crystal that serves as a waveguide for the IR beam (Fig. 1). The IR beam is directed in such a way that it hits the crystal/sample interface several times. Each time, the evanescent wave penetrates a few microns into the sample, so that the signal that reaches the detector contains information about the absorbance of the sample. Since the penetration depth is limited to a few microns, very good contact between the sample and the crystal is required. For soil samples, this can be achieved by working with samples close to water saturation (Linker et al., 2004; Linker et al., 2005; Linker et al., 2006; Shaviv et al., 2003). Unfortunately, water exhibits very strong absorbance bands in the mid-IR range, which may distort or hide bands of interest.

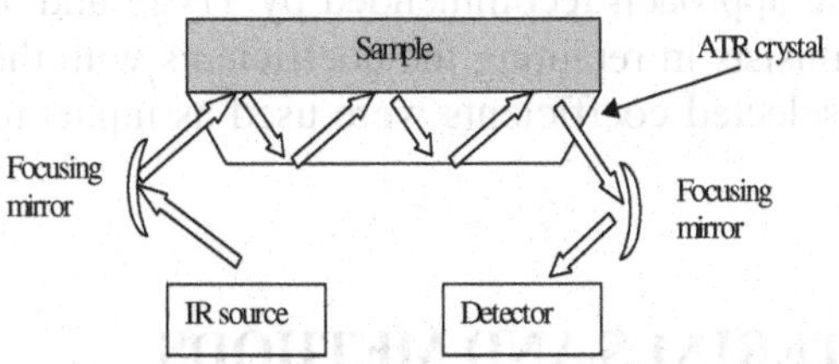

Figure 1. Schematic description of ATR spectroscopy

Changwen et al. (Changwen et al., Changwen et al.,) recently suggested the use of photoacoustic (PAS) mid-IR spectroscopy for soil identification. Photoacoustic spectroscopy is based on absorption-induced heating of the sample, which produces pressure fluctuations in a surrounding gas. These fluctuations are recorded by a microphone, and constitute the PAS signal (McClelland et al., 2001). The major advantage of photoacoustic spectroscopy is that it is suitable for highly absorbing samples, such as soils, without any special pre-treatment.

Regardless of the method used to obtain the spectra, mathematical processing is required for automated soil classification. Due to the high dimensionality of the spectra, this typically involves a data-reduction stage followed by some classification tool. The most commonly used method for data reduction is principal component analysis (PCA) (Jolliffe, 1986) that performs a linear decomposition of the data. This approach was used by Linker et al. (Linker et al., 2006) and Chanwen et al. (Changwen et al.,) for ATR and PAS soil spectra, respectively. Wavelet transform is another data-reduction method that is becoming increasingly popular for spectrum analysis (Walczak et al., 1997; Trygg et al., 1998; Ehrentreich et al., 2002; Liu et al., 2004; Figueiredo et al., 2007). The main feature of wavelet decomposition is that the resulting coefficients contain information about both the location and shape (sharpness) of the spectral bands. With respect to spectroscopy, three types of wavelet-based methods have been investigated (continuous wavelet, discrete wavelet and wavelet packet transform), and details relative to each method can be found in the literature (Walczak et al., 1997; Figueiredo et al., 2007; Leung et al., 1998; Jahn et al.,). Wavelet transformation by itself does not produce a compressed representation of the original data, and data reduction is achieved by eliminating the wavelet coefficients that do not contain valuable information. This is a non-trivial task and various approaches have been reported in the literature, such as eliminating all "small" coefficients using either simple thresholding (Ehrentreich et al., 2002; Liu et al., 2004; Figueiredo et al., 2007; Leung et al., 1998), mutual information (Alsberg et al., 1998) or genetic algorithms (Depczynski et al., 1999; Zhang et al., 2003).

The present paper presents a method based on wavelet decomposition and a neural network classifier for classification of both ATR and PAS spectra.

In this study, the approach recommended by Trygg and Wold (Trygg et al., 1998), which consists in retaining the coefficients with the highest variance, was used. The selected coefficients were used as inputs to a neural network (NN) classifier.

2. MATERIALS AND METHODS

2.1 Sample preparation and spectroscopic measurements

Details concerning sample preparation and spectroscopic measurements can be found in (Linker et al., 2006) and (Changwen et al. Changwen g et al.) for ATR and PAS, respectively, and only crucial information is recalled here. The ATR measurements were conducted with 202 samples close to water saturation, representative of five soil types commonly encountered in Israeli agriculture (Table 1). For the PAS measurements, 160 air-dried samples of the same types of soils were used. Although these samples were not strictly identical to the ones used for the ATR study, they had very similar characteristics and belonged to the same soil types (details not shown, see (Changwen et al.).

Table 1. Properties of the soils used for ATR measurements

Soil type denomination in text	Clay content (%)	$CaCO_3$ content (%)	Organic matter content (%)
Grumosol	50-70	5 -25	1.1-1.3
Loess	15-30	10-30	0.8-1.1
Rendzina	40-55	35-45	0.8-1.1
Hamra	5-35	0-1	0.5-0.8
Terra Rosa	45-70	<1	1.0-1.4

2.2 Data analysis

2.2.1 Pre-processing of spectra

The ATR spectra were smoothed using a second-order Savitzky-Golay filter with a 15-point window, and the method developed by Linker et al. (Linker et al., 2004) was applied for water-subtraction and baseline correction. The corrected spectra were then normalized so that the integral of the spectra would be equal to one. For the photoacoustic spectra, only smoothing (using a first-order Savitzky-Golay filter with a 25-point window) and normalization was applied.

2.2.2 Data reduction

Data reduction was achieved using the discrete wavelet transform with a Coiflet mother-wavelet. The approach recommended by Trygg and Wold (Trygg et al., 1998) was used to determine which of the resulting wavelet coefficients contained most of information. According to this method, the coefficients were sorted according to their variances and only the N coefficients with the highest variances were retained. The procedure is depicted schematically in Fig. 2 and the main steps are (1) wavelet transformation of each spectrum, (2) concatenation of the wavelet coefficients, (3) calculation of the coefficient variances and (4) extraction of the coefficients with the largest variances.

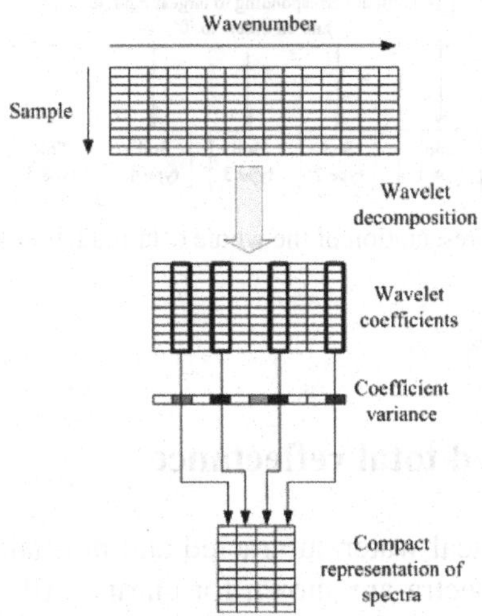

Figure 2. Schematic description of the wavelet-based procedure for data reduction

2.2.3 Classification

Feedforward neural networks (NN) with sigmoid activation functions were used as non-linear classifiers (Haykin et al., 1999) The inputs of the classifiers consisted of the wavelet coefficients selected at the previous stage. The NN had five outputs, corresponding to the five soil classes included in the study. A "winner-takes-all" approach was used, according to which the identified type was that of the output node with the largest value. All the classifiers were calibrated using half of the data (chosen randomly) and validated with the remaining data.

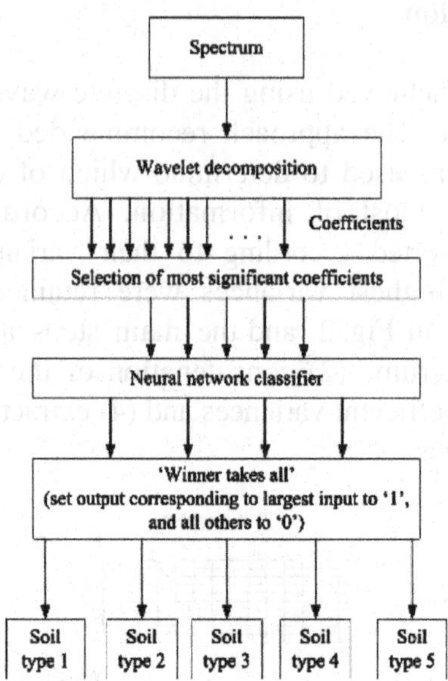

Figure 3. Schematic representation of the whole data-reduction & classification procedure

3. RESULTS

3.1 Attenuated total reflectance

Figure 4 shows typical water-subtracted and normalized spectra. For each soil type, only five spectra are shown for clarity. All the soils have various absorbance bands in the 800-1200 cm-1 interval, which are centered at 870, 915, 1025, 1110 cm-1. In addition, the calcareous soils have a strong absorbance band centered at 1440 cm-1 that corresponds to calcium carbonate (left frames in Fig. 4). Comparison of these results with Fig. 5 in (Changwen et al.,) (which was based on non-normalized spectra) shows that within-type variability was greatly reduced by normalization.

Three types of classifiers were investigated: (1) based on the 800-1200 cm-1 interval only, (2) based on the 1250-1550 cm-1 interval only, and (3) based on the whole 800-1550 cm-1 interval. In each case numerous classifiers based of various levels of wavelet decomposition, number of coefficients used as NN input and number of NN hidden nodes, were tested.

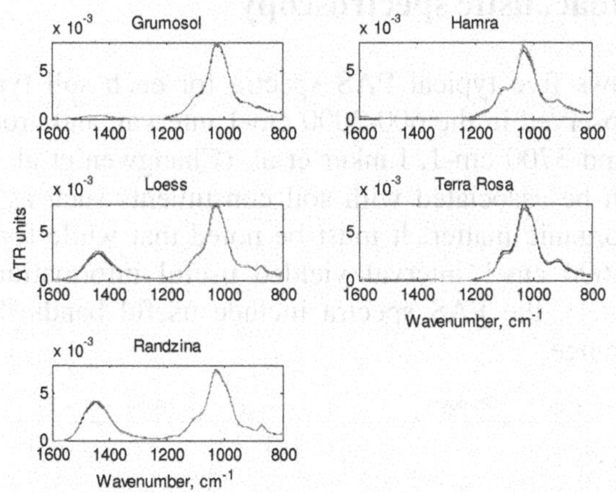

Figure 4. Typical ATR soil spectra after water subtraction, baseline correction and normalization

Table 2 presents the results obtained for the NNs that had the highest classification rate for the validation samples. Regardless of the spectral interval used, very good results are obtained and all the validation samples are correctly classified except some Hamra samples that are classified as Terra Rosa. Such misclassification is not surprising due to the very high similarity of the spectra of both soils (Fig. 4, right frames). However, as pointed out by Linker et al. (Linker et al., 2006), this is not a serious shortcoming since in practice these two soils are not found in the same regions. Furthermore, the water contents of the saturated pastes of these soils are very different (due to difference in clay content), so that it is possible to ensure perfect classification by adding the paste water content to the NN inputs (not shown).

Table 2. Classification results based on ATR spectra (validation spectra only)

Spectral interval	1250-1550 cm-1	800-1200 cm-1	800-1600 cm-1
Decomposition level	5	5	5
Number of coefficients selected	4	5	5
Number of NN hidden nodes	4	4	4
	Percentage correct classification		
Grumosol	100	100	100
Loess	100	100	100
Randzina	100	100	100
Hamra	91	91	96
Terra Rosa	100	100	100

3.2 Photoacoustic spectroscopy

Figure 5 shows five typical PAS spectra for each soil type. Numerous bands can be observed in the 600-2000 cm-1 interval, and around 2550 cm-1, 2900 cm-1 and 3700 cm-1. Linker et al. (Changwen et al.,) showed that these bands can be associated with soil constituents such as clays, carbon carbonate and organic matter. It must be noted that while for ATR spectra only the 800-1600 cm-1 interval yielded useful information (due to the presence of water), the PAS spectra include useful bands throughout the whole spectral range.

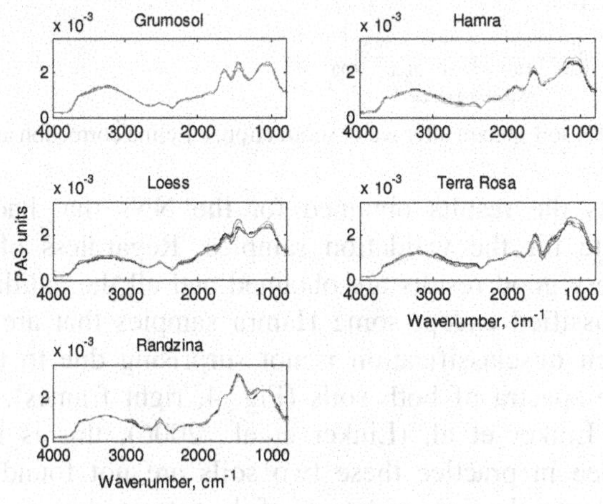

Fig. 5. Typical normalized PAS soil spectra

The results of the classification procedure are summarized in Table 3. Again, various classifiers were tested, and only the best ones are reported in the Table. The best results were obtained when using the whole spectrum, in which case only a few Hamra samples were incorrectly classified.

Table 3. Classification results based on photoacoustic spectra (validation spectra only)

Spectral interval	800-4000 cm-1	800-2700 cm-1	800-2300 cm-1
Decomposition level	5	10	10
Number of coefficients selected	10	20	20
Number of NN hidden nodes	4	4	4
	Percentage correct classification		
Grumosol	100	100	100
Loess	100	100	93
Randzina	100	100	100
Hamra	96	95	96
Terra Rosa	100	100	85

4. CONCLUSIONS

Mid-infrared attenuated total reflectance and photoacoustic spectroscopy both appear to be very promising techniques for rapid analysis of soil samples. The main limitation of the ATR approach is that it requires samples close to water saturation. As a result, the spectral range that yields useful information is rather limited. By comparison, PAS measurements are conducted with air-dried samples, and useful absorbance bands are observed throughout the whole spectrum. The bands are also sharper and more clearly defined than the ATR ones.

Both types of spectra can be used for soil classification, which requires data reduction and classification. A new method based on wavelet decomposition and neural network classification has been developed, which results in correct classification of over 95% of the validation samples.

REFERENCES

Alsberg B. K., Woodward A. M., Winson M. K., Rowland J. J. and Kell D. B., Variable selection in wavelet regression models, Analytica Chimica Acta, Vol. 368, 1998, pp. 29-44.

Ben-Dor E. and Banin A., Near-infrared analysis as a rapid method to simultaneously evaluate several soil properties, Soil Science Society of America Journal, Vol. 59, 1995, 364-372.

Changwen D., Linker R. and Shaviv A., Characterization of soils using photoacoustic spectroscopy, Applied Spectroscopy (Accepted)

Changwen D., Linker R. and Shaviv A. Soil identification using mid-infrared photoacoustic spectroscopy. Submitted to Geoderma

Daniel K. W., Tripathi N. K. and Honda K., Artificial neural network analysis of laboratory and in situ spectra for the estimation of macronutrients in soils of Lop Buri (Thailand), Australian Journal of Soil Research, Vol. 41, 2003, pp. 47-59.

Depczynski U., Jetter K., Molt K. and Niemoller, A., Quantitative analysis of near infrared spectra by wavelet coefficient regression using a genetic algorithm, Chemometrics and Intelligent Laboratory Systems, Vol. 47, 1999, pp. 179-187.

Ehrentreich F., Wavelet transform applications in analytical chemistry, Analytical and Bioanalytical chemistry, Vol. 372, 2002, pp. 115-121.

Figueiredo dos Santos R. N., Galvao R. K. H., Araujo M. C. U. and Cirino da Silva E., Improvement of prediction ability of PLS models employing the wavelet packet transform: A case study concerning FT-IR determination of gasoline parameters, Talanta, Vol. 71, 2007, pp. 1136-1143.

Gerzabek M. H., Antil R. S., Kogel-Knabner I., Knicker H., Kirchmann H. and Haberhauer G., How are soil use and management reflected by soil organic matter characteristics: A spectroscopic approach, European Journal of Soil Science, Vol. 57, 2006, pp. 485-494.

Haberhauer G. and Gerzabek M. H., DRIFT and transmission FT-IR spectroscopy of forest soils: an approach to determine decomposition process of forest litter, Vibrational Spectroscopy, Vol. 19, 1999, pp. 413-417.

Haberhauer G., Rafferty B., Strebl F. and Gerzabek M. H., Comparison of the composition of forest soil litter derived from three different sites at various decompositional stages using FTIR spectroscopy, Geoderma, Vol. 83, 1998, pp. 331-342.

Haykin S., Neural networks: A comprehensive foundation, Prentice-Hall, 1999.

Jahn B. R., Linker R., Upadhyaya S. K., Shaviv A., Slaughter D. C. and Shmulevich I., Mid infrared spectroscopic determination of soil nitrate content. Biosystems Engineering, Vol. 94, 2006, pp. 505-515.

Janik L. J., Merry R. H. and Skjemstand J. O., Can mid infrared diffuse reflectance analysis replace soil extractions, Australian Journal of Soil Research, Vol. 38, 1998, pp. 681-696.

Jolliffe I. T., Principal Component Analysis, Springer-Verlag, New York, 1986.

Leung A. K., Chau F. T., Gao J. B. and Shih T. M., Application of wavelet transform in infrared spectrometry: spectral compression and library search, Chemometrics and Intelligent Laboratory Systems, Vol. 43, 1998, pp. 69-88

Linker R., Kenny A., Shaviv A., Singher L. and Shmulevich I. FTIR/ATR nitrate determination of soil pastes using PCR, PLS and cross-correlation, Applied Spectroscopy, Vol. 58, 2004, pp. 516-520.

Linker R., Shmulevich I., Kenny A. and Shaviv A., Soil identification and chemometrics for direct determination of nitrate in soils using FTIR-ATR mid-infrared spectroscopy, Chemosphere, Vol. 61, 2005, pp. 652-658.

Linker R., Weiner M., Shmulevich I. and Shaviv A., Nitrate determination in soil pastes using FTIR-ATR mid-infrared spectroscopy: Improved accuracy via soil identification, Biosystems Engineering, Vol. 94, 2006, pp. 111-118.

Liu Y. and Brown S. D., Wavelet multiscale regression from the perspective of data fusion: new conceptual approaches. Analytical and Bioanalytical chemistry, Vol. 380, 2004, pp. 445-452.

McBratney A. B., Minasny B., Viscarra Rossel R. A., Spectral soil analysis and inference systems: A powerful combination for solving the soil data crisis, Geoderma, Vol. 136, 2006, pp. 272-278.

McCarty G. W., Reeves III J. B., Follett R. F. and Kimble J. M., Mid-infrared and near-infrared diffuse reflectance spectroscopy for soil carbon measurement, Soil Science Society of America Journal, Vol. 66, 2002, pp. 640-646.

McClelland J. F., Jones R. W. and Bajic S. J., Photoacoustic Spectroscopy, In Handbook of Vibrational Spectroscopy (Volume II), J. M. Chalmers and P. R. Griffiths, Eds., Wiley & Sons, 2001.

Nguyen T. T., Janik L. J. and Raupach M., Diffuse reflectance infrared Fourier transform (DRIFT) spectroscopy in soils studies, Australian Journal of Soil Research, Vol. 29, 1991, pp. 49-67.

Shaviv A., Kenny A., Shmulevich I., Singher L. Reichlin Y. and Katzir A., IR fiberoptic systems in situ and real time monitoring of nitrate in water and environmental systems, Environmental Science & Technology, Vol. 37, 2003, pp. 2807-2812.

Trygg J. and Wold S., PLS regression on wavelet compressed NIR spectra, Chemometrics and Intelligent Laboratory Systems, Vol. 42, 1998, pp. 209-220.

Viscarra Rossel R. A. and McBratney A. B., Soil chemical analytical accuracy and costs: Implications from precision agriculture, Australian Journal of Experimental Agriculture, Vol. 38, 1998, pp. 765-775.

Viscarra Rossel R. A., Walvoort D. J. J., McBratney A. B., Janick L. J. and Skjemstad J. O., Visible, near infrared, mid infrared or combined diffuse reflectance spectroscopy for simultaneous assessment of various soil properties, Geoderma, Vol. 131, 2006, pp. 59-75.

Walczak B. and Massart D. L., Noise suppression and signal compression using the wavelet packet transform, Chemometrics and Intelligent Laboratory Systems, Vol. 36, 1997, pp. 81-94.

Zhang X., Jin J., Zheng J. and Gao H., Genetic algorithms based on wavelet transform for resolving simulated overlapped spectra, Analytical and Bioanalytical Chemistry, Vol. 377, 2003, pp. 1153-1158.

NONDESTRUCTIVE TESTING SYSTEM FOR EGGSHELLS BASED ON DSP

Lihong He[1], Dejun Jiang[1], Lan Liu[1], Jingang Liu[2,*]
[1] *Department of Mechanical and Electrical Engineering, Hunan Institute of Engineering, Xiangtan, China, 411104*
[2] *Department of mechanical and automotive engineering, Hunan University, Changsha, China, 410082*
** Corresponding author, Address: Room 533, 17th Dormitory, Hunan University, Changsha, 410082, P. R. China, Tel: +86-731-2855111, Fax: +86-732-2855114, Email: wellbuild@126.com*

Abstract: In order to check and eliminate the cracked eggs quickly and exactly in the process of egg products, a nondestructive testing system for eggshells based on digital signal processor (DSP) was established. The system utilized a TMS320VC33 DSP as a hardware platform, and has many functions such as controlling knock-equipment, sampling and processing the acoustics signals, judging the cracked-shell eggs, communicating with the host computer etc. In order to improve the detecting accuracy, 6 characteristic parameters were extracted in cepstrum domain besides the common used characteristic parameters in frequency domain as candidate distinguishing factors. Step discriminant was used to optimize the combination of these characteristic parameters, then established the optimum discriminant functions of normal eggshells and cracked eggshells respectively according to the step discriminant result. The accurate recognition ratio for cracked eggs is up to 95.5 percent, and the system can check 12 eggs per sec, which satisfied the requirements of the process of egg products.

Keywords: nondestructive testing, acoustic signal, cepstrum, power spectrum

1. INTRODUCTION

In egg processing plants, culling the cracked-shell eggs is an important working procedure, which is largely done by handwork currently. However, the quality of eggs detected by handwork cannot be guaranteed due to variable detection accuracy that results from difference in workers'

He, L., Jiang, D., Liu, L. and Liu, J., 2008, in IFIP International Federation for Information Processing, Volume 259; Computer and Computing Technologies in Agriculture, Vol. 2; Daoliang Li; (Boston: Springer), pp. 1147–1154.

experience, emotion and physical capability. Furthermore, the workers' working conditions and labor intension is too bad. So it is very urgent to find a scientific, applicable and fast detecting method for eliminating cracked-eggs.

At present, there are two different automatic techniques employed in quality detection and sorting of eggs: machine vision and image analysis; and mechanical stiffness measurements. The accuracy of the machine vision method relies on the resolution of the camera, the sorting algorithm and the type of defect. Machine vision inspection works excellently for dirty shells, broken shells and odd shapes, however, detection of small cracks is more difficult (Goodrum et al., 1992; Patel et al., 1998; Nakano et al., 2000; Garcia-Alegre et al., 2000). The mechanical stiffness measurements were based on the measurement and analysis of the mechanical behavior of the eggshell (Nakano et al., 2001). Ketelare et al. developed a method that eggshell crack detection was based on the analysis of the acoustically measured frequency response of an egg excited with a light mechanical impact on different locations on the eggshell equator. This method allowed a crack detection level of 90% and a false reject level of less than 0.5% (Ketelare et al., 2000). J. Wang et al. also developed an experimental system to generate the impact force, measure the response wave signal and analyze the frequency spectrum for physical property detection of eggshell (Wang et al., 2004).

In recent years, acoustic testing technique has been developed and seemed to be a promising method in some nondestructive testing system (Duprat et al., 1997; Cho et al., 2000). The same concept was applied herein to the case of detecting and sorting the eggs. In this paper we established an experimental nondestructive testing system for eggshells based on DSP. In the system, acoustic signals of normal eggs and cracked eggs were collected, followed by analysis of the differences between the two groups in power spectrum and cepstrum. The developed hardware platform and the introduction of cepstrum in identification process improved the detection accuracy, ratio, reliability and convenience effectively.

2. EXPERIMENTAL EQUIPMENTON

The nondestructive testing system for eggshells was designed with a DSP to enhance the detection rate and accuracy. The developed experimental hardware platform should have many functions such as controlling knock-equipment, sampling and disposaling the acoustics signals, identifying the cracked-shell eggs, communicating with the PC etc. In order to achieve the above-mentioned performances, The hardware was designed as several sub-systems: electromotor-drive sub-system, the signal-amplify and filter subsystem, the signals' starting point identification sub-system, the A/D convertor sub-system, the CPLD logic-control sub-system, the power source sub-system, the main processor sub-system, and the interface sub-system which communicated with the PC etc. The systematic structure figure is shown in Fig. 1.

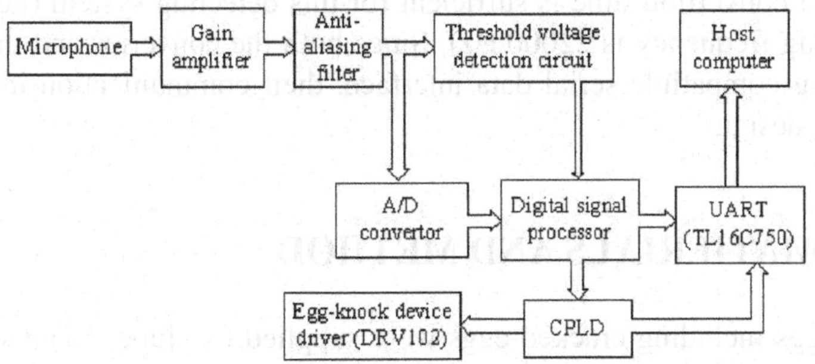

Fig. 1. Schematic diagram of the experimental equipment

Egg-knock device is composed of a solenoid with a return spring. A rubber hammer joints with the armature of the solenoid. The rubber hammer will knock the eggshells when drive current flows through the solenoid coil, and will return rapidly when the drive current is shut off. In order to prevent the rubber hammer break the eggshell the drive current should be shut off at once controlling by the relative program when the eggshell makes a sound.

Since the detecting system must deal with vast floating-point operation and should have high real time performance, TMS320VC33 was selected as the main processor, which performs great efficiency, high speed, low cost and easy exploiture. It is the most perfect DSP in the occasion of floating-point operation. The system doesn't need expand the external memory because it has an additional 1M bits of on-chip SRAM, which is sufficient for detecting calculation. FUM9750BP was selected as the acoustic sensor of the system, which is the single-point electrets microphone whose sensitivity is -51±3dB; frequency response is 100-16,000 Hz. INA217 is the low noises and low distortion audio frequency preamplifier which magnified the low-amplitude signals. A two-order butterworth low pass filter was designed with an operation amplifier OPA134 to filter random noise and baseline drift signals output by the microphone in this work. The cutoff frequency is 10,000 Hz according to the effective bandwidth of the acoustic signal made by the eggshell (Wen et al., 2002).

In order to reduce the calculation amount of signal processing and realize real time process, the system adopted the threshold circle to check the signals' starting point, which successfully resolve the contradiction between DSP interior memorizer is finite and extracting effective signal by software must use a lot of memory resources. When the objective signal is larger than the threshold voltage, the high speedy comparator TL714 will output low level voltage, and then the system considers that the effective signal is coming, and the sampling process is activated at once. The system used the component ADS7813 as the analog-to-digital convertor, which is a 16-Bit 20*us* A/D converter with an input multiplexer, sample/hold, clock, reference, and serial data interface. As the maximum convertion time of the convertor is 20*us*, the convertion time is sufficient for this detecting system (note that the sampling frequency is 22000 Hz). Since both the convertor and the DSP have timing compatible serial data interface, their communication interface is easily to design.

3. MATIERIALS AND METHOD

Fresh eggs including cracked eggs were supplied by Hubei Xiantao Food Corporation in China. The extent of eggshell crack in the eggs was varied from broken shells to small cracks. In this study, the rubber hammer knocked once for each eggshell controlled by a program running on DSP. The audio signal made by the eggshell was amplified; filtered and then sampled by the signal gathering program. It was found that the sampling frequency of 22000 Hz was optimal for accurate sampling after comparing the results obtained under different sampling number, and that the effective signal lasted no more than 4*ms*, so 128 point data were sampled for once knocking. In order to find obvious characteristic parameters the digital signal processing program transformed the sampled signal into frequency domain and cepstrum domain. Power spectrum and cepstrum profiles of normal eggs and cracked eggs were compared, from which several characteristic parameters can be retrieved for eggshell crack detection. At last, step discriminant was used to optimize the combination of these characteristic parameters, then establish the optimum distinguishing functions of normal eggshells and cracked eggshells respectively according to the step discriminant result.

4. EXTRACTION OF OBVIOUS CHARACTERISTIC PARAMETERS

4.1 The power spectrum characteristic parameters

When the rubber hammer knocked the eggshell the eggshell made a sound responding to the pulse excitation. The power spectrum can straightly reflect the acoustic signals' energy distribution in the frequency domain. The power spectrum was gained by 128-point Fast Fourier Transform (FFT) algorithm method in this work. According to the property of Discrete Fourier Transformation (DFT), the power spectrum of the acoustic signals is even symmetry on the 64 point; therefore, the paper only analyzed the former 64 point of it. The power spectrum of the typically normal egg's acoustic signal and the typically cracked egg's acoustic signals were shown in Fig. 2. The x-coordinate is the signals' sampling point and the y-axis is the amplitude of the sampling point. Six obvious characteristic parameters were extracted from frequency domain, they were: the first and second peak value in the high frequency band x_1 and x_2; the area of power spectrum x_3; the energy ratio of the low frequency band x_4; the energy ratio of the low frequency band x_5; the difference between the energy ratio of the low frequency band and the medium frequency band x_6. When the egg was knocked the force maybe uneven, in order to eliminate the influence on the detecting accuracy caused by the uneven knocking force, we chose the energy ratio but not the energy as the characteristic parameter. The definitions of the parameter were shown in Table 1, in every formula, p_i was the amplitude of the power spectrum, and $Peak_k$ was the value of the kth power spectrum peak in the high frequency band.

Table 1. Definitions of each characteristic parameter form frequency domain

x_1	x_2	x_3	x_4	x_5	x_6
$\max p_i$ $33 \le i \le 63$	$\max Peak_k$ $k \ne 1$	$\sum_{i=0}^{63} p_i$	$\frac{\sum_{i=7}^{15} p_i}{x_3}$	$\frac{\sum_{i=36}^{44} p_i}{x_3}$	$x_4 - x_5$

4.2 The cepstrum characteristic parameters

When the eggshell is knocked, it should make a sound; the acoustics theory thinks that the acoustic signals are the eggshell's response to the excitation signal $u(n)$, and which can be considered as the outputs of a linear system $h(n)$ excited by the excitation signal $u(n)$. Determined by the eggshell's structure, the linear system $h(n)$ apparently contains the more information of the eggshell.

The theoretical analyses shows that the cepstrum of the acoustic signals $\hat{c}(n)$ is the sum of the cepstrum of the eggshell's impulse response sequence $\hat{h}(n)$ and the cepstrum of the excitation signal $\hat{u}(n)$. The cepstrum of the eggshell's impulse response sequence $\hat{h}(n)$ is a two sided sequence and its amplitude attenuated rapidly when the sampling point n increases. The cepstrum of the excitation signal $\hat{u}(n)$ is a periodic sequence and its amplitude is not zero only when $n \ne kN_p$, but at the other point its amplitude is zero. Thus it is highly effectual that the cepstrum of the eggshell's impulse response sequence $\hat{h}(n)$ is extracted using the short time-window. The cepstrum of the acoustic signals is shown in Fig. 3. The x-coordinate is the signals' sampling point in the cepstrum domain and the y-axis is the amplitude of the sampling point.The cepstrum figure was weighed by the constant-coefficient 128 and the origin value wasn't included because it was so great that its display will hide other datum. Because the energy of the cepstrum $\hat{h}(n)$ centralized near the origin, the former six cepstrum values including the origin were retrieved as six characteristic parameters in cepstrum domain:

$$x_{i+7} = \hat{c}(i), \ (i=0, 1, \ldots, 5) \tag{1}$$

The above mentioned characteristic parameters x_1–x_{12} were selected as candidates distinguish factors for differentiating normal eggs and cracked eggs.

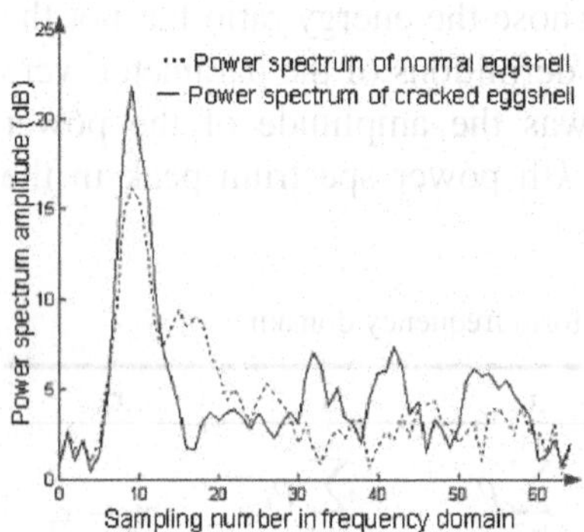

Fig. 2. Power spectral density of the egg

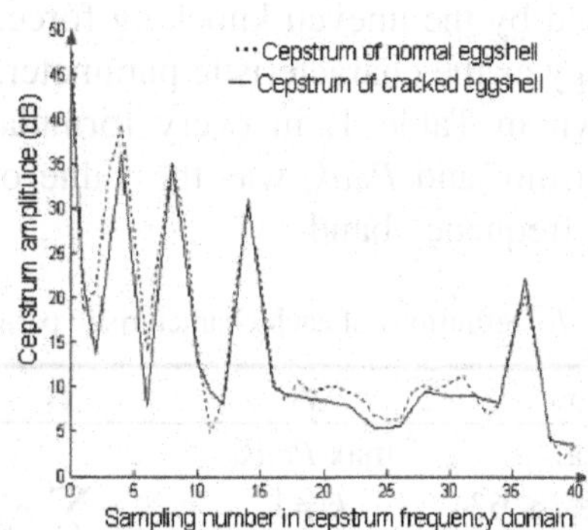

Fig. 3. Cepstrum of the sound given by an egg

5. STEPWISE DISCRIMINANT ANALYSIS OF EGGSHELL CRACK DETECTION

Twelve characteristic parameters had been extracted as candidate distinguishing factors. However, some characteristic parameters may be a nonsignificant or reverse influence on the discriminant accuracy, in this study the step-by-step discriminatory method was adopted. The stepwise process of disriminant analysis consisted of identifying the optimal subset of features, as well as the optimal linear discriminating functions of this subset that minimized the pooled expected risk of misclassification (PERM).

Experiments were carried out according to the methods stated in the previous part and a set of data from 200 normal eggs and 150 cracked eggs were collected. However, the data were too large to be listed here. The differentiation models of normal eggs and cracked eggs using SAS were developed. Results from the SAS outputs were shown in Table 2.

Table 2. Course STEPDISC

Step	Entered	Removed	Partial R-square	F Value	Pr>F
1	x_{11}	-	0.6780	136.84	<.0001
2	x_4	-	0.3165	33.94	<.0001
3	x_{10}	-	0.0739	5.03	0.0285
4	x_{12}	-	0.1111	7.75	0.0071
5	x_6	-	0.0935	6.29	0.0148
6	x_5	-	0.0309	1.91	0.1716
7		x_5	0.0309	1.91	0.1716

Differentiation function of cracked eggs is as follows:

$$\begin{aligned} G_{zh} &= -81.9501 + 2.84802x_4 - 1.86661x_6 \\ &\quad + 3.77679x_{10} + 0.36412x_{11} + 6.88054x_{12} \end{aligned} \tag{2}$$

Differentiation function of normal eggs is as follows:

$$\begin{aligned} G_{zp} &= -103.48 + 2.11697x_4 - 1.42783x_6 \\ &\quad + 0.30921x_{10} + 1.75619x_{11} + 10.6156x_{12} \end{aligned} \tag{3}$$

To further evaluate the preciseness of the models, detection was carried out in a much larger sample in which 1120 normal eggs and 980 cracked eggs were included. Detection data were generated and collected upon execution of the differentiation system. Among the 1120 normal eggs, 64 were wrongly identified as cracked eggs; so the average detection accuracy of the good egg model was 94.2%. On the other hand, among the 980 cracked eggs, 44 eggs were wrongly identified as good eggs; so the average detection accuracy of the cracked egg model was 95.5%.

6. CONCLUSIONS

In this study, a nondestructive testing system for eggshells based on digital signal processor (TMS320VC33) is established. Some obvious characteristic parameters extracted from frequency domain and from cepstrum domain were utilized to distinguish the eggshell status. The accurate recognition ratio for cracked eggs of this system is up to 95.5 percent, and the system can check 12 eggs per sec, which satisfied the requirements of the process of egg products.

The discrimination errors for cracked eggs arise mainly from the fact that not all the impact can happen just at or quite near the cracking spot by the egg-knock device. The further experimental results show that the accuracy can be improved by increasing the times of impact or detecting different spots by rotating the egg. But other issues may also come up together with the time increase in impact, such as elongation of the data processing period and subsequent reduction in efficiency of the system.

ACKNOWLEDGEMENTS

This study has been funded by the Department of Science of Huan Institute of Engineering (project number: 120641). The author also thanks Prof. Youxian Wen in Huazhong Agriculture University for introducing active modulation to us and for providing much valuable guidance.

REFERENCES

Goodrum J W, Elaster R T. Machine vision for crack detection in rotating eggs. Transaction of the ASAE, 1992, 35: 1323-1328

Patel V C, McClendon R W, Goodrum J W. Color computer vision and artificial neural networks for the detection of defects in poultry eggs. Artificial Intelligence Review, 1998, 12: 163-176

Nakano K, Sasaoka K, Ohtsuka Y. A study on nondestructive detection of abnormal eggs by using image processing. Asian Federation for Information Technology in Agriculture, 2000: 345-352

Garcia-Alegre M C, Ribeiro A, Guinea D, et al. Eggshell defects detection based on color processing. SPIE 2000 Electronic Imaging Conference, 2000

Duprat F, Grotte M, Pietri E, et al. The acoustic impulse response method for measuring the overall firmness of fruit. Journal of Agricultural Engineering Research, 1997, 66: 251-259

Wang J, Jiang R J, Yu Y. Relationship between dynamic resonance frequency and egg physical properties. Food Research International, 2004, 37: 289-294

Ketelaere B D, Coucke P, Baerdemaeker J D. Eggshell crack detection based on acoustic resonance frequency analysis. Journal of Agricultural Engineering Research, 2000, 76: 157-163

Wen Y, Wang Q, Zong W, et al. Study on crack detection of duck eggs [J]. Huazhong Agricultural University Transaction, 2002, 26: 285-287 (in Chinese)

Cho H K, Choi W K, Paek J H. Detection of surface cracks in shell eggs by acoustic impulse method. Transactions of ASAE, 2000, 43: 1921-1926

Nakano K, Usui Y, Motonaga Y, et al. Development of non-destructive detector for abnormal eggs. Workshop on Control Applications in Post-Harvest and Processing Technology, 2001: 71-76

FARMERS' ACCESS TO INTERNET INFORMATION: PATHWAYS, INTERESTS AND COST

A Typical Survey in Southern Hebei Province of China

Gubo Qi[1], Haimin Wang[1], Ting Zuo[1,*]

[1] *College of Humanitites and Development, China Agricultural University, Beijing, China, 100094*

[*] *Corresponding author, Address: College of Humanities and Development, China Agricultural University, 2 Yuanmingyuan West Road, Beijing, 100094, P. R. China, Tel:+86-10-62731319, Fax: +86-10-62731027, Email: zuoting@cau.edu.cn*

Abstract: Set in six rural communities in southern Hebei Province of China, the present study focuses on rural farmers' access to, interests about and management of internet information. With the perspective of farmers, this empirical study attempts to understand the potential effects of information technology on agricultural information system and to capture its development prospect as well. The research findings indicate that 1) how the surveyed rural residents gain access to internet information and to what extent the internet as media is acceptable to farmers are dependent on such factors as individual education background, age, family income, as well as outside interventions; 2) innovative information technology plays a significant role in improving local livelihoods and accelerating local development; 3) poverty reduction through information technology has laid positive but limited impact towards rural residents' understanding, learning and utilization of the internet.

Keywords: farmers' access, innovative information technology, poverty reduction

1. BASIC DATA OF THE SURVEYED INDIVIDUALS AND HOUSEHOLDS

Along with more and more open and consistent market and some agencies and companies' promotion on information technology in rural area, there are also a certain number of townships, villages and rural residents that have

Qi, G., Wang, H. and Zuo, T., 2008, in IFIP International Federation for Information Processing, Volume 259; Computer and Computing Technologies in Agriculture, Vol. 2; Daoliang Li; (Boston: Springer), pp. 1155–1168.

already been successfully accessing through telephone lines to internet to search for marketing information and even to release the product information (Li, 2003). Out of those cases, UNDP has initiated the "Project on Poverty Reduction through Information Technology in Rural China" in 12 villages of 5 counties since 2000. The project has so far witnessed preliminary achievements as expected (Wang, 2003). This empirical research attempts to understand the potential effects of information technology on agricultural information system and to capture its development prospect as well, with focusing on farmers' perspectives.

The typical survey was conducted in six villages in Wuan County of Hebei Province, including Menwangzhuang village, Shangdian village, Qianboshan village and Qianqu village of Huoshui Township, and Pushang village and Xiazhuang village of Paihuai township. Among them, Menwangzhuang village, Shangdian village and Pushang village belong to project villages whereas their neighboring villages: Qianqu village, Qianboshan village and Xiazhuang village are all non-project villages. The entire survey sample contains 189 households (individuals). 95 respondents, inclusive of 37 man and 58 women are from project villages, taking up 50.3% of the total. The rest 94 respondents being consisted of 55 men and 39 women are from non-project households, holding up about 49.7%. In general, the sample consists of 92 male participants (48.7%) and 97 female ones (51.3%).

Considering annual income of surveyed household with average 4 family members- RMB 8728.1 Yuan, the surveyed villages are relatively with higher living standard than the average standard in rural China. Off-farm work, wages and small business are three main income sources contributing to 78% of total income. There is only 20.3% of total income coming from agricultural production.

Figure 1 shows the structure of households' expenditure. Expenses of housing, education, medicine and the medical service, and food make up 72.99% of the total expenditure. As to the facilities, TV set and telephone are popularly used by 97.9% and 67% households respectively, while few households can afford to the computer. Some even hold the view that computer is useless in rural areas. When it comes to entertainment, 71.4% of the respondents reported "watching TV", 12.4% responded "visiting relatives" and 8.6% chose "playing cards/mahjong". The most frequently chosen TV channels includes CCTV-1, CCTC-7, Hebei Provincial

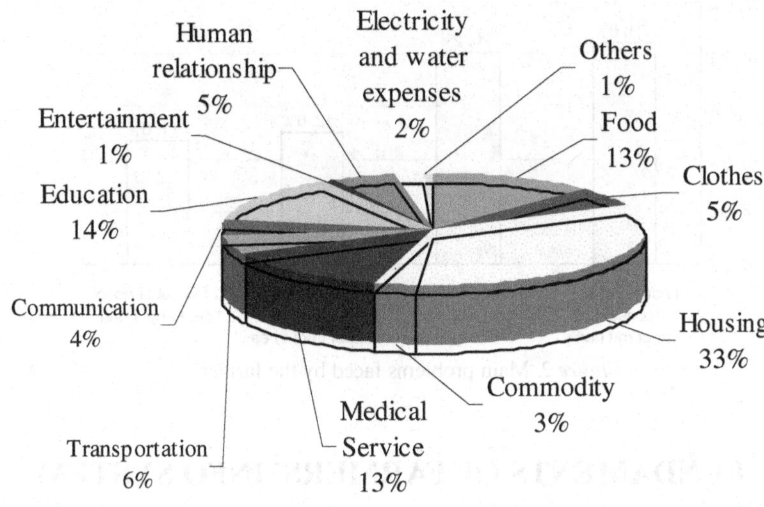

Figure 1. Expenditure structure of the households

Television, Henan Provincial Television, etc. and the most frequently watched programmers are reported as Weather Forecast, Focus Interview, News, TV Series, Agricultural Extension programmers and so on. Some households who have showed strong interest in innovative agriculture technology watch frequently the CCTV-7 for new technology and have managed to apply in reality to yield fair results.

To better understand local situations and local people needs, this survey explored the main problems faced by the farmers and their expected assistance. Insufficient access to information, insufficient technology, insufficient training opportunities and difficulties of getting loan took largest proportions, as being showed in Figure 2. As to assistance, 97 persons (51.9%) of the surveyed demand "more financial support", 49 respondents (26.2%) request "better production technology/breeds", and 26 respondents (13.9%) mentioned "more market information". It can be concluded that rural residents have strong demands for market information, apart from financial support and technical assistance. If new information and technology generated from internet network can be applied in agricultural production, it will be surely of great significance and benefit to local farmers.

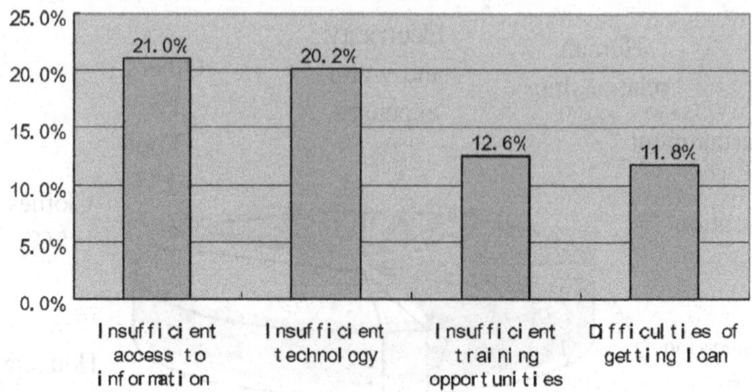

Figure 2. Main problems faced by the farmers

2. FUNDAMENTS OF FARMERS' INFO SYSTEM

2.1 Information Farmers Interested in

A total of 188 valid samples were gathered for this analysis, out of which 94 samples are for the project villages and 94 for non-project villages. The questionnaire contains 13 categories of information. It is found out that 45.7% of the investigated are interested in information of new technology wherein there is no significant difference between that of project villages (48.8%) and that of non-project villages (51.2%), and that 12.8% of the investigated are interested in where to get jobs wherein the percentage of project and non-project villages are fifty-fifty. Immediately after the top-ranking two categories is information on market price of quotidian livelihood goods, day-to-day knowledge about medical care, health and education, and basic knowledge on laws. The summing-up of the three categories show that 60.9% of the investigated are interested on new technology, 33.3% interested on market demands for products, 33.2% interested on day-to-day knowledge (of which 17.8% of the total investigated chose this kind of knowledge as the third most interesting). Taking new technology as the top priority suggest a fairly strong desire of farmers for technological information, and the fact that as many as 12.8% of those searched for needed information on employment opportunities is a demonstration of the outward migration trend of the surplus labors from the rural areas.

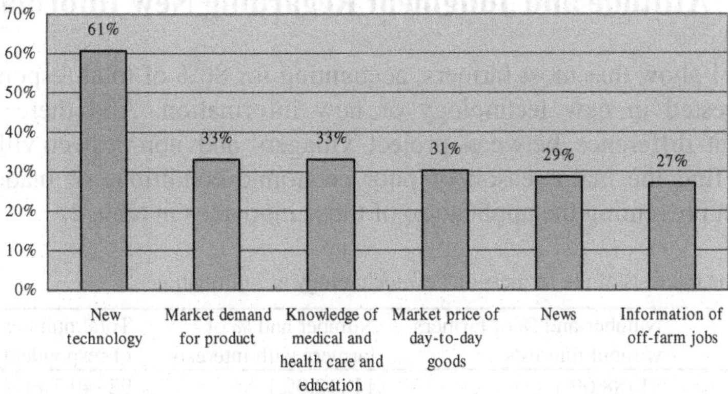

Figure 3. Information that the farmers are interested in

2.2 Sources and Channels of Acquiring Information

Relatives, friends and TV broadcasting constitute the primary sources for people to receive information, while books, newspapers and magazines play certain role in serving that purpose as well. In terms of other sources of obtaining information, the effect of Villagers Committees and their loudspeakers is deemed much poorer than expected, only as little as 3.2% of the respondents consider official institutions in charge of disseminating technology at the township level as useful, and merely 5.9% replied favorably for the township governments. The internet usage is poor: only 1.6% of the surveyed have ever searched for information on their own, only 2.7% have ever got help from the materials disseminated by the information centers, 6.4% of them have ever searched for information with the help of the people working at the information centers. That is to say, farmers seldom used internet and their ability to use it is inadequate. All the three who replied that they ever "searched the internet independently" and all the 12 who said "yes" to "ever searched for information with the help of the people working at the information centers" are from the project villages. In addition, four out of those five who were affirmative that they used "materials printed out by information centers" are from the project villages. Presumably most of the investigated had access to internet through the info centers established under the project of internet poverty alleviation projects. As 100% of those who ever "searched the internet independently" or "searched for information with the help of the people working at the information centers" are from the project villages, and as many as 80% of those used "materials printed out by information centers" are from the project villages, there is no doubt that establishing information centers plays a positive role in promoting information awareness.

2.3 Attitude and Judgment Regarding New Information

Table 1 show that most farmers, accounting for 86% of total respondents, are interested in new technology or new information. And there is not significant difference between project villagers and non-project villagers. We can find the main reason of poor economic conditions or inadequate education preventing the application of those innovates in table 2.

Table 1. The interests of the farmers on new technology or information

Village	Number and % of farmers without interests	Number and % of farmers with interests	Total number and % of respondents
Project	81 (88.0%)	11 (12.0%)	92 (49.7%)
Non-project	79 (85.0%)	14 (15.0%)	93 (50.3%)
Total	160 (86.0%)	25 (14.0%)	185 (100%)

Table 2. Reasons on their no interests on new technology or information

Village	The number and % of answers of 'could not apply due to fund limitation'	The number and % of answers of 'could not learn due to low education level'	The number and % of answers of 'Fear of failure'	The number and % of answers of 'other reasons'	Total number and % of respondents
Project	3 (27.3%)	6 (54.6%)	0 (0%)	2 (18.1%)	11 (45.8%)
Non-project	3 (23.1%)	3 (23.1%)	1 (7.6%)	6 (46.2%)	13 (54.2%)
Total	6 (25%)	9 (37.5%)	1 (4.2%)	8 (33.3%)	24 (100%)

As to the sources of those information that the respondents care most, relatives and neighbors was ranked as the first one out of 16 information sources listed by the farmers, and the following ranks are as following: TV program, newspapers and magazines, books, village broadcasting, village committee, telephone call, radio, relatives doing off-farm work outside, children studying outside, information center staff's help, township government, township technology institutes, materials distributed by the information center, law consultation departments, and internet browser, etc. Although through internet browser was ranked as the last one, both project village and non-project village have farmers who started adopting this kind of source. And in project village, there were 6% respondents using internet browser independently, which is 4.4 points higher than that of non-project village. Relatives and TV programs are still the ones they trusted most and they think most useful.

3. BRIEFING ON FARMERS' USE OF INTERNET INFORMATION TECHNOLOGY

3.1 Use of Internet by Farmers

Internet can now be witnessed in very corner of the cities all around the country and it has substantially improved the efficiency of all urban sectors. However, to what extent has the internet influenced the rural area? 76 participating households (41.1%) report that one of the family members is capable of using internet, whereas some reflect that more than one family member can make it. It seems that quite a lot of rural residents in survey area are able to use computer, which has laid a solid foundation for information network construction in rural areas.

The proportion of households capable of using computer in project village is much higher than that in non-project villages. Specifically, 49 households (51.6%) of the households in project villages are able to operate computers. In non-project villages, 27 responding households (30%) were using computers, mainly influenced by the project villages. It can hence be concluded that poverty reduction through information technology has obtained obtained its preliminary effects on advancing the rural information network.

Out of 113 respondents, 68.4% learned basic computer use skills at school, 22.4% acquired from the information center, and 5.3% from the computer training courses. These are the three major approaches for the local people to learn the computer. Ever since the implementation of the information poverty alleviation, the Science and Technology Bureau in Wu'an County has organized quite a few computer internet training workshops, each of which were attended by quite a large number of local residents. From all these results, it may be concluded that information center has played a preliminary but positive role in helping local farmers learn and use internet.

Who are the computer users and what characteristics do they have? The research reveals that among all of the computer users, "my son can" and "my daughter can" have occupied 29.3% and 26.7% of the total. "Both son and daughter can" accounts for 34.7%. It is observed that most of the computer users are youths, with higher education background. Some of them are even registered students, who have learned to use computers in school. Besides that, the proportion of the male computer users is higher than their female counterparts. Also, most attendees of the computer training workshops are man whereas women usually merely took a curious look and left.

To verify the correlation between household income and the use of computers, the result shows that the higher the family income is, the more possibility in using computers. For instance, no one earning below RMB 1000 per year is reported to be able to use the computer. At the income ranges of RMB "1000-3000", "3000-6000", "6000-10000", "10000-20000" and above 20000, the ratio of computer use amount up from 19.0%, 35.6%, 48.0%, 63.2% and to 84.6% respectively. Along with the increase of the income, the proportion of the family who can use the computer is gradually going up, which demonstrates the obvious correlation between family income and computer use.

3.2 Farmers' Cognition Towards Internet Information

182 valid samples altogether have responded to this aspect, inclusive of 94 (51.6%) from project village and 88 (48.4%) from the non-project village. Within project village, 72 (76.6%) of the survey reported that they know information can be searched from internet whereas 22 (23.4%) interviewees provided negative answer. Contrast in non-project villages, there were 46 (52.3%) participants knowing and 42 (47.7%) unknowing about the fact that internet is source of information. Such an observable difference between non-project households and project households has to certain extent reflected the progress brought alongside with the poverty reduction project.

When answering the question "How do you know that internet is one of the sources of information?" 32 (44.4%) out of the 72 respondents from project village referred to the information center, 14 (19.4%) referred to the TV, and 13 (18.1%) for other ways. In the non-project village, 30 (65.2%) of the 46 surveyed referred to the TV, 9 (19.6%) referred to their children who were perusing their study outside the village, and 6 (13%) for other ways, like knowing it from neighbors, friends, broadcast and so on. All of these have shown that the information center built by the project has contributed a lot for farmers to get new information from outside.

In addition, we have carried out a correlation analysis over "household income" and the "cognition about the internet information" and no significant correlation has been traced. The ratio between the known person and the unknown person in each income segment is nearly 1:2. Similarly, no significant correlation has been identified between gender and the cognition of internet information in that out the 182 respondent to this question, 58 women (31.9%) and 60 (33%) gave the positive answer. However, when it comes to education background, it is observed that the cognition rate increases along with the improvement of education level.

3.3 Use of Internet by Surveyed Households

We have chosen 30 households respectively from the project village and the non-project village to compare who have earlier access to internet information. None of the 30 respondents randomly selected from non-project village had ever used internet whereas 10 out of the 30 households from project village had ever used internet, with 8 of them started as early as 2002 and 2 in 2003. The project has urged the villagers to accept such a new object as internet and it is deemed to have brought about changes to the village. It yet needs to be alert that there is still a long way to go before the rest 2/3 of the villagers started to accept and use this new object.

81 individuals responded to the question that whether they know the relative knowledge before accessing to internet. 56 answered no and 25 answered yes. 8 (32%) claimed to get the information through project training and 6 (24%) learned from the book, 6 (44%) reported other ways. To get a further understanding of the specific situation, we have done the comparison analysis regarding whether they know that they can get the information from the internet and whether they know the relative knowledge before accessing the internet. 68 out of the 118 households who believe that internet is an information source have participated this session, with 26 households claiming to know the relative knowledge before accessing and 42 do not. Among the other 64 households who don't know that internet is an information source, only one out of the 13 respondents claims to know the relative knowledge before accessing while the other 12 do not. Therefore, we can infer there is a great positive correlation between whether they can get information from internet and whether they know the relative knowledge. It is necessary to make the farmer household know that they can get the information they want from the internet before publicizing the internet knowledge in the rural area. Only in this way, it can inspire the farmer households' enthusiasm to learn the internet. Whereas, we should see that in the 68 households who know that they can get the information from the internet, there are only 26 households who know the internet knowledge, which has occupied 38.2% of the whole number. The other 61.8% still don't know about the internet knowledge. So if we want to get ground the internet knowledge widely, there is a long way to go before all farmer households know that they can get the knowledge from the internet.

4. ROLES OF THE INFORMATION CENTER

4.1 Approaches to Access Internet Information and Various Information Gained

Responding to the "approaches to access internet information" are 52 households, with 18 (34.6%) reported that they search for internet information by themselves while 26 (50%) turned to others (mainly the information messenger at the info center) for help and the last one (1.9%) gained related knowledge from the printouts at the info center. Though the village information center is currently the main place for villagers to log on the internet but the most favored venue is at home. 17 (53.1%) of he 32 respondent prefers to access internet at home, 7 households (21.9%) prefers independent search of information at information center and 6 households (18.8%) likes dependent search of information at information center.

There are 48 persons who have answered the question of "what types of information have you gained from the internet". The first three options include the new technology, common sense knowledge in every day life, the price of the production materials, and the last three options are the product purchase venue, the living standard of the outsiders, the daily goods' price information. In terms of the use of information center, no particular trend has been identified. The frequency of info center visit range from once to 200 times, according to the 45 respondents. The duration in info center also depends, although most of them report that each visit cost less than two hours. When comparing the distance to info center with the frequency of info center visit, we find that 33 out of the 36 respondents, that is 91.7% of them, dwell a five-minute-walk distance away from info center,

From gender perspective, no obvious distinction has been diagnosed between gender and frequency of info center visit. Nor is there great correlation between education background and frequency in visiting info center. However, a close examination reveals that those 44 individuals who frequently visit the information center consist of 13 (29.5%) people with primary schooling, 18 (40.9%) with junior high education background, 6 (13.6%) with above senior high education and two illiterate (4.5%). It can be explained that the visitors to info centers are mostly students from primary and high schools, who have more curiosity and time to access internet.

4.2 Internet Information and its Usage

Among the 35 responding households, 17.8% use internet mainly for sending E-mail, 35.9% for playing games, chatting and watching the film,

15.4% for reading the news and the other 30.8% for other purposes including typing exercises and listening to music. From the analysis above, we can see that there are some wrong understandings on using the internet. Most of farmers have taken the internet as a kind of tool for pastime. Action needs to be taken to update their understanding and change their behavior so that internet can play a more positive role for their economic activities and income generation.

46 out of the 48 respondents regard it helpful to access internet information, while the rest two remain pessimistic on this. The 46 positive respondents prioritized the top two functions of internet to include "broadening the mind" (24 people, i.e. 50%) and "providing of useful market information" (8 people, i.e.16.7%)

When it comes to the impact of internet on daily life, 26 (57.8%) out of the 44 respondents reported that internet had increased their contact with outside world, 10 (22.2%) reflected that internet had helped them in improving income, 5 (11.1%) have reported other impacts. Despite such a small sample, it can be inferred that farmers have noticed internet can bring about more information. Yet more efforts need to be done to strengthen the impact of internet in rural areas in an all-round way.

In terms of the difficulties in using internet to get information, 43 responding households prioritize the top three challenges to include awkward operation (15 people), insufficient computers (10), and not knowing how to locate needed information (5). It can be inferred that local farmers' poor knowledge about internet has added to the difficulty for them to access internet. When exposed to such a new invention as internet, it is very hard for farmers to effectively respond.

When asked the question whether there is some data that can not be found in the internet", 43 (84.3%) out of the 51 interviewees in total reflected that internet is a better information source, 5 (9.8%) believed that internet is no different from other information sources, and the rest 3 (5.9%) households commented that internet can not compete with other means of information channels. It is also learned that these who are in favor of internet think that internet information is more swift and comprehensive, whereas these who are against internet actually had no experience in accessing internet information.

When responding to needed assistance in information access, 46 respondents prioritized their needs to include more training (27 households), more advice from professionals (9 households) and better access to internet (4 households). It can then be inferred that lack of internet related knowledge and shortage of internet professionals in rural areas have constitute the two major constraints for local farmers to access internet.

4.3 Cost of Internet Access and its Management

93 households answered the question "what kinds of equipments are needed for accessing internet? 65 persons (69.9%)'s gave negative answers while 28 (30.1%) expressed that they know some of them. Among these 28 persons, 21 believed that a computer is necessary for accessing internet, 20 persons know that a phone line is needed, and 8 persons know that the modem should be used. However, 32 (61.5%) of all 52 respondents did not even know how much money is needed for buying these equipments. For the rest 20 people who claimed to know exactly how much money is needed, they have provided diversified answers regarding to the actual prices of these equipment.

A closer examination indicates that 64 out of the above mentioned 93 respondents come from the project village, occupying 68.8% of the whole. Within the project household sample are 25 positive respondents (26.9%) and 39 negative respondents (41.9%). In contrast, the 29 respondents from the non-project village have provided 3 positive answer (3.2%) and 26 negative one (26%). It can be observed from these figures that there exists great disparity between project villages and non-project villages in this regard, in that households form project village have more exposure to computers and internet.

The same is true when drawing comparison between project villages and non project villages in terms of the cognition towards the needed equipment for internet connection. Among the 52 respondents, 45 households are in the project village, including 19 (36.5%) answering yes, 26 (50%) answering no. For the rest 7 from non-project village, only one surveyed claim to know while the other 6 remain ignorant in the regard.

When asked the question "would you like to pay for the internet service in the future once the provision of free service is terminated?" 90 farmer households have given the answer, among whom, 37 (41.4%) persons are reluctant to pay and 53 (58.9%) persons respond positively. When it comes how much one would like to pay, 42 (87.5%) out of 48 respondent regarded it reasonable to pay for less than 3 RMB per hour and 19 (39.6%) prefer 1 RMB per hour, 12 (25%) would like to pay 2 RMB per hour, 11 (22.9%) choose 3 RMB per hour.

It, however, shall be point out that the 29 households are from non-project villages where there are no free internet service. So their willingness to pay deserves a second thought. Out of the 61 households from project village, 40 expressed their willingness to pay while 21 insisted on free service. Further interview indicates that 81.25% of these who are willing to pay actually would like to pay less than 3 yuan per hour for the service. According to our analysis, there is no significant correlation between household income and the willingness to pay, nor is there significant correlation between the frequency of visiting information center and the willingness to pay.

46 out of the 81 responding households, that is 56% of the total, had got plans to buy computer, whereas the rest 35 households (43.2%). For these who have planed to buy computers, it is learned that 23 (65.7%) households will take action once they have earned enough money, 7 (20%) households would not buy it until their child has grown up, and the rest 5 households have provided other plans. In this case, no significant correlation has been identified between household income and the wiliness in buying the computer.

When drawing a comparison between project village and non-project village, we find that those residents from project village have much higher will in buying computer. Specifically, 20 out of the 39 respondents from the project village were going to buy computers, accounting for 51.3%, leaving 19 respondents (48.7%) with no specific plan to buy computers. In non-project village, 15 out of the 42 respondents were going to buy computers, accounting for only 35.7%, leaving 27 respondents (64.3%) with no specific plan to buy computers.

5. CONCLUSION

As with the technical and economic development in rural areas, internet has become one of the available sources of information. The case in Hebei province has provided with very good implications. Through internet, farmers are capable of getting information about policy, market, technology and so on, to well support their own livelihood development. In the process of accessing internet information, there is high potential in encountering information filtering and false information. In this case, Info centers and information programmers shall be in place to provide due assistance. Meanwhile, it needs to be noticed that internet resources as a public good requires huge financial input. Otherwise, most farmers themselves would be unable to afford the internet services. It is hence suggested that internet development shall be enlisted as one of the most important public goods for the nation's New Countryside Construction Program.

ACKNOWLEDGEMENTS

The present study, especially the field work has been supported by the projects of UNDP and MOST while the Science and Technological Bureau of Wu'an County has also contributed to the organization of the field survey. The authors take full responsibility in explaining the views in this article.

REFERENCES

Li, Xiaoyun, Zuo, Ting. 2002 status of rural China. China Agricultural University Press. 2003, Beijing: 23–24

Rural department of State Development Research Center. Rural Public Service System in China. China Development Press. 2003

Wang, Haimin. Analysis of the mechanism of poverty reduction through internet and its impacts, 2003, Beijing: 5–6

SPECTRUM CHARACTERISTICS OF COTTON CANOPY INFECTED WITH VERTICILLIUM WILT AND INVERSION OF SEVERITY LEVEL

Bing Chen[1], Keru Wang[1,2], Shaokun Li[1,2,*], Jing Wang[3], Junhua Bai[1,2], Chunhua Xiao[1], Junchen Lai[1]

[1] *Key Laboratory of Oasis Ecology Agriculture of Xinjiang Bingtuan, Shihezi University, Research Center of Xinjiang Crop Yield. Shihezi University, shihezi 832000, xinjiang*

[2] *Institute of Crop Science, Chinese Academy of Agricultural Sciences, The National Key Facility for Crop Gene Research and Genetic Improvement, NFCRI, Beijing, 100081, P. R. China*

[3] *College of Resources and Environment, North west Science and Technology University of Agriculture and Forestry, Yang ling, Shan xi 712100*

[*] *Corresponding author, Address: P.O. Box 121, Chinese Academy of Agricultural Sciences & The National Key Facility for Crop Gene Research and Genetic Improvement NFCRI, Institute of Crop Science, Beijing, 100081, P. R. China, Tel: +86-10-68918891, Email: lishk@mail.caas.net.cn*

Abstract: Verticillium wilt of cotton is one of the diseases of cotton with extensive occurrence and maximal harming in our country even in the world. Hyper spectrum remote sensing with the fine spectrum information has becoming the efficient method to monitor the verticillium wilt of cotton. The research was conducted in Xinjiang, the largest cotton plant region of China. The paper used data which was collected both canopy spectrum infected with verticillium wilt and SL (severity level) in the year 2005–2006, the quantitative correlation were analyzed between SL and canopy reflectance spectrum, derivative spectrum. The tested results indicated that spectrum characteristics of cotton canopy infected with verticillium wilt had better regularity with the increase of SL in different periods and varieties. Spectrum reflectance increased in visible light region (620–700 nm) with the increase of the SL, which inverted in near-infrared region, and extreme signification in 780 – 1300 nm. When SL got 25%, cotton canopy infected with verticillium wilt could be used as a watershed and diagnosed index in an early time. The tested results also indicated there were evident characteristics of first derivative spectrum in these SL, it changed significantly in red edge ranges (680–760 nm) with different SL, red edge swing decreased, and red edge position equal moved to

Chen, B., Wang, K., Li, S., Wang, J., Bai, J., Xiao, C. and Lai, J., 2008, in IFIP International Federation for Information Processing, Volume 259; Computer and Computing Technologies in Agriculture, Vol. 2; Daoliang Li; (Boston: Springer), pp. 1169–1180.

the blue. The results indicated that 1001–1110 nm and 1205–1320 nm were selected out as sensitive band regions to SL of canopy. These some inversion models for estimating cotton canopy infected with verticillium wilt all reached the most significantly level. At last, the results suggested that different spectrum characteristics of cotton canopy infected with verticillium wilt were obvious, the first derivative spectrum ($FD_{731\ nm}$–$FD_{1317\ nm}$) will invert the cotton canopy SL accurately, and it may be used to forecast the position of cotton canopy infected with verticillium wilt in quantitatively.

Keywords: cotton; verticillium wilt; canopy spectrum; SL; inversion mode

1. INTRODUCTION

Verticillium wilt of cotton, which is widly- outbreak, strong- epidemic and highly- of outbreak, is one of the diseases of cotton with extensive occurrence and maximal harming in our country even in the world. Traditional monitoring method of verticillium wilt was that people investigated and sampled in field, which was limited time-consuming, hard-sledding, so prevention and cure of disease were effected to some extent. In this paper, spectrum characteristics of cotton canopy infected with verticillium wilt were analyzed, and inversion modes of SL were established. The studying results will provide theoretic support for further monitoring cotton verticillium wilt in large area by aviation and spaceflight remote sensing. As shown in paper (Huang et al. 2006; Ma et al. 2004), due to plant diseases and insect pests, some canopy spectrum parameters, such as LAI (leaf area index), biomass and coverage etc., had happened change, leading canopy spectrum reflectance correspondingly changed too, which made possibly fast monitoring occurrence and development of verticillium wilt of cotton in large area by remote sensing technology.

Some domestic and overseas scholars have had some correlate experiments on characteristics of canopy spectrum of plant diseases and insect pests, and made a great deal of achievements (Hamed Hamid Muhammed, 2005; Zhang et al. 2005; Bravo et al. 2003; Huang et al. 2003; Wu et al. 2002; Liu et al. 2002; Qiao et al. 2002; Chen, 2007). However, there have been few studies on surveying crop disease (especially to plant disease types) by remote sensing (Luo et al. 1997; Kefyalew Girma et al. 2005). When cotton was harmed by verticillium bacterium, the spectrum of cotton plant and remote sensing image will put up some especial diagnosis spectrum characteristics, which had different spectrum characteristics between single leaf and canopy, at present, however there are little report about them, it has become a unsubstantial technique on monitoring cotton growth situation by remote sensing. What are spectrum characteristics of cotton canopy infected with verticillium wilt? How to recognize cotton of disease and inverst SL of disease by remote sensing? So far, these studying have had little systemic reportion. Elucidating above mentioned questions, for disclosing spectrum mechanism of cotton canopy infected with verticillium wilt, monitoring district and trend of verticillium wilt, instructing reasonable distribution with breed resisted have very important meaning.

2. MATERIALS AND METHODS

2.1 Design of experiments

The experiment was conducted in cotton region of Shihezi in Xinjiang in the 2005-2006. The small area experiment was conducted in tentative cotton disease garden of institute of Key Laboratory of Oasis Ecology Agriculture in Shihezi university (44°18′N, 86°03′E). The soil type was characterized as a soil-grey desert loam with organic matter content 1.93%, alkali-hydrolysis nitrogen 77.4mg·kg^{-1}, rapidly available phosphorus 93mg·kg^{-1} and rapidly available potassium 315mg·kg^{-1} in 20cm depth, former stubble plant was cotton. The selected cultivars in this experiment were some cotton varieties with different plant types and same growing periods, for example, Xin Luza 7(XLZ-7), Xin Luza 8(XLZ-8), Xin Luza 13(XLZ-13), Xin Luza 24(XLZ-24), and Zhong Mian 36(ZM-36). The plot area was 42.5m^2 with three repeats, the plant density was 240000 plants per ha, and row spacing was 60cm+30cm, the seeding time was the 20rd April, plant mode was on-film order programmer with under-film drip irrigation. Irrigation carried out under-film irrigation with 3300m^3•hm^2, fertilizer amounts were pure nitrogen 300 kg•ha^{-1}, 150kg•ha^{-1}P_2O_5, 75 kg•ha^{-1}K_2O. Besides, the plot had not weed, the other treatments were applied according to high yield cultivation mode in the local region. The field area experiment was conducted in cotton region of Shihezi, locating 19 company at 143 regiment, seed station at 147 regiment and 11 company at 148 regiment, where all had occurred continuously verticillium wilt in large area in the past years. Cultivars in this experiment were the same as small area experiment. Besides; some new cotton cultivars were selected, for example 602 and 4432. The soil type was characterized as a soil-grey desert loam with middle fertility, and the seeding time date was on the last ten-day of April, other was the same as small area experiment.

2.2 DI (Disease index) investigation and SL classification

DI was investigated with five point method, which was, symmetrical five points were selected in treatment plot to investigate DI, and ten cotton plants (about 1 m^2) were selected out in very one point after DI was investigated. SL of single plant was divided into five grades, namely, normal single plant was 0 grade; numbers of leave with disease on single plant had not exceed 25% as 1 grade; numbers of leave with disease on single plant between 25%and 50% as 2 grade; numbers of leave with disease on single plant had exceed 50% as 3 grade; single plant had died or almost died as 4 grade. After finishing investigating, single plant numbers and grade were registered in every point, and calculated formula of DI with tested canopy as follows (Zhang et al. 2005):

$$DI = \frac{\sum (X * f) \times 100}{n * \sum f} \tag{1}$$

In the formula, *x* denoted grade value from every grade; *n* denoted grade number from highest grade; *f* denoted plant numbers from every grade. Then, according to different DI with tested canopy, SL of tested canopy were divided into five grades, namely, normal (b0): DI=0; mild degree (b1): DI was between 0 and 25; moderate degree (b2): DI was between 25 and 50; severe degree (b3): DI was between 50 and 75; most severity degree (b4): DI was between 75 and100.

2.3 Hyper spectrum reflectance data collections

Hyper spectrum reflectance measurements were taken under clear sky conditions from 11:30 to 14:00 (Beijing local time) using an ASD Field spec Pro FR 2500 spectrometer (Analytical Spectral Devices, Boulder, CO, USA) fitted with a 512 spectrum bands, operating in the 350–2500 nm spectrum region with a sampling interval of 1.4 nm between 350 and 1000 nm, and 2 nm between 1000 and 2500 nm, and with spectrum resolution of 3 nm between 350 and 1000 nm, 10 nm between 1000 and 2500 nm. A panel radiance measurement was taken with 25° field of view before and after the canopy measurement by two scans each time, and scans time of each sample was 0.2s, sample of spectrum reflectance was measured five times in every plot, finally, spectrum reflectance value of each plot was acquired by averaging. Different SL of cotton canopy infected with verticillium wilt was measured by the method that the sensor probe of the spectrometer was taken vertically from a height of 1.5m above canopy, as a standard, the solar radiation spectrum value was calibrated by using the standard white board (40cm*40cn, $BaSO_4$) before every measurement canopy of spectrum. In the year 2005-2006, cotton canopy infected with verticillium wilt was measured in different growth stages, for example, peak budding period (20 June), peak flowering period (19 July, 21 July), peak bolling period (3 August, 12 August), peak opening period (4 September).

2.4 Data analysis methods

Using view spec program software (made ASD company) managed primitive canopy reflectance spectrum data, we obtained reflectance spectrum data and reflectance curve, at the same time, the first derivative spectrum method was selected with matlab 7.01 software, and calculated formula of the first derivative spectrum as follows (wang et al., 2005):

$$\rho'(\lambda_i) = \frac{[\rho(\lambda_{i+1}) - \rho(\lambda_{i-1})]}{2\Delta\lambda} \tag{2}$$

In the formula, λ_i denoted wave length value from $_i$ wave band; $\rho\ (\lambda_i)$ denoted spectrum value from λ_i wave length; $\Delta\lambda$ denoted space from λ_{i-1} to λ, which was decided by sampling interval of spectrum. The correlation between original spectrum, first derivative spectrum and SL was analyzed by Excel 2003, wave band with spectrum characteristic and combination parameters from wave band were extracted from them (Huang et al. 2003; Gu, 2003). On the other hand, SPSS 10.0 software, was applied in statistics analysis, number of samples used in establishing model and testing were 89 and 73 respectively.

3. RESULTS AND ANALYSES

3.1 Total spectrum characteristics on cotton canopy infected with verticillium wilt

The results indicated that spectrum reflectance of cotton canopy had obvious diversity between normal cotton and disease cotton with verticillium wilt (Figure 1, 2). The value of spectrum reflectance of cotton canopy infected with verticillium wilt was higher than normal in visible light wave band, and it had little influence by cotton growing periods. The value was lower than normal canopy in near-infrared wave band, and it had little influence by cotton growing periods and varieties.

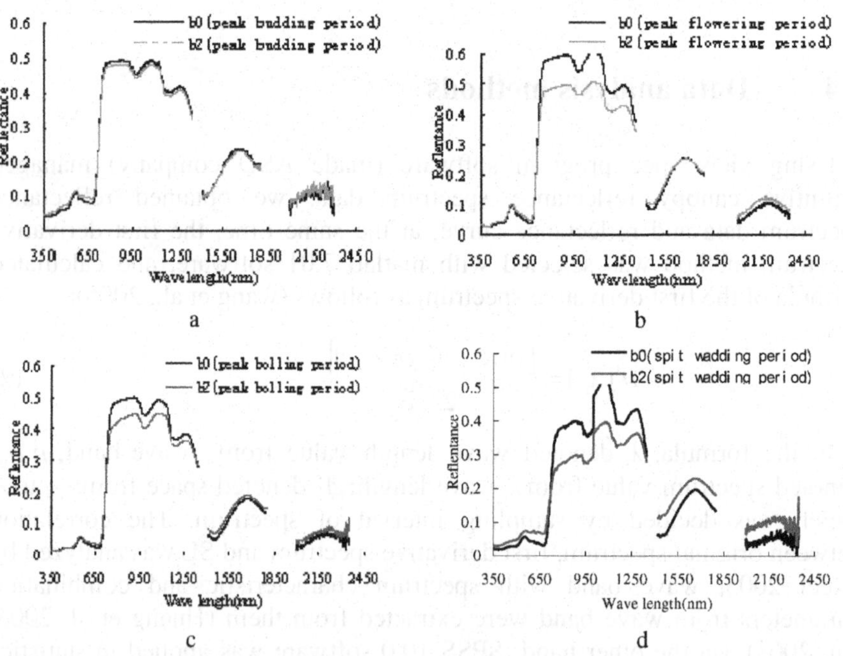

Fig. 1. Spectrum graph between the disease and CK treatment at different growth stages of cotton (XLZ-7)

3.2 Spectrum characteristics of cotton canopy verticillium wilt with different SL

The results indicated from analyzing canopy spectrum of different disease degrees (Figure 3), spectrum reflectance raised in visible light region (620–

700 nm) with the increase of the SL, however, it inverted in near-infrared region, spectrum reflectance fallen down with the increase of the SL, and extreme signification in 780–1300 nm namely, spectrum reflectance of normal canopy (b0) was the highest in near-infrared region, mild degree canopy (b1) took second place, most severity degree canopy (b4) was the lowest. Canopy spectrum from different varieties and different growth stages had homologous change trend, but spectrum reflectance value was different. Consequently, spectrum characteristics difference of cotton canopy with different SL can be used to recognize SL of verticillium wilt.

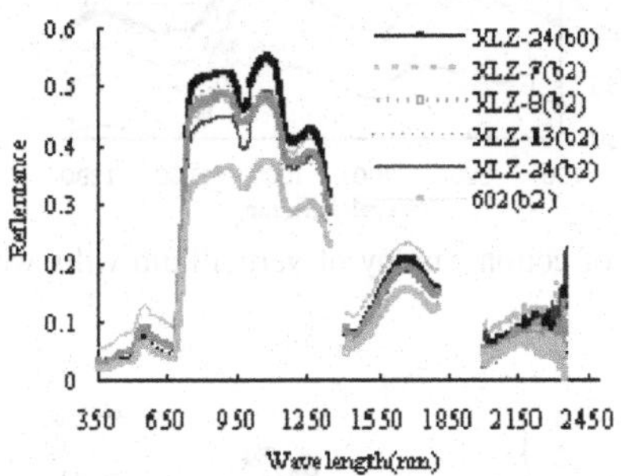

Fig. 2. The reflectance spectrum curve of varieties cotton canopy of verticillium wilt in peak bolling stage

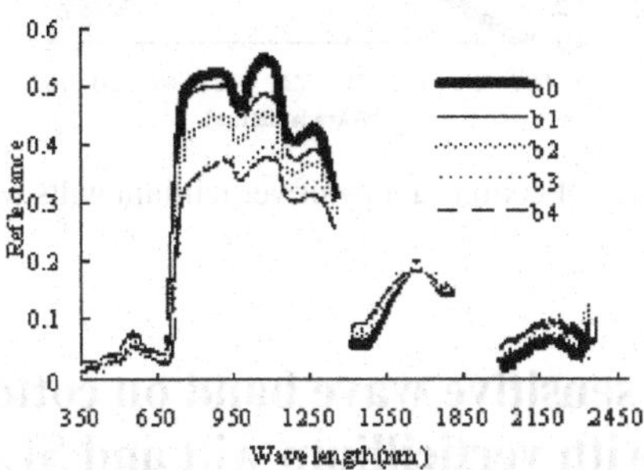

Fig. 3. The reflectance spectrum curve of cotton canopy of verticillium wilts with different SL

In near-infrared wave band (780-1300 nm), discrepancy of canopy spectrum reflectance with different SL was obvious between diseases and normal (Figure 4). With the SL increasing, discrepancy of canopy spectrum reflectance was increased generally, when SL got b1, canopy reflectance of diseases was lower by 3% than normal, but when SL got b2 (25%), it lower 8% than normal. So b2 (25%) could be used as a watershed and diagnosed index in early time.

This study indicated (Figure 5), wave bands, which located the red edge region in first derivative spectrum, changed biggest, namely, maximum of the first derivative spectrum and corresponding wavelength between 680 nm and 760 nm. As for normal canopy, the red edge of cotton canopy infected with verticillium wilt with different SL all had "double swings" phenomenon, and "double swings" descended at the same time, besides, red edge position all moved to the short-wave direction obviously, namely, red edge swing decreased, and red edge position happened "blue movement".

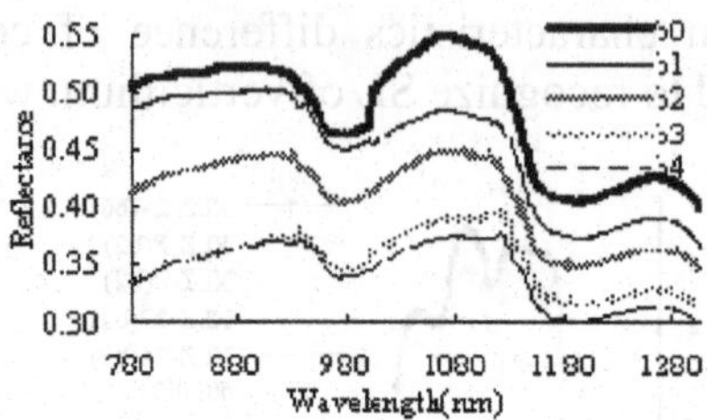

Fig. 4. The spectrum curve of cotton canopy of verticillium wilts with different SL in 780–1300 nm

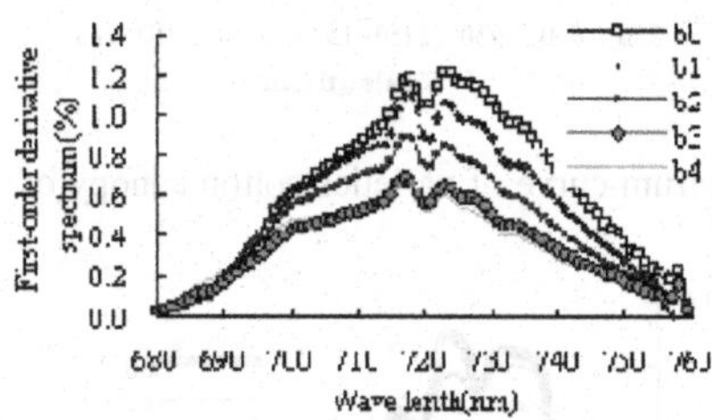

Fig. 5. The red edge of cotton canopy of verticillium wilts with different SL

3.3 Selecting sensitive wave band on cotton canopy infected with verticillium wilt and SL invertion

The spectrum reflectance curve shape of cotton canopy which infected with verticillium wilt with different SL was likeness. However, the depth value of band swing and valley was different; namely, cotton canopy spectrum infected with verticillium wilt had certain regular (Figure 3). Except for effect on atmosphere and water, the sensitive wave bands were selected in 400–1350 nm, 1400–1800 nm and 1950–2350 nm. After the correlation between SL of cotton infected with verticillium wilt and the canopy spectrum reflectance was analyzed, we got correlation coefficient curve (Figure 6). The result indicated, the correlation between spectrum reflectance and SL was positive in 620–700 nm (visible light region), and was negative in 780–1300 nm (near-infrared area), which showed consistent with spectrum reflectance, the spectrum reflectance rose in visible light region, and descended in near-infrared region with SL aggravation, and the Correlation between spectrum reflectance and SL had arrived at the best significantly level in 733–1350 nm bands, so they were selected out as sensitive band region to canopy SL of verticillium wilt, and 1001–1110 nm and 1205–1320 nm were selected out as spectrum sensitive wave band of them. Due to the correlations between spectrum reflectance and SL was maximum at 806 nm and higher at 690 nm, 1455 nm, they were selected out, and base on their combinations, estimation models of canopy infected with verticillium wilt with SL were established. At the same time, estimation models were tested. As shown (Table 1), the wave band combinations model, which was established by R806 nm – R1455 nm, had higher estimative accuracy, the RMES was the least (0.743).

Table 1. The inversing model and test for colony severity level of cotton with verticillium wilt

Wave Band Parameter (λ)	Diagnosing model (sl)	Determination coefficients (R^2)	Root mean square error (RMES)	Correlative coefficients (r)
R_{806nm}	sl=-11.64x + 7.0722	0.675**	0.799	0.798**
R_{806nm}- R_{690n}	sl=-11.66x + 7.0722	0.675**	0.805	0.789**
$R_{806\ nm}$ -R_{1455nm}	sl=-10.588x + 5.6764	0.690**	0.743	0.824**
R_{806nm}/R_{1455nm}	sl=-0.5309x + 4.7848	0.558**	0.910	0.733**
(R_{806nm}- R_{690nm})/(R_{806nm}+ R_{690nm})	sl=-8.5564x + 7.4851	0.537**	0.901	0.747**
FD_{731nm}	sl=-509.42x + 5.3279	0.715**	0.750	0.846**
FD_{1317nm}	sl=2184.4x + 5.4427	0.627**	0.804	0.818**
FD_{731nm}–FD_{589nm}	sl=-484.88x + 5.2745	0.713**	0.753	0.844**
FD_{731nm}–FD_{1317nm}	sl=-438.41x + 5.5722	0.741**	0.730	0.853*

Note: significance at the *P*>0.01 confidence interval. Number of samples used in establishing model and testing are 89 and 73 respectively.

After the characteristics of first derivative spectrum on cotton canopy infected with verticillium wilt were analyzed, the correlation between SL of disease and the first derivative spectrum was present (Figure 7). Because the correlation between first derivative spectrum and SL was higher at 589 nm, 731 nm and 1317 nm, they were selected out and base on their combinations, estimation models of canopy infected with verticillium wilt with SL were established, at the same time, those models were tested. As shown (Table 1), the wave band combinations model, which was established by FD731 nm – FD1317 nm, showed higher estimate accuracy, the RMES was the least (0.735). Therefore, we drown an conclusion that using first derivative spectrum, SL of cotton canopy infected with verticillium wilt could be estimated. Moreover, the testing result of model indicated, the b value was not zero in regression model, and possible reason was that

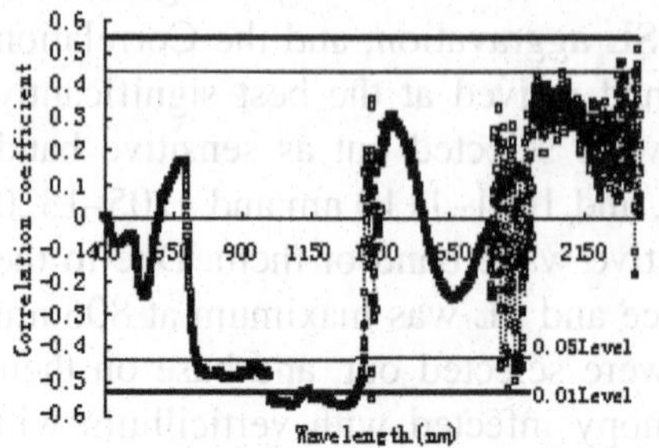

Fig. 6. Correlation between the cotton canopy of verticillium wilt different SL and spectrum reflectance

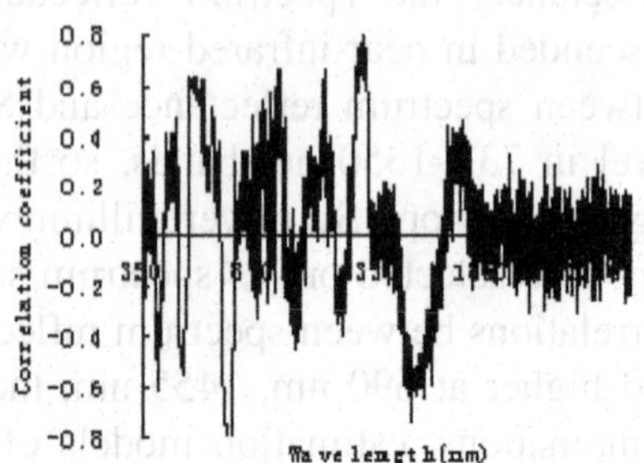

Fig. 7. The correlation analyses between the first derivatative spectrum datas and SL

establishing and testing model samples were come from different years and varieties, and then certain system error was inevitable.

4. CONCLUSIONS AND FUTURE WORKS

From the above analysis, the main results can be described as follows:

(1) Spectrum characteristic on cotton canopy infected with verticillium wilt had better regularity with the increase of SL in different periods and varieties. With the aggravation of the SL, spectrum reflectance rose at visible light region (620–700 nm), and declined at near-infrared region, and however, it was special signification at 780–1300 nm.

(2) When SL got 25%, cotton canopy infected with verticillium wilt could be used as a watershed and diagnosed index in early time.

(3) The characteristics of first derivative spectrum on verticillium wilt indicated, wave bands, which located red side district (680–760 nm) changed significantly, from normal to mild to most severity disease, red edge swing decreased, while the red edge position equal moved to the blue, presented special spectrum characteristics of disease.

(4) The results indicated that 1001–1110 nm and 1205–1320 nm can be selected out as sensitive bands region to monitor SL of canopy. Base on above study, some inversion models, which estimated canopy infected with verticillium wilt, were established, and those models in which the first derivative spectra at $FD_{731\ nm}$–$FD_{1317\ nm}$ would improve accuracy the SL of cotton canopy, and it may be acted as the best recognized model to SL of verticillium wilt.

(5) Future work is analyses various stresses, including disease bring diversities on physiology, bio-chemical mechanism in cotton and its respond with canopy parameters of verticillium wilt on the spectrum, establish higher precision spectrum identification models and canopy parameters inversion modes.

ACKNOWLEDGEMENTS

This work was supported by National High Technology Research and Development Program of China (863 programmes 2006AA103A302, 2006AA10Z207) and Key Laboratory of Oasis Ecology Agriculture of Xinjiang Bingtuan open task.

REFERENCES

Bravo, C, Moshou, D, West, J McCartney, A. 2003, Ramon H. Early disease detection in wheat fields using spectral reflectance. Biosystems Engineering, 84, 137-145.

Chen Pengcheng. 2006, Application of groung-based hyperspectral remote sensing to detect spider mite. (In Partial Fulfillment of the Requirements for the Degree of Postgraduation), Xinjiang, Shi hezi university, 14-15 (in Chinese with English abstract)

Guo Ni. 2003, Vegetation index and its advances. Arid Meteorology, 21, 71-75 (in Chinese with English abstract)

Hamed Hamid Muhammed. 2005, Hyperspectral crop reflectance data for characterizing and estimating fungal disease severity in wheat. Biosystems Engineering, 91, 9-20.

Huang Lin, Zhang Xiaoli, Shi Ren. 2006. Current status and problems in monitoring forest damage caused by diseases and insects based on remote sensing. Remote sensing information, 71-75 (in Chinese with English abstract)

Huang muyi, Wang jihua, Huang wenjiang, Huang yide, Zhao chunjiang, Wan anmin. 2003, Hyperspectral character of stripe rust on winter wheat and monitoring by remote sensing. Transactiongs of The CSAS, 19, 154-158 (in Chinese with English abstract)

Kefyalew Girma, Mosali, J, Raun, W R, Freeman, K W, Freeman, K.W, Martin, L.K, Solie, J.B, Stone, M L I, 2005, Dentification of optical spectral signatures for detecting cheat and ryegrass in winter wheat. Crop Science, 45, 477-485.

Liu Liangyun, Huang Muyi, Huang Wenjiang, Wang Jihua, Zhao Chunjiang, Zheng Lanfen, Tong Qingxi. 2004, Monitoring stripe rust disease of winter wheat using multi-temporal hypersoectral airborne data. Journal of Remote Sensing, 8, 276-281 (in Chinese with English abstract)

Luo Xiuling, Zhou Hesheng, Xue Qin, Zhang Changhong, Wang Zhigang, Zhou Xu. 1997, Satellite multiangular remote sensing for distinguishing areas harmed by mice and insects in the grassland. Journal of remote sensing, supple, 1, 212-219 (in Chinese with English abstract)

Ma Jianwen, Han Xiuzhen, Ha Sibagan, Wang Zhigang, Yan Shouxun, Dai Qin. 2004, Remote sensing new model for monitoring the east Asina migratory locust infections based on its breeding circle. Journal of remotes sensing, 8, 370-377 (in Chinese with English abstract)

Qiao Hongbo, Jian Guiliang, Zhou Yafei, Cheng Dengfa. 2007, Influence of Fusavium wilt to differenct resistance cultivars on spectrum of cotton. Cotton science, 19, 155-158 (in Chinese with English abstract)

Wang Xiuzhen, Li Jianlong, Tang Yanlin. 2004, Approach the action of derivative spectral for determining agronomic parameters of cotton. Journal of Huanan Agriculture University, 25, 17-21 (in Chinese with English abstract)

Wu Suwen, Wang Renchao, Chen Xiaobin, Shen zhangquan, shi zou. 2002, Effect of rice leaf blast on spectrum reflectance of rice. Journal Of Shanghai Jiaotong University, 20, 73-84 (in Chinese with English abstract)

Zhang Kaiqiang, Li Sshengjian, Lu Shenbing, Luo Hua, Wu Yinghua, Li Xiaobing, Wu Anping, Qing Jiugang. 2004. Fusarium and verticillium wilt of cotton were investigated in plain of jianghan in 2004. Cina cotton, 31, 38-39 (in Chinese)

Zhang Minghua, Qin Zhihao, Liu Xue. 2005, Remote sensed spectral imagery to detect late blight in field tomatoes. Precision Agriculture, 6, 489-508.

THE DESIGN AND IMPLEMENTATION OF SUGAR-CANE INTELLIGENCE EXPERT SYSTEM BASED ON EOS/MODIS DATA INFERENCE MODEL

Zongkun Tan [1,2,*], Meihua Ding [1,2], Xin Yang [1,2], Zhaorong Ou [1,2], Yan He [1,2], Zhaomin Kuang [1,2], Huilin Chen [3], Xiaohua Mo [4], Zhongyan Huang [5]

[1] *Remote Sensing Application and Test Base of National Satellite Meteorology Centre, Nanning, China, 530022*

[2] *GuangXi Institute of Meteorology, Nanning, China, 530022*

[3] *HaiNan Province Meteorological Administration, Haikou, China, 570203*

[4] *Zhangjiang City Meteorological Administration, Zhangjiang, China, 524001*

[5] *YunNan Province Meteorological Administration, Kunming, China, 650034*

* *Corresponding author, Address: GuangXi Institute of Meteorology, Nanning, 530022, P. R. China, Tel: +86-771-5875207, Fax: +86-771-5865594, Email: tanzongkun@163.com*

Abstract: One of the major problems in the real time decision of agricultural intelligence expert system is how to be obtained the real time information of crops growth and its close relation environment data. As result, the extraction of crops planting areas and their spatial distribution and their growth variety, especially when the natural disaster arises, such as drought, its spatial distribution and crops suffer from harmful degree have become the extraordinary important factors of the real time decision in agricultural intelligence expert system.

In order to be obtained the real time information of crops growth and its close relation environment data. In the first place, this paper presents an automatic approach to the sugar-cane planting areas and its spatial distribution and growth and classification of drought extraction for mixed vegetation and hilly region, more cloud using moderate spatial resolution and high temporal resolution EOS/MODIS data around Guangxi province, south of China. Next, the framework and the method for knowledge expressing and inference mechanism of the real time decision of sugarcane intelligence expert system are proposed. Finally, the information of sugarcane planting area and

Tan, Z., Ding, M., Yang, X., Ou, Z., He, Y., Kuang, Z., Chen, H., Mo, X. and Huang, Z., 2008, in IFIP International Federation for Information Processing, Volume 259; Computer and Computing Technologies in Agriculture, Vol. 2; Daoliang Li; (Boston: Springer), pp. 1181–1191.

sugarcane growth variety and sugar-cane drought distribution are carried out by using multi-phase EOS/MODIS data and weather forecast are used in the sugarcane intelligence expert system, the mechanism combines concern rectangle and inference based on the produce type inference, and makes good use of their advantages.

Keywords: RS, Sugarcane, Intelligence Expert System, Inference diagnosis, Real time decision

1. INTRODUCTION

Agricultural intelligence expert system is one of the pervasive and the persuasive of modern agricultural technology in recent decade, China. Till 2006 year, there are more than 200 agricultural intelligent systems have been used in 29 provinces and cities in China, which involved 25 kind species crops and domestic animal and birds and marine lives etc. Simulate model and the rule knowledge of produce have been using in agricultural intelligent expert system, the people who have based on the localization of the tradition idea and experience knowledge to solve the agricultural management problems for long time have been broken. However, the agricultural management is an extraordinary complex system, which is impacted by climatic and soil and other biology, as well as manpower inference and interaction between individual and colony of biology. In addition, the agricultural management is faced the biology organism which growth with time change. Consequently, the integrity and reliability of model capability and the construct level of expert knowledge database contribution directly influence on the whole of capability of agricultural intelligent expert system. The integrity and reliability and real timely and of completely agricultural base database, especial the foresee of the environment factors variety of corps growth, which have become the key factor of deciding the agricultural intelligent expert system whether provide real time serves. These days, however, the agricultural expert systems of implementation in Guangxi province have had lack the decision capability of according crops growth and environment factors variety, because the information of the real time environment data and the information of crops growth are difficult to be obtained.

In view of Guangxi province is the biggest region of sugar-cane planting in China, and the sugar is extraordinary important material in our country. In this paper, Guangxi province in southern China was selected as the study area. The objectives of this research were to: (1) find out the way of identify and extraction of sugar-cane planting spatial distribution information in hilly, mixture vegetation and cloudy in southern, China; and (2) the retrieve a suitable monitor model for sugar-cane growth and drought based on EOS/MODIS data, and validate the model with estimated sugar-cane planting areas and the model with forecasted annual yield based on remote sensing; and (3) set up the model of sugar-cane intelligence expert system based on EOS/MODIS data.

2. METHODS

2.1 Study area

The study area is located in Guangxi province, south of China. It latitude is 20°54′–26°23′N and longitude is 104°29′–112°04′E. Its total area is 236700.0 km^2. It belongs to monsoon region of south subtropical zone and north tropical zone without four clearly demarcated seasons of spring, summer, autumn and winter. The climate here is hot and humid in summer and warm and dry in winter. In 2006/2007 year crushing season, sugar-cane planting area approach 853000hm^2, the yield of sugar over 7.08×10^9KG in Guangxi province, which sugar yield approach 65% of the whole country.

2.2 Data acquisition

In this paper, EOS/MODIS imageries obtained from on December 1, 2002 to on December 31, 2006 were required; whose path/row is 1151/1230. When the data were obtained, more than one county without cloud or little cloud images were selected, then they were synthesized one or more images in the same month.

2.3 Methods

Due to the relationship between vegetation indices calculated by different algorithms, reflectance of bands and field measurements of NDVI, we can retrieve NDVI using EOS/MODIS data. With this relationship, a NDVI retrieval model for study area can be established. A specific flow chart of retrieval technique is shown as Fig. 1.

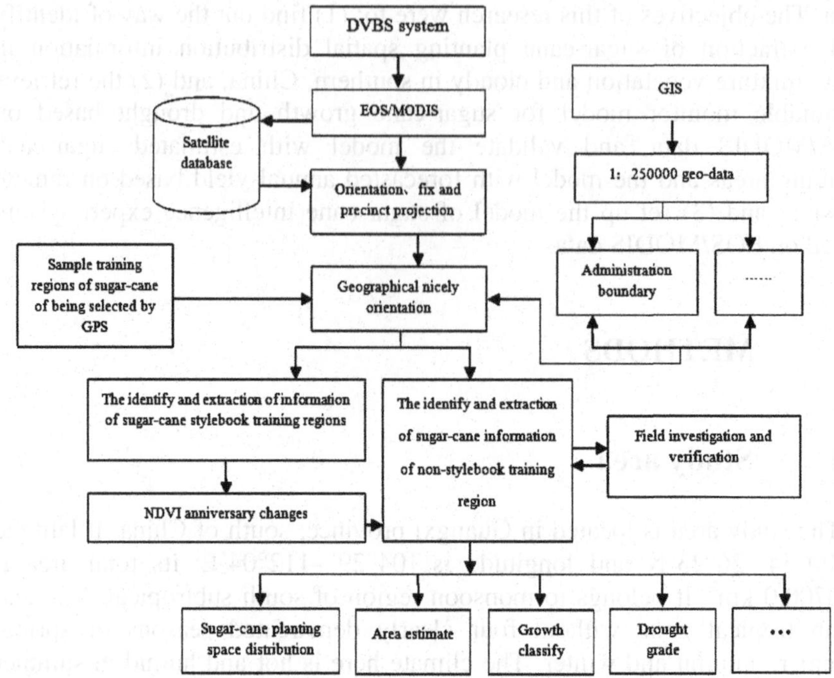

Fig. 1. The flow chart of the identify and extraction of sugar-cane planting space distribute information based on EOS/MODIS data

Based on the above flow chart of technique, detailed steps are described as follows:

Inversing reflectance for EOS/MODIS imagery

The objective of atmospheric correction for EOS/MODIS data is to attain related parameters which can indicate the vegetation inherent properties of the region. Since the remotely sensed image was affected by reflective solar energy, solar elevation, zenith angle, the thickness of aerosol and the bidirectional scattering due to the mutual influence of ground environment factors, we should take into account both atmospheric and bidirectional scattering to obtain accurate ground reflectance. Because the parameters of atmospheric profile based on measurements data or standard atmospheric profile were not established in China, in this paper we adopted international standard parameters of atmospheric profile to correct EOS/MODIS image.

Obtaining characteristic parameters of vegetation

Due to the chlorophyll and inner architecture of foliage, a special reflective spectrum of vegetation foliage was formed like intensive absorption in the red waveband and intensive reflection in the near infrared waveband. By the reflectance difference varied in the red and the near infrared waveband, we can calculate related parameters that indicate the conditions of vegetated surfaces, such as normalized difference vegetation index (NDVI) which is a simple, effective and experiential measurement to vegetation activity. To some extent, NDVI indicates the vegetation information of status and succession.

NDVI, a parameter (range -1–1) denoting the ground vegetation coverage can be derived from the reflectance in the red and the near-infrared wavebands. This equation reads:

$$NDVI = \frac{\rho NIR - \rho R}{\rho NIR + \rho R}$$

From the equation we can see that in the water area and roadway area and city or town area, theirs value of NDVI are below 0 or approach constant value in different seasons. But for the land surface with cover foliage, NDVI ranges from 0.1 to 0.7. NDVI has been applied in many fields, such as land cover or change, vegetation and environment change, net primary productivity and the assessment of crop yield.

Sample training regions of sugar-cane of being selected by GPS

To the same foliage, its value of NDVI is various with its growth process. As result, the values of NDVI between foliages are diversity in different seasons. In order to mastery the spectrum characteristic of sugar-cane and distinguish sugar-cane from many kinds of foliages, some sample training regions of sugar-cane (the area must be bigger than 7 hm^2) in different county of Guangxi were selected by GPS (the Global Positioning System).

The identify and extraction of sugar-cane planting information based on EOS/MODIS data

In the first place, the values of NDVI of sample training regions of sugar-cane during the main growing seasons were calculated. As result, we could find the variety trend of curves of sugar-cane in different regions being consistent (Fig. 2).

For the sugar-cane, its growth lasting more 8 to 12 months, and the main of crops is sugar-cane in Guangxi province during winter, and the areas of sugar-cane cover are reduced during crush season. When the crush season was over, the values of sugar-cane planting areas approached 0.

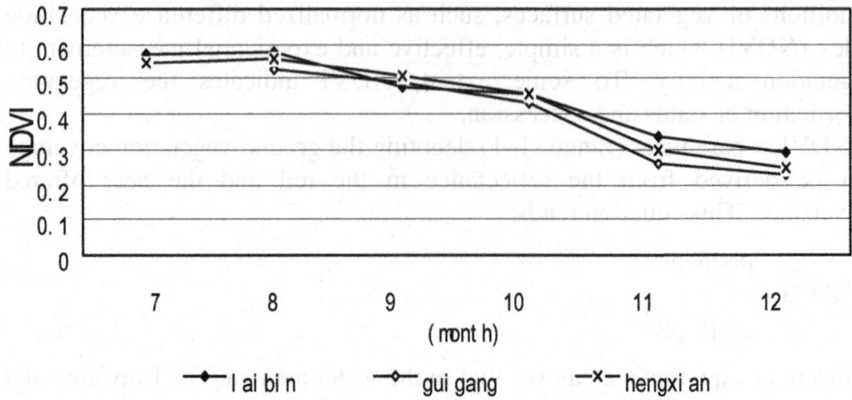

Fig. 2. The curse of NDVI of sugar-cane in different regions

Fig. 3. The imagine of sugar-cane planting and its spatial distribution based on EOS/MODIS in Guangxi province, 2006

And the same time, corn and rice and soybean, theirs growth (from sowing to harvest) are general lasting 3 or 4 months. The south subtropical zone and north tropical zone forest growth lasting more than 12 months, but its value of NDVI anniversary approach constant. Consequently, the curves of NDVI variety in different foliages during the main growth seasons are difference. We can use the Maximum likelihood to extract the information of sugar-cane planting and its spatial distribution through the calculation of multiple-phase MODIS-NDVI from different foliages in Guangxi province. The result shows that the information of sugar-cane planting and its spatial distribution in 2006 were clearly in remote sensing imagine (fig. 3). The survey of field also showed that the information of sugar-cane planting based on multiple-phase EOS/MODIS data was highly reliable and truth.

3. THE FRAMEWORK AND FUNCTION OF SUGAR-CANE INTELLIGENCE EXPERT SYSTEM BASED ON RS

Compare with other agricultural intelligence expert system, the sugar-cane intelligence expert system based on RS certain consists of following sects: User connects and server and inference diagnosis and expert's knowledge database and environment database, which close relate sugar-cane growth and development, such as weather and water and soil and fertilize etc. it must be point out that the crop growth simulation model in agricultural expert system is replaced by the information of sugar-cane monitoring based on remote sense. Its flow chart as Fig. 4.

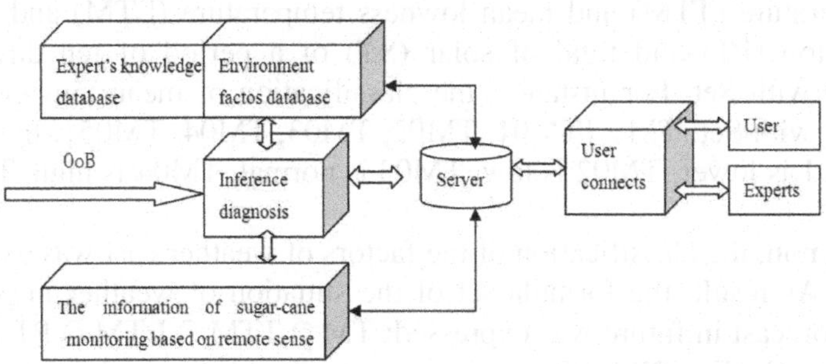

Fig. 4. The framework of the sugar-cane intelligent expert system based on RS

4. SUGAR-CANE EXPERT'S KNOWLEDGE EXPRESS BASED ON RS

According the value of sugar-cane NDVI and its yield data and field survey in different years, the growth of sugar-cane was divided by four steps: bad and average and good and better, which remarked for GG={G01,

G02, G03, G04}. At the same time, the majority disasters which affect on sugar-cane yield and sugar yield, such drought and frost, were remarked for ZZ={Z01, Z02, Z03, …}. The losing degree which were caused by nature disasters was divided by light and middle and severity and extremely: FF={F01, F02, F03, F04}. In addition, the disasters from occur to die out were divided by arise and continue and aggravate and relax and relieve, its remarked for XX={X01, X02, X03, X04, X05}.

Since sugar-cane certain planting in dry land without irrigation in our country, the yield and sugar contain of sugar-cane are close relation with weather condition, especial the precipitation total and its distribution in different months. As result, the weather data and weather forecast information must be imported when we are constructing expert knowledge rules. The climatic factors estimated index as following:

$\leq \bar{x}i - 2\delta i$	lower (less)
$(\bar{x}i - 2\delta i,\ \bar{x}i - 0.5\delta i)$	low (little)
$(\bar{x}i - 0.5\delta i,\ \bar{x}i + 0.5\delta i)$	average
$(\bar{x}i + 0.5\delta i,\ \bar{x}i + 2\delta i)$	high (much)
$\geq \bar{x}i + 2\delta i$	higher (more)

In demand side,

$\bar{x}i$ is the value of a period of ten days average in many years of climatic factor.

δi is the mean square error of the climatic factor historical data list.

In the first place, the classification of mean temperature (TM) and mean top temperature (TTM) and mean lowness temperature (LTM) and total of precipitation (PT) and total of solar (ST) of a period of ten days were expressed with Set. For instance, the classification of mean temperature is expressed with Set: TM= {TM01, TM02, TM03, TM04, TM05}. In demand side, TM01 is lower, TM02 is low, TM03 is normal, TM04 is high, TM05 is higher.

In common, the classification of the factors of weather cast was expressed with Set. As result, the formula set of the situation of weather in past and weather forecast in future was expressed: $TM \cap TTM \cap LTM \cap PT \cap ST \cap TF \cap TTF \cap LTF \cap PF \cap SF$.

From what has been discussed above, the weather and climate survey in past years and the information of sugar-cane remote sensing monitor were synthesis, the corresponding agricultural management rules have been formed requirement of service object.

5. THE INFERENCE MECHANISM OF SUGAR-CANE INTELLIGENCE EXPERT SYSTEM BASED ON RS

The inference mechanism of sugar-cane intelligence expert system based on RS was equally following the principle of the disaster first. The management of agriculture or enterprise should adopt the measure of avoid and Anti- and relief disaster, and could take no account of the influence of the other factor when the monitoring the sugar-cane have already suffered the influence of different degree disaster or the disaster will arise forecast. Its inference process as follows:

(1) The user was distinguished first when he or she entered the sugar-cane intelligence expert system, and the system demanded the user to choose sugar-cane region which he or she concerned.

(2) According to the user's choice, the system will automatically link the data of sugar-cane remote sensing monitoring and will show that the newest imagine of sugar-cane remote sensing monitoring whether surfer the disaster which the region of the user chose. The system will give the conclusion of anti- and relief measures from the expert knowledge database to the user when the disaster arises.

(3) The system indexes the data of weather forecast information and choices the latest forecast information of region where the user chose, and judgment the disaster emergence or not. If the disaster emergence, the system will give the conclusion of the disaster prevention, reduce the disaster for user.

6. THE REALIZATION OF THE REAL TIME DECISION FUNCTION OF SUGAR-CANE INTELLIGENCE EXPERT SYSTEM BASED ON RS

The influence sugar cane grows the growth's disaster from take place to die out could mean with following relation: XX·RZ. In demand side,

Light degree Middle degree Severity degree Extremity degree

$$RZ = \begin{bmatrix} Z1F01 & Z1F02 & Z1F03 & Z1F04 \\ Z2F01 & Z2F02 & Z2F03 & Z2F04 \\ Z3F01 & Z3F02 & Z3F03 & Z3F04 \end{bmatrix}$$

XX={X01, X02, X03, X04, X05}

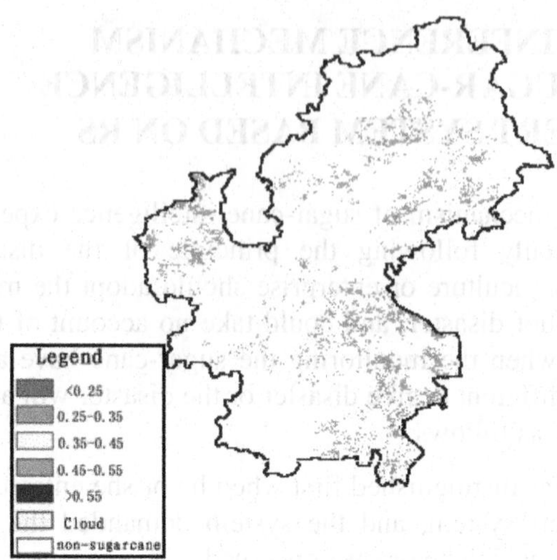

Fig. 5. The sugar-cane drought classification in Congzuo city of Guangxi province, China. Oct, 2006

Fig. 5 is the image of remote sensing which the sugar-cane drought classification and distribution in Congzuo city of Guangxi province, Oct, 2006. For the whole city, we could find that the drought degree was diversity in different villages. There for, the villages of minimum administration district were carved up in imagine of remote sensing in order to the sugar-cane intelligent system could identify the information of sugar-cane. In the nature village, we regulated the most of pixels which suffer the same disaster grade to representative the whole village suffer disaster status, and its values of disaster grade was set 1, and the rest grade was set 0. Because of weather disaster grade forecast is only, if the result of forecast shows that the disaster would arise, its result of forecast will be set 1, if the result of forecast shows that the disaster would not arise, its result of forecast will be set 0. So we could calculate out the value of XX·RZ. According to the expert knowledge databases and identify the code, the system would give out the accurate diagnosis conclusion immediately to user.

7. CONCLUSIONS AND DISCUSSIONS

Based on the above study and analysis, some conclusions can be drawn as follows:

(1) It is an effective way to extraction of sugar-cane planting spatial distribute information by taking some sample training regions of sugar-cane in different county.

(2) The information of sugarcane planting area and sugarcane growth variety and sugar-cane drought distribution are carried out by using multi-phase EOS/MODIS data and weather forecast are used in the sugarcane intelligence expert system, the mechanism combines concern rectangle and inference based on the produce type inference, and makes good use of their advantages.

ACKNOWLEDGEMENTS

This study has been funded by China National 863 Plans Projects (Contract Number: 2001AA115360), It is also supported by the China Meteorological Administration new technology extend project (Contract Number: CMATG2006M42), Sincerely thanks are also due to Guangxi Climate center and National Satellite Meteorology Center for providing the data for this study.

REFERENCES

Cheng Qian, Huang J F. 2003. Analyses of the correlation between rice LAI and simulated MODIS vegetation indices, Red Edge Position [J]. Transactions of the CSAE, 19(5):104-108.

Yang Bangjie, Pei Zhiyuan, Zhang Songling. 2001. RS-GIS GPS-based agriculture condition monitoring systems at a national scale [J]. Transactions of the CSAE, 17(1):154-158.

Kontoes C, Wilkinson G G, Burril A. et al. 1993. An Experimental System for the Integration of GIS Data in Knowledge Based Image Analysis for Remote Sensing of Agriculture [J]. International Journal of Geographical Information Systems, 7(3):247-262.

Murthy C S, Raju P V, Badrinath K V S. 2003. Classification of Wheat Crop with Multi-temporal Images: Performance of Maximum Likelihood and Artificial Neural Networks [J]. INT. J. Remote Sensing, 24(23):4871-4890.

Zhao M S, Fu C B, Yan X D et al. 2001. Study on the relationship between different ecosystem and climate in China using NOAA/AVHRR data. Acta Geographica Sinica, 56(3):287-296. (in Chinese)

Zheng Y R, Zhou G S. 2002. A forest vegetation NPP model based on NDVI. Acta Phytoecologica Sinica, 24:9-12. (in Chinese)

RESEARCH ON KINEMATICS SIMULATION OF PARAMETERIZED MECHANISM BASED ON VISUALIZATION IN SCIENTIFIC COMPUTING

Gang Zhao, Liangxi Xie*

College of Mechanical Automation, Wuhan University of Science and Technology, Wuhan, Hubei, 430081, P. R. China

* *Corresponding Author, Address: P.O. Box 242, College of Mechanical Automation, Wuhan University of Science and Technology, Wuhan Hubei, 430081, P. R. China, Tel: +86-27-62423171, Email: lx_tse@163.com*

Abstract: It is fundamental for the kinematics simulation and mechanism design to apply the visualization in scientific computing on the simulation research of 3D mechanism. It is also the basis of independent development for 3D CAE software without any copyright limit. This paper offered a parameterized kinematics simulation template for 3D mechanism based on the visualization in scientific computing, in which the open graphics library is applied by MFC compiler. This paper illuminates the fundamental principle of the simulation development applying the OpenGL technology. The principle of OpenGL, the procedure structure, the working flow of OpenGL, the concept of program designing and the code implementation based on VC++ is demonstrated and an application approach for mechanism simulation is developed under the MFC environment.

Keywords: visualization in scientific computing, MFC compiler, OpenGL, parameterized, kinematics simulation for 3D mechanism

1. VISUALIZATION IN SCIENTIFIC COMPUTING APPLIED IN KINEMATICS SIMULATION

Visualization in Scientific Computing was first present in a report (Computer Graphics, 1987, Vol. 6) submitted to US National Science Foundation in 1987 by B. H. McCormick. It is goal to show the evidence

Zhao, G. and Xie, L., 2008, in IFIP International Federation for Information Processing, Volume 259; Computer and Computing Technologies in Agriculture, Vol. 2; Daoliang Li; (Boston: Springer), 1193–1203.

between the data content and the graphics by an organic synthesis, which is formed by a mass of data obtained by digital computation or experiment according to its physical background. It is advantageous to totally grasp the designing evolution, discover the internal physical principle, enrich the scientific research and shorten the research period for Visualization in Scientific Computing. In an application of kinematics simulation by the visual technology, the problem is how to construct a physical model based on a data set, how to set up a visual model and implement a concrete graphics technology (Pu, 2004).

The visualization technology first transforms the abstract description of data to a description of graphic element and forms a visual physical model constituted by graphic elements such as vertex, curve, surface and solid. And then plots the visual model by the plot technology based on computer graphics, which extract the common properties of geometric elements to implement the data visualization, graphic parameterization, graphic property and property parameterization. Graphic parameterization includes the parameter control of geometric element, lighting model, perspective spot and projection type. Graphic property includes color, transparence, optical property of material and mapping texture. The property parameter and graphic parameter form the inter-visual computing model. In the plot technology, surface plot is basic to form the graphic element by vertex, curve and geometric contour based on them. Compared with this plot technology, solid plot distinguish itself from surface plot technology, which directly generate a solid contour by solid element, not an intermediate surface.

Under the principle of tomography, the visualization technology decreases the model dimensions and factually simulates the data field of solid by correct method offered by computer graphics. Solid data field exits in 3D space of research system with the time elapse. Visualization of solid data field is a mapping from data set to geometric property. It extracts solid data about space time according to the need. And that data with variable types is extracted when data is displayed.

2. IMPLEMENTATION OF MECHANISM KINEMATICS SIMULATION BY MFC COMPILER

The setup of parameterized model is a need for graphic intercourse on mechanism analysis and kinematics research. By several control parameter settings and the analysis for 3D parameterized graphic model, the best effect on geometric entity, the pertinence analysis of parameters, the parameter dependence on objects and the harmony between objects is obtained. All these parameters are the man-machine interface. It is hard to set up a visual parameterized model for kinematics simulation by that professional software such as ProE, Solidworks, ADAMS and so on, which is limited by the compiled code or the copyright. Therefore by the implementation of open graphic library in MFC compiling environment, it is relatively easy to develop a system of kinematics simulation, all copyrights reserved.

Generation of 3D graphic for solution is the object-oriented basic for the simulated objects by OpenGL. The question is transformed to the object and scene processed by OpenGL. Scene is constructed by geometric and graphic elements such as vertex, curve and surface, and complex scenes need some better algorithms for surface generation. By a set of image process for objects and its internal elements, such as material definition, lighting, polygon antialiasing, fairing, transparence and image synthesis, texture mapping, atomization etc, anticipated realistic effect is obtained after the geometric solid is generated in scene.

Because it is a software interface for special graphic device, a program language compatible with C++ grammar for visualization technology, OpenGL is advantageous to visualization in scientific computing. As a software interface for graphic device, OpenGL is composed of hundreds of commands and functions. These commands allow the users to instruct the 2D or 3D geometric object and operate object for frame rendering. OpenGL is a command set to operate the graphic hardware and retain a mass of geometric information. Due to the network transparence, OpenGL can transfer the graphic information to remote client servers or display devices, even the shared operation with other systems. OpenGL offers an enhanced capability of plot to ensure the reliability of popular graphic application, which is applied in the coming generation fields of medical imaging, geography information, prospect for petroleum, climate simulation and amusement animation, etc (He, 1994).

It is a technology question to choose, describe and plot the graphic elements when 3D mechanism is constructed. Fundamental graphic element is the most minimum figure unit in a plotting operation. An ideal description of fundamental graphic element shall be expressed by a flexible data structure, a correct data operation and mechanism recognition. Consequently, the ideal graphic element according with the need is the various components and the kinematic pairs. Considering the complexity of component contour and the diversity of kinematic pair, the fundamental graphic element RRR, RRP and RPR group is applied to construct the complicated mechanism of second grade group (Wang, 2001) (Cheng, 2003).

3. STRUCTURE OF OPENGL

As the software interface about graphic hardware, the principle function of OpenGL is to convert the 2D or 3D objects data to image frame buffer. Those solids are composed of a series of vertexes describing the geometric attributes of objects and the pixels describing the images. OpenGL convert that solid data to pixel data and form a final display in frame buffer of image. The fundamental functions of OpenGL include the core function of OpenGL, implement library function, auxiliary library function, windows professional function and Win32 API function. The fundamental structure of OpenGL includes GLYPHMETRICSFLOAT, LAYERPLANEDESCRIPTOR, PIXELFORMATDESCRIPTOR, POINT-FLOAT.

Flow chart of OpenGL includes the operation and command of graphic element, graphic control, interpreted language of OpenGL command, fundamental operation of OpenGL, as the following Fig. 1.

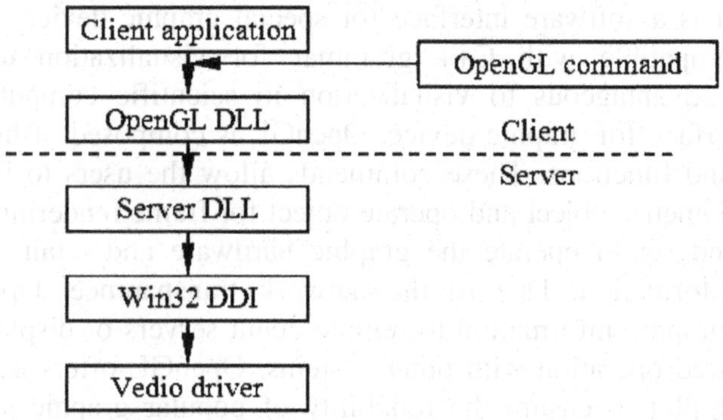

Fig. 1. Flow chart of openGL on windows

Interpreted mechanism of OpenGL command is based on client/server mode, in which the request from client application is interpreted by kernel server of OpenGL. Based on the client/server mode, it is convenient for various users on various computers to share the service offered by other servers. That's to say the structure of OpenGL is transparent to network.

Library function of OpenGL is packed in the dynamic link library file opengl32.dll. OpenGL function access published by client application is first operated by file opengl32.dll. After transferred to server, it is processed by file winsrv.dll and then transferred to DDI, finally to video display driver.

User's command is divided into two parts since it entered OpenGL. One part plots certain geometric solid and the other instructs the program how to operate solids at various stages. Many commands may be listed in display list as queuing messages waiting for operation. An effective method for curve and surface approximation is offered by the polynomial function, whose input value is offered by valuator. Then the geometric element described by vertex is operated. At this stage, vertex is transformed and rendered, graphic element is clipped to perspective space and the raster display is ready. Raster generates the frame buffer address for a series of images and the 2D expression of graphic element. The generated result is called substrate. Each substrate is fit to operate the single substrate before the frame buffer is changed. These operations include the buffer update according to the field depth, various tests and image synthesis, shield operation for substrate color and saved color, the logical operation and fade effect for substrate (Yan, 1995).

Operation of pixel data makes the pixel, displacement and image to be raster displayed after pixel operation. Because the graphic element of OpenGL is expressed by vertexes, it is convenient to operate data vertex by vertex, assemble the fundamental units by them, and then form substrate by raster display. As for the pixel data, the results are saved in texture memory in order to access pixel information from texture memory for raster display.

Display list of OpenGL is factually a function group, which is saved for OpenGL operation in future. The valuator of OpenGL is factually composed of many special functions, which generates the coordinate of vertex, normal, texture, color and transfers operational result to execute module, by a polynomial mapping about one or two variables (Ouyang, 2004). Valuator interface offers a template to generate curves and surfaces based on OpenGL. Compared with complex NURBS interface, the valuator interface in OpenGL is more advantageous when polynomial valuator is efficiently applied to express the non-NURBS curves and surfaces and special surface attributes without the NURBS transform (Yu, 2001).

Finally, the raster includes two part, geometric and physical mapping. Geometric operation turns graphic element to 2D image, physical operation computes the color and depth information of each vertex. Therefore, a raster of graphic element is completed by two steps. Firstly, the pixel occupied by graphic element in an integral raster of window coordinate is determined. Secondarily, the value of color and depth is computed for each pixel. The computed results are conferred to next step of OpenGL for update of correct units in frame buffer.

To operate graphic by OpenGL and finally plot 3D scene on screen, the main steps include:

(1) Format pixel. Format important OpenGL information such as plot style, color mode, color bit, depth bit, etc.
(2) Set up geometric model. Set up and demonstrate the 3D model by mathematics and fundamental graphic elements.
(3) Set scene. Determine the location of object in 3D space and set the perspective correctly.

(4) Effects setting. Set material of object and lighting condition.
(5) Raster display. Convert the information of scene and color into the pixel information displayed on screen.

4. VC++ IMPLEMENTATION OF KINEMATICS SIMULATION FOR 3D PARAMETERIZED MECHANISM

4.1 Concept of Modularized Program Designing

System frame of real-time simulation for mechanism is implemented by OpenGL as the following Fig. 2.

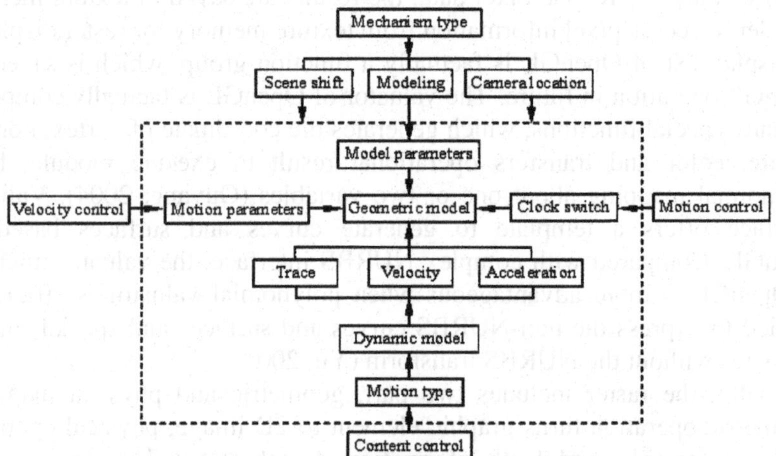

Fig. 2. OpenGL implementation of kinematics simulation for 3D mechanism

Considering the concept of modularized program designing, mechanism type is determined by writing a simple conditional sentence in the window drawing function void CSimulationView::OnDraw(CDC* pDC). With the increase of mechanism types, program adds the corresponding head files and case condition in function OnDraw(). It is no need to modify the main frame of the program. Object-oriented Concept of program designing is implemented overall. The application codes for choice of mechanism type in function OnDraw() are listed as follows,

```
void CSimulationView::OnDraw(CDC* pDC)
{ CSimulationDoc* pDoc = GetDocument();
  ASSERT_VALID(pDoc);
  switch (mechanism)
```

```
  { case Planer:
       DrawScene();
       break;
    ......
    default:
       OnFileNew();
       break;
    }
}
```

4.2 Generation of Geometric Solid

OpenGL offers a number of drawing functions for basic geometry solid. Under the VC++ compiling environment, a head file to draw the basic solid for the corresponding mechanism type is added into project. This head file cites the file glaux.h from the open graphic library. Consequently, the plot function can be directly loaded from library file. For the kinematics simulation of crank rocker mechanism, the head file GeoPlaner.h loads the library file gluCylinder() to draw rods, whose application codes are listed as following,

```
void DrawBar(UINT nLists, GLfloat m_radius, GLfloat m_length)
{ GLUquadricObj* quadObj;
  glNewList(nLists, GL_COMPILE);
  quadObj=gluNewQuadric();
  gluQuadricDrawStyle(quadObj, GLU_FILL);
  gluQuadricOrientation(quadObj, GLU_INSIDE);
  gluQuadricNormals(quadObj, GLU_SMOOTH);
  switch (nLists)
  { case 1: case 5:
      gluCylinder(quadObj, m_radius, m_radius, m_length, 10, 1);
      break;
  ... ...
  }
  glEndList();
  gluDeleteQuadric(quadObj);
}
```

Only the plot function for basic solids is offered by OpenGL, so it is relatively difficult to construct a complex model. In order to construct various complex solids efficiently, 3DMAX is introduced as the professional 3D graphic software (Zhang, 2003). It is an efficient approach to construct the complex geometric solid by 3DMAX and control or convert these solids by OpenGL. At first, construct a complex solid by 3DMAX and then extract the space coordinate for vertexes of 3D solid and the material information to form a C++ binary source file by model transform utility named 3D Exploration. Domination program reads the corresponding binary file by file interface to generate a display list dominated by OpenGL when the certain 3D object needs to be displayed. The file interface of 3D model includes a read interface function and a generation function of display list. Users substitute specified material for model material and implement an intercourse control of object materials by generation of display list interface function (Zhu, 2000).

4.3 Implementation of Motion Simulation

Motion parameters are used to locate the motion objects, which is based on the object coordinate and local coordinate in OpenGL. The number of motion parameters is equal to freedom. The motion parameters control the motion speed by man-machine interface.

Absolute motion is implemented by a rational mathematics model based on a specified geometric model. Motion object in 3D space is located by motion parameters in real-time simulation. OpenGL implementation of absolute motion is demonstrated as the Fig. 3.

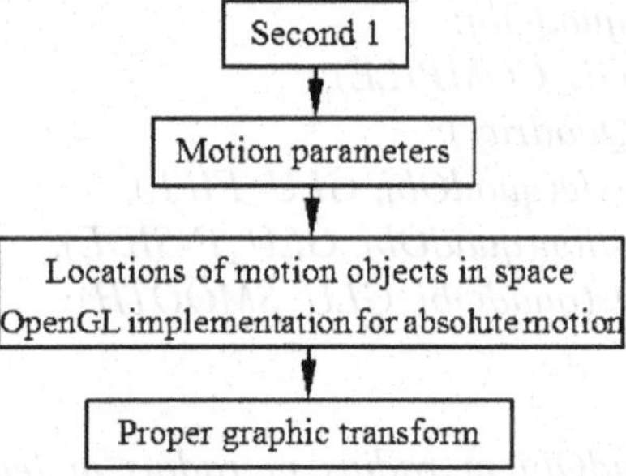

Fig. 3. OpenGL implementation of absolute motion for mechanism

OpenGL convert the absolute motion into relative motion of objects about the dynamic reference frame by its certain expression. There are two ways to get it. One is to take advantage of a relative coordinate directly. The other is to demonstrate a relative motion based on an existed absolute motion. Obviously, the second way is more efficient for a complicated motion mechanism. OpenGL implements a relative motion as the following Fig. 4. When dynamic reference frame restores to the second $t = 0$, the graphic transform corresponding to the control parameters at this second locates all motion objects in virtual space from top source. Consequently, various complex relative motions are simulated easily.

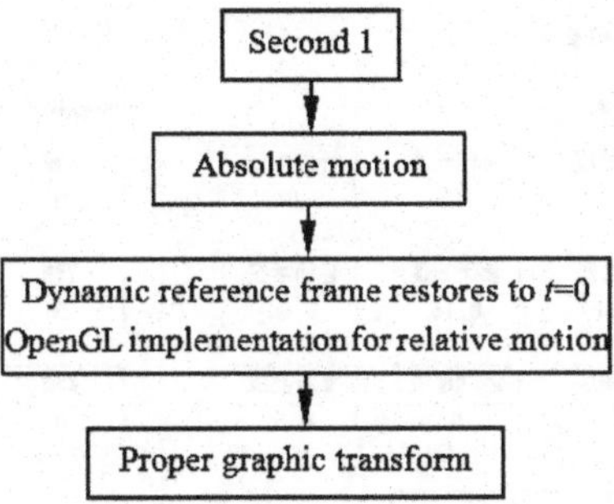

Fig. 4. OpenGL implementation of relative motion for mechanism

Whatever absolute motion or relative motion, time engine for mechanism motion is implemented by the following codes,

```
void CSimulationView::OnTimer(UINT nIDEvent)
{ m_rotateAngle+=m_speedFactor;
  if (m_rotateAngle>360.0f)
  {  m_rotateAngle=0.0f;
  }
  Invalidate(FALSE);
  CView::OnTimer(nIDEvent);
}
```

4.4 Performance of Man-machine Interaction

Based on OpenGL, man-machine interaction of kinematics simulation embodies some aspects such as model setup, motion control, content control, perspective control, camera location, scene shift, etc. Dialog box in simulation of crank rocker mechanism is shown as the Fig. 5. In order to understand the motion of a certain object or a local detail, it is necessary to observe the absolute motion, relative motion, motion trace, velocity and acceleration, etc. in various perspectives. Perspective parameters are determined by variables m_perspectiveX, m_perspectiveY and m_perspectiveZ, which can be given in dialog box or real-time controlled by cursor message series.

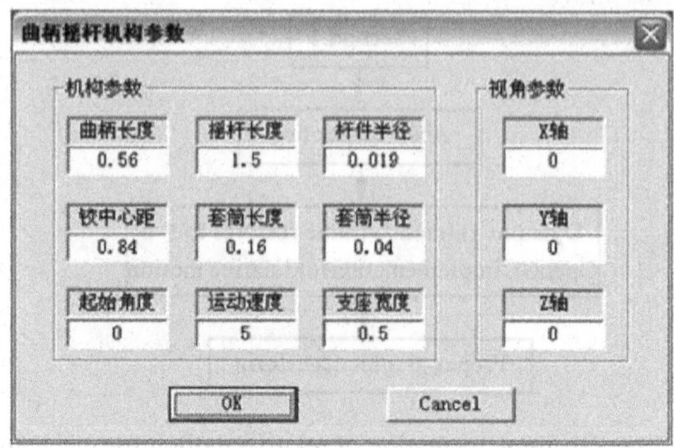

Fig. 5. Dialog box of simulation parameter input

4.5 Effect of Interface and Professional Splash

It is not enough just to simulate the motion of mechanism. In order to obtain the better effect closer to the real, it is necessary to render the interface image by functions such as lighting effect function EnableLighting(), background effect function DrawBackground(), coordinate display function DrawBackground(), velocity control function OnSpeed(), image size control function OnSize (UINT nType, int cx, int cy), function of sound effect and motion control OnControlMove(), etc. Splash effect is completed by class CSplashWnd to make the simulation software more professional. It shows the running interface as the Fig. 6.

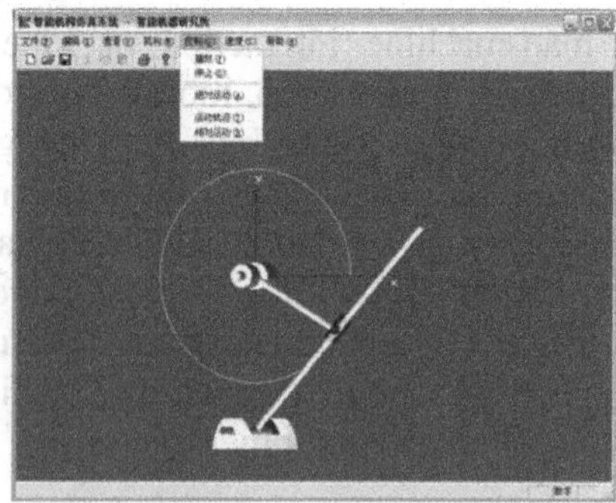

Fig. 6. Interface of parameterized kinematics simulation of 3D mechanism

5. CONCLUSIONS

When the simulation software is completed, some professional tools such as Installshield are used to archive and encrypt the program files. The obtained installation package is up to the trade standard. Now, the simulation software has had the fundamental elements belonging to the professional software. It can efficiently set up the mechanism, simulate the motion, plot the trace of any point on mechanism, output any kinematic variables for analysis adjust the angle of view and zoom the field of view, observe the details about the motion properties. The flexible manipulation, the better vision effect, the real and reserved copyrights and so on, are the advantages of the simulation software.

A kinematics simulation template for parameterized 3D mechanism is constructed by the open graphic library and the open MFC compiler based on the visualization in scientific computing. Many attempts are benefit to the independent development for software of kinematics simulation and mechanism construct in this paper. Due to being the fundament of virtual reality and manufacturing information technology, visualization in scientific computing is efficiently applied in the mechanism simulation, the mechanism construct, the kinematics analysis and the dynamics field.

REFERENCES

Pu Zhixin, Gu Yanfeng, Li Ying, Realization of NC lathe-turning simulation software based on OpenGL [J], Machine Tool & Electric Apparatus, 2004; (2): 14-16

He Xiaoli, Du Ying, Visual C++ MFC class library design & user manual [M], the first edition, Beijing: Xueyuan Press, 1994; 223-229

Wang Chengzhi, Huang Kaixuan, Chen Feiyan, Research on mechanism drawing and animation system based on computer visualization [J], Machinary Design & Manufacture, 2001; (1): 24-25

Cheng Chonggong, Hu Guanyi, Dai Juan, The research of programme method for kinematics analysis based linkage group [J], Journal of Changsha University, 2003; 17 (4): 27-30

Yan Wen, C++ graphic program design — C++ interface and graphic program instance [M], the first edition, Beijing: Science Press, 1995; 177-192

Ouyang Zhen, Di Ruikun, Qin Feng, Research and exploitation of NC simulation control system based on OpenGL [J], Machine Tool & Hydraulics, 2004; (5): 66-68

Yu Bin, Liu Rongzhong, Research on simulation system for NC manufacturing based on OpenGL [J], Journal of Sichuan University (Engineering Science Edition), 2001; 33 (5): 16-19

Zhang Wei, Li Yibing, Hu Yuanzhi, Construction of interactive virtual 3D accident scene based on OpenGL [J], Journal of Highway and Transportation Research and Development, 2003; 20 (2): 108-111

Zhu Qidan, Su Peng, Zhang Weiming, Application of the visual modeling method for 3D graph based on OpenGL [J], Techniques of Automation and Applications, 2000; 19 (1): 27-29

TOWARDS DEVELOPING A WEB-BASED GAP MANAGEMENT INFORMATION SYSTEM FOR CUCUMBER IN CHINA

Jianping Qian[1], Ming Li[2], Xinting Yang[1,*], Xuexin Liu[1], Jihua Wang[1]

[1] *National Engineering Research Center for Information Technology in Agriculture Beijing, 100097, P. R. China*

[2] *College of Information and Electrical Engineering China Agricultural University Beijing, 100083, P. R. China*

[*] *Corresponding author. Room 307, Beijing Agricultural Science Mansion, Building A, Banjing, Haidian district, Beijing, 100097, P. R. China, Tel: +86-10-51503476; Fax: +86-10 -51503476; Email address: yangxt@nercita.org.cn*

Abstract: The integrated management of Good Agricultural Practices (GAP) for cucumber (*Cucumis sativus* L.) plays a key role in guaranteeing the high quality and safety of cucumber production. This paper describes an attempt to develop a web-based GAP management information system for cucumber. Through applying the system methodology and mathematical modeling technique to analyzing the dynamic relationships between control point of GAP and produce process of cucumber, self-assessments algorithm was established. Based on the algorithm, Web-based GAP management information system for cucumber (Cucumber-GAPS) with the functions of data manage, criterion query, GAP self-check, model manage and so on was designed with the web structure of Browse/Server, and it has been applied successfully in China.

Keywords: cucumber; Good Agricultural Practices (GAP); self assessment; management information system

1. INTRODUCTION

With the requirement and emphasis on quality, safety and sustainability of farm produce, it has been placed new demands for the development and adoption of Quality Assurance Systems, such as Good Agricultural Practices

Qian, J., Li, M., Yang, X., Liu, X. and Wang, J., 2008, in IFIP International Federation for Information Processing, Volume 259; Computer and Computing Technologies in Agriculture, Vol. 2; Daoliang Li; (Boston: Springer), pp. 1205–1212.

(GAP) (Linus, 2003). GAP are defined as general practices to reduce microbial food safety hazards in the cultivation, harvesting, sorting, packing and storage operations for fresh fruits and vegetables, and should be developed in a stepwise manner based on the risk associated with individual fruits and vegetables and the scientific data available (De Roever, 1998; Pabrua, 1999). Countries over the world are recognizing its importance (Adriano et al, 2006). The Euro_Retailer Produce Working Group and Good Agricultural Practices (EUREPGAP) started as an initiative of the retailers in 1997 to harmonize GAP for all sources of supply; According to the Food and Drug Administration (FDA), GAP was an offshoot of the Food Safety Initiative/Action plan developed in 1998 (Howard and Gonalez, 2001).

In China, the Agro-Food Safety Program which is basically a guarantee system on standards, monitoring and certification was implemented in 2001. This program also controls measures for inputs and environmental protection. China's Good Agricultural Practices (CHINAGAP) was established in December 2005 and enforced in May 2006 (Wu and Liang, 2007). It includes 11 standards such as Terminology, all farm base control points and compliance criteria, crops base control points and compliance criteria, combinable crops control points and compliance criteria, fruit and vegetable control points and compliance criteria and so on (GB/T 20014.1-20014.11-2005).

Relative information systems serving for GAP are few according to literature review. Based on the principle of export vegetable safety production process controlling, Hazard Analyses Critical Control Points (HACCP) and Good Manufacturing Practice (GMP), the development of Intelligent Decision Support System for Export Vegetable Safety Production based on HACCP was reported (Shen and Huang 2005). The lack of such information represents a gap in our knowledge of GAP information systems, thus a web-based CHINAGAP Management Information System for Cucumbers in China has been designed.

2. ANALYSIS OF CUCUMBER PRODUCE PROCESS BASED ON GAP

GAP, are recommended, in order to sustain best practices for farming, i.e. soil and water management, crop production, storage, waste disposal etc. China has moved to benchmark national Good Agricultural Practices (CHINAGAP) with EUREPGAP (Nigel and Elmé, 2005). From figure 1, the cucumber produce process based on CHINAGAP included eight sections or blocks: (1) seed and nursery, (2) site selection, (3) crop production, (4) harvest and transport, (5) storage, washing and treatment, (6) environment, waste & pollution, operator health, safety & welfare, (7) documentation, traceability and quality system requirements, (8) general regulations.

Figure 1 shows the total produce process of cucumber is covered. The first linked to the farm in which the crop is grown, until the moment when the crop is harvested, after which the recording is linked to batches or lots and the produce handling site. The General Regulation which sets out the rules by which the standard will be administered. Control Points and Compliance Criteria Protocol (CPCC) including in (1)-(7) is the standard with which the farmer must comply, and which gives specific details on how the farmer complies with each of the scheme requirements.

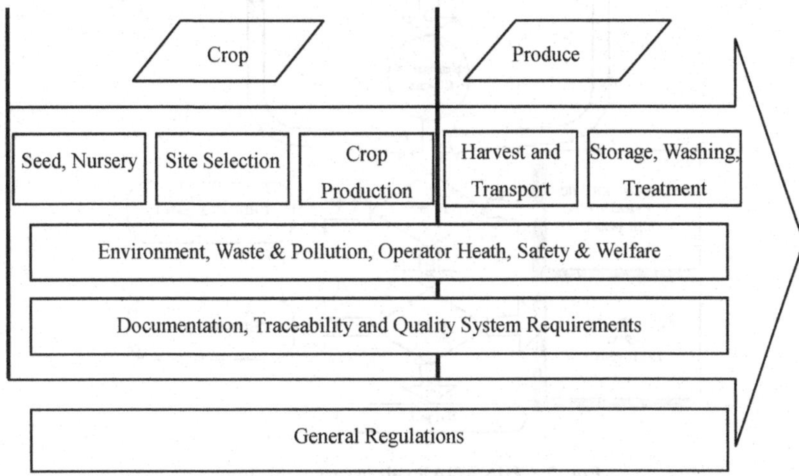

Fig. 1. Produce process based on CHINAGAP for cucumber

3. ESTABLISHMENT SELF-ASSESSMENTS ALGORITHM

The checklist used for inspection of producers, which contains all the control points and must be used during inspection by the Certification Bodies (CB). The checklist can also be used by the producer/group when performing the self-assessments.

Compliance with CHINAGAP Fruit and Vegetables consists of three types of control points, that the applicant is required to undertake in order to obtain CHINAGAP recognition; Grade 1 (MAJOR MUSTS). Grade 2 (MINOR MUSTS) and Grade 3 (RECOMMENDATIONS), and must be fulfilled as follow Fig. 2.

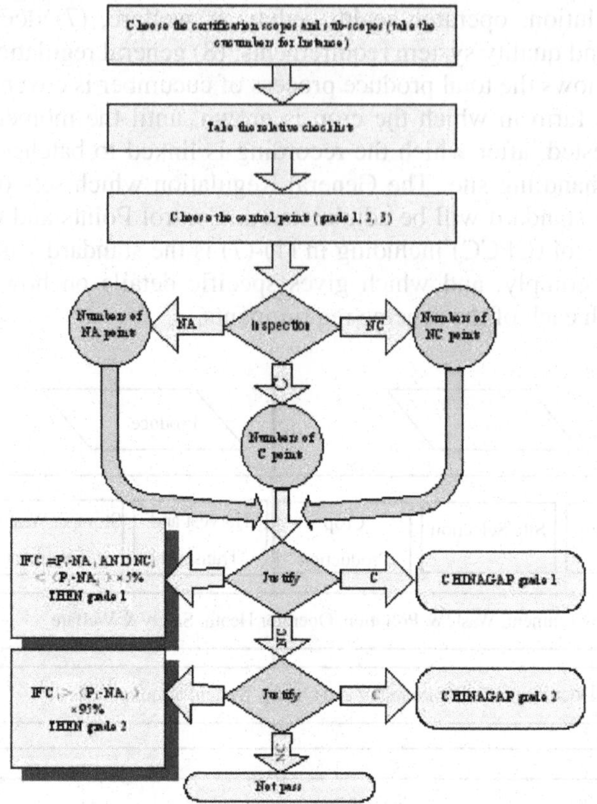

Fig. 2. CHINAGAP self assessment flow
NA: Not Applicable; NC: Not Compliance; C: Compliance

4. SYSTEM DESIGN

4.1 System Architecture

GAP Management Information System for Cucumber is composed of four layers: user browse layer, service layer, .Net framework and base layer (Fig. 3). The first layer is a standard browser, such as Internet Explorer or Netscape, can transfer request to business service layer, receive and analyze html document from server. The second layer is web server with Microsoft Internet Information Server 6.0. The third layer consists of business facade layer and business rules layer. The business façade layer is the separate layer which insulates user interface and varied business and the business rules layer includes login, data access and so on. The fourth layer consists of database layer, standard base and model layer including environment evaluation model, early warning model of cucumber disease, balanced fertilization model and self-assessments model.

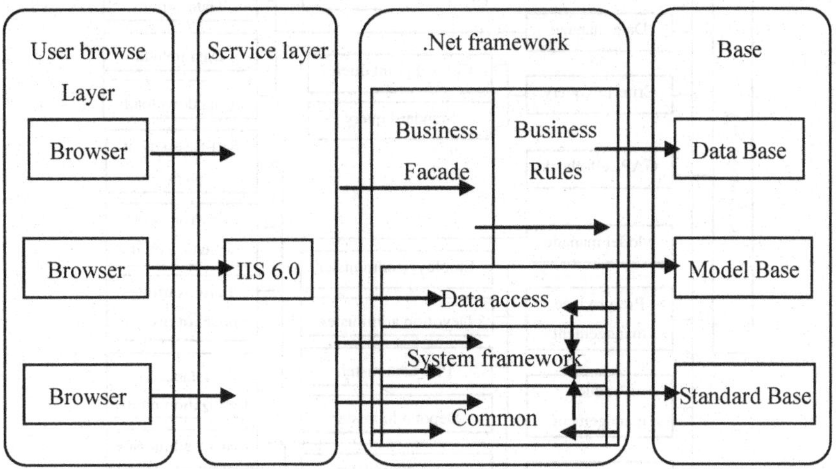

Fig. 3. Architecture of cucumber-GAPS

4.2 System Function

The system architecture (Fig. 4) is described below.

System Setting

Two functions of base setting and user setting is included. The former indicates enterprise basal information setting, data server address setting and so on. The latter indicates user information addition, user information edition and purview distribution.

Data Manage

Data import, data export, data upload and data download is provided in the system. The data of excel format and access format can be import and export. Data upload and download between server and customer is by web service.

Criterion Query

Criterion query consists of control point query including different phase of seed, plant, fertilize, plant prevent and standard query including Free-pollution Products, Green-Food and Organic-Food standard. Two query matters which are singleness query and assorted query are provided. When user inputs the query condition, the query result is showed.

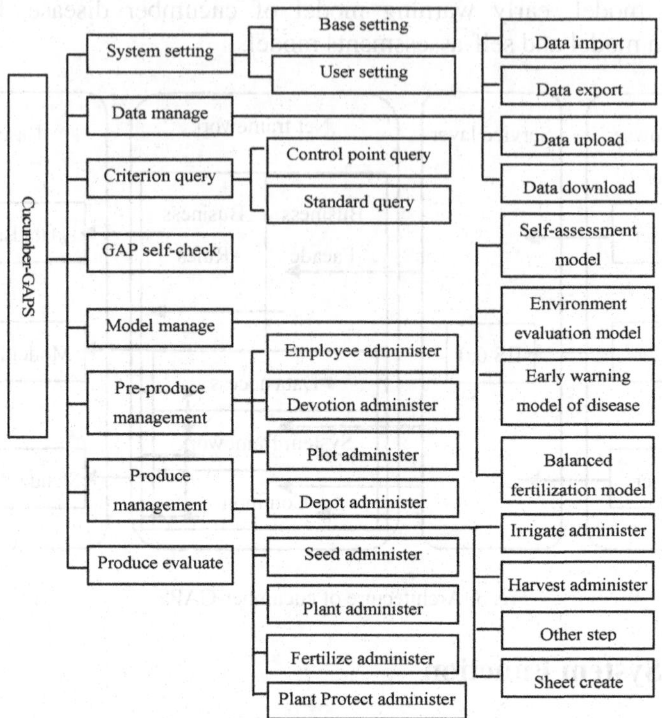

Fig. 4. Function of cucumber GAPS

GAP Self-check

The result whether the enterprise is accord with the criterion of GAP can be indicated according to self-check with GAP Management Information System for Cucumber. A check-list containing 95 questions attributed three grades was drawn up according to Good Agricultural practice (GB/T 20014.1-20014.11) and, distributed in ten sections or blocks: (1) varieties and rootstocks, (2) site history and site management, (3) soil and substrate management, (4) fertilizer use, (5) irrigation, (6) crop protection, (7) harvesting, (8) produce handling, (9) worker welfare and (10) environmental issues.

Model Manage

Four kinds of model are managed in the system. Self-assessment model is established on basis of self-assessment algorithm, and other models are based on the theory of risk analysis (Jacob, 2006). The system provided the management of risk assessment, risk management and risk communication.

Pre-produce Management

Employees administer, devotion administer, Plot administer and depot administer are provided in the system. The information of employee is added, delete and edit in the part of employee administer. The information of source, specification, amount and purchaser about seed, pesticide and fertilizer is managed in the part of devotion administer.

Produce Management

The crucial control points including seed, plant, fertilize, irrigation, plants protect and harvest are selected and supervised according to GAP in the system. For example, in harvest administer, when the date of harvest is input to the system, the hint whether exceeds the safety interval schedule is showed. The producer can adjust the date of harvest in order to produce safety farm products according to the hint. The function of GAP form table print is provided in this part.

Produce Evaluate

Executive conditions of varied control point are evaluated via comparing actual produce records with GAP criterions. Improved measures are afforded for the sake of advance the produce safety.

5. CONCLUSION

GAP is used widely as a management tool for ensuring farm produce safety, but an exercisable software system for GAP is reported rarely. In this paper, by applying the system methodology and mathematical modeling technique to analyzing the dynamic relationships between control point of GAP and produce process of cucumber, self-assessments algorithm was established. Based on the algorithm, Cucumber-GAPS with the functions of data manage, criterion query, GAP Self-check, model manage and so on were designed with the web structure of Browse/Server.

ACKNOWLEDGEMENTS

It is an outcome of Hi-Tech Research and Development Program of China (No: 20060110Z2002) and National Key Technology R&D Program (No: 2006BAD10A08).

REFERENCES

Adriano G C, Sergio A C, Maria C A. Good agricultural practices in a Brazilian produces plant. Food Control, 2006, 17: 781-788

De R C. Microbiological safety evaluations and recommendations on fresh produce. Food Control, 1998, 9: 321-347

GB/T 20014. 1-20014.11-2005. Good agricultural practice (in Chinese)

Howard L R, Gonzalez A R. Food safety and produce operation: what is the future? Hortscience, 2001, 36: 33-39

Jacob M. Risk analysis-an essential modern approach to food safety. Environmental Policy and Practice, 1996, 6: 91-96

Linus U O. Traceability in agriculture and food supply chain: a review of basic concepts, technological implications, and future prospects. Food, Agriculture & Environment, 2003, l.1 (1):101-106

Nigel G, Elmé C. Options for the development of National/Sub-regional Codes of Good Agricultural Practice for Horticultural Products Benchmarked to EuropGAP. Http://www.unctad.org/trade_env/test1/meetings/eurepgap/EurepGAP_benchmarking_UNCTAD_November-NG.pdf. September 2005

Pabrua F. Good agricultural practices: methods to minimize microbial risks. Diary, Food and Environmental Sanitation, 1999, 19(7): 523-526

Shen G R, Huang D F. Intelligent decision support system of export vegetable safety production based on HACCP. Agriculture Network Information, 2005, 11:18-20 (in Chinese with English abstract)

Wu R, Liang Y. GAP management system and software development for Chinese Crude Drugs. Beijing: Chinese Medical and Pharmaceutical Science Press. 2007, p. 29 (in Chinese)

MAIZE PRODUCTION EMULATION SYSTEM BASED ON COOPERATIVE MODELS

Shijuan Li[1], Yeping Zhu[1,*]

[1] *Agricultural Information Institute, Chinese Academy of Agricultural Sciences, Beijing, China, 100081*

[*] *Corresponding author, Address: Library 311, Agricultural Information Institute, Chinese Academy of Agricultural Sciences, No. 12 Zhongguancun South Street, Beijing, 100081, P. R. China, Tel: +86-10-68919652-2342, Fax: +86-10-68919886-2339, Email: zhuyp@mail.caas.net.cn*

Abstract: Based on the maize_ecophysiological characteristics, the maize developping process-based cooperative models including growth model,_developmental phase models, water balance model and nitrogen balance model etc. was built combined with the basic data such as variety characteristics, weather data, soil level and cultivation management with the technology support of system engineering method, crop simulation and computer. On the basis of cooperative models, this paper further constructed Maize Production Emulation System (MPES) with several additional functions such as determining variety characteristic parameters, deciding the planting design, simulating maize phenology stages and production features, warning of the nitrogen leaching in advance, simulating the water and nitrogen deficit degree and maize growth three-dimensional display. The system reproduces the maize production process in digital form. MPES was test through actual experiment, and the results verified its strong mechanism and prediction performance as well as its universal adaptation.

Keywords: maize; cooperative models; simulation; emulation system

1. INTRODUCTION

Beginning in 1960s, crop simulation model had been developed from infantility to maturation, from experiential form to mechanism explanation, from theory model to practical application. As the core of modern

Li, S. and Zhu, Y., 2008, in IFIP International Federation for Information Processing, Volume 259; Computer and Computing Technologies in Agriculture, Vol. 2; Daoliang Li; (Boston: Springer), pp. 1213–1221.

agricultural production management and agricultural resource optimizing management, it is the foundation of precision agriculture and information agriculture implemented now in china. In America, Holand, England and Australia, It has been developed different advanced crop simulation system, respectively. Among them DSSAT developed by American contains many famous crop model, such as CERES, GROPGRO etc; APSIM developed by Australian integrates deferent models into a public platform, and can simulate crop, vegetable and weed production systems; SUCROS developed by Dutch comprises the practical models, MACROS, BACROS and WOFOST. Furthermore, the EPIC can be used to study effects of the climate change on agriculture. Though these models have been developed into powerful system, only the minority of them were applied to agricultural production successfully. At present, the basic role of crop simulation model is to predict yield, and it is also applied to research on world food supplies, ecology agriculture, regional yield prediction, soil resource management, effect evaluation on agriculture of environment and social economic etc.

With the development of information technology and the increase of social requirement, many countries have diversified their crop growth model, for example adding new technology (GIS, GPS, RS) or other models to improve its generality, veracity and operation. InfoCrop model evolved from several models could simulate the effects of weather, soils, agronomic management and major pests on crop growth, yield, soil carbon, nitrogen and water, and greenhouse gas emissions (Aggarwal et al., 2005). The Dynamic North Florida Dairy Farm Model (DyNoFlo) incorporated livestock model, manure N mode and crop model (Cabrera et al., 2006). The old edition for ORYZA were integrated to one rice model ORYZA2000, which simulates the growth and development of rice under conditions of potential production and water and nitrogen limitations (Bouman et al., 2006). A crop-parasitic weed interactions model was designed and implemented under the framework of the Agricultural Production Systems Simulator (APSIM), and became the first module to reflect biology treat in APSIM (Grenz et al., 2006). At the same time, many concerned researches appeared in China. Nanjing agricultural university developed WebGIS—based system for agricultural spatial information management and assistant decision-making, which combined knowledge model with GIS and realized the functions as cropping system evaluation, ecological distribution of farm products, estimation of cropping potential production etc (Liu Xiaojun et al., 2006). According to the studies on WebGIS and ES, National Engineering Research Center for Information Technology in Agriculture constructed Crop Management Geography Information System including database, model library and repository (Li Xiang et al., 2003).

On the basis of past researches (Yan Dingchun, 2005) and the deep analysis on maize growth and development discipline, the author applied system engineering, computer technology and model method to design maize cooperative models and built Maize Production Emulation System (MPES). It realizes the digital management for maize production, and will establish foundation for digital agriculture-based agricultural informatization.

2. SYSTEM DESIGN AND STRUCTURE

2.1 Development environment and system design

Using programming language Visual C++ and Visual Basic 6.0 to design system interface, MPES was developed on the operation system of Windows 2000 server. Database was built in Access 2000, and forms and graphs output was carried out by Teechart control unit. The models had been integrated according to COM standard by programming language Visual C++.

Cooperative models groupwares compose the main content and central work of the MPES, and the groupwares include maize growth model, development model, water balance model, N balance model, maize quality model, economic analysis model and pest forecast model, furthermore, it include variety parameter test module and weather creation module. Basic parameters database saves variety information, soil data and agronomic management. Weather data day by day are kept in weather database. The simulation and predicting results of models were transported into result database. Interface functions implement the data transfer between models and databases.

2.2 System structure

Based on the past studies, we collected related literatures and agronomic expert information in a large scale, then designed maize cooperative models in accordance with maize growth and development discipline, and combined the models with corresponding database and repository, and constitute Maize Production Emulation System (MPES). The system consists of cooperative models, database and interface etc. Next its components will be elaborated.

Following figure 1 shows the detail system structure.

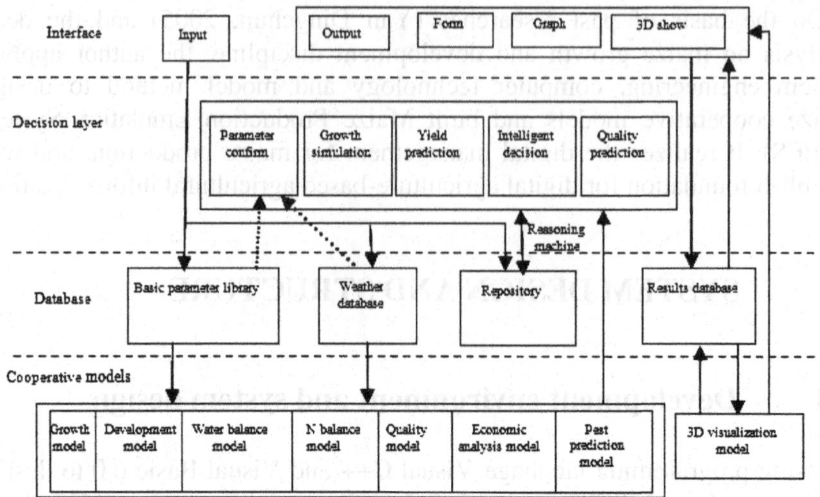

Figure 1. The system structure for maize production emulation system

2.2.1 Databases

Including basic parameter database, weather database, repository and results database.

Basic parameter database consists of the information about location, fertilizer type, fertilizer management, irrigation management, cultivation variety, phenology data, soil texture and soil parameter, among which phenology data, as actual experiment data, was used to determine the genetic characteristic parameters for certain variety; soil parameters contain former crop remains information, water content before sowing, and organic matter percent, volume weight, NH_4-N, NO_3-N and PH of every soil layer.

The datum of daily highest temperature, lowest temperature, rain and solar radiation or sunlight time input by user are saved in weather database. The qualitatively technology knowledge about maize cultivation can not be analyzed quantitatively through models. Thereby, they are expressed with production rule and concluded into repository, which solves the practical problems by means of reasoning machine according to the input and aim of users. Results database stores the simulation results from cooperative models and decision layer, and the tables are set on the basis of the output menus.

2.2.2 Cooperative models

Based on the past modeling experiences, agronomic knowledge and experimental data, deep analysis on maize growth and development discipline, and applying system engineering, computer technology and model method, we quantified maize growth and its effective factors to several models which realized valid cooperation. The following gives details of the main cooperative models, respectively.

(1) Maize growth model

It computes LAI, Light Interception, photosynthesis, and dry matter production and distribution in maize, and calculates N uptake and distribution, described by N content and percent for plant and every organ. Firstly, daily potential accumulation of dry matter is estimated according to photosynthesis effective radiation. Secondly, considered the effects of temperature, water and N on photosynthesis, Daily actual accumulation of dry matter is calculated. In every development stage, maize plants has different growth center and distribution rule of dry matter. For example, at grain filling stage, photosynthate was allocated priority to grain, which is the growth center. If intraday photosynthesis matter is not enough for grain intraday increase, all the newly photosynthate was transported to grain, and temporary dry matter stored in stem was reallocated to grain to meet the demand of grain growth.

(2) Maize development model

This model divided maize development into 8 stages: from sowing to germination, from germination to seeding emergence, from seeding emergence to juvenile stage, from juvenile stage to jointing stage, from jointing stage to silking stage, from silking stage to beginning of grain filling, from beginning of grain filling to physiological maturity. Temperature, water, photoperiod and genetic parameter restrict the replacement of development stages. Genetic parameter can be input by user or decided automatically by parameter determination program in system.

(3) Water balance model

Based on water movement rules, soil water status and maize absorption characteristic, Water balance model simulates field potential and actual evapotranspiration. The model considered soil from surface to two meters depth, and divided it into 10 layers. Based on weather data and basic parameter input by user, after counting the soil water restricting parameters, model calculates water leakage, runoff, soil evaporation, plant transpiration and root water absorption. Priestly-Taylor equation and SCS Curve Number Method are adopted to simulate potential evapotranspiration and runoff, respectively. Water deficit index calculated by this model affects directly daily accumulated value of maize dry matter and increases amount in LAI. Model contains a water-saving module which offers irrigation management on the basis of irrigation target set by user.

(4) N balance model

This model mainly simulates the processes of N mineralization and fixation, N losses and uptake by maize. Corresponding with the soil layers in water balance model, it counts and renews N content of every layer in terms of the water movement up and down, and the N deficit index calculated by the model affects directly daily accumulated value of maize dry matter and LAI. Firstly, the mineralization amount of organic matter in soil is calculated, and N (NH_4-N and NO_3-N) released from fertilizer should be added at the data of fertilized. Secondly, model computes nitrification and denitrification, N deficit index, N uptake amount by maize plant and organs (leaf, stem, ear and grain). Finally the amount of nitrate leached to the second meters depth soil layer with water movement is counted, and evaluated possible effects of the leached nitrate on groundwater.

(5) Grain quality model

Combining past research achievement or results and the relative literatures collected in a large scale, we analyzed the effects on main maize quality (protein, starch and fat) of variety trait, weather, cultivation management and nutrition, and got the algorithm with Logistic equation to draw up the relation between quality and factors such as density, days after grain-filling, water and nutrition, then built maize quality model, which integrated with maize growth model very well.

(6) Economic analysis model

This model offers optimal variety, sowing date and density when user choose a target variety such as variety with highest yield, largest ratio of output to input, highest net income and lowest input. It can calculate annual total input, output and net income to offer decision-making support for user in terms of the relative input.

(7) Pest prediction model

With the increasing study on crop simulation model, more and more researcher attach importance to pest prediction. Some information system may be used to pest investigation and prediction in America and Europe. Holland and Germany developed special wheat plant protection decision-making system EPIPRE and PRO-PLANT (Forrer, H.R. et al, 1992; Frahm, J. et al, 1993). MPES considered the major maize pest-maize borer, and use nerve network prediction method to built model. After study training according to effect factors set by user, and integrating weather data and plant density, the model can simulate the occurrence law of maize borer.

(8) 3D visualization model

3D visualization model mainly simulates the maize growth based on agronomic shape knowledge, image and 3D animation technology, and the simulation data in results database. It makes it easy to observe and predict maize growth, development and yield formation directly. The methods such as rigid body kinematics and flexible system were applied to model construction in order to realize the functions such as zoom, growing longer, growth, horizontal movement, rotation, color change and total shape change.

2.2.3 Decision layer

Applying the above maize cooperative models and repository, decision layer implements maize growth simulation, yield and yield components prediction, maize quality prediction and intelligent decision-making. In addition, genetic parameter determination module creates the characteristics needed by maize development model automatically on the basis of basic parameter database and weather data.

2.2.4 System interface

User operates system or models on the interface, which includes data input and results output. System has friendly and beautiful interface, and it's easy to be operated and understood. The parameter input unit reduced the amount of parameters that is uneasy to get for user. Simulation results are showed with form and graph, and system can show the daily data of more than 40 indexes. 3D visualization show simulates the shape changes of seed, root, stem, leaf, spike and female flowers from sowing to maturity, and the corresponding content for soil water, NO3-N and NH4-N are displayed on the right screen. Fig. 2 is one picture for 3D maize visualization show.

3. FUNCTIONS OF MPES AND DEMONSTRATION ANALYSIS

MPES simulates maize production of different varieties under different location and yield levels with one day as time step. System has the following functions: 1) It simulates yield and yield components, main quality formation, dynamic change of soil water and N, water and N uptake and utilization by maize, N leaching, and shows the simulation results with form and chart directly. 2) It gives anticipative target yield and quality, variety choose, sowing date and density determination, fertilizer and water management according to user's requirement and biological environment of decision location. Based on simulation model, system analyzes maize variety, water and nutrition status, weather resource and offers assistant decision-making, such as the suggestions on optimal variety, sowing date and density, the amount and time of irrigation and fertilizer. 3) Maize 3D visualization shows maize growth vividly, and it is the real reflect of simulation results from cooperative models.

Using actual experiment data under deferent water conditions in 1996 and 1997 in Hebei province, we calibrated and validated Maize Production Emulation System based on cooperative models. Results indicated the system has good prediction performance and strong applicability. The RMES of yield and biomass predicted by system were 426.3kg/hm-2, 477.5 kg/hm-2 and 1029.8 kg/hm-2, 1356.0 kg/hm-2 respectively. The relative RMES were 6.78%, 5.55% and 7.60%, 7.50%, all less than 10%. Fig. 3 and Fig. 4 show the correlation between prediction value and observation value for yield and biomass.

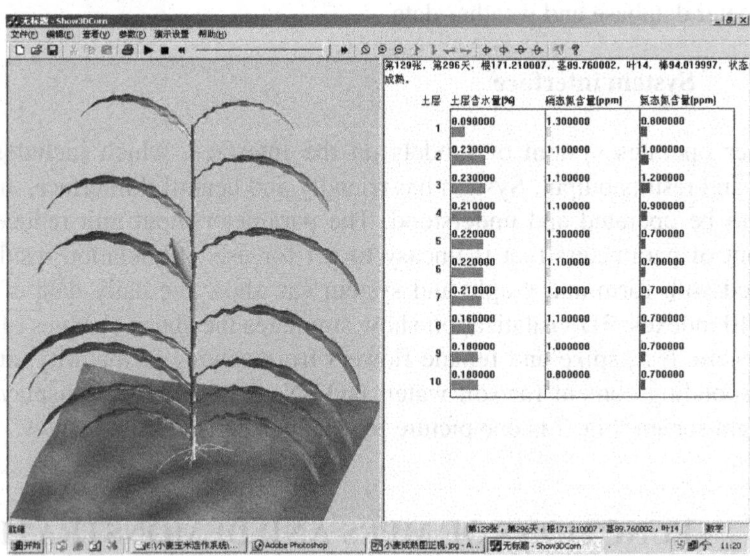

Figure 2. Maize 3D visualization show

ACKNOWLEDGEMENTS

This research was supported by Digital Agriculture Program of State High-tech Research and Development Project of China (No. 20060110Z2059), and by National Scientific and Technical Supporting Programs Funded by Ministry of Science and Technology of China (2006BAD10A06).

REFERENCES

Aggarwal, P.K., et al. InfoCrop: a dynamic simulation model for theassessment of crop yields, losses due to pests, and environmental impact of agro-ecosystems in tropicalenvironments. I. Model description. Agr. Syst. 2005

Bouman, B.A.M., et al. Description and evaluation of the rice growth model ORYZA2000 under nitrogen-limited conditions, Agricultural Systems, 2006, 87: 249-273

Forrer, H.R., Experiences with the cereal disease forecast system EPIPRE in Switzerland and prospects for the use of diagnostics to monitor the disease state. Brighton Crop Protection Conference, Pests and Diseases, 1992, (2): 711-720

Frahm, J., et al. PRO_PLANT - a computer-based decision-support system for cereal disease control. Bulletin-OEPP. 1993, 23(4): 685-69

Grenz, J.H., et al. Simulating crop–parasitic weed interactions using APSIM: Model evaluation and application, Europ. J. Agronomy 2006, 24: 257-267

Li Xiang, et al. Study on the Crop M anagement Inform ation System Based on W ebGIS and ES, Acta Agriculturae Boreali-Sinica, 2003, 18(2): 106-109 (in Chinese)

Liu Xiaojun, et al. WebGIS-based system for agricultural spatial information management and aided decision-making, Transactions of the Chinese Society of Agricultural Engineering, 2006, 22(5): 125-129 (in Chinese)

Victor, E., et al. An integrated North Florida dairy farm model to reduce environmental impacts under seasonal climate variability, Agriculture, Ecosystems and Environment, 2006, 113: 82-97

Yan Dingchun, et al. A Knowledge Model- and Growth Model-based Management System For Corn. Proceeding of the Third International Symposium on Intelligent Information Technology in Agriculture (ISIITA), 2005, pp. 14-16, Beijing, China.

THE AGRICULTURE APPLICATION PROSPECT OF IGIS

Xiaoyang Cui[1], Kaimeng Sun[1], Dingcun Yan[1], Yeping Zhu[1,*]
[1] *Agricultural Information Institute, Chinese Academy of Agricultural Sciences, Beijing, China, 100081*
* *Corresponding author, Address: Library 311, Agricultural Information Institute, Chinese Academy of Agricultural Sciences, No. 12 Zhongguancun South Street, Beijing, 100081, P. R. China, Tel: +86-10-68919652-2342, Fax: +86-10-68975172, Email: zhuyp@mail.caas.net.cn*

Abstract: The article expatiates the conception, definition and application status of GIS and IGIS and discusses the two systems to form agriculture application IGIS combined with three GIS development tools including Arcinfo, MapInfo and SuperMap.

Keywords: GIS, IGIS, SuperMap

1. INTRODUCTION

We discuss GIS, AI and IGIS as followings:

1.1 Geography Information System (GIS)

1.1.1 Conception

It is a computer application software system with those strong functions to collect, store, edit, manage, describe and analyze space data. Because of it's powerful management and analysis of space data and drawing ability, It is used widely to the research and application of description and expression of space elements and correlative relations.

Cui, X., Sun, K., Yan, D. and Zhu, Y., 2008, in IFIP International Federation for Information Processing, Volume 259; Computer and Computing Technologies in Agriculture, Vol. 2; Daoliang Li; (Boston: Springer), pp. 1223–1230.

GIS is the integration of computer science, geography, measurement science and cartology. It is difficult to give a definite definition. Because of it's wide coverage, it can be given different definitions from different angels. It can usually definite GIS from four different ways. First we definite from function, GIS is a system to collect, store, check, handle, analyze and show geography data. Second we definite from application, we can divide GIS into different application system according to the different GIS application fields such as soil information system, city information system, layout information system and space decision support system etc. (Fan Hong, 2002) Third we can definite from toolbox, GIS is a collection of tools to collect, store, query, change and show space data. The definition emphasizes the tool to handle geography data provided by GIS. Finally we can definite from database, GIS is such a database system whose data has space order and provide an operation collection to deal with those data to answer the query of space entity in database Fig. 1.

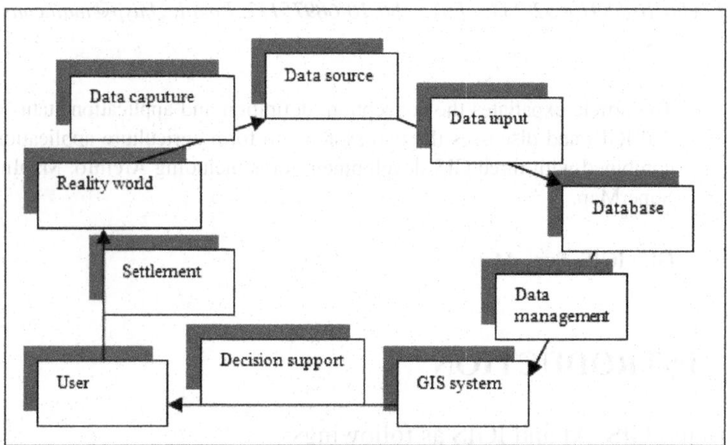

Fig. 1. GIS function structure diagram

1.1.2 Application trend

The current application trends of GIS represent is as following aspects (Lv Xin, 2002):

(1) Integration
(2) Intelligence
(3) Modularization
(4) Network
(5) Subassembly
(6) Standardization

1.2 AI (Artificial Intelligence)

1.2.1 Definition

It is an important research field of computer science. Professor Nilsson of AI research center of Stanford University thinks: "AI is a science about knowledge – a science about how to express knowledge and how to obtain and use knowledge". (Sun Chengming, 2004) Professor Winston of MIT University indicates: "AI is to research how to use computer to do the intelligent work which can be done only by man in the past". Those definitions reflect the basic thoughts and scale of AI study.

1.2.2 Applications

Currently, the main application fields of AI include:

(1) Expert System. The system stimulates the procedure to solve a sophisticated problem which can be solved only by mankind experts.
(2) Repository. To store, process and manage the mankind's knowledge and to dispose and apply the knowledge according to the need.
(3) Decision support system. To process assistant decision of mankind's activity by means of stimulation and reasoning through model and knowledge (Hu Cun, 1998).
(4) Natural language comprehension. To comprehend the mankind's nature language to realize the direct communication between mankind and computer to promote the broad application of computer.
(5) Intelligent robot. It is the robot which has feeling, identification and policy decision function.
(6) Mode recognition. To stimulate the functions of hearing and seeing of mankind to identify the sound, image, scene and character.
(7) Auto program design. To implement the automation of program design with the computer to achieve the program's validation and integration.

1.3 Intelligent Geography Information System (IGIS)

IGIS should be studied firstly from space analysis intelligence, artificial neural network and inheritance arithmetic put a doable way to space information intelligence.

1.3.1 The status quo of IGIS

Currently, GIS is used successfully to the fields including resource management, establishment management, city and district layout, population and commerce management, traffic and transportation. The traditional GIS have strong data input, storage, search and display ability. With the day and day going deep into those fields's application, we encounter many things of experiment of some determinative questions to make traditional functions of GIS unable to satisfy the demand; this requires combining GIS with the very popular AI to exert each merit to solve the problems perfectly. The nineties last century, many domestic and abroad scholars begin to study the IGIS and obtain many successful applications. The southern Africa scholar Veldic Vado and Han City university Chulmin Jun apply independently to the city's ground address selection with the integration of Expert System and Geography Information System and study the integration method of GIS and ES, GIS and math model. Professor David Lanter of California University studied particularly the insufficiency existing in GIS, and import AI into GIS to reinforce GIS to make it more practical and flesh. The scholar Leonid Stoimenov of Jugoslavia develops the alarm prediction system based on repository and GIS (Wu Xincai, 2000).

Moreover, there are many researches applying to the intelligent study of GIS and this makes many productions. Domestically, those studies are just in primary stage and many scholars publish some articles to discuss the integration problem of AI and GIS. In a case of electrify network distribution, transportation and trouble diagnose, the Orient Electron co., ltd combined the Expert System and Geography Information System to apply. Zhao-Shi peng of Northeast Normal University applies the integration technology of GIS and ES in the evaluation of mud and stone flow. Wang-Shi Chen, Ma-Sheng Zhong of Jilin University have applied to combine the GIS and ES to predict the mine production and developed a set of unique theory for many years.

We can conclude from above statement that IGIS research has a full application coverage and vast prospect research field. But it's research is still in primary stage Fig. 2.

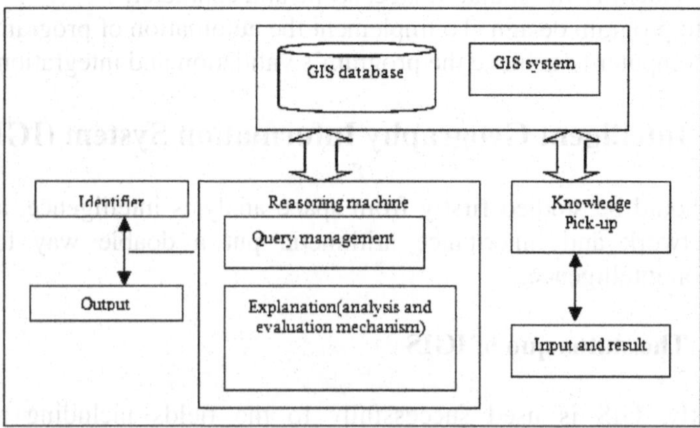

Fig. 2. A IGIS system prototype

2. ARCINFO, MAPINFO AND SUPERMAP

2.1 Arcinfo

It is developed by American ESRI (American Environment System Research Institute), It is a professional GIS (geography information system) platform software with abundant function and includes the following diversified and advanced functions:

(1) The function to input and edit data. We can acquire data from digital instrument, graph scanner, graph transportation to edit graph and property.

(2) Data transportation and integration. We can transform those standard data format and support the relation database which accords with SQL standard.

(3) Basic GIS functions. Map projection and projection transformation, data maintenance and management, cushion and fold analysis.

(4) Space data and property query to show corresponding graph including grid graph display and management.

(5) Geography data management. We can manage a large distributed database using info database or ArcSDE.□

(6) It provides the interface design tool and the system second development tool. It uses aml language, MO module base and VC, VB which support industry standard as main development tools.

(7) Data output. It provides functions to make high quality digital map and report forms.

(8) It supports edition management and business transaction.

2.2 MapInfo

It is desktop geography information software of American MapInfo Corporation, It is a desktop solution scheme to visualize data and map information.

2.3 Supermap

SuperMap GIS is a new generation large geography information system platform developed by Beijing SuperMap Geography Information Technology co., ltd depending on the technology advantage of CAS and based on technology innovation to fulfill different kind of customer's need of all trades. SuperMap GIS 5.2 series productions are the newest edition of

SuperMap GIS. It supports the incorporation of storage structure, usage and management of space data and statistics subject, and supports the application and development of multi layers structure based on B/S structure (Xun Fanlun, 2002).

2.3.1 Database engine and module structure

(1) SuperMap SDX+ 5.2 —Space database engine which supports huge space data management.
(2) SuperMap Objects 5.2 —Full module GIS development platform suitable to large professional application system.
(3) SuperMap IS .NET 5.2 —Internet GIS development platform used to build large network GIS service.
(4) eSuperMap 5.2 —Embeded GIS development platform suitable to moving terminal device.
(5) SuperMap Deskpro 5.2 —Popular desktop GIS software used to geography space data disposal and analysis model.
(6) SuperMap Express 5.2 —Desktop GIS software used to edit and handle geography space data.
(7) SuperMap Viewer 5.2 —Tool software used to browse geography space data(Zhao Peng, 2003).

3. AGRICULTURE APPLICATION OF IGIS

There are those agriculture applications combined with the aforesaid IGIS secondly developed by GIS development platform at present:

(1) Wild moving data collection system eFieldSurvey based on the palm computer.
(2) Agriculture climate district-division geography information system.
(3) Heilongjiang forest fire information and fire loss evaluation system.
(4) Fangshan forest industry basic space database management system.
(5) Forest fireproofing and resource geography information system.
(6) Extract agriculture farm geography information system.

I will give a brief introduction to the exact agriculture farm geography information system:

The system uses IGIS, GPS and intelligent agriculture technology to adjust and control agriculture production information management and farm devotion to increase benefit, avoid resource waste and reduce environment pollution. The farm geography information system of the Geography Institute of CAS is developed cooperatively by the Geography Institute of

CAS and Beijing SuperMap Geography Information Technology co., ltd. The system adopted full module GIS software SuperMap Objects as development platform using its characters easy to integrate with other systems and development advantage. Combining farm GIS and farm ration computation module system, it provide a strong tool for quantificational study farm ecology system and the improvement of farm unit production (Smith T R, 2004).

(7) Corn ecology which is suitable to district division and, responding corn seed selection.

(8) Ascertain of the match-series scheme suitable to every cooked area.

(9) Prediction of the yield.

(10) Forecast the agriculture nature disaster.

4. SUMMARIZATION

GIS is developed rapidly, It provides a good basic platform for our agriculture science and research worker to develop agriculture intelligent geography information system (IGIS), The article gives a detailed discussion to the GIS's concept, definition and development trend and discusses the agriculture application of GIS, In our country, though the research and application make some progress, it is basically in the primary stage. I wish the article have an active function to promote the research, development, popularization and application of IGIS to a deep domain (Robinson V B, 2003).

ACKNOWLEDGEMENTS

The article is sponsored by the Center-level Commonweal-Quality Science and Research Academy and Institute Basic Science and Research Operation Fee Special Item Capital Item-the Country Science and Technology Support Item (No. 2006BAD10A06).

REFERENCES

2004. Weather Publishing Company, 53-66

Fan hong, Zhan Xiaoguo. 2002. Arcinfo application and development

http://www.supermap.cn/gb/application/0ssgl/JTGPS/minhangjj/

Hu Cun. 1998. GIS's application in agriculture information management and district-division. Tianjin Agriculture Science, 4, 34-37

Lv Xin. 2002. Geography information system and its application in agriculture

M. Wooldridge, N. Jennings, and D. Kinny. 2000. The Gaia Methodology for Agent-Oriented Analysis and Design. Journal of Autonomous Agents and Multi-Agent Systems, 3(3).

Robinson V B. 2003. Expert systems and geographical information system. Review and Prospects, Journal of Surveying Engineering, 6, 92-95

Smith T R. 2004. A knowledge-based geographical information system. International Journal of Geographical Information Systems, 1, 116-118

Sun Chengming. 2004. The agriculture application and evolvement of geography information system [J], Shanghai Agriculture Transaction, 3, 18-21, technology (emended edition). Wuhan University Publishing Company, 97-113

The brief introduction of civil aviation economy information GIS application system.

Wu Xincai, Guo Linlin. 2000. The development status and prospect of geography information system. Computer Engineering and Application, 4, 8-9

Xun Fanlun. 2002. Artifical intelligence and computer's application in agriculture modernization. Agriculture Modernization Research, 5,123-126

Zhao Peng. 2003. Intelligent geography information system's development and research. Computer Development, 2, 37-40

DEVELOP WHEAT GROWTH MODEL MULTI-AGENT SYSTEM WITH MASE METHOD

Shengping Liu[1], Yue E[1], Yeping Zhu[1,*]

[1] *Agricultural Information Institute, Chinese Academy of Agricultural Sciences, Beijing, China, 100081*

[*] *Corresponding author, Address: Library 311, Agricultural Information Institute, Chinese Academy of Agricultural Sciences, No. 12 Zhongguancun South Street, Beijing, 100081, P. R. China, Tel: +86-10-68919652-2342, Fax: +86-10-68919886-2339, Email: zhuyp@mail.caas.net.cn*

Abstract: The advent of multi-agent systems has brought together many disciplines and given us a new way to look at intelligent, distributed systems. However, traditional ways of thinking about and designing software do not fit the multi-agent paradigm. Based on agent theory, technology research and multi-agent environment, this paper describes agent oriented programming software development method – MaSE and analyzed wheat growth model. Through requirement analysis, system design and system modeling, this paper builds manage agent, fertilize agent, irrigate agent, weather agent, soil agent, growth agent and simulation agent and develops wheat growth multi-agent system.

Keywords: MaSE, Multi-Agent System, Wheat Growth Model, AOP, Agent Cooperation

1. INTRODUCTION

Agent and multi-agent system technology have given us a new way to look at distributed systems and provided a path to more robust intelligent applications (Wooldridge et al., 2000). The advent of multi-agent system has brought together many disciplines in an effort to build distributed, intelligent, and robust applications. In our research, we combine with wheat growth model technology, agent technology, distributed computing technology, artificial intelligence technology, database technology and MAS

Liu, S., E, Y. and Zhu, Y., 2008, in IFIP International Federation for Information Processing, Volume 259; Computer and Computing Technologies in Agriculture, Vol. 2; Daoliang Li; (Boston: Springer), pp. 1231–1242.

modeling technology. We have been developing wheat growth model multi-agent system with multi-agent system engineering method and multi-agent environment platform.

In this research, we view agents as a specialization objects. Agents coordinate their actions via conversations to accomplish individual and community goals instead of objects whose methods that are invoked directly by other objects. At the same time, agents can communicate each other with message mode through MAGE platform. In this way, we can design and build intelligent and non-intelligent agents with the same framework. Actually, wheat growth model multi-system includes 10 independent agents, which can offer different function and accomplish wheat growth simulation via MAGE platform.

This paper is structured as follows. We begin by describing the wheat growth model simulation, MaSE in Sections 2 and 3. Section 4 and 5 presents how wheat growth model multi-agent system should be modeled and realized. Section 6 concludes and presents some ongoing work.

2. WHEAT GROWTH MODEL SIMULATION

Plant growth models with various degrees of complexity have been elaborated for the use of plant growers and breeders, ranging from simple statistical regression models to models with a complicated mechanism, capable of simulating anything from a simple process to the behavior of the whole plant (Harnos, 2006). In addition to experimental research, many scientists also use simulation models to gain a better understanding and description of environmental stress effects, such as high temperature, drought stress, and their effects on the development and yield of plants grown at elevated atmospheric CO_2 concentration.

In recent decades more than 70 wheat models have been published internationally, including AFRCWHEAT2, Ceres-Wheat and SIRIUS. This paper uses Ceres-Wheat, which can be used to simulate the growth and yield of wheat under different environments. The model simulates the effects of variation in weather, crop genotypes, soil properties and crop management practices. The simulation of growth and yield is based on the quantification of phasic development; photosynthesis; respiration, morphogenesis; growth; biomass accumulation and partitioning; extension growth of leaves, stem, roots and grain; soil water extraction and plant nitrogen status. However, the effects of insect pests, diseases and natural calamities, such as wind and hailstorm damage, are not accounted for in this model.

3. MULTI-AGENT SYSTEMS ENGINEERING (MASE) METHOD

The Multi-agent Systems Engineering (MaSE) is a general purpose methodology for developing multi-agent systems that is founded on the basis software engineering principles. MaSE divides the development process into two major phases: the analysis phase and the design phase. The analysis phase consists of the following stages: capturing goals, applying use cases, and refining behavior. The design phase consists of the following stages: creating agent classes, constructing agent communication, assembling agent classes, and system design (Scoot, 2001). Fig. 1 presents the development process proposed by MaSE.

The purpose of the analysis phase is to provide a set of roles whose tasks meet the system requirements, i.e., specifying what the system should do. According to MaSE, the analysis phase consists of the following stages: capturing goals, applying use cases, and refining behavior. In capturing goals stage the system goals are being elaborated specified from the system point of view and not from the user point of view (DeLoach, 2006). In applying use cases stage the system use cases are being specified. In refining behavior stage the system functional decomposition is determined.

The purpose of the design phase is to specify the way the system-to-be should behave and be constructed. That means, specifying how the system will achieve its goals. The design phase consists of the following stages: creating agent classes, constructing agent communication, assembling agent classes,

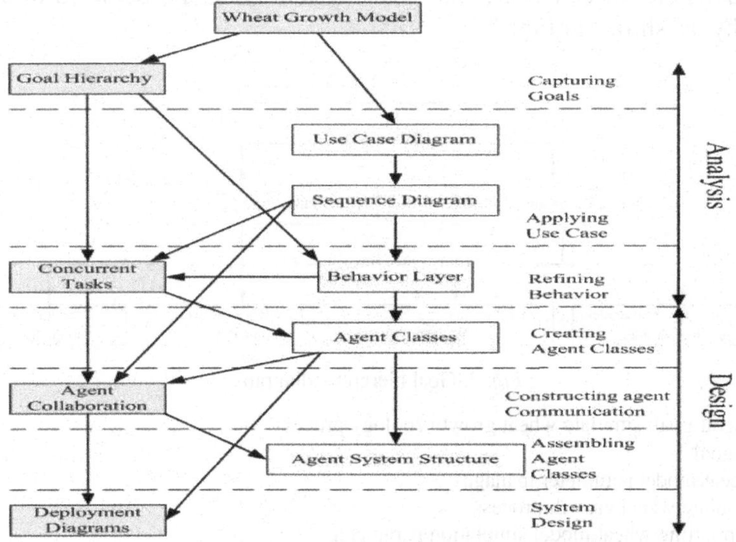

Fig. 1. MaSE developments stages

and system design. In creating agent classes' stage, the overall multi-agent system architecture in terms of agent and the conversations among them is determined. In constructing agent communication stage, the designer defines the coordination protocols between agent couples. In Assembling Agent stage the internal architecture of the agents is being specified. System design stage is aim at depicting the physical system architecture and the distribution of the various agent classes' instances within that architecture.

4. MAS ANALYSIS AND DESIGN

By using MaSE method and MAGE it is possible to develop wheat growth model MAS. In this paper we analyze and design system with six steps. The steps presented here are: capturing goals, applying use cases, refining roles and behaviors, creating agent classes, assembling agent, system deployment.

4.1 Capturing Goals

The first step is capturing goals, in MaSE, a goal is an abstraction of a set of functional requirements. The stage of capturing goals comprises two sub-stages: identifying the goals and structuring them in a hierarchy, in terms of goal and their sub-goals is being constructed.

Goals are identified by distilling the essence of the set of requirements. Wheat growth model multi-agent system includes one general goal and two sub-goals. Each goal will be associated with behaviors and agent classes that are responsible for satisfying that goal. System goals are depicted in a goal hierarchy, as shown in Fig. 2.

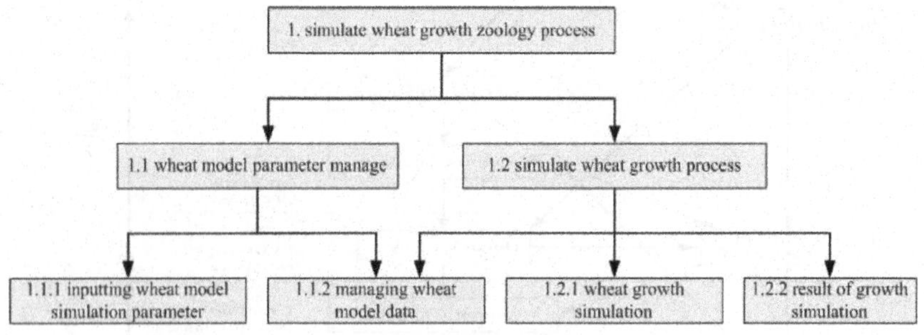

Fig. 2. Goal hierarchy diagram

- General goal: simulate wheat growth zoology process
- Sub-goals:

1.1 wheat model parameter manage
1.2 simulate wheat growth process
1.1.1 Inputting wheat model simulation parameter
1.1.2 Managing wheat model data
1.2.1 Wheat growth simulation
1.2.2 Result of growth simulation

4.2 Applying Use Cases

Applying use cases is divided into two sub-stages: the creation of use cases and the creation of the sequence diagrams. A use case is a set of interactions and describes the general system behavior. The transformation from the use cases specification to sequence diagrams is straightforward; each entity becomes a role and information passing becomes an event.

Table 1. List of system use cases

Number	Use Cases
1	Wheat Parameter Input
2	Wheat Model Simulate
3	Simulate Result
4	Model Data Manage

Use cases are drawn from the system requirements and described in Table 1. To help determine the actual communications required within a multi-agent system, the use cases are restructured as sequence diagrams, as shown as in Fig. 3 and Fig. 4.

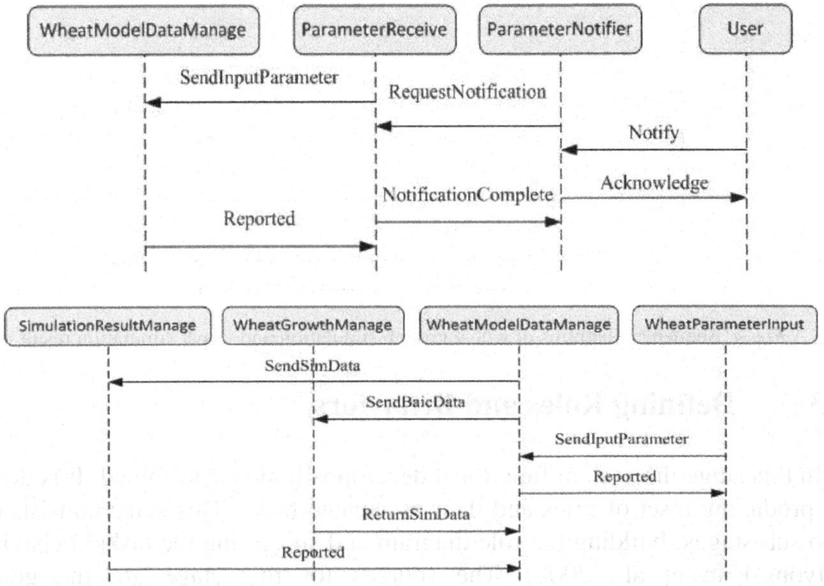

Fig. 3. Sequence diagrams of wheat model simulation parameter and wheat model data manage

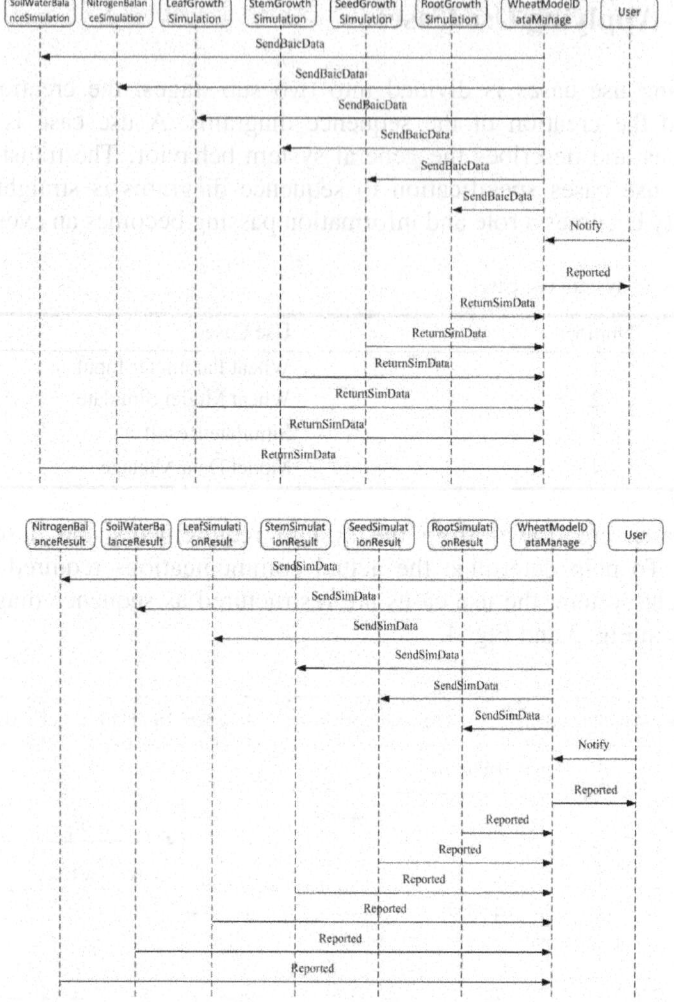

Fig. 4. Sequence diagrams of wheat growth simulation and wheat simulation result

4.3 Defining Roles and Behaviors

In this stage the system functional decomposition is determined. It is done by producing a set of roles and their associated tasks. This stage consists of two sub-stages: building the role diagram and specifying the tasks' behavior (Myong-Hun et al., 2006). The sources for that stage are the goals determined in the first stage and the sequence diagrams created in the second stage. A role can be derived from the roles determined during the sequence diagram creation or can be formed directly from the goals hierarchy. In addition, each role should have tasks that realize its goals. The task model in MaSE are specified using a finite state automaton and is called system role model diagram. Fig. 5 illustrates system role model diagram that shows systems' roles and behaviors.

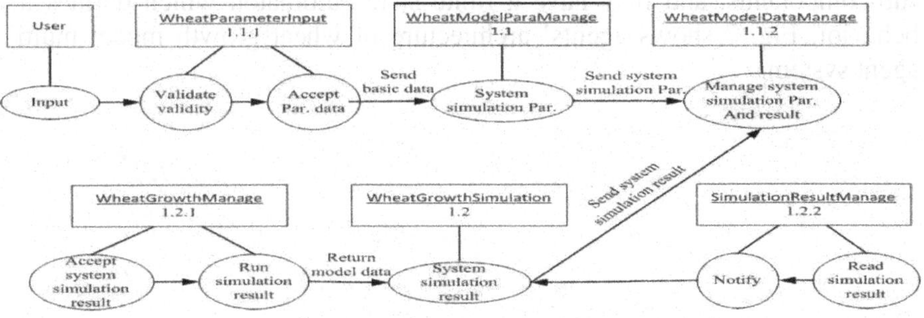

Fig. 5. System roles model

4.4 Creating Agent Classes

In this stage, the overall multi-agent system architecture in terms of agent and the conversations among them is determined. Agents classes are created from the roles define in analysis phase by assigning roles to agents (Bellifemine et al., 2006). Each agent is associated with at least one role. In addition to the agent classes, the conversations among them are also specified utilizing the protocols defined in the analysis phase.

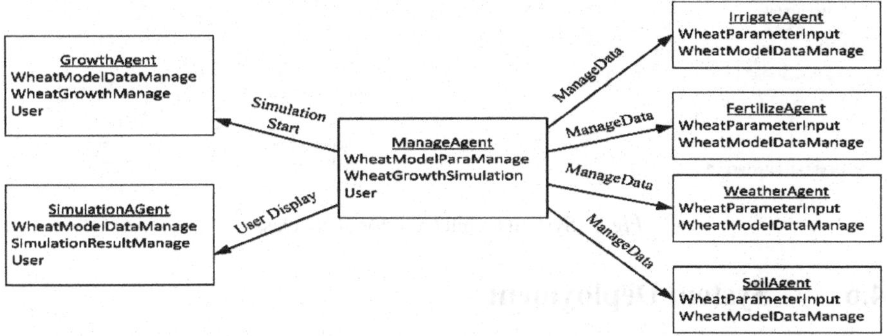

Fig. 6. System agent classes diagram

The Wheat growth model MAS mainly defines 7 agent classes: GrowthAgent, SimulationAgent, ManageAgent, IrrigateAgent, FertilizeAgent, WeatherAgent, SoilAgent, as show in Fig. 6.

4.5 Assembling Agent

In this stage the internal architecture of the agents is being specified. We can use its own architecture to build an agent (e.g., BDI) or convert the tasks from the previous stage into components. The agent architecture consists of the components and the relationships among them (Collis et al., 2006). These components can be specified recursively, i.e., a component may have sub-components, and may have a finite state automaton which defines its behavior. Fig. 7 shows agents' architecture of wheat growth model multi-agent system.

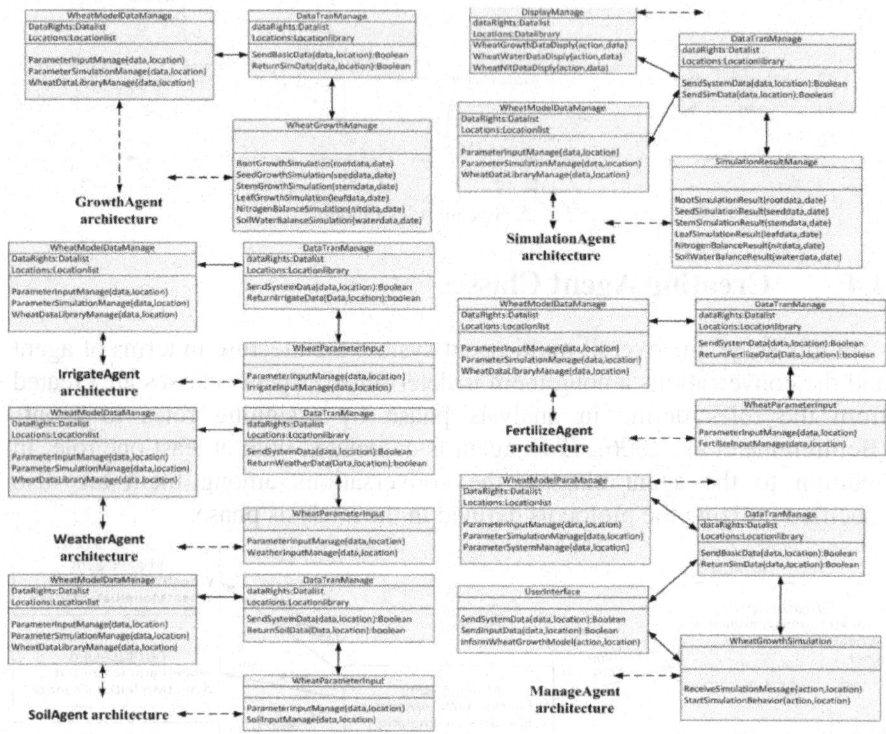

Fig. 7. System architecture for agent class

4.6 System Deployment

This stage is aim at depicting the physical system architecture and the distribution of the various agent classes' instances within that architecture. The agents in a Deployment Diagram are actual instances of agent classes from the Agent Class Diagram. Since the lines between agents indicate communications paths, they are derived from the conversations defined in the Agent Class Diagram as well.

System deployment diagram includes 10 agents, thereinto, Weather Agent, Fertilize Agent, Irrigate Agent and Soil Agent run in the same physical node. Manage Agent, Irrigate Agent and Soil Agent run in the same physical node, as shown in Fig. 8.

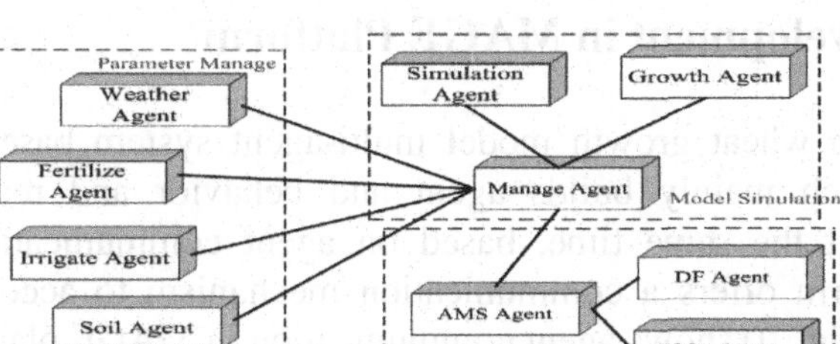

Fig. 8. System deployment diagram

5. WHEAT GROWTH MODEL MAS REALIZATION

5.1 System Framework

Based on the analysis and design above, we obtain the following holistic framework of multi-agent system. The framework is composed of four layers: GUI Layer, Application Service Layer, Agent Platform Layer and Resource Layer. In GUI Layer, users input data for the model, feed back operation result of simulation model, and manage the model operation via GUI Layer. Application Service Layer mainly provides functions such as system data application service and simulative operation of wheat growth model, as well as interacts with agents in GUI Layer. Agent Platform Layer is an agent through which MAGE platform manage multi-agent system, and consist of agent management system agent, directory service agent and agent that manages message transfer. Resource Layer is made up of some information contents about wheat model, contents related to database and several parts of the growth model, supplying service for system operation. The four layers above together constitute the framework of wheat growth model MAS.

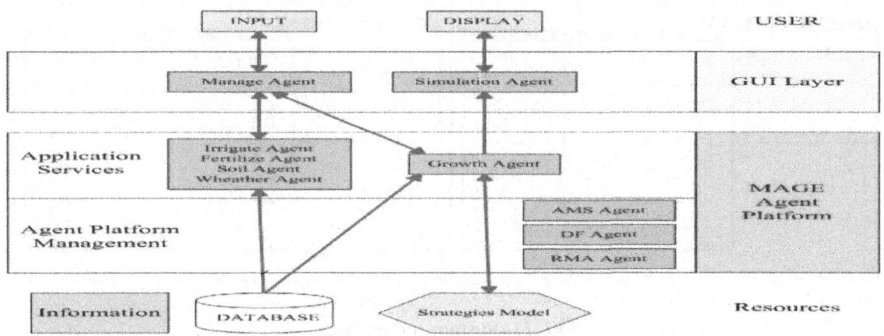

Fig. 9. System framework diagram

5.2 Development in MAGE Platform

We develop wheat growth model multi-agent system based on MAGE platform, which mainly builds agent and behavior and realizes system framework. At the same time, based on agent communication language, MAGE platform offers a communication mechanism to accomplish agent cooperation. Fig. 10 shows agent communication in MAGE platform.

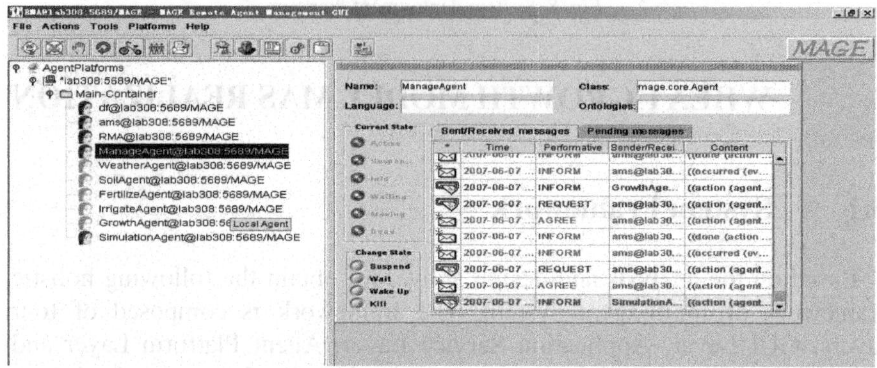

Fig. 10. Agent communication in MAGE

This paper we develop wheat growth model multi-agent system in terms of MAGE, which includes requirement analyzing, system designing, agent generating and system implementing.

5.3 System Realization

Fig. 11 shows some picture of final system interface, which realizes wheat growth model from Data Input, Run Model and Result Display.

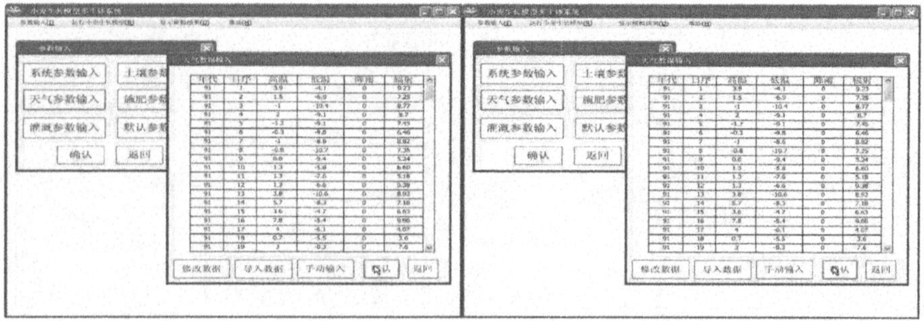

Fig. 11. System screenshot

6. CONCLUSIONS AND FUTURE WORKS

The Multi-agent Systems Engineering methodology is a seven-step process that guides a designer in transforming a set of requirements into a successively more concrete sequence of models. By analyzing the system as a set of roles and tasks, a system designer is naturally lead to the definition of autonomous, pro-active agents that coordinate their actions to solve the overall system goals.

From our research on MaSE and MAGE, we have learned some lessons.

(1) Based on the Research and analysis of the mechanism of the wheat growth model, and used for reference of the advanced research achievements at home and abroad, this paper constructs a wheat growth model, in which photosynthesis as the core, dry matter accumulation and wheat morphological development as the main content.

(2) This paper develops the wheat growth model using Agent oriented software method, analyzes the requirement of wheat growth multi-agent system, designs system object hierarchical structure diagram, use case diagram, sequence diagram, system role model, agent class, system agent class, agent class architecture, system deployment diagram, system framework diagram, and then realizes the system development and construction. The study carries out a new develop method for exploding similar crop model.

(3) This paper builds agent for every wheat grow phase, uses MAGE as developing environment, and makes up Manage Agent, Fertilize Agent, Irrigate Agent, Weather Agent, Soil Agent, Growth Agent and Simulation Agent for wheat growth model that can be reused., then realizes the cooperate working mechanism between agents and completed the wheat growth model process, using message transmission platform of MAGE and Agent Communication Language.

ACKNOWLEDGEMENTS

This research was supported by Digital Agriculture Program of State High-tech Research and Development Project of China (No. 20060110Z2059), and by National Scientific and Technical Supporting Programs Funded by Ministry of Science and Technology of China (2006BAD10A12).

REFERENCES

DeLoach, S.A. 2006. Engineering Organization-based Multiagent Systems. LNCS Vol. 3914, Springer, 109-125.

F. Bellifemine, A. Poggi, and G. Rimassa. 2001. Developing Multi-Agent Systems with JADE. In C. Castelfranchi and Y. Lespérance, editors, Intelligent Agents VII. Agent Theories, Architectures, and Languages – 7th. International Workshop, ATAL-2000, Boston, MA, USA, July 7-9, 2000, Proceedings, Lecture Notes in Artificial Intelligence. Springer-Verlag, Berlin.

F. Bousquet, et al. 2002. Multi-agent systems and role games: collective learning processes for ecosystem management. In Complexity and ecosystem management: The theory and practice of multi-agent systems, pages 248-285. Edward Elgar.

J. C. Collis, D. T. Ndumu, H. S. Nwana and L. C. Lee. 1998. The ZEUS agent building toolkit. BT Technol. J. Vol. 16 No. 3 July.

M. Wooldridge, N. Jennings and D. Kinny. 2000. The Gaia Methodology for Agent-Oriented Analysis and Design. Journal of Autonomous Agents and Multi-Agent Systems, 3(3).

Myong-Hun Chang and Joseph E Harrington Jr. 2006. Agent-based models of organizations. In Handbook of Computational Economics II.

N. Harnos. 2006. Applicability of the AFRCWHEAT2 wheat growth simulation model in Hungary. APPLIED ECOLOGY AND ENVIRONMENTAL RESEARCH 4(2): 55-61.

Scott A. DeLoach and Mark Wood. 2001. Developing Multiagent Systems with agentTool. Intelligent Agents VII. Agent Theories, Architectures, and Languages – 7th International Workshop, ATAL-2000, Boston, MA, USA, July 7-9, 2000, Proceedings, Lecture Notes in Artificial Intelligence. Springer-Verlag, Berlin.

WATER QUALITY MANAGEMENT IN INTENSIVE AQUACULTURE IN CHINA

Chengxian Yu [1,2], Bin Xing [1,2], Liying Xu [1,2], Daoliang Li [1,2,*]

[1] *College of Information and Electrical Engineering, China Agricultural University, Beijing, China, 100083*

[2] *Key Laboratory of Modern Precision Agriculture System Integration, Ministry of Education, Beijing, China, 100083*

[*] *Corresponding author, Address: P.O. Box 121, EU-China Center for Information & Communication Technologies, China Agricultural University, 17 Tsinghua East Road, Beijing, 100083, P. R. China, Tel: +86-10-72736717, Fax: +86-10-62737679, Email: li_daoliang@yahoo.com*

Abstract: Intensive aquaculture is adopted by more and more fish farms in China. This paper focuses on the typical water monitoring and controlling methods in intensive fish culture in Chinese fish farms. The paper illustrates the main factors in fish culturing and the way to control them: dissolved oxygen (DO), temperature, PH, particulate matter and ammonium.

Keywords: intensive aquaculture, fish culture, water quality control

1. INTRODUCTION

The history of aquaculture in China goes back at least 2500 years, but from the year 1978 when the open-policy and economic reformations have been adopted, China's aquaculture entered a new era. Then the aquaculture in China developed quickly, and the total output of aquaculture in China reached 5.102×10^7 tons in 2005, 70% and even higher share of total aquatic production in China (National Bureau of Statistics, 2005).

The traditional poly-culture and integrated fish farming has been well known in the world, but it has been challenged by the culture of high value species. For example, land based farming, such as tank culture for flounder

Yu, C., Xing, B., Xu, L. and Li, D., 2008, in IFIP International Federation for Information Processing, Volume 259; Computer and Computing Technologies in Agriculture, Vol. 2; Daoliang Li; (Boston: Springer), pp. 1243–1252.

is popularized in northeast China. The culture of high value fish is a rapidly expanding field, both in research and industrial aquaculture. A driving force behind this growth is the inherently high value placed upon the products in the marketplace.

The growth of aquaculture has been much contributed by the scientific advancement, such as hatchery techniques, water quality control, breeding technologies, and disease control. Intensive fish farming has a high production per unit area, but on the other hand, large fish farms generate large amounts of particulate organic waste, as well as soluble-inorganic excretory waste (Ackefors and Enell, 1994). This will not only lead to the crack of water quality but also cause seriously environment problems if not properly managed. For this reason, an increased awareness of the consequences of interaction between intensive fish farming and the environment has emerged (Iwama, 1991). To satisfactorily manage the scale of enrichment from fish farming and to ensure that the ecological change does not exceed predetermined levels, monitoring and recycling of water should be regarded as part of a large management framework.

Typical intensive fish culture is conducted in recycling aquaculture systems (RAS). These systems have practical applications in commercial aquaculture hatcheries, in water quality monitoring, in water quality control of fish tanks, and in fish breeding. Water is typically filtered and recycled when there is a specific need to minimize water replacement, to maintain water quality conditions which differ from the supply water, or to compensate for an insufficient water supply. RAS for holding and growing fish have been used by fisheries researchers for more than three decades in western countries but only recent years in China. These systems link the fish culturing tank to the controller of the RAS, which provides for removal of particulate matter, biological filtration, buffering of pH, temperature control, DO control, and water management. As such, "dirty" water flows out of the fish culturing tank, through the RAS for cleaning and sterilization, and returns "clean" to the fish culturing tank as needed. The use of RAS technology enhances bio-security and increases environmental and hydrodynamic control, maximizing production survival and system reliability.

In a typical RAS, there are many water quality variables important for different kind of fishes, but in this paper, only some water quality variables important for most fishes are described. They are removal of particulate matter, biological filtration, buffering of pH, temperature control, DO control, and water management. These processes can be achieved by several separate simple composite units and by several interconnected components, more details will be demonstrated in this paper.

2. MATERIALS AND METHOD

2.1 DO control

DO is by far the most important chemical parameter in aquaculture. If DO levels in a water body drop too low, fish and other organisms will not be able to survive. Different aquatic plants may impact on DO levels in various ranges.

Low-dissolved oxygen levels are responsible for more fish kills. Fish require oxygen for respiration like humans. The amount of oxygen consumed by the fish is a function of its size, feeding rate, activity level, temperature, and the type of the fish. Small fish may consume more oxygen than large fish because of their higher metabolic rate. Each 18oF increase in temperature will cause double metabolic rate of the fish, with oxygen requirement much higher.

The amount of oxygen that can be dissolved in water decreases at higher temperatures (Table 1), but decreases with increases in altitudes and salinities. In combining this relationship of decreased solubility with increasing temperatures, it can be seen why oxygen depletion are so common in the summer when higher water temperatures occur.

Table 1. Oxygen dissolved in water at different temperatures

Temperature (°C)	DO (salinity 32)	
	mL/L	mg/L
0	8.21	11.74
5	7.23	10.34
10	6.44	9.21
15	5.79	8.28
20	5.26	7.52
25	4.81	6.88
30	4.43	6.33

In conventional fish culture, when summer comes, it may be necessary to supply supplemental aeration to maintain adequate levels of dissolved oxygen. Whereas in recirculation systems the farmer must supply 100 percent of the oxygen needed for the fish and beneficial nitrifying bacteria.

To obtain good growth, fish must be cultured at optimum levels of dissolved oxygen. A good rule of thumb is to maintain DO levels at saturation or at least 4mg/L. If dissolved oxygen levels become less than 4mg/L, it can place undue stress on fish; and if DO levels drop to less than 0.6mg/L it can result in death (Knowledge of aquaculture).

Fish are not the only consumers of oxygen in aquaculture systems; bacteria and zooplankton consume large quantities of oxygen as well.

Decomposition of organic materials (algae, bacteria, and fish wastes) is the single greatest consumer of oxygen in aquaculture systems (Figure 1). Problems encountered from water recycling systems usually stem from excessive ammonia production in fish wastes. Consumption of oxygen by nitrifying bacteria that break down toxic ammonia to non-toxic forms depends on the amount of ammonia entering the system.

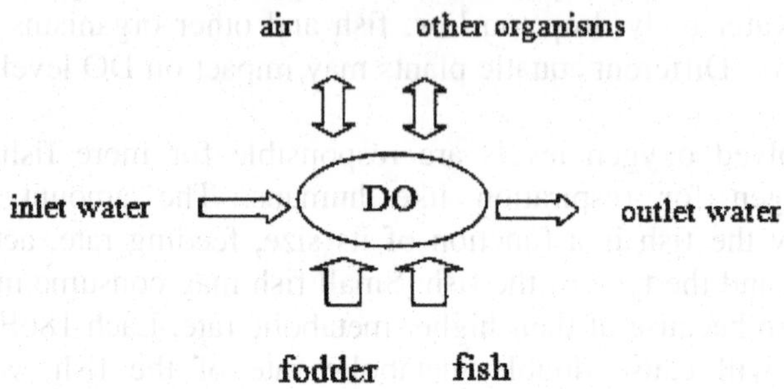

Figure 1. Factors to affect DO in water

Oxygen enters the water primarily through direct diffusion at the air-water interface and through plant photosynthesis. Because of the lack of photosynthesis in indoor water recycling systems, mechanical means of aeration is the only alternative for supplying oxygen to animals cultured in these systems. In RAS oxygen depletions can be calculated, but as predictions may be misleading, these systems should never be substituted for actual measurements. Several forms of mechanical aeration are available to the fish farmers. The general categories are:

Paddlewheels
Venturia pumps
Agitators
Liquid oxygen injection
Impellers
Airlift pumps
Air diffusers

Which aeration should be used depends on the actual environment of the fish farm.

2.2 Temperature control

After oxygen, water temperature may be the single most important factor affecting the welfare of fish. Fish are cold-blooded organisms and they keep approximately the same temperature as their surroundings. The activity, feeding, growth, and reproduction of all fishes are all affected by water temperature because metabolic rates in fish double for each 18oF rise in temperature.

Ideally, species selection of fish farms should be based in part on the temperature of the water supply. Because any attempt to match a fish with less than their ideal temperatures will involve energy expenditures for heating or cooling. This added expense will subsequently increase production costs, so fish farms from different areas usually select different species of fish for the reason of temperature.

The amount of dissolved gases (oxygen, carbon dioxide, nitrogen, etc.) is also determined by water temperature. When the water gets cooler, the gas becomes more soluble. And temperature plays a major role in the physical process called thermal stratification. Water has a high-heat capacity and the density qualities are not unique. In spring, water temperatures are nearly equal at all pond depths. As a result, dissolved gases, nutrients, and fish wastes are evenly mixed throughout the pond. As the days become warmer, the surface water becomes warmer and lighter while the cooler and denser water forms a layer underneath. Because of the different densities between the two layers of water, circulation of the colder bottom water is prevented. In natural fish tanks, dissolved oxygen levels will decrease in the bottom layer as photosynthesis and contact with the air is reduced. The already low oxygen levels are further reduced through decomposition of waste products, which settle to the pond bottom, so farmers use mechanical aeration to circulate the water.

Summer stratification is a greater problem for fish cultured in deeper farm ponds. Localized dissolved oxygen depletion poses a very real problem to fish farmers and stratification may last for several weeks. This condition may develop into a major fish kill when sudden summer rains occur. These rains will cool the warmer upper layer of water enough to allow it to mix with the oxygen poor layer below. Decomposing materials in the oxygen-poor layer are again mixed evenly throughout the pond, resulting in an overall reduction in the dissolved oxygen level. Fish previously able to avoid the oxygen depleted layer are now susceptible to low-dissolved oxygen syndrome and possibly death. To avoid this, there are usually three ways: mechanical aeration, temperature converter, mix warmer water with the tank water by circulating (Figure 2).

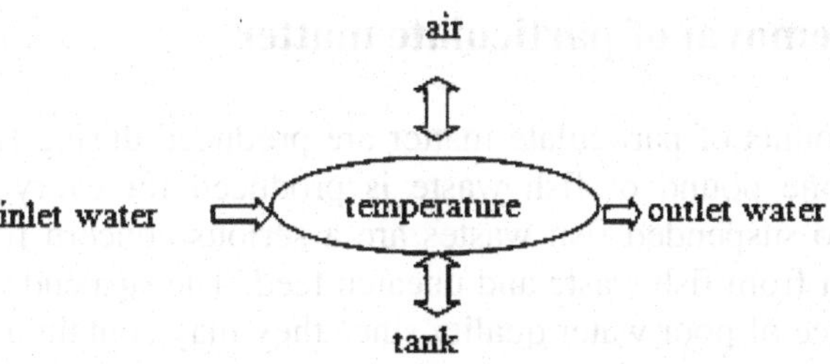

Figure 2. Factors to affect temperature of water

2.3 Buffering of pH

In intensive aquaculture ponds, fish metabolism and bacterial nitrification are the main formation of acids.

The quantity of hydrogen ions (H+) in water will determine if it is acidic or basic. A value of 7 is considered neutral and values below 7 are considered acidic; above 7, basic. Most fish can tolerate a pH range of 5-10; however, a range of 6.5-8.5 is preferred for most aquaculture species.

A buffering system to avoid wide swings in pH is essential in aquaculture. Without some means of storing carbon dioxide released from plant and animal respiration, pH levels may fluctuate in ponds from approximately 4-5 to over 10 during the day. Figure 3 shows the reason for the change of PH in culture tanks. In recycling systems constant fish respiration can raise carbon dioxide levels high enough to interfere with oxygen intake by fish, in addition to lowering the pH of the water. To replace lost alkalinity and sustain the buffering capacity of water, one of the most important substance is carbonate (CO_3^{2-}), in the form of limestone, bicarbonate of soda, or other common sources is added. Often, bio-filter media (oyster shell) or some other components of the system (concrete tanks) serve as a source of carbonates. Depending on the species cultured, frequent monitoring of water hardness, alkalinity and pH are required. To avoid fluctuate of pH levels, carbonate substance or recycling water is required.

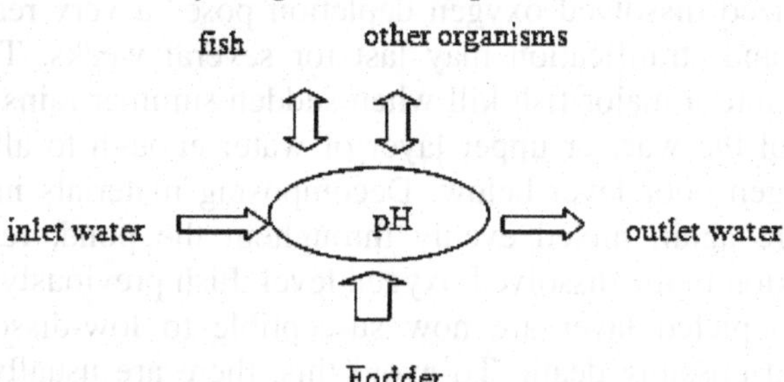

Figure 3. Factors to affect pH in water

2.4 Removal of particulate matter

Large amounts of particulate matter are produced during fish production. As a rule, one pound of fish waste is produced for every pound of fish produced. So suspended fish wastes are a serious concern for RAS. Solids mainly result from fish waste and uneaten feed. The suspended solids can be a major source of poor water quality since they may contain up to 70 percent of the nitrogen load in the system. As they not only irritate the fish's gills, cause several problems to the biological filter, and contribute a portion of the oxygen demand and toxic ammonia in RAS systems, but also promote the growth of bacteria that produces–rather than consumes ammonia, they must be concentrated for removal.

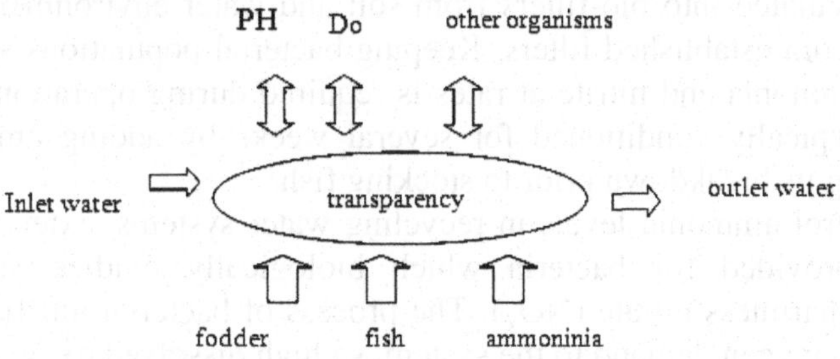

Figure 4. Factors to affect transparency in water

This can be accomplished in the following ways (Figure 4): by having the "dirty" water in settling basins for some time with reduced water turbulence; by medicine which can concentrate the particulate matter to deposits; by mechanical filtration through porous material such as sponge, sand or gravel. Solids that accumulate will gradually be mineralized (broken down) by bacterial action and their volume reduced. The most commonly used method is the settling basins and the mechanical filters. All of the methods require regular cleaning since they are prone to clogging when dirty.

2.5 Biological filtration

In RAS, there are many ways ammonia be produced. The most important way is that fish and other aquatic organisms release their nitrogenous wastes primarily as ammonia excreted across the gill membranes. Also, urine, solid wastes, and excess feed have undigested nitrogen fractions, which are additional sources of ammonia. Ammonia is toxic to fish and can result in poor growth and lower resistance to disease.

Fish excrete ammonia and lesser amounts of urea into water as wastes. Two forms of ammonia occur in aquaculture systems, ionized and un-ionized. The un-ionized form of ammonia (NH_3) is extremely toxic while the ionized form (NH_4^+) is not. Both forms are grouped together as "total ammonia" (Figure 5).

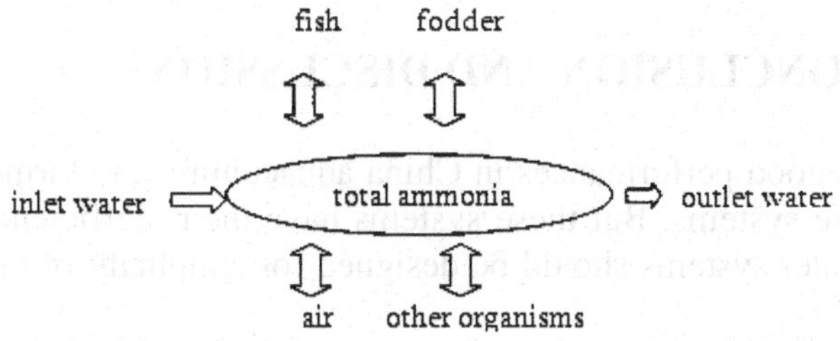

Figure 5. Factors to affect total ammonia in water

To reduce the amount of total ammonia, bio-filters which keep bacteria in them are usually used. The bacteria responsible for these reactions can be easily inoculated into bio-filters from soil and water environments, or with material from established filters. Keeping bacterial populations sufficient to remove ammonia and nitrite at rates is required during operation. So a bio-filter is typically conditioned for several weeks by adding ammonia and monitoring its breakdown prior to stocking fish.

To control ammonia levels in recycling water systems, extensive surface area is provided for bacteria which biologically oxidize ammonia to relatively harmless nitrate (NO_3^-). The process of bacterial nitrification adds additional oxygen demand to the system, so high dissolved oxygen levels are required. Bacterial nitrification is a two-stage process resulting first in the transformation of ammonia to nitrite (NO_2^-), then a further oxidation of nitrite to nitrate. Nitrite is also toxic to fish at low concentrations; hence, both reactions must occur for successful bio-filtration. Through biological processes, toxic ammonia can be degraded to harmless nitrates.

Media for bio-filters can be virtually any substrate which provides maximum surface area for bacterial growth: oyster shell, gravel, nylon netting, plastic rings, and sponge foam pads are among popular choice. In designing bio-filters, the principal concerns should be maximum surface area for bacterial growth, high dissolved oxygen levels, uniform water flow through the filter, sufficient void space to prevent clogging, and proper sizing to ensure adequate ammonia removal capability. As filter surface area, fish density, and water flow are all important considerations, required size of a bio-filter is difficult to predict.

Nitrification is typically carried out in a variety of systems, which can be grouped into six general types: submerged filters, trickling filters, reciprocating filters, rotating biological contactors (RBC), rotating drums, and fluidized bed reactors (Wheaton et al., 1991). A number of comparative studies of these systems have been conducted and RBCs were found to give the best performance with respect to specific ammonia removal efficiency (ammonia removal per surface area per time). However, due to operating problems with RBCs, often related to shaft and bearing failures, their commercial use has been limited.

3. CONCLUSION AND DISCUSSION

RAS have good performances in China aquaculture, and farmers benefit a lot from these systems. But these systems have their inefficiencies as well. Recycling water systems should be designed for simplicity of operation and economic feasibility. Sufficient time must be allowed for conditioning of the bio-filter prior to introducing fish. Ammonia and nitrite concentrations must be checked frequently. Dissolved oxygen should be sustained above 60% of saturation and periodically verified. Alkalinity, hardness, and pH need to be measured and adjusted, if necessary, at regular intervals. Filters should be inspected and cleaned as required. Frequent monitoring of the performance of the recycling system will allow the manager to improve and refine its operation over time. As the factors mentioned above, a new kind of RAS is required, which will automatically monitor these water quality parameters, auto control the equipments to adjust water quality parameters to the proper level. And all the parameter level will be computed in a model base on the knowledge of separate kind of fish (Figure 6). This will come true in three years in Fengze Company in Lijin, Dongying of China.

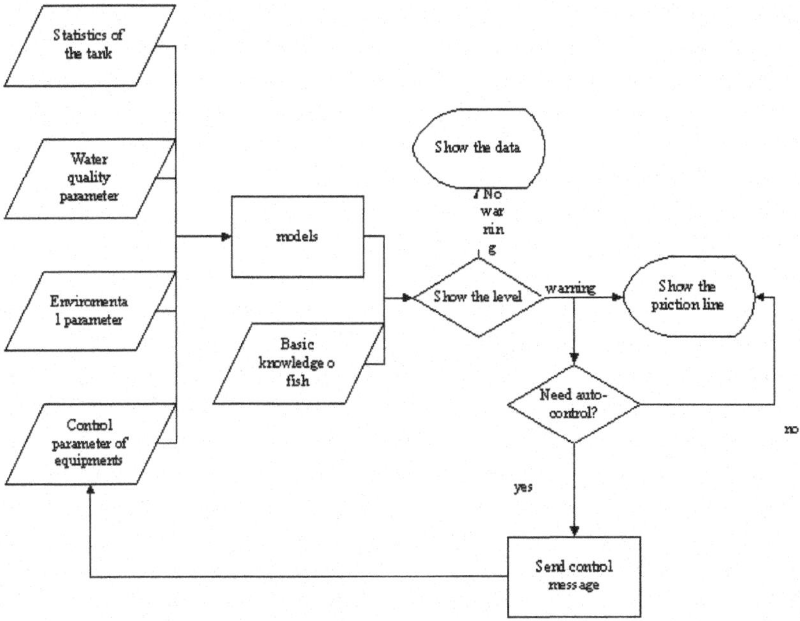

Figure 6. Structure of the system

ACKNOWLEDGEMENTS

We thank the members from Southern Regional Aquaculture Center for their data. And thank Fengze Company for their help of offering us the opportunity to do some research in their company.

REFERENCES

Ackefors, H., Enell, M., 1994. The release of nutrient and organic matter from aquaculture systems in Nordic countries. J. Appl. Ichthyol. 10, 225-241.

Iwama, G.K., 1991. Interactions between aquaculture and the environment. Crit. Rev. Environ, Contr. 21, 177-216.

National Bureau of Statistics, China Statistical Yearbook, China Statistics Press, 2005.

Knowledge of aquaculture http://www.instrument.com.cn/netshow/SH100886/C15717.htm

Wheaton, F., Hochheimer, J. and Kaiser, G.E., 1991. Fixed film nitrification filters for aquaculture. In: D.E. Brune and J.R. Tomasso (Editors), Aquaculture and Water Quality. World Aquacult. Sot., Baton Rouge, LA, pp. 272-303.

AN EFFICIENT SEGMENTATION METHOD FOR MILK SOMATIC CELL IMAGES

Heru Xue[1,*], Shuoshi Ma[2], Xichun Pei[1]
[1] *College of Computer and Information Engineering, Inner Mongolia Agricultural University, Hohhot, China, 010018*
[2] *College of Machinery and Electrical Engineering, Inner Mongolia Agricultural University, Hohhot, China, 010018*
[*] *Corresponding author, Address: College of Computer and Information Engineering, Inner Mongolia Agricultural University, 306 Zhaowuda Road, Hohhot, China, 010018, P. R. China, Tel: +86-471-4309237, Fax: +86-471-4306865, Email: xuehr@imau.edu.cn*

Abstract: The accurate segmentation of milk somatic cells in microscope images may contribute to development of a successful system that automatically analyzes, detects and counts cells in microscope images. We present a method for milk somatic cell Segmentation. Our approach is based on segmentation of subsets of bands using mathematical morphology followed by the fusion of the resulting segmentation “channels”. For color images, the band subsets are chosen as RG, RB and GB pairs. The segmentation in 2D color spaces is obtained using the watershed algorithm. These 2D segmentations are then combined to obtain a final result using a region split-and-merge process. Milk somatic cell images are segmented, and background, nucleus and cytoplasm can be extracted correctly. The most important feature of this method is the improved performance.

Keywords: milk somatic cell, color image segmentation, mathematical morphology, image fusion

1. INTRODUCTION

Most Color image segmentation is basically a 3D image histogram clustering. Many methods for color image segmentation on 3D histogram have been developed so far. However, clustering a 3D histogram can be

Xue, H., Ma, S. and Pei, X., 2008, in IFIP International Federation for Information Processing, Volume 259; Computer and Computing Technologies in Agriculture, Vol. 2; Daoliang Li; (Boston: Springer), pp. 1253–1258.

expensive because of huge amount of data involved. One remedy is to project the 3D color space into a lower dimensional space such as 2D (Kurugollu et al., 2001) or even 1D space (Cheng et al., 2000). In this paper, the method we present is partly based on the 2D morphological clustering method (Xue et al., 2003).

A typical milk smear consists of somatic cell, milk creaminess, lactoproteid, debris and so on. The goal of segmentation is to locate the milk somatic cells and to mark their nucleus and cytoplasm regions. Numerous methods have been proposed for digital cell images of peripheral blood (Comaniciu et al., 1999).

The above cell segmentations need to be accomplished in 3D space. Hence they have poor performance, especially in running time and memory consumption. In this paper, we present an improved approach for milk somatic cell color image segmentation.

2. SEGMENTATION SCHEME

2.1 Segmentation in 2D color spaces

Fig. 1 shows the procedure of the proposed segmentation scheme. The first step is to apply the morphological segmentation process (Géraud et al., 2001).

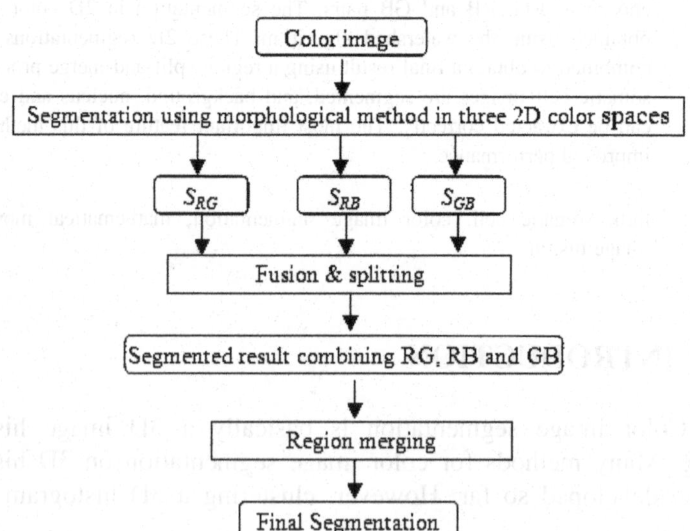

Fig. 1. The flowchart of the proposed approach

Instead of computing a 3D histogram in the 3D RGB space, we compute these 2D histograms, respectively in the 2D color spaces RG, RB, and GB. For instance, $H_{RG}^{(0)}$ is a 2D histogram of input image I which is only based on the red and green components.

From the three 2D histograms $H_{RG}^{(0)}$, $H_{RB}^{(0)}$, and $H_{GB}^{(0)}$, we output these segmented images S_{RG}, S_{RB}, and S_{GB} using watershed algorithm (Géraud et al., 2001).

2.2 Fusion of segmentation results

None of these three 2D segmentations is good enough to be the final resulting segmentation of the original color image. So a fusion process of segmentation results is required.

2.2.1 Fuzzy matching degree between two classes

Let us symbolize by T a 2D projection (T can be RG, RB, or GB), l being a class label, we denote by $S_T^{(l)}$, the set of points assigned to class l in the segmentation S_T. In other words:

$$S_T^{(l)} = \{p \mid S_T(p) = c_T^{l_T(p)}\} \tag{1}$$

A partial similarity degree between both corresponding point sets is defined as:

$$\mu_{T_1 \to T_2}^{(l_1,l_2)} = \frac{card(S_{T_1}^{(l_1)} \cap S_{T_2}^{(l_2)})}{card(S_{T_1}^{(l_1)})} \tag{2}$$

Then, we get a fuzzy matching degree:

$$\mu_{T_1,T_2}^{(l_1,l_2)} = \mu_{T_1 \to T_2}^{(l_1,l_2)} \oplus \mu_{T_2 \to T_1}^{(l_1,l_2)} \tag{3}$$

where: $\oplus$ can be any fuzzy T-norm operator. Here, we use "max" operator.

2.2.2 Region splitting

Given a criterion $m \in (0,1)$, if $\mu_{T_1,T_2}^{(l_1,l_2)} > m$, we split set $S_{T_1}^{(l_1)}$ into two subsets $S_{T_1}^{(l_1)} - S_{T_1}^{(l_1)} \cap S_{T_2}^{(l_2)}$ and $S_{T_1}^{(l_1)} \cap S_{T_2}^{(l_2)}$.

The splitting process is iterative. For instance, we can take the order as illustrated in Fig. 2 in the case of RGB images.

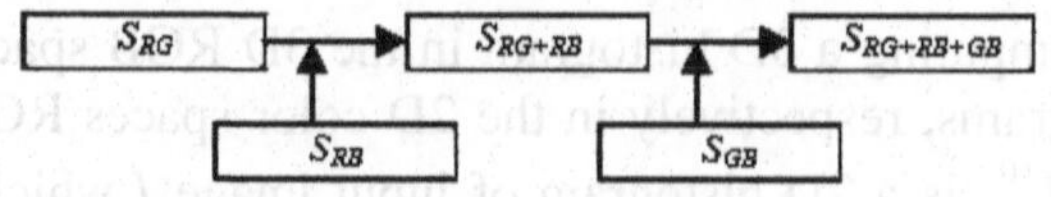

Fig. 2. Diagram of splitting process

The splitting process, while combining information from these classifications, separates nearly all classes that overlap in 2D color spaces.

2.2.3 Region merging

For input RGB image *I*, the result after the splitting process is an over-segmentation $S_{RG+RB+GB}$ of *I*. Therefore, a region merging is necessary. In the merging algorithm, the regions with the smallest color distance are first merged.

Here, we perform the algorithm in CIE L*a*b* color space which has more uniform perceptual properties than other spaces (Cheng et al., 2000).

3. MILK SOMATIC CELL IMAGE SEGMENTATION

Large numbers of milk somatic cell images have been processed by our segmentation scheme. Fig. 3 shows the segmentation results of an image of a milk smear.

As can be shown, the background, cytoplasm and nucleus are correctly separated, even while the images don't have very good contrast between the background and the cytoplasm of the cell images. We utilize the fact that the nucleus has minimum gray value (g) for locating the milk somatic cells,

Where:

$$g = \frac{R+G+B}{3} \quad (4)$$

Where R, G and B are the center of each class.

We have employed a large variety of milk somatic cell color images in our experiments. Some results of the proposed approach are shown in Fig. 3. The corresponding segmentation method using automatic morphological approach in 3D space (Géraud et al., 2001) is also tested for the comparison between the 3D method and our 2D method. The 3D method is based on 3D histogram and no fusion is needed. The experiments show that the somatic cell segmentations using the two methods are both correct. Furthermore, the execution time for the algorithm based on 2D space is less than that in 3D space.

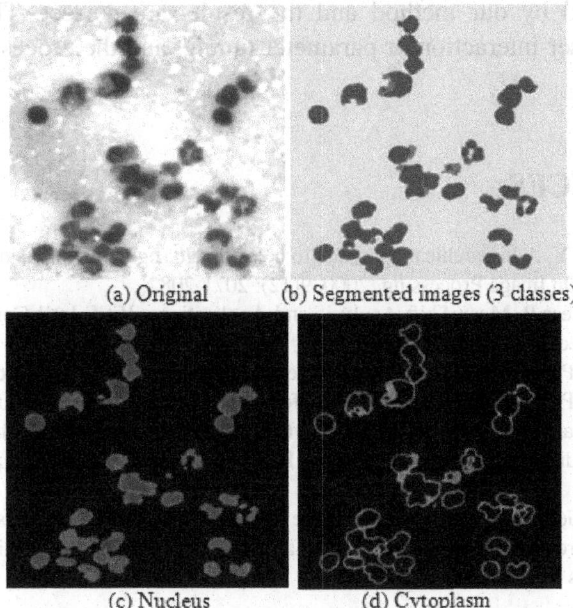

Fig. 3. Segmentation results of milk somatic cell images

Some experimental results for the comparative performance between the 2D method proposed in this paper and the 3D method proposed by Géraud (2001), are listed in Table 1.

The experimental results indicate that both results of the 3D method and the ones obtained from the 2D method are correct. However the latter is more efficient, particularly regarding running time.

Table 1. Comparison between 2D and 3D methods

Images	Running time (s)	
	3D	2D
1	58.13	5.91
2	83.10	6.09
3	101.05	8.91
4	78.43	11.67
5	88.40	6.04
6	63.83	5.42

4. CONCLUSION

In this paper, an efficient somatic cell image segmentation method has been presented. The method is based on morphological classification in 2D color space and a fusion technique. Large mount of milk somatic cell images are segmented by our method and the results are correct. This approach requires no user-interaction or parameter tuning, and the processing speed is fast.

REFERENCES

Cheng H D, Sun Y. A hierarchical approach to color image segmentation using homogeneity, IEEE Trans. On Image Processing, 2000, 9(12): 2071-2082.

Comaniciu D, Meer P. Mean Shift Analysis and Applications, IEEE Int'l Conf. Comp. Vis., Kerkyra, Greece, 1999: 1197-1203.

Géraud T, Strub P Y, Darbon J. Color image segmentation based on automatic morphological clustering, in Proc. IEEE International Conference on Image Processing, 2001: 70-73.

Kurugollu F, Sankur B, Harmanci A E. Color image segmentation using histogram multithresholding and fusion, Journal of Image and Vision Computing, 2001, 19(13): 915-928.

Xue H, Pei X, Géraud T, Ma S. Color image segmentation using fuzzy sets and fusion, in Conference Proceedings of the Sixth International Conference on Electronic Measurement & Instruments (ICEMI'2003), 2003: 72-75.

RESEARCH ON REMOTE CROP PRODUCTION MANAGEMENT SYSTEM BASED ON CROP SIMULATION MODELS AND WEBGIS

Jianbing Zhang [1,*], Yeping Zhu [2], Liping Zhu [1]
[1] *Department of Computer Science and Technology, China University of Petroleum-Beijing, Beijing, China, 102249*
[2] *Institute of Agricultural Information, Chinese Academy of Agricultural Sciences, Beijing, China, 100081*
[*] *Corresponding author, Address: Department of Computer Science and Technology, China University of Petroleum-Beijing, 18 Fuxue Road, Changping, Beijing, P. R. China, 102249, Tel: +86-10-89733006, Email: zhangbing153@yahoo.com.cn*

Abstract: In order to simulate growth and yield of crop in field scale, crop simulation models always require a lot of initial parameters. WebGIS is a powerful tool for obtaining, managing and analyzing spatial data on the Web. With the support of WebGIS, crop models based on web services can be easily run on the web by remote client and provide decision information and measures for crop production. In this paper, a remote crop production management system based on crop models and WebGIS is discussed. After the input variables for the crop model are obtained by WebGIS tool, the model will be run and the database will be updated with the results of the simulation. Spatial analysis and representation of crop model input data and output results can be done effectively on the web. Both the input data for the model and outputs from the model can be displayed on a map.

Keywords: crop model, WebGIS, crop production

1. INTRODUCTION

Crop production, farm management and decision making are influenced by a lot of uncertain factors, such as biological and technological ones. Some important decisions in agricultural production, such as irrigation application

Zhang, J., Zhu, Y. and Zhu, L., 2008, in IFIP International Federation for Information Processing, Volume 259; Computer and Computing Technologies in Agriculture, Vol. 2; Daoliang Li; (Boston: Springer), pp. 1259–1265.

dates and amount, fertilizer application dates and amount, the choose of crop variety are depended heavily on the existing knowledge base of current environment conditions like soil and climate, water resource (Wang Shiqi, 1998). Cropping inputs such as fertilizer are applied at varying rate across a field in response to variations in crop needs. Spatial variability of crop production such as different soil conditions, weather conditions and water conditions needs different agricultural production management practices within a target-region.

Crop model simulates growth and yield of crop in field scale and can be used to provide useful field information for farmers. Due to the complexity of crop-environment system, the limit of agricultural data collection and some technology restrictions, crop model techniques have not been widely applied in agricultural production. Researchers have used crop models for many years, but have had limited success in packaging these complex models in a framework that make them easy for producers to use.

Web-based applications provide an efficient and powerful way for delivering crop models to crop producer. An important issue in crop simulation model is that the basic units (water, soil and weather data) have a spatial distribution. WebGIS is a powerful tool for spatial data management on the web. With the support of WebGIS tool, crop models can be easy for agricultural producers as well as policy makers to use via the Internet.

This paper describes the application of integration WebGIS with web-based crop models for crop production management. An integrated system of WebGIS and crop modeling helps people know the impact of differences between input and output spatially from one place to other, provide better management information from productivity and profitability viewpoint quickly. The design and implementation of the remote crop production management system are also introduced in this paper.

2. CROP SIMULATION MODEL

Crop model simulates growth and yield of crop in field scale, and requires a lot of initial conditions and parameters as input data such as soil, water, climate data etc. There are many crop models nowadays in the world. But few can be run on the web. In order to implement remote crop production management, we need the model to be run on the web.

Intelligent Wheat and Corn Management System is the result of many years' research in wheat and corn simulation by Yeping zhu, Shiqi Wang in CAAS (Chinese Academy of Agricultural Science). The research was supported by Chinese Academy of Agricultural Science and National 863-306 project of China.

The wheat and corn models simulate growth and yield of crop in field scale, and they also require initial conditions and parameters. The inputs for the model include environmental conditions (soil type, daily maximum and minimum temperature, rainfall, and solar radiation) and management practices (variety, row spacing, plant population). The simulation result includes daily growth of vegetative, stage, and water and nitrogen stress. Some decision information can also be given to assist producers in their management.

In order to be used on the web environment, wheat and corn crop models are both packaged with web service and XML technology. Web service is a technology that designed to support interoperable Machine to Machine interaction over a network. Web services are frequently just Web APIs that can be accessed over a network, such as the Internet, and executed on a remote system hosting the requested services. By using web services technology, the wheat and corn models can be run on the web by remote client. The data exchange format is XML.

3. WEBGIS

WebGIS is a powerful tool for obtaining, managing and analysing spatial data on the Internet. WebGIS holds the potential to make distributed geographic information available to a very large worldwide audience.

It is possible to add GIS functionality to web-based crop model applications by WebGIS tool. It makes data access including model inputs and model results more conveniently. Both the input data for the model and outputs from the model can be displayed on a map. It helps effective use of GIS for the spatial analysis and representation of input data and output results of crop model. By this way, we can get a more clear impression on the spatial distribution of those environment elements and simulation results.

GeoBeans software is WebGIS software which is developed by network technology department, Institute of Remote Sensing Application, CAS. In our research, we use GeoBeans Map Server for map function and model data management.

4. DESIGN AND IMPLEMENTATION OF REMOTE CROP PRODUCTION MANAGEMENT

4.1 Design of the system

In order to make the system more robust, flexible, we adopt the popular three-tier architecture in our system design. The architecture of the system is composed of three layers: the user interface layer, the application logic layer and the database layer. The architecture makes our development and integration work easily and efficiently.

Before users run the model on the Internet, the crop variety and simulation location must be pointed out. For each cell of the grid where remote client lives on, WebGIS tool will extract the input variables (climate, crop, soil, …) for the crop model from the corresponding data layers, then the model will be run and the database will be updated with the results of the simulation.

The workflow and architecture of the system are as follow:

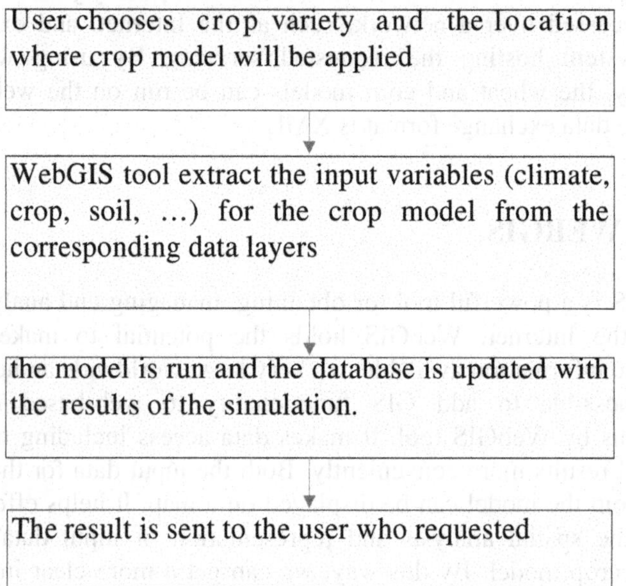

Fig. 1. Showing the work flow of the system

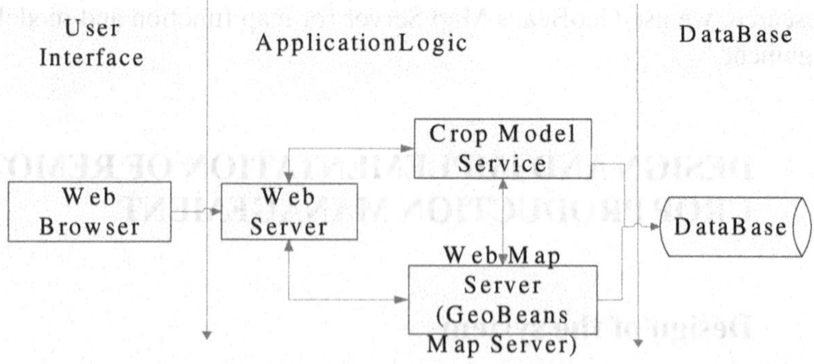

Fig. 2. Showing the architecture of the system

4.2 Implementation

The remote crop management system is implemented by the integration of WebGIS and crop simulation model. The communication between WebGIS service and crop model services is the key part of the system. Web services, XML technology are applied to the communication. XML is exchange data format.

After connecting WebGIS and crop models successfully, users can run the model easily through internet in their convenience. The delivering crop models to crop producer has achieved in our system. It helps crop producer a lot to know the impact of different measurement in crop production.

Some of running results of the system are as follow:

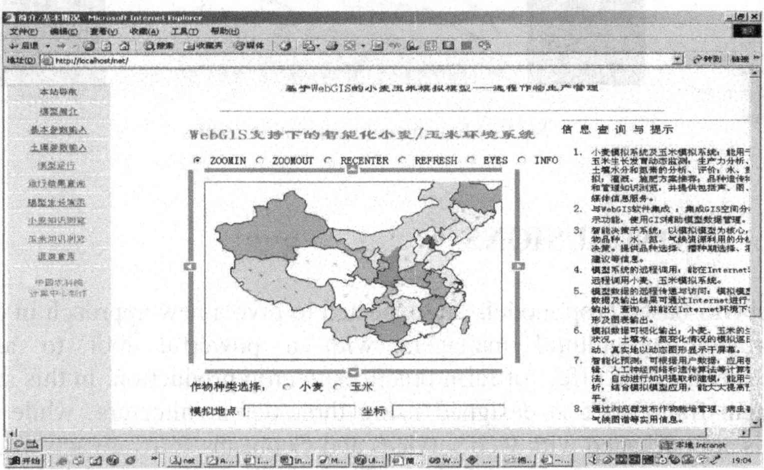

Fig. 3. Showing the main window of the system

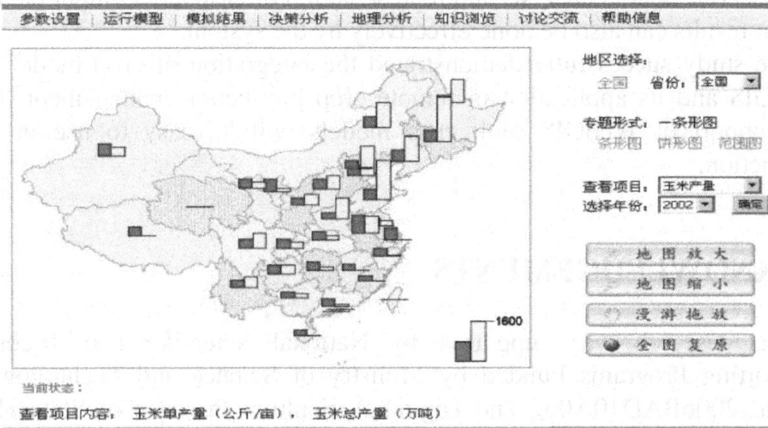

Fig. 4. Showing simulation results on the map

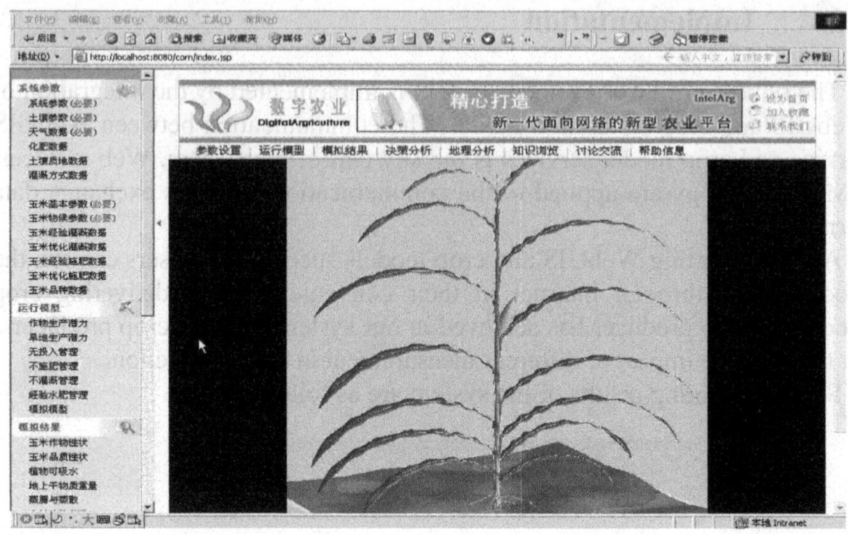

Fig. 5. Showing corn simulation results

5. CONCLUSIONS

WebGIS based crop models are expected to give a new approach in order to provide agricultural managers with a powerful tool to assess simultaneously the effect of farm practices to crop production. In this study, a system framework is designed using three-tier architecture, while crop models are packaged with web services and Geobeans software acts as a pre and post processor for model data. The graphical display of inputs and outputs of wheat and corn model (CAAS) are also supported by Geobeans Map Server. Spatial analysis and representation of crop model input data and output results can also be done effectively by the system.

The study successfully demonstrated the integration of crop model with WebGIS and its application in remote crop production management. With the support of WebGIS tool, crop models will be easy to use in crop production.

ACKNOWLEDGEMENTS

This research was supported by National Scientific and Technical Supporting Programs Funded by Ministry of Science and Technology of China (2006BAD10A06), and Digital Agriculture Program of State High-tech Research and Development Project of China (No. 20060110Z2059).

REFERENCES

Engel T. AEGIS/WIN: A computer program for the application of crop simulation models across geographic areas [J]. Agron. J. 1997, 89(6):919-928.

Pan Xuebiao. Study on the Spatial Distribution and Variation of Cotton Production in Counties of China Based on GIS. Scientica Agricultura Sinica, 2003, 36(4):382-386.

Pan Yuchun, Wang Jihua. WebGIS-based system for crop quality monitoring and planting optimization. Transactions of the Chinese Society of Agricultural Engineering, 2004, 20(6):120-123.

Paz, JO, WD Batchelor. Web-based soybean yield simulation model to analyze the effects of interacting yield-limiting factors variations of winter time air pollution concentrations in the city of Graz, Austria. Environmental Monitoring and Assessment, 2001, 65:79-87.

Wang hong, Wu Shuan, Tang zhenghong. Crop productivity model and its application. Chinese Journal of Applied Ecology, 2002, 13(9):1174-1178.

Wang Shiqi Zhu Yeping etc. A System Framework Based on Crop Simulator. Modeling for Crop-Climate-Soil-Pest System and Its Applications in Sustainable Crop Production 1998, 6.

FORECAST RESEARCH OF CROP WATER REQUIREMENTS BASED ON FUZZY RULES

Jianbing Zhang[1,*], Yeping Zhu[2], Feixiang Chen[3]
[1] *Department of Computer Science and Technology, China University of Petroleum-Beijing, Beijing, China, 102249*
[2] *Institute of Agricultural Information, Chinese Academy of Agricultural Sciences, Beijing, China, 100081*
[3] *College of Information, Beijing Forestry University, Beijing, China, 100083*
[*] *Corresponding author, Address: Department of Computer Science and Technology, China University of Petroleum-Beijing, 18 Fuxue Road, Changping, Beijing, P. R. China, 102249 Tel: +86-10-89733006, Email: zhangbing153@yahoo.com.cn*

Abstract: This paper put forward the idea of producing fuzzy rules by genetic algorithms based on Takagi-Surgeon Fuzzy Logic System from the dataset of multi-dimension climate data and crop water requirements, and establishing the fuzzy model to predict crop water requirements. The forecast model was tested and the result showed that it was an effective way to forecast crop water requirements by fuzzy rules model.

Keywords: Fuzzy logic system, Genetic algorithms, Crop water requirements

1. INTRODUCTION

The forecast of crop water requirements is importance for irrigating forecast. In order to make plans for water use in large irrigation districts, we need to predict crop water requirements and their changes in a period of time later. Many technologies had been applied in predicting for crop water requirements, such as regression analysis, gray prediction, Artificial Neural

Zhang, J., Zhu, Y. and Chen, F., 2008, in IFIP International Federation for Information Processing, Volume 259; Computer and Computing Technologies in Agriculture, Vol. 2; Daoliang Li; (Boston: Springer), pp. 1267–1273.

Network (Renato Silvio, 1998; Zhang bing, 2000). But the results were not satisfied. Today the research of complex system has been changed from constructing precise mathematics models to constructing fuzzy models based on intelligent technology. Fuzzy logic technology is called the core technology of the 21st Century (Dou zhengzhong, 1996), it had achieved many remarkable success in industry and many high technological fields. Fuzzy logic system provides better forms of rule expression, and fuzzy technology provides reasoning logic like human thought. Genetic algorithms are adaptive heuristic search algorithm premised on the evolutionary ideas of natural selection and genetic, and they represent an intelligent exploitation of a random search within a defined search space to solve problems (Zhou Ming, 1999). Based on the combination of genetic algorithms and fuzzy logic technology, we put forward the idea of producing fuzzy rules automatically by genetic algorithms, then establishing the fuzzy model of the system. From the dataset of multi-dimension climate and crop water requirements, the fuzzy rules of crop water requirements were extracted out successfully and the model of crop water requirements and climate factors was established and tested for the forecast of crop water requirements (green pepper). The test showed better results in predicting water requirements. The method can be also used in modeling other agricultural complex system.

2. TAKAGI-SUGENO FUZZY LOGIC SYSTEM

Fuzzy logic systems have been successfully applied to a number of scientific and engineering problems during recent years. The expression of fuzzy rules and the reasoning in fuzzy systems are flexible. They provide flexible architecture for modeling nonlinear systems. In fact, fuzzy models are only one of mathematic expression forms of fuzzy rules and reasoning. Takagi-Surgeon Fuzzy logic systems are based on local linear function and the global nonlinear is implemented by blending the subsystems' models (Wang shitong, 1998).

The main feature of Takagi-Surgeon fuzzy models can be grouped under the form of the following steps that also point out their operating mode:

Firstly, the input space is decomposed into subspaces;

Then, within each subspace, the system model can be approximated by simpler models, in particular linear ones;

Finally, the global fuzzy model in the state-space is derived by blending the subsystems' models in terms of the weighted average of rule contributions.

Fuzzy inference systems are composed of a set of IF–THEN rules. A TS fuzzy model has the following form of fuzzy rules:

$$R^i : IF \quad x_1 \ is \ A_1^i, x_2 \ is \ A_2^i, \ldots\ldots, x_k \ is \ A_k^i,$$

$$THEN \quad y^i = p_0^i + p_1^i . x_1 + \ldots + p_k^i . x_k$$

where R^i means that the sequence number of rule is i; $x_1, x_2, \ldots, x_k$ are inputs of fuzzy system; $A_1^i, A_2^i, \ldots, A_k^i$ are fuzzy subsets, their membership function is general piecewise continuous function or Gauss function; y^i is the output of rule i; p_j^i (j=0,1,…,k) is called conclusive parameter. If there are n pieces of rules in the fuzzy model, and the input values is X= $(x_1, x_2, \ldots, x_k)^T$, the overall output of the fuzzy model is the weighted average value of y^i:

$$Y = \sum_{i=1}^{m} w^i y^i \Big/ \sum_{i=1}^{m} w^i$$

The computing formula of w^i is:

$$w^i = \prod_{j=1}^{k} u_{A_j^i}(x_j)$$

where $u_{A_j^i}(x_j)$ is the membership function for fuzzy set.

3. GENETIC ALGORITHMS

GAs are global optimization and parallel search method. The basic concept of GAs is designed to simulate processes in natural system necessary for evolution, specifically those that follow the principles first laid down by Charles Darwin of survival of the fittest. As such they represent an intelligent exploitation of a random search within a defined search space to solve a problem.

Algorithm is started with a set of solutions (represented by chromosomes) called population. Solutions from one population are taken and used to form a new population. This is motivated by a hope, that the new population will be better than the old one. Solutions which are selected to form new solutions (offspring) are selected according to their fitness – the more suitable they are the more chances they have to reproduce. The algorithms are described as follow:

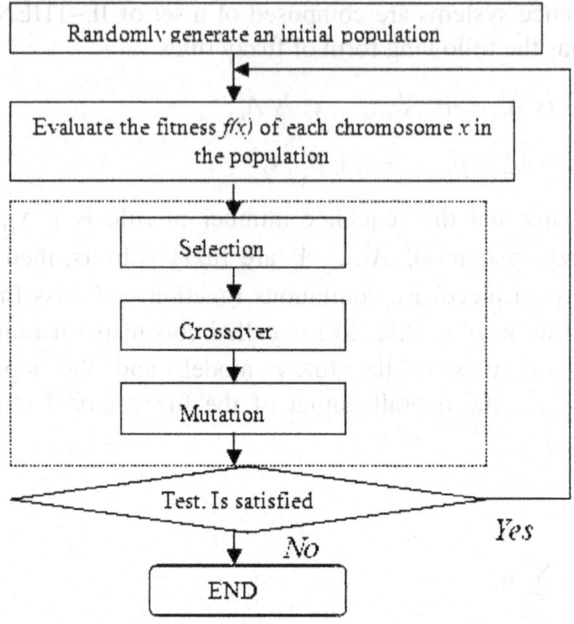

Fig. 1. The process of GA

As a computational analogy of adaptive systems, GAs are modeled loosely on the principles of the evolution via natural selection, employing a population of individuals that undergo selection in the presence of variation-inducing operators such as mutation and recombination (crossover). A fitness function is used to evaluate individuals, and reproductive success varies with fitness.

GAs are used for solving parameters optimization of fuzzy logic systems. Finding the appropriate structure and parameters of fuzzy logic systems automatically is implemented by using genetic algorithms.

4. APPLICATIONS

A crop (green pepper) water requirements fuzzy model is established to test the method. The research data is from the reference (Renato Silvio, 1998). The dataset is the 30 continuous water requirements data, which is observed during the growth process of green pepper in 1994 (florescence). The data is filtered by regression analysis. Y is crop water requirements value ($mm.d^{-1}$); X1 is radialization data ($W.\ m^{-2}.d^{-1}$). X2 is value of relative humidity. The research data is shown in Table 1:

Table 1. Crop water requirements and climate data (Year 1994)

Date(Month-Day)	Original Data			Standardized Data	
	Y	X1	X2	x1	x2
5-4	1.33	2023	92.35	0.308409	1.225619
5-5	3.16	5870	76.55	0.894890	1.015930
5-6	3.67	5560	77.15	0.847630	1.023893
5-7	1.73	3175	84.15	0.484033	1.116793
5-8	4.29	7611	73.66	1.160308	0.977576
5-9	4.92	7931	71.54	1.209092	0.949440
5-10	4.03	6116	67.75	0.932393	0.899141
5-11	5.15	7754	67.41	1.182108	0.894629
5-12	4.29	6131	66.60	0.934680	0.883879
…	…	…	…	…	…
5-31	5.43	8140	73.68	1.240955	0.977841
6-1	3.98	6273	82.02	0.956328	1.088525
6-2	2.86	4822	82.51	0.735121	1.095028

X1, X2 are the results of standardized data of X1, X2. The model is established using the data from 5-4 to 5-25. The data from 5-26 to 6-2 is used for model verification. The data of x1 and x2 is used for system input, Y is the output of the system. The membership function is used Gaussian function $u(x) = e^{-k(x-c)^2}$. The forecast result of fuzzy model after training is as follow:

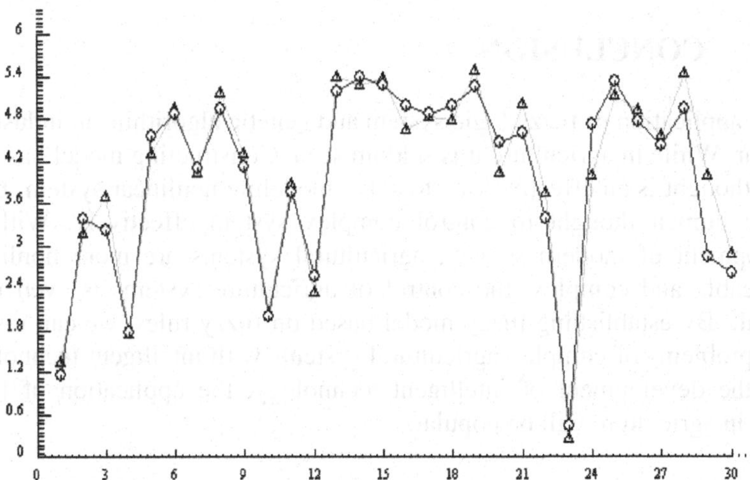

Fig. 2. The Comparison between output of model and actual data: Circle points represent actual value; triangle points represent the output of fuzzy model

The output of model is close to actual values, and most data error is less than 10%. The result of 22 training data is satisfied, and the later 8 forecast result is matched with actual situation. Nine fuzzy rules are obtained by training sample data using fuzzy logic system and GAs:

Table 2. The nine fuzzy rules

	radicalization: low	radicalization: middle	radicalization: high
relative humidity: low	Y=0.060000* X2/75.35	Y=0.772525*X1/6559.47+ 3.391403	Y=2.138495*X1/6559.47+0.080402* X2/75.35+2.639984
relative humidity: middle	Y= 1.043489	Y=3.479416*X1/6559.47+ 0.239196*X2/75.35	Y=3.939482*X1/6559.47+0.049539* X2/75.35
relative humidity: high	Y= -0.273271* X1/6559.47	Y=-8.340000*X1/6559.47+ 2.790065*X2/75.35+ 6.025024	Y=7.458415* X1/6559.47

To the research district, if radicalization and relative humidity are low, crop water requirements are direct ratio with relative humidity, Y = 0.060000*X2/75.35. If radicalization is low and relative humidity is middle, Y = 1.043489. The other rules also have certain meanings. We can see that a distinct difference between the fuzzy model and forecast model of artificial neural network is that we can get fuzzy rules by using fuzzy logic system. From the result of model running, our model is established the relationships of crop water requirements and climate factors effectively based on 9 fuzzy rules, and the forecast test showed the result was satisfied. An advantage of fuzzy models is that a complex task can be done by several simple fuzzy rules.

5. CONCLUSION

The application of fuzzy logic system and genetic algorithms in industry is popular. While in agriculture it is seldom seen. Constructing model based on fuzzy thought is an effective way to solve modeling nonlinear system. It can imitate human thought to control complex system effectively. With the development of modern society, agricultural systems are more nonlinear, changeable and complex, the control of agriculture systems is even more difficult. By establishing fuzzy model based on fuzzy rules, we can resolve those problems of complex agricultural systems with intelligent technology. With the development of intelligent technology, the application of fuzzy model in agriculture will be popular.

ACKNOWLEDGEMENTS

This research was supported by National Scientific and Technical Supporting Programs Funded by Ministry of Science and Technology of China (2006BAD10A06), and Digital Agriculture Program of State High-tech Research and Development Project of China (No. 20060110Z2059).

REFERENCES

Cao xianbing, Zhuang zhenquan, A method of automatically produce fuzzy rule based on genetic algorithm, Pattern recognition and artificial intelligence 1997, 10(2):171-175.

Dou zhengzhong, Fuzzy logic technology is core technology of 21st century, Application research of computers 1996, 13(4)8-12.

Renato Silvio da Frota Ribeiro. Fuzzy logic based automated irrigation control system optimized via neural networks [D]. American: The University of Tennessee, 1998.

Wang shitong, Neuro-Fuzzy System and its application. Beihang University Press, Beijing 1998, 6.

Zhang Bing, Yuan qishou. Model for predicting crop water requirements by using L-M optimization algorithm BP neural network. Transactions of the Chinese society of agricultural engineering 2000, 20(6): 73-76.

Zhang jianbing, Zhu yeping Forecasting research of diseases and pests based on fuzzy rules. System Sciences and Comprehensive Studies in Agriculture 2000, 16(4): 283-285.

Zhou ming. The Theory and application of genetic algorithm. National Defence Industry Press 1999.

RESEARCH ON THE TRANSFORMATION OF DISPLAY FORMAT FOR WEB INFORMATION

Sufen Dong, Guifa Teng [*], Dan Wang, Yan Hu, Fang Wang, Shuhui Chang
School of Information Science and Technology, Agricultural University of Hebei, Baoding, China, 071001
[*] *Corresponding author Address: School of Information Science and Technology, Agricultural University of Hebei, No. 289 Ling Yu Si Street, Baoding, 071001, China, Tel: +86-312-7521807, Fax: +86-312-7521807, Email: tguifa@hebau.edu.cn*

Abstract: As so far, most of the information on the internet are designed for computer display, which can achieve the desired effect at 800*600 or higher resolution, but the normal TV's resolution is much lower, so the web page looks abnormally on TV. Furthermore, it also can't be browsed by phone and PDAs as PC does. Based on the analysis of the display principles of TV and computer, a new method which combines the following two methods together is proposed for the low resolution, thumbnail and relayout based on width adjustment. Web pages are divided into navigation page and content page. Thumbnail is corresponded to the former pages; the latter ones are adjusted through the format transformation with width adjustment algorithm. The web pages can be made more adaptive for TV.

Keywords: format transformation, thumbnail, width adjustment, relayout

1. INTRODUCTION

Internet is being used widely in many fields. However, at present on the comparatively lower development area in agriculture. The distribution of the information on the web has many restrictions. One main reason is that many countrymen can't afford the expensive personal computer. Comparatively the TV's possession rate is very high in countryside, especially in China.

Dong, S., Teng, G., Wang, D., Hu, Y., Wang, F. and Chang, S., 2008, in IFIP International Federation for Information Processing, Volume 259; Computer and Computing Technologies in Agriculture, Vol. 2; Daoliang Li; (Boston: Springer), pp. 1275–1282.

How to take full advantage of the abundant information resources on the internet for these areas is becoming a critical problem for their development. The imbalance of the economy among different areas makes many kinds of connection to the internet to solve the "last kilometer" an inevitable choice (Fig. 1), such as TV to internet, mobile to internet etc. In the countryside of China, TV has already basically popularized the new means of using TV plus set top box to access the internet is much more suitable for the specific areas. It will have a broader development space. However, as so far, most of the information on the internet is designed for computer, which can achieve the desired effect at 800*600 or higher resolution. But the PAL pattern TV's resolution is 720*576, lower than the required level. Surfing on the internet by this kind of device can't reach the ideal effect and even mis-layout or mess occurs. This is difficult to distribute this kind of connection. There are two scenarios for transformation for TV (Set Top Box): one is to rewrite web pages for TV, this is specific web pages are compiled for TV. Another is to transform web pages for TV. Thus they can be made to display on TV normally. The preceding method suits to write new pages. As far as the magnanimous existing network resources are concerned, rewriting the page will waste time and energy, and even impossible to do. However, the latter one can make the web pages adapt to display on TV. So it is an effective method to make full use of the network resources. Based on the second method a solution is presented, it has an important realistic meanings and application values.

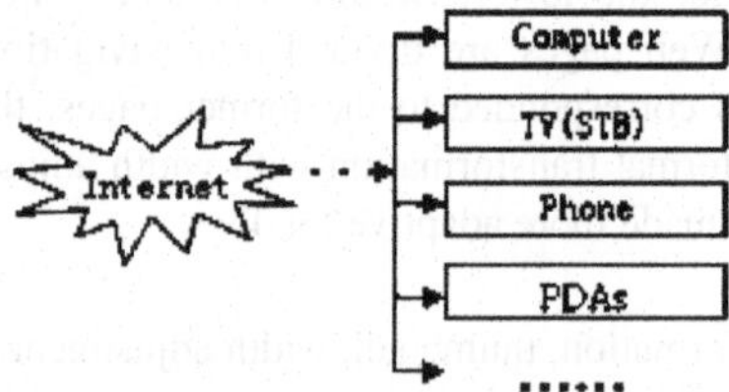

Fig. 1. Connections to internet

2. DIFFERENCES BETWEEN TV AND PC

The differences between TV and PC consist of resolution, color, display mechanism and the EMS memory. Thereinto, what TV resolution is lower than that of PC's is the main difference between them. TV resolution is image quality guide line during REC transmission and display, and the connatural screen structure on representing the degree of image particularity that is the scan format of a single image signal and the pixel standard of the device. Generally it is expressed as horizontal-pixel × vertical-pixel. As to the scan format of 720×576 PAL TV, its horizontal pixel is 720 points which depends on the maximal resolution of the cathode ray tubes (CRT). So if it wants to display the 720×576 resolution accurately, the display needs at least 720 horizontal pixels and 576 scan lines. Furthermore, TV resolution is also limited to the scan format of receiving TV signal, which usually uses 625 scan lines (625 line, 50 frames, interleaved mode, in China). But computer's display resolution is only limited to the pixel standard of the hardware (Shen, 2001). At present, there are many kinds of resolution standards: 800*600, 1024*768, 1280*1024 and so on.

In addition, their differences in color make the bright color look brighter on TV, prone to make the audience's vision weary if watching for a long time. So the Web TV pages had better use the dismal background. The differences in display mechanism make the static text less clear viewing on TV. What's more, the small EMS memory of the STB (Set Top Box) is not suitable for storing the plethoric content pages.

The differences of the users' habits are distance and interactive.

(1) Distance: TV audiences are used to watch TV program from a long distance but the PC users prefer to sit in front of the computer. So the small typeface is not suitable for viewing on TV.

(2) Interactive: TV audiences are accustomed to obtain information, not to carry out lots of interactive operation. Therefore, it should avoid using the multilayer navigation tree and the screen filled fully with graphical interfaces components in the pages (Shen, 2001; Lu et al., 2000).

3. REQUIREMENT FOR WEB TV PAGES

According to the analysis above, the requirements for Web TV pages could be concluded as follows:

(1) The navigation information of a web site should be displayed at the top-page, and it had better display within one screen to reduce the scroll operation, which will be convenient for the user to locate.

(2) Color: Web pages usually use white as their background, but this will make the screen look distortion on TV, and the boundary of the page will appear arc. Thus, Web TV page had better use dark background and bright foreground (Lu et al., 2000).

(3) Page width: There are many kinds of TV resolution. The TV screen doesn't support horizontal scroll bar, so the web page should be dynamically adjusted according to the screen resolution, otherwise the exceeded area could not be displayed.

4. SOLUTIONS

4.1 System framework

The system function structure is shown in Fig. 2. It is composed of TV (client, shown as I), Set Top Box (page transformation, shown as II) and internet (shown as III). First, a parameter to represent the screen resolution is set to the STB by the client, and stored in STB. Then an URL can be input to send a http request (shown as step (1) Fig. 2), which received by STB and transmitted to the internet (shown as step (2)). The internet responds to it and sends the requested html page to STB (shown as step (3)). When STB receives this html source file, it will transform it into Web TV page according to the parameter set in advance and return it to the TV client (shown as step (4)) to display on TV normally. The format transformation will be explained in the following paragraphs.

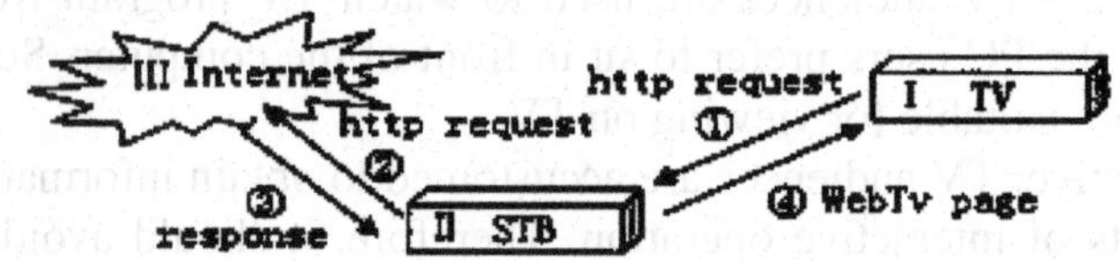

Fig. 2. Transformation process of web TV page

4.2 System flow

The system flow chart is shown in Fig. 3. Firstly, an URL is input, then a html source file will be received and preprocessed by STB. And it is changed into a well formed grammar tree through accidence analysis and syntax analysis. Afterwards, the page type is judged. If it is a navigation page, the system will invoke thumbnail algorithm to change web page into a thumbnail and if it is a content page, to invoke the page width adjustment algorithm to relayout the page. Finally, the page accorded with the TV display standard is achieved.

4.3 Distinction of web pages

Web pages are divided into two sorts: navigation page and content page. The navigation pages usually provide hyperlink information and doesn't contain many specific contents. So how to accelerate the user's browsing speed is a critical factor to this kind of pages. On the contrary, the content

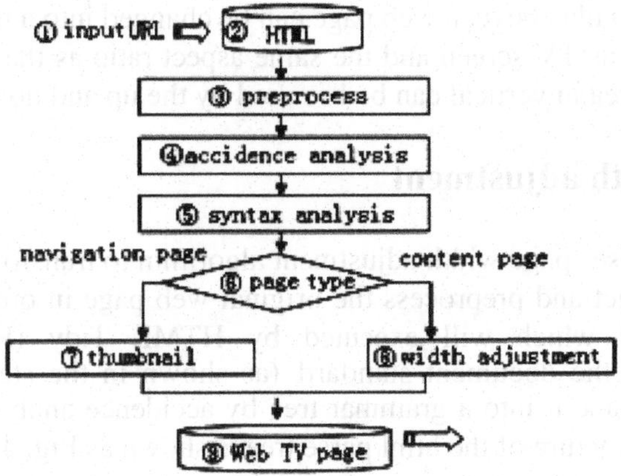

Fig. 3. System flow of page transformation

page mainly provides some specific contents. The main difference between the two pages is the hyperlink rate, with which they can be divided through the page distinction algorithm and the proportion of the navigation words (Yue et al., 2003).

4.4 Thumbnail

Thumbnail algorithm is used to transform navigation pages. Firstly, a http request is submitted, then received by STB, and the scaling can be calculated according to the preset of screen resolution parameter. At the same time, STB will access the DOM tree of the original page through web server and get the information which contains the content and location elements from the DOM tree, then send them together to the client. Thus, html elements will correspond to the thumbnail coordinates (Wobbrock et al., 2002; Yoshikawa et al., 2006). When user chooses a certain part of a thumbnail, the system will take out the corresponding html code and transform it into an appropriate html page and send it to the client.

Calculation of the thumbnail size: To define the screen width and height as ScreenWidth (*SW*), ScreenHeight (*SH*); the thumbnail width and height as ThumbnailWidth (*TW*), ThumbnailHeight (*TH*) and the web page width and height as WebPageWidth (*WPW*), WebPageHeight (*WPH*) individually. Due to the TV screen doesn't support the horizontal scroll bar, the thumbnail width should be set as SW, when the navigation page is transformed into a thumbnail (Wang, 2006). The formulas are shown as follows:

$$TW=SW \quad (1)$$

$$TH= (SW \times WPH)/WPW \quad (2)$$

Based on formula above, a web page can be changed into a thumbnail with the same width as TV screen and the same aspect ratio as the original page. The excessed area in vertical can be browsed by the up and down keystroke.

4.5 Width adjustment

The system uses page width adjustment algorithm to transform the content page. First, to get and preprocess the original web page in order to gain the html document which will executed by HTML Tidy (Raggett, 2003) according with the document standard (as shown in the step (3) Fig. 3), afterwards to parse it into a grammar tree by accidence analysis and syntax analysis. The structure of the html parse tree is shown as Fig. 4.

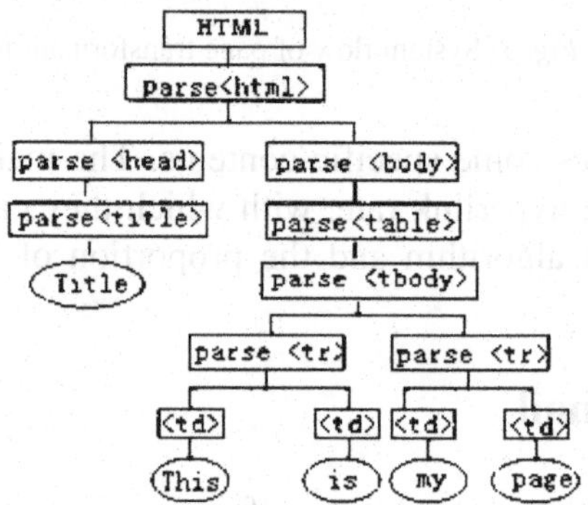

Fig. 4. Structure of HTML parse tree

It is composed of html root note and tag attribute notes. As the page structure is the essential factor that has to be considered, and the content in web page is usually stored in tables (Li, 2006), the table note will be disposed as follows:

(1) Firstly, to find the table note in the html dom tree (Sun et al., 2004).

(2) To get information which united as table, and invoke width adjustment algorithm to adjust the page width.

(3) Width adjustment algorithm.

a. To invoke table adjustment algorithm to adjust the table width.

b. When a table is filled fully with an image, and it exceeds the screen width, then to invoke image adjustment algorithm to scale down it, as shown in the step (1) Fig. 5.

c. In web TV pages, when the table contains both text and image on one side, if the width of the text by the side of the image is less than the preset parameter N (N is set as the least width of text area according to the user's browsing habits), the image has to be scale down. Then the text in the table is realigned, as shown in the step (2) Fig. 5.

d. In web TV pages, when the table contains both text and image on both sides, if the width of the text between the images is less than the preset parameter N, the image has to be scale down. Then the text in the table is realigned, as shown in the step (3) Fig. 5.

e. When a table is filled fully with text, width adjustment algorithm is invoked to realign the text, as shown in the step (4) Fig. 5.

f. Other type of layout in a table can also be done according to the above principles.

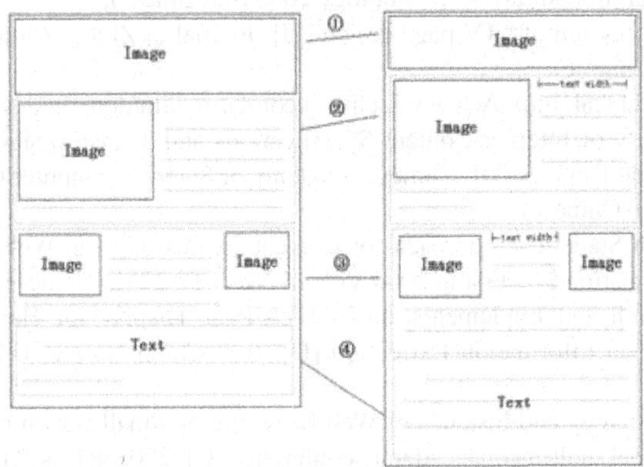

Fig. 5. Transformation of PC page into web TV page

5. CONCLUSION

In development countries, especially in China TV has already basically popularized in countryside. The new means of using TV plus Set Top Box to access the internet is springing up. However, as for the differences between TV and computer in display resolution, the abundant information on the internet can't be fully utilized. A solution is proposed to transform web pages for TV in this paper. This makes it very convenient for the rural to make full use of information on the internet, and promote the informatization in agriculture. Of course, this research is just a trial. It should be further studied and verified in the future research.

ACKNOWLEDGEMENTS

This study has been supported by Scientific Research Funds for the non-life discipline and emerging discipline of Agricultural University of Hebei (Contract Number: FSZ200618). Many thanks to the leaders and colleagues of our university for their supports and help.

REFERENCES

Dave Raggett. http://www.w3.org/People/Raggett/tidy/.

Hiroaki Yoshikawa, Osamu Uchida. Web Content Transducing System for Cellular phones [J]. proceeding of the Adwanced international conference on telecommunications and internarional Conference on Internet and Web Application and Services (AICT/CIW 2006).

Jacob O. Wobbrock, Jodi Forlizzi, Scott E. Hudson, Brad A. Myers. WebThumb: Interaction Techniques for Small-Screen Browsers [J]. UIST'02 2002 205-208.

Li J. Research and Implement of Network Information Extraction Oriented to The Mobile Platform [D]. Harbin Institute of Technology 2006 (in Chinese).

Lu C, Cheng Y. Discussion of "TV page design" [J]. Journal of Zhong Zhou University No. 2 2000 67-69 (in Chinese).

Nodes have a hierarchical http://www.w3schools.com/htmldom/dom_nodes.asp.

Shen X F. The Study of Internet Content Specifications and Transformation Tools Suitable for Several Digital Devices [D]. Chinese Academy of Science computer technical research institute 2001 (in Chinese).

Sun C J, Guan Y. A Statistical Approach for Content Extraction from Web Page [J]. Journal of Chinese Information Processing 2004 Vol. 18 No. 5 17-22 (in Chinese).

Wang Q B. Research and Implementation of Web Page Display for Handhold Intelligent Terminal Based on Information Extraction [D]. East China Normal University 2006 (in Chinese).

Yue W, Wang Y. Strategy and Instance of Web Browsing on Small-screen Devices [A]. Proc. Of the 12th national multimedia academic conference [C]. 2003 460-468 (in Chinese).

AN OPTIMIZATION GENETIC ALGORITHM FOR IMAGE DATABASES IN AGRICULTURE

Changwu Zhu[1], Guanxiang Yan[2], Zhi Liu[3], Li Gao[1,*]

[1] *Department of Computer Science, Hua Zhong Normal University, Wuhan 430079, China*

[2] *School of Information Management, Wuhan University, Wuhan 430072, China*

[3] *Wuhan Junxie Shiguan School, Wuhan 430079, China*

[*] *Corresponding author, Tel: 027-62265691, Email: lgao@mail.ccnu.edu.cn*

Abstract: Data Mining is rapidly evolving areas of research that are at the intersection of several disciplines, including statistics, databases, pattern recognition, and high-performance and parallel computing. In this paper, we propose a novel mining algorithm, called ARMAGA (Association rules mining Algorithm based on a novel Genetic Algorithm), to mine the association rules from an image database, where every image is represented by the ARMAGA representation. We first take advantage of the genetic algorithm designed specifically for discovering association rules. Second we propose the Algorithm Compared to the algorithm in, and the ARMAGA algorithm avoids generating impossible candidates, and therefore is more efficient in terms of the execution time.

1. INTRODUCTION

The image databases contain an enormous amount of information, and it is becoming more and more complex while its size continues to grow at a remarkable rate. So it can be exceedingly difficult for users to locate resources that are both relevant to their information needs and high in quality. Vast numbers of images have accumulated on the Internet and in entertainment, agriculture, education, and other multimedia applications.

Zhu, C., Yan, G., Liu, Z. and Gao, L., 2008, in IFIP International Federation for Information Processing, Volume 259; Computer and Computing Technologies in Agriculture, Vol. 2; Daoliang Li; (Boston: Springer), pp. 1283–1289.

Therefore, how to mine interesting patterns from image databases has attracted more and more attention in recent years (G. Chen, 2002).

Many additional algorithms have been proposed for association rule mining. Also, the concept of association rule has been extended in many different ways, such as generalized association rules, association rules with item constraints, sequence rules etc. Apart from the earlier analysis of market basket data, these algorithms have been widely used in many other practical applications such as customer profiling, analysis of products and so on (Gaoli, 2003).

Genetic Algorithm (GA) is one self-adaptive optimization searching algorithm. GA obtains the best solution, or the most satisfactory solution through generations of chromosomes' constant evolution includes the reproduce, crossover and mutation etc. operation, until a certain termination condition is coincident (K. Koperski, 1995).

Association rules mining Algorithm Based on a novel Genetic Algorithm (ARMAGA) is an optimal algorithm combing GA with ARMA.

In this paper we first take advantage of the genetic algorithm designed specifically for discovering association rules. Second we propose a novel spatial mining algorithm, called ARMAGA, Compared to the algorithm in (G. Chen, 2002), and the ARMAGA algorithm avoids generating impossible candidates, and therefore is more efficient in terms of the execution time.

2. ASSOCIATION RULES

Definition 1 Confidence

Set up $I=\{i_1, i_2, i_m\}$ for items of collection, for item in $i_j(1 \le j \le m)$, $(1 \le j \le m)$ for lasting item, $D=\{T_1, T_N\}$ it is a trade collection, $T_i \subseteq I(1 \le i \le N)$ here T is the trade.

Rule $X \to Y$ is probability that $X \bigcup Y$ concentrates on including in the trade.

The association rule here is an implication of the form $X \to Y$ where X is the conjunction of conditions, and Y is the type of classification. The rule $X \to Y$ has to satisfy specified minimum support and minimum confidence measure (Shijue Zheng, 2006).

The support of Rule $X \to Y$ is the measure of frequency both X and Y in D

$$S(xy) = |xy|/|D| \tag{1}$$

The confidence measure of Rule $X \to Y$ is for the premise that includes X in the bargain descend, in the meantime includes Y

$$C(x \to y) = S(xy)/S(x) \tag{2}$$

Definition 2 Weighting support

Designated ones project to collect I = {i_1, i_2, i_m}, each project i_j is composed with the value w_j of right {$0 \leq wj \leq 1, 1 \leq j \leq m$}. If the rule is $X \rightarrow Y$, the weighting support is

$$S_w(xy) = \frac{1}{k} \sum_{i \in xy} w_j S(xy) \tag{3}$$

And, the *K* is the size of the Set *XY* of the project. When the right value *Wj* is the same as I_j, we calculating the weighting including rule to have the same support.

3. GENETIC ALGORITHM (GA)

Genetic Algorithm (GA) is a self-adaptive optimization searching algorithm. GA obtains the best solution, or the most satisfactory solution through generations of chromosomes constant evolution includes reproduction, crossover and mutation etc.

Here is the general description of this problem:

$$F(x) = a \times S(x) + b \times C(x) \tag{4}$$

a, b is constants, $a \geq 0$, $b \geq 0$, $S(x)$ is the support, and $C(x)$ is the confidence.

4. ASSOCIATION RULES MINING BASED ON A NOVEL GENETIC ALGORITHM

4.1 Encoding

This paper employs natural numbers to encode the variable A_{ij}. That is, the number of the lines of every range in the matrix A in which the element 1 exists is regarded as a gene. The genes are independent of each other. They are marked by $A_1, A_2 \ldots A_j, \ldots, A_n$, in which $A_j \in [1, m], j \in [1, n]$ and A_n may be a repeatedly equal natural number.

When the distributive method at random is employed to produce the initial population comprised of certain individuals, the population must be in a certain scale in order to achieve the optimal solution on the whole. The best way is the generated *M* individuals randomly that the length is N, then the chromosome bunch encoded by the natural number is calculated as the initial population.

4.2 The Fitness

Formula (3) is properly transformed into:

$$F(xy) = W_s \times \frac{S(xy)}{S_{\min}} + W_c \times \frac{C(xy)}{C_{\min}} \tag{5}$$

Here, $W_c+W_s=1$, $W_c \geq 0$, $W_s \geq 0$, S_{min}, is minimum support, and C_{min} is minimum confidence.

4.3 Reproduction Operator

Reproduction is the transmission of personal information from the father generation to the son generation. Each individual in each generation determines the probability that it can reproduce the next generation according to how big or small the fitness value is. Through reproducing, the number of excellent individuals in the population increases constantly, and the whole process of evolution head for the optimal direction. We are adopting roulette selection strategy; each individual reproduction probability is proportion to fitness value.

1) Compute the reproduction probability of all the individuals

$$P(i) = \frac{f(i)}{\sum_{i=1}^{M} f(i)} \tag{6}$$

2) Generate a number r randomly, r=random [0, 1];

3) If $P(0)+ P(1)+ \ldots + P(i-1)<r< P(0)+ P(1)+ \ldots + P(i)$, the individual i is selected into the next generation.

4.4 Crossover Operator

Crossover is the substitution between two individuals of the father generation that is to generating new individual. The crossover probability *Pc* directly influences the convergence of the algorithm. The larger *Pc* is the most likely is the genetic mode of the optimal individual to be destroyed. However, the over-small of *Pc* can slow down the research process (Wu zhaohui, 2005). Here is the definition of the crossover operator:

Computing crossover probability *Pc*

$$Pc = \begin{cases} 0.9 - \dfrac{0.3(f(x) - \overline{f(x)})}{f_{\max}(x) - \overline{f(x)}} & f(x) \geq \overline{f(x)} \\ 0.9 & f(x) \prec \overline{f(x)} \end{cases} \tag{7}$$

In which, $f_{max}(X)$ is the maximum fitness value of the population, $\overline{f(X)}$ is the average fitness value of the population.

4.5 Mutation Operator

The role of the mutation operator lies in that it enables the whole population to maintain a certain variety through the abrupt change of the mutation operator when a local convergence occurs in the population. The selection of the mutation probability P_m is the vital point because it influences the action and performance of the ARMNGA. If P_m is over-small, the ARMNGA will become a pure random research. Here is the definition of the mutations operator, computing the mutation probability P_m .

$$P_m = \begin{cases} 0.1 - \dfrac{0.009\,(f(x) - \overline{f(x)})}{f_{\max}(x) - \overline{f(x)}} & f(x) \geq \overline{f(x)} \\ 0.1 & f(x) \prec \overline{f(x)} \end{cases} \tag{8}$$

In which, $f_{max}(X)$ is the maximum fitness value of the population, $\overline{f(X)}$ is the average fitness value of the population.

4.6 Termination Condition

When the matching error ε≈0 or the condition is not coincident, the process will naturally stop.

5. EXPERIMENTS ON SYNTHETIC DATA

To check the research capability of the operator and its operational efficiency, such a simulation result is given compared with the GA in [2], The platform of the simulation experiment is a Dell power Edge2600 server (double Intel Xeon 1.8GHz CPU,1G memory, Redhat Linux 9.0).

We first compare the performance of our proposed method with the algorithm in Fig. 1 shows the runtime vs. the size of an image for both algorithms, where the size of the image varies for the synthetic dataset. As the size of the image increases, the runtimes of both algorithms decrease; nevertheless, the runtime of the ARMAGA algorithm does not change very much.

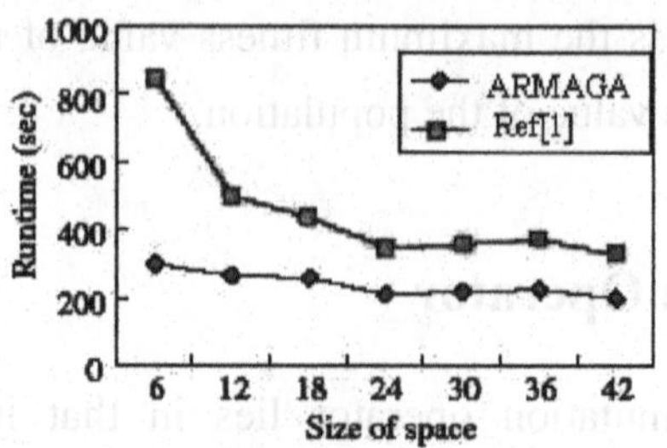

Fig. 1. Runtime vs. image size

Fig. 2 shows the runtime vs. number of objects for both algorithm, where the number of objects varies from 25 to 100 for the synthetic dataset. Since the average size of process and number of transaction are both fixed, the average support for the item sets decreases as the number of objects increases. Thus, the runtimes of both algorithms decrease slightly when the number of objects increases. Nevertheless, our proposed algorithm is faster than the algorithm in.

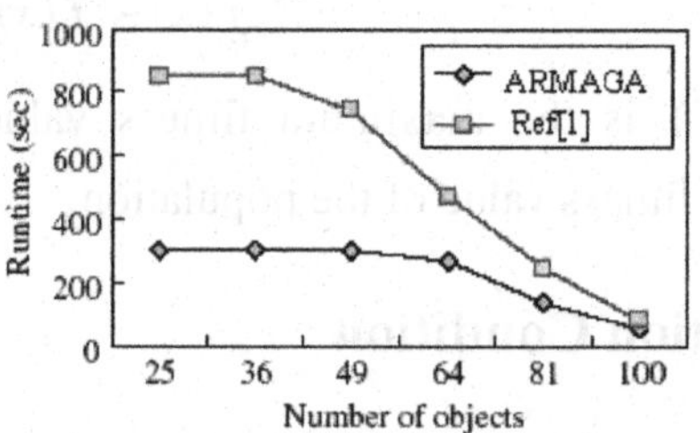

Fig. 2. Runtime vs. number of objects

From Fig. 1 and Fig. 2, we can educe that ARMAGA has a higher convergence speed and more reasonable selective scheme which guarantees the non-reduction performance of the optimal solution. Therefore, it is better than GA and ARMA through the theoretic analysis and the experimental results.

6. CONCLUSION

The image data mining in agriculture is a newly researching hot point in database area. But general data mining get knowledge from large quantities of data. We propose an Association rules mining based on a novel Genetic Algorithm, designed specifically for discovering association rules. We compare the results of the ARMAGA with the results of (G. Chen, 2002), and, it is better than GA and ARMA through the theoretic analysis and the experimental results.

REFERENCES

G. Chen, Q. Wei. Fuzzy association rules and the extended mining algorithms, Information Sciences 147 (2002) pp. 201–228.

Gao Li, Li Dan, Dai Shangping. A mining Algorithm of constraint based association rules, journal of Henan University Vol. 33 (2003) pp. 55–58.

Koperski, K., J. Han. Discovery of spatial association rules in geographic information databases, in: Proc. of International Symposium on Advance in Spatial Databases, SSD, LNCS, Vol. 951, Springer Verlag, 1995, pp. 47–66.

Hu, P.Y., Y.L. Chen, C.C. Ling. Algorithms for mining association rules in bag databases, Information Sciences 166 (2004) pp. 31–47.

Shijue Zheng, Wanneng Shu, Li Gao. Task Scheduling Using Parallel Genetic Simulated Annealing Algorithm, 2006 IEEE International Conference on Service Operations and Logistics, and Informatics Proceedings June 21–23, 2006, Shanghai, pp. 46–50.

Wu zhaohui. Association rule mining based on simulated annealing genetic algorithm, Computer Applications Vol. 25 (2005) pp. 1009–1011.

CONSTRUCTION AND APPLICATION BASED ON COMPRESSING DEPICTION IN PROFILE HIDDEN MARKOV MODEL

Zhijian Zhou [1,2], Daoliang Li [2,3], Li Li [2,3], Zetian Fu [2,3,*]

[1] *College of Science, China Agriculture University, Beijing, China, 100083*

[2] *Key Laboratory of Modern Precision Agriculture System Integration, Research Ministry of Education, Beijing, China, 100083*

[3] *College of Information and Electrical Engineering, China Agricultural University, Beijing, China, 100083*

[*] *Corresponding author, Address: P.O. Box 209#, China Agricultural University, East Campus, Beijing, 100083, P. R. China, Tel: +86-10-62736323, Email: fzt@cau.edu.cn*

Abstract: A method to express Profile Hidden Markov Model (Profile HMM) parameters with compressing matrix is presented, which is obtained by imposing the characteristics of both the state transfer and the character output in the Profile HMM.

Keywords: Profile HMM, Multiple sequence alignment, Bioinformatics

1. INTRODUCTION

The Profile HMM is composed by a Markov chain including matching, insertion, delete states, and an observable stochastic process namely as observation chain. The state chain depicts the transfer relationship among different state. The observation chain represents the statistical association between the state and observation. Generally, the state in Markov process cannot be observed directly. It can only be understood through the observable process. The Profile HMM model was introduced to bioinformatics by Krogh (1994), and now was widely used in the Alignment of biological sequence.

Zhou, Z., Li, D., Li, L. and Fu, Z., 2008, in IFIP International Federation for Information Processing, Volume 259; Computer and Computing Technologies in Agriculture, Vol. 2; Daoliang Li; (Boston: Springer), pp. 1291–1297.

A one grade Profile HHM with matching range as L is illustrated as Figure 1. This is a linear model. It progress only one direction from left to right. There are 3L+1 system states in the system namely as matching (M), insertion (I), and delete (D) respectively. For convenience in coding the program, two extra states, say as beginning (M_0) and ending (M_{L+1}), are involved in the model. They do not output any character to influence the model activity. The character set depends on the concerning object. For example, there are four characters in DNA sequence denoted by A, G, C, and T, and twenty characters for amino acid sequence. In the Figure 1, rectangle, diamond, and circle denote matching, insertion and delete state respectively. The arrow connected different states indicates the state transfer relation and direction. In an ascertain model, the transfer probability and character release between different state is wholly determinate.

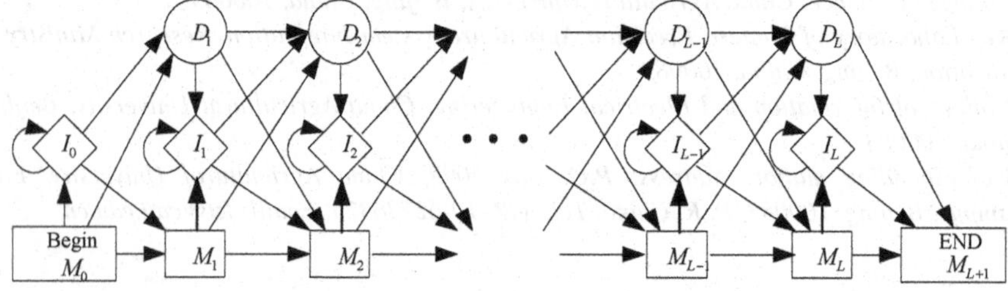

Figure 1. Model of one grade Profile HHM with matching range as L

2. COMPRESSION FORMS OF PROFILE HMM PARAMETERS

Based on the stochastic processes theory, a Profile HHM can be totally ascertained if $\lambda = \lambda(S, \Omega, \boldsymbol{A}, \boldsymbol{B}, \boldsymbol{\pi})$ is known. Where S denotes as state set, Ω is observation set, A is state transfer probability matrix, B is character output probability matrix, and π is initial state probability distribution function, respectively. Suppose a Profile HMM has matching range as L. Then the number of elements for S is $N = 3L+1$. If two extra state, beginning and ending stats, are also included, the number of elements for S become as $N = 3(L+1)$. So the order of state transfer matrix A is $3(L+1)\times 3(L+1)$. It can be found from Figure 1, at any state such as M_l, three front states as inputs of the state are at the most. They locate at same layer, say as l-1. This means only three occasions, $M_{l-1}M_l$, $I_{l-1}M_l$ and $D_{l-1}M_l$, can turn up when the state transfer from M_{l-1} to M_l. Similarly, only three occasions,

M_lI_l, I_lI_l, D_lI_l and $M_{l-1}D_l$, $I_{l-1}D_l$, $D_{l-1}D_l$ for I_l and D_l state occur respectively. Remembering none of character output for delete state, the state transfer probability matrix as order as $3(L+1)\times 3(L+1)$ can be compressed to $9\times(L+1)$ without lost any information:

$$\overline{\mathbf{A}}_{9\times(L+1)} = \begin{pmatrix} a[M_0M_1] & a[M_1M_2] & \cdots & a[M_{l-1}M_l] & \cdots & a[M_{L-1}M_L] & a[M_LM_{L+1}] \\ a[I_0M_1] & a[I_1M_2] & \cdots & a[I_{l-1}M_l] & \cdots & a[I_{L-1}M_L] & a[I_LM_{L+1}] \\ 0 & a[D_1M_2] & \cdots & a[D_{l-1}M_l] & \cdots & a[D_{L-1}M_L] & a[D_LM_{L+1}] \\ a[M_1I_1] & a[M_2I_2] & \cdots & a[M_lI_l] & \cdots & a[M_LI_L] & 0 \\ a[I_1I_1] & a[I_2I_2] & \cdots & a[I_lI_l] & \cdots & a[I_LI_L] & 0 \\ a[D_1I_1] & a[D_2I_2] & \cdots & a[D_lI_l] & \cdots & a[D_lI_L] & 0 \\ a[M_0D_1] & a[M_1D_2] & \cdots & a[M_{l-1}D_l] & \cdots & a[M_{L-1}D_L] & 0 \\ a[I_0D_1] & a[I_1D_2] & \cdots & a[I_{l-1}D_l] & \cdots & a[I_{L-1}D_L] & 0 \\ 0 & a[D_1D_2] & \cdots & a[D_{l-1}D_l] & \cdots & a[D_{L-1}D_L] & 0 \end{pmatrix}_{9\times(L+1)}$$

where $S=\{M_0, I_0, M_1, I_1, D_1, M_2, I_2, D_2\cdots, M_L, I_L, D_L, M_{L+1}\}$ are state set, and $a[X_iY_j]$ represents the one step transfer probability from X_i to Y_j. The elements sign as "0" means no corresponding state transfer, and elements in l column denote the probability of one step state transfer ending at l layer. Let $\overline{\mathbf{A}}_{9\times(L+1)} \equiv (\overline{a}_{ij})_{9\times(L+1)}$, $\overline{a}_{ij} = a[X_dY_j]$, $1\le i\le 9$, $1\le j\le L+1$, then following relations can be obtained,

$$\text{(1) } i \bmod 3 = \begin{cases} 1, & X = M \\ 2, & X = I \\ 0, & X = D \end{cases} \tag{1}$$

$$\text{(2) } (d,Y) = \begin{cases} (j-1,M) & if\ 1\le i\le 3 \\ (j,I) & if\ 4\le i\le 6 \\ (j-1,D) & if\ 7\le i\le 9 \end{cases} \tag{2}$$

$$\text{(3) } \overline{a}_{31} = \overline{a}_{91} = \overline{a}_{4(L+1)} = \cdots = \overline{a}_{9(L+1)} = 0 \tag{3}$$

$$\text{(4) } a_{ij} + a_{(i+3)(j-1)} + a_{(i+6)j} = 1,\ i=1,2,3;\quad j=2,3,4 \tag{4}$$

Similarly, let $\Omega=\{\omega_1, \omega_2,\cdots, \omega_K\}$ is observation character set, then character output probability matrix $\boldsymbol{B}_{K\times(3L+1)}$ can be compressed as

$$\overline{\boldsymbol{B}}_{K\times(2L+1)} = \begin{pmatrix} b_0^I[\omega_1] & b_1^M[\omega_1] & b_1^I[\omega_1] & b_2^M[\omega_1] & b_2^I[\omega_1] & \cdots & b_L^M[\omega_1] & b_L^I[\omega_1] \\ \cdots & \cdots & \cdots & \cdots & \cdots & \cdots & \cdots & \cdots \\ b_0^I[\omega_K] & b_1^M[\omega_K] & b_1^I[\omega_K] & b_2^M[\omega_K] & b_2^I[\omega_K] & \cdots & b_L^M[\omega_K] & b_L^I[\omega_K] \end{pmatrix}_{K\times(2L+1)} \tag{5}$$

where $b_j^X[\omega_k]$ is the probability of output ω_k at the state X_j. Let $\overline{\boldsymbol{B}}_{K\times(2L+1)} \equiv (b_{ij})_{K\times(2L+1)}$, $b_{ij} = b_d^X(\omega_i)$, $1 \le i \le K$, $1 \le j \le 2L+1$, following relations can be obtained,

$$j \bmod 2 = \begin{cases} 1, & (X,d) = (I, \dfrac{j-1}{2}) \\ 0, & (X,d) = (M, \dfrac{j}{2}) \end{cases} \tag{6}$$

Therefore a profile HMM can be simplified by the compression state transfer probability matrix with order of 9× (L+1) and the compression character output probability matrix with order of K× (2L+1).

3. FORWARD ALGORITHM

Let us consider the observation sequence $O = (o_1, o_2, \cdots, o_T)$. The matching range is L in Profile HMM (λ). Based on compress state transfer probability matrix $\overline{\boldsymbol{A}}_{9\times(L+1)}$ and character output probability matrix $\overline{\boldsymbol{B}}_{K\times(2L+1)}$, forward algorithm can be obtained.

Definition 1 Let $\alpha_l^X(t) = P(o_1, o_2, \cdots, o_t, \text{ end of } X_l \mid \lambda)$, $X = M, I, D$ be the probability when part sequence $O_t = (o_1, o_2, \cdots o_t)$ output in X_l state at l ($1 \le l \le L$) layer. Then $(\alpha_l^M(t), \alpha_l^I(t), \alpha_l^D(t))^T$ is the probability vector for l layer, denoted as $\boldsymbol{\alpha}_l(t)$.

Definition 2 Let $\varphi(X_r) = (a[M_{r-1}X_r], a[I_{r-1}X_r], a[D_{r-1}X_r])^{\mathrm{T}}$, $X = M, D$ be a column vector composed by one step transfer probability from state X_{r-1} to X_r, and $\varphi(I_q) = (a[M_qI_q], a[I_qI_q], a[D_qI_q])^{\mathrm{T}}$ also be a column vector composed by one step transfer probability from state I_{q-1} to I_q. Thus $\varphi(M_p) = (\overline{a}_{1p}, \overline{a}_{2p}, \overline{a}_{3p})^{\mathrm{T}}$, $\varphi(I_q) = (\overline{a}_{4q}, \overline{a}_{5q}, \overline{a}_{6q})^{\mathrm{T}}$ and $\varphi(D_r) = (\overline{a}_{7r}, \overline{a}_{8r}, \overline{a}_{9r})^{\mathrm{T}}$, if compress state transfer probability matrix is $\overline{\boldsymbol{A}}_{9\times(L+1)}$ and character output probability matrix is $\overline{\boldsymbol{B}}_{K\times(2L+1)}$.

Now the forward algorithm for Profile HMM can be express as,

1) Initiation

$$\boldsymbol{\alpha}_0(0) = (1,0,0)^{\mathrm{T}}$$

$$\boldsymbol{\alpha}_0(t) = (0,0,0)^{\mathrm{T}},\ t = 1, 2, \cdots, T+1 \tag{7}$$

$\boldsymbol{\alpha}_l(0) = (0,0,0)^{\mathrm{T}}, l = 1,2,\cdots,L+1$

2) Recursion calculation

$$\boldsymbol{\alpha}_l(t) = \begin{pmatrix} \boldsymbol{\alpha}_{l-1}^{\mathrm{T}}(t-1)\varphi(M_l)b_l^M(o_t) \\ \boldsymbol{\alpha}_l^{\mathrm{T}}(t-1)\varphi(I_l)b_l^I(o_t) \\ \boldsymbol{\alpha}_{l-1}^{\mathrm{T}}(t)\varphi(D_l) \end{pmatrix} \quad t=1,2,\cdots,T, \quad l=1,2,\cdots,L \tag{8}$$

where $b_d^M(o_t) = b_{i(2d)}$, $b_l^I(o_t) = b_{i(2d+1)}$, when $o_t = w_i$.

3) Ending

Thus

$$\alpha_{L+1}^M(T+1) = \alpha_L^M(T)a_{1(L+1)} + \alpha_L^I(T)a_{2(L+1)} + \alpha_L^D(T)a_{3(L+1)} \tag{9}$$

and the probability is

$$P(O \mid \lambda) = \alpha_{L+1}^M(T+1) \tag{10}$$

4. APPLICATION OF PROFILE HMM

An example of compress state transfer probability matrix and character output probability matrix for known multiple sequence comparison is shown below. Suppose DAN sequences have be aliment as Table 1. Matching states locate in first, second, and sixth column. Thus m is 5, T is 6 and L is 3. The corresponding Profile HMM framework is depicted in Figure 2.

Table 1. Alignment of multiple DNA sequence

	1	2	3	4	5	6
bat	A	G	-	-	-	C
rat	A	-	A	G	-	C
cat	A	G	-	A	A	-
gnat	-	-	A	A	A	C
goat	A	G	-	-	-	C

In order to avoid zero in probability calculation, pseudo counting is adapted. For example, character A appears 4 times in first column, the probability to output A for this state is $b_1^M(A) = \frac{4+1}{4+4} = 0.625$. Character G does not show up and the probability to output G is $b_1^M(G) = \frac{0+1}{4+4} = 0.125$.

Table 2 list character output probability at the state in which state transfer real occurs. As the same rule used, the compress state transfer probability matrix can be obtained as matrix (11).

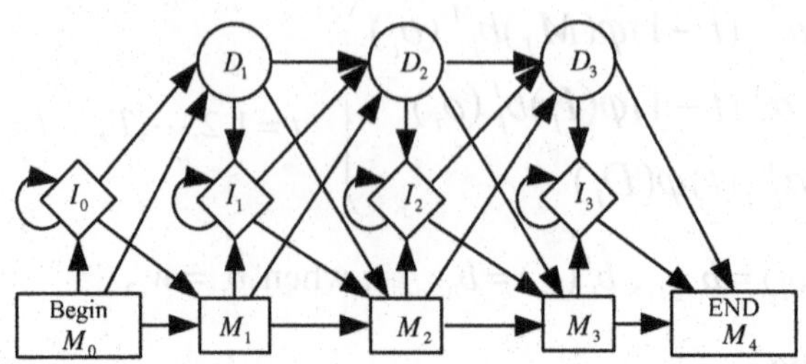

Figure 2. Profile HMM frameworks for the case of three states

Table 2. Frequency of character output

	$b_1^M(\omega_k)$	$b_2^M(\omega_k)$	$b_2^I(\omega_k)$	$b_3^M(\omega_k)$
A	0.625	0.143	0.333	0.125
G	0.125	0.571	0.333	0.125
C	0.125	0.143	0.111	0.625
T	0.125	0.143	0.222	0.125

$$A=\begin{pmatrix} 0.625 & 0.571 & 0.500 & 0.833 \\ 0.333 & 0.333 & 0.500 & 0.500 \\ 0.000 & 0.250 & 0.200 & 0.667 \\ 0.143 & 0.333 & 0.167 & 0.000 \\ 0.333 & 0.167 & 0.500 & 0.000 \\ 0.250 & 0.600 & 0.333 & 0.000 \\ 0.250 & 0.286 & 0.167 & 0.000 \\ 0.333 & 0.333 & 0.333 & 0.000 \\ 0.000 & 0.500 & 0.200 & 0.000 \end{pmatrix} \tag{11}$$

Suppose $O = AGC$ is a new observation sequence. Thus the probability that the sequence can be observed by using the model as shown above is $P(O|\lambda) = 0.0373$

In practical application, the observation sequence can be classed according to its probability for different model λ_i that established by corresponding training data.

ACKNOWLEDGEMENTS

This work supported by the grant (10631070) from the National Natural Science Foundation of China.

REFERENCES

Gong Guang-lu, Qian Min-Ping, 2004, Application stochastic process course and stochastic model in intelligent calculation, Qinghua Publisher, Beijing.

Katoh A, Mian I S, Haussler D., 1994, A hidden Markov model that finds genes in E.coli DNA. Nucleic Acids Research, 22:4768-4778.

Krogh A, Brown M, Mian I S, Sjolander K, and Haussler D., 1994, Hidden Markov models in computational biology: applications to protein modeling, Journal of Molecular Biology, 235:1501-1531.

Wang Yu-Fei, Shi Ding-Hua, 2006, Bioinformatiic intelligent arithmetic and its application, Chemistry industry Publisher, Beijing.

WHEAT GRAIN QUALITY FORECASTING BY CANOPY REFLECTED SPECTRUM

Wenjiang Huang*, Jihua Wang, Xiaoyu Song, Chunjiang Zhao, Liangyun Liu

National Engineering Research Center for Information Technology in Agriculture, P.O. Box 2449-26, Beijing, 100097, China

* *Corresponding author, Address: P.O. Box 2449-26, Beijing, 100097, P. R. China, Tel: +86-10-51503676, Fax: +86-10-51503750, Email: yellowstar0618@163.com*

Abstract: Advanced site-specific determination of grain protein content by remote sensing can provide opportunities to optimize the strategies for purchasing and pricing grain, and to maximize the grain output by adjusting field inputs. Field experiments were performed to study the relationship between grain quality indicators and foliar nitrogen concentration. Foliar nitrogen concentration at the anthesis stage is suggested to be significantly correlated with grain protein content, while spectral vegetation index is significantly correlated to foliar nitrogen concentration around the anthesis stage. Based on the relationships among nitrogen reflectance index (NRI), foliar nitrogen concentration, and grain protein content, a statistical evaluation model of grain protein content was developed. NRI proved to be able to evaluate foliar nitrogen concentration with a coefficient of determination of $R^2 = 0.7302$ in year 2002. The relationship between measured and remote sensing derived foliar nitrogen concentration had a coefficient of determination of $R^2 = 0.7279$ in year 2003. The results mentioned above indicate that the inversion of foliar nitrogen concentration and the evaluation of grain protein content by NRI are surprisingly good.

Keywords: Winter wheat (Triticum aestivum L), Canopy reflected spectrum, Grain protein

Huang, W., Wang, J., Song, X., Zhao, C. and Liu, L., 2008, in IFIP International Federation for Information Processing, Volume 259; Computer and Computing Technologies in Agriculture, Vol. 2; Daoliang Li; (Boston: Springer), pp. 1299–1301.

1. INTRODUCTION

Wheat (Triticum aestivum L) is one of the main grain crops in Northern China. It is important to evaluate wheat grain quality before the harvest. However, evaluating grain quality before it is ripe is difficult using current methods. Remote sensing can potentially rapidly determine the grain quality condition of crops over large areas. The objective of this paper was to determine a credible and applicable method to evaluate winter wheat foliar nitrogen concentration and grain protein content by in situ canopy reflected spectrum.

2. MATERIALS AND METHODS

The experiment was conducted at Beijing Xiaotangshan Precision Agriculture Experimental Base, in Changping district, Beijing (40°10.6′N, 116°26.3′E) from 2001–2002 and 2002–2003.

3. RESULTS

Grain protein indicators such as grain hardness, protein content, wet gluten content and dry gluten content. Grain starch indicators such as flour peak viscosity, trough, breakdown, final viscosity, and peak time. The correlation coefficients among grain hardness, flour final viscosity, and flour peak time were significantly positive, with correlation coefficients of 0.396 and 0.498. Correlation coefficients between grain sedimentation value and flour final viscosity, flour peak time and flour trough were highly significantly negative, which were respectively -0.616, -0.652, and -0.399. They did not reach robust correlations at 5% significance level among protein content, dry gluten content (grain protein indicators), flour trough, flour breakdown, flour final viscosity, and flour peak time.

Grain bread quality indicators include developing time, stability, elasticity, volume, and total score. The correlation coefficients among wet gluten, dry gluten content, sedimentation value (grain protein quality indicators), bread elasticity, and volume (grain bread quality indicators) were significantly negative. The correlation coefficients among wet gluten, dry gluten content, sedimentation value, total score, and specific volume were significant positive. They did not reach robust correlations at 5% relativity level between protein content, bread elasticity, volume, total score, and specific volume.

The foliar nitrogen concentration at the anthesis stage was thus highly correlated to grain protein content. If the foliar nitrogen concentration at the anthesis stage could be monitored by remote sensing technology, the grain protein content could be evaluated. Foliar nitrogen concentration at anthesis stage is suggested to be significantly correlated with grain protein content, while spectral vegetation index is significantly correlated to foliar nitrogen concentration around anthesis stage. Based on the relationships among nitrogen reflectance index (NRI), foliar nitrogen concentration, and grain protein content, a statistical evaluation model of grain protein content was developed. Our results reveal good agreement between the measured foliar nitrogen concentration and NRI, with a coefficient of determination of $R^2 = 0.7302$ (n = 240), which was very significantly positive (r(0.01,240) = 0.181). The model of the relationship between foliar nitrogen concentration and NRI was validated using data from 2003. The coefficient of determination between remote sensing derived and the measured foliar nitrogen concentration from 2003 is 0.7279, which is extremely significant.

This study showed robust correlations between NRI and foliar nitrogen concentration, suggesting that NRI is a promising indictor to predict winter wheat grain protein for winter wheat. The NRI proved to be able to evaluate foliar nitrogen concentration with a coefficient of determination of $R^2 = 0.7302$ using the data from 2002. The relationship between foliar nitrogen concentration that was measured and derived by remote sensing had a coefficient of determination of $R^2 = 0.7279$ in 2003. The correlation between the grain protein content that was measured and that was derived from in situ canopy-reflected spectrum was $R^2 = 0.7661$ in 2003.

Our study contributes towards developing optimal procedures for predicting wheat grain quality through the analysis of canopy-reflectance spectrum data before the harvest of large areas. Using the results of this paper, we are developing some simple instruments with selected bands of sensitivity. For example, we are developing optical camera lens and sensors that focus on 570nm and 670nm, which could be placed on agricultural machines traveling in the field. This would allow for on-site and non-sampling modes of crop growth monitoring, fertilizing, and water guidance without a priori knowledge. Such portable instruments can also estimate grain quality at the anthesis stage by the regression models of this paper.

According to the results of this paper, the relationships among NRI, leaf nitrogen concentration, and grain protein content were significant. A model for evaluating grain protein was established based on the transfer principle of foliar nitrogen concentration, which made it possible to optimize crop nitrogen management for grain quality.

Our results indicate that evaluating grain quality indicators using NRI spectral index is surprisingly good.

ACKNOWLEDGEMENTS

This work was subsidized by the National High Tech R&D Program of China (2007AA10Z201, 2006AA10Z203), it was also supported by the State Key Laboratory of Remote Sensing Science (KQ060006), the program from Ministry of Agriculture (2006-G63). The authors are grateful to Mrs. Zhihong Ma, Mr. Weiguo Li and Mrs. Hong Chang for data collection.

GEO-REFERENCED SPATIAL MULTIMEDIA APPLICATION FOR AGRICULTURAL RESOURCE MANAGEMENT

Hui Liu[1] , Gang Liu[1] , Zhijun Meng[2]

[1] *Key Laboratory of Modern Precision Agriculture System Integration Research, Ministry of Education, China Agricultural University, Beijing 100083, China Corresponding author, Email: pac@cau.edu.cn*

[2] *National Engineering Research Center for Information Technology in Agriculture, Beijing, P. R. China*

Abstract: Agricultural resource management is one of important technologies in agricultural application. The new technology, geo-referenced spatial multimedia has revolutionized the data collection, storage, management and display for agricultural resource management. This paper presents geo-referenced spatial multimedia technology including the system architecture, data collection procedures and data processing based on VMS system of Red Hen Company, and then introduced the application for agricultural resource management.

Keywords: GPS, GIS, Geo-referenced spatial multimedia, precision agriculture

1. INTRODUCTION

In recent years, more and more information technologies such as Global Positioning Systems (GPS) and Geographic Information Systems (GIS) are applied in agricultural resource management for digital data collection, storage, manipulation and display. Although many GIS have been successfully implemented, it has become quite clear that two-dimensional maps cannot precisely present multidimensional and dynamic spatial phenomena. Geo-referenced

Liu, H., Liu, G. and Meng, Z., 2008, in IFIP International Federation for Information Processing, Volume 259; Computer and Computing Technologies in Agriculture, Vol. 2; Daoliang Li; (Boston: Springer), pp. 1303–1306.

spatial multimedia is a new technology that merges photography and geography to help users understand space by experiencing abstraction and reality through the integration of maps and images.

Some researchers have reported their applications of geo-referenced spatial multimedia. Ayers et al. (2000) used the video mapping system for field data collection in Lorry State Park, and generated geo-referenced information, which provided visual source and observation for management of the park in terms of timing. Liu et al. (2000) explored the application of the GPS/GIS and VMS in agricultural machinery.

This paper presents geo-referenced spatial multimedia technology based on video mapping systems (VMS) of Red Hen Company, and then introduces VMS applications in agricultural resource management.

2. MATERIAL AND METHODS

2.1 Geo-referenced spatial multimedia technology

Video mapping systems (VMS) developed by Red Hen Systems is a comprehensive spatial multimedia system including data collection devices and data processing software. System architecture can be described as three-layer framework – hardware layer, management layer and application layer.

The hardware layer is for data collections. In the first method, a camera or a video and a GPS receiver work separately, there is no cable between them. But it is necessary to log GPS track and take still image or video at the same time and same place. In another method, a specific device is used for data collections of merging for position and media to achieve data merging.

After data collection, these multi-resources data are imported to the management layer, MediaMapper, which is data processing software. MediaMapper software automatically merges these source media with spatial information from GPS receivers, building an interactive media map. MediaMapper can also export into GIS format and HTML file.

Based on the processed spatial media data, differential areas can design their application system according to differential needs such agricultural management, city facilities management and environment protection.

2.2 Applications in agricultural resource management

Many high technology of precision agriculture including GPS, GIS, ES, variable management have been applied in our demo farm. The interesting positions such as farm infrastructure, sampling points, pest occurrence could be imported and displayed in the digital map for precision management, whilst most of these positions were also photographed. But the interesting positions could not be related with their photos in the past. We require a means of associating GPS data with pictures and videos. Geo-referenced spatial multimedia mapping technology becomes the ideal selection. The objective of our application was to summarize the utilization of GPS, GIS and VMS to obtain geo-referenced video and images for agriculture resource management, and to export the spatial data into ArcGIS for analysis and management.

The image or video and their locations for agricultural resource data had been collected in two ways of post-processing integration and real-time merging.

A Nicon digital camera and a Garmin portable GPS receiver were selected for data collection of post-processing integration. It was necessary to calibrate time of camera and GPS receiver before collection. The simplest way for time calibration was taking photo for UTC time display of the receiver with the camera. If using video camera, a video for UTC time display could be taken for time calibration. It is sure to make the photo or video for time calibration distinguishable because of it is important for data processing. GPS receiver was configured to output GGA, RMC sentences. Then data collection was beginning. Taking photos at the interesting location while logging GPS track. It is sure to keep the same step while operating the camera and receiver for data matching during processing.

In real-time merging method, a Sony digital video camera and a VMS device which had inner GPS module for positioning were selected. It was necessary to make sure to connect devices correctly. The video camera and the VMS device were connected using the microphone (MIC) connector. After the antenna of GPS module connected and the satellites locked, it was time to take video of agricultural resource data when recording real-time NMEA GGA and RMC strings on the videotape.

2.3 Results

GPS receiver, digital camera, video camera and VMS device were connected the computer separately for data input. MediaMapper provided a data import wizard, which made the operation easier. After data processed, users could add the layers of background digital map and set the options for display resulted in a geo-referenced spatial multimedia map.

The data of agricultural application were exported into ArcGIS of ESRI Company in SHP file for management and analysis after the extension tool setup for the ArcGIS environment. Figure 1 shows a sprinkler of the farm in ArcView. The data were also created and exported in HTML format. Figure 2 shows a HTML file. We put HTML files onto the HTTP server system and provided consultation via Internet.

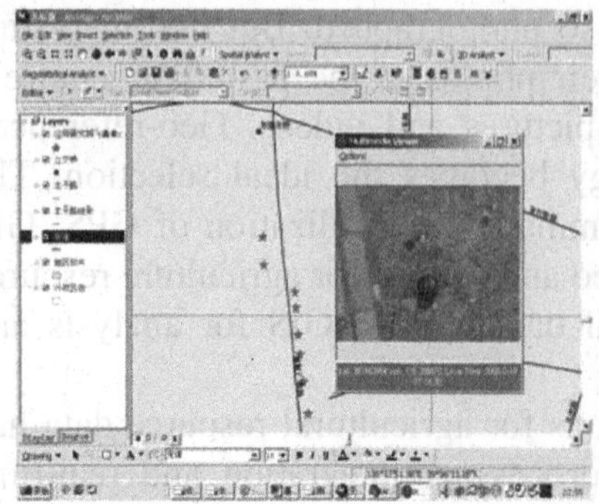

Figure 1. ArcView display of spatial multimedia data

Figure 2. HTML files of spatial multimedia data

3. CONCLUSION

In agriculture, geo-referenced spatial multimedia could be applied not only in resource management such as geo-referenced data collection in fields, machines and facilities management in farms, but also in crop analysis, pest and disease prevention and weeds monitoring. This new technology should highly improve visual display and management efficiency in agricultural activities, and help people to find enjoyment during farming.

REFERENCES

Ayers, P., Juhua Liu, and J. Hocheder. GPS-based video mapping for nature resource management. Proceedings of the Second International Conference on Geospatial Information in Agricultural and Forestry. pp. I-464-468. Disney's Coronado Springs Resort, Lake Buena Vista, Florida. January 10-12, 2000.

Liu, Juhua, P. Ayers, and N. Geng. Database Modeling of Agricultural Machinery Safety and Production Based on Integrating GPS/GIS and VMS Technology. Presented at 2000 ASAE Annual International Meeting Sponsored by ASAE Milwaukee, Wisconsin, USA July 9-12, 2000.

DEVELOPMENT OF DGPS GUIDANCE SYSTEM FOR AGRICULTURAL MACHINERY

Zhijun Meng[1], Wenqian Huang[1], Hui Liu[2], Liping Chen[1], Weiqiang Fu[1]
[1] *National Engineering Research Center for Information Technology in Agriculture, Beijing, P. R. China Corresponding author, Email: mengzj@nercita.org.cn*
[2] *Key Laboratory of Modern Precision Agriculture System Integration Research, Ministry of Education, China Agricultural University, Beijing, 100083*

Abstract: The purpose of this research was to develop a kind of light bar parallel guidance system based on DGPS for small agriculture machinery. This system consists of field computer, light bar and DGPS. The field computer is developed using PC/104 CPU module, running on which the embedded real-time operation system was customized by a special development tool. Single chip and CAN controller was selected for guidance light bar design, which use a row of indicator LEDs to guide drivers. Field computer and guidance light bar communicate with each other by using CAN bus. The guidance software running on the field computer was designed in object-oriented manner by integrating embedded GIS development kit. Field-testing and experiments have been done to evaluate the accuracy of the system with Trimble AgGPS 132 under different speeds, application width and light bar sensitivity, and the average of offline distance was 0.4517 meter. Results show this kind of guidance system is suitable for small-size machinery and can meet the need for parallel swath guidance for different kinds of field operation.

Keywords: DGPS, Agricultural Machinery, parallel guidance

1. INTRODUCTION

Agricultural machinery are usually driven in parallel swath pattern during the operation of seeding, fertilizing, spraying and seedbed preparing, so the machines' moving track will affect the operation directly. In order to avoid overlap or skip in the operation, the scriber, furrow, ditch and crop rows are

Meng, Z., Huang, W., Liu, H., Chen, L. and Fu, W., 2008, in IFIP International Federation for Information Processing, Volume 259; Computer and Computing Technologies in Agriculture, Vol. 2; Daoliang Li; (Boston: Springer), pp. 1307–1310.

used as guidance in traditional methods. Operators must focus their attention on driving to be sure of the tractor's moving in parallel pattern with the help of this guidance. However, guidance effect is dependent on the proficiency of operators, especially when using a large-scale machine with a wide application width at a high speed (O'Connor, 1996).

GPS guidance and machine vision guidance are the two main promising ways. With the development of precision agriculture, GPS equipments with high accuracy are used widely. Guidance based on GPS and operation path layout has many advantages such as high reliability and less time limitation. Recently, there are many commercial products of Light-bar Guidance System. Researchers do many researches and experiments on the light-bar guidance precision. Buick and Lang researched and analyzed the precision and efficiency between foam marker method and DGPS light-bar guidance system through experiments in 1998 (Buick, 1998).

The purpose of the present study was to develop a kind of light bar parallel guidance system based on DGPS for small agriculture machinery. Field-testing and experiments have been done to evaluate the accuracy of the system with the use of a Trimble AgGPS under different speeds, application width and light bar sensitivity.

2. SYSTEM DESIGN

This parallel swathing guidance system consists of field computer, light bar guidance aids and DGPS. Field computer for agriculture machinery is designed for controlling, displaying and data storage during the process of parallel swathing guidance, variable rate application of fertilizer and chemicals, field information collection etc. This field computer is designed based on PCM-3350, ADVANTECH PC/104 CPU module, which with GX1-300 MHz onboard processor and low power, fangless performance. PCM-3350 supports Compact Flash SSD solution, 18-bit TFT LCD, touch panel, some standard communications ports. In order to make field computer support CAN bus communication, PCM 3680 CAN adapter card was selected for this field computer. PCM 3680 support CAN interface memory address of CAN interface can be changed by address switch.

Windows CE .NET was selected as embedded real time operation system (RTOS) for this field computer. Windows CE .NET is 32-bit, real-time, multitasking OS, which is highly componentized, scalable and has wide variety of CPU support. It combines an advanced, real-time operating system with powerful tools for rapidly creating the next generation of smart, connected, and small-footprint devices and Windows CE for Automotive and Industrial Automation. Some configuration and development of Wince operation system have been done to meet the field computer hardware environment by using Platform Builder. Platform Builder is a Configuration and debugging tool for deploying an operating system.

In order to avoid overlap or leaving gaps in the field, the light-bar guidance system is used to help the operator to drive the machine moving along the desired parallel passes, which can improve the operation quality and efficiency. W77E58 C51 microcontroller of Winbond Company, the independent CAN controller SJA1000 and CAN transceiver PCA82C250 of Philips Company are adopted, which realize the CAN communication. The high-speed optocouplers 6N137 are used for voltage isolation between input and output and watchdog timer X5045P for independent protection for microcontroller. Since the variable rate controller and the light-bar guidance unit are in the same CAN local area network, the guidance unit can receive the deviation data from the controller using the special communication protocol. The 35 LEDs and the LCD can display the deviation in an obvious way. The 74ALS373 and 74F377 are used for address logic and LED control signal output respectively.

In order to meet the need of parallel guidance, the application should provide the functions such as parallel guidance display, vehicle motion control and application data management. This application has to deal with lots of control, operation, object and data. Following object-oriented programming thought, above data and objects concerned in a parallel swath guidance application procedure could be abstracted into different classes. Workspace class was used to manage all of data and objects, which is created when parallel swath guidance application project begins. This object organizes the whole application procedure, creates relevant objects such as logging object, field object, vehicle object and so on. Guidance data communication protocol prescribes the data format for field computer and guidance light bar. Current version of this protocol includes navigation sentence, GPS status sentence, navigation status sentence and machine travel status sentences. This protocol can be extended according to system need. Navigation protocol between the central computer and the stimulation light bar should be defined so as to the navigation message the sender and receiver identify string. A typical navigation string is shown as following: $L, L, 2, *hh, CR/LF.

3. FIELD TEST

Field test was accomplished in September of 2005 on National Precision Agriculture Demonstration Farm at suburb of Beijing. The vehicle model TD724 was an agricultural tractor made in China, Trimble AgGPS132 DGPS receiver was used for the vehicle positioning. Considering vehicle speed, swath width and light bar sensitivity, 9 experiments were conducted totally. There are no ground references for guidance in the testing field. Track points, track line and swath coverage data was recorded during the test. test Recorded data was processed with ArcMap by using layer overlap, intersection and clip tools. A program was designed for getting the distance between the actual track point and the ideal swath. The maximal value of offline distance was 0.6128 meter and the minimal value was 0.3021 meter among the results. And the average offline distance was 0.4517 meter. The field computer and light bar worked well in the test. The guidance system could be proven to have good performance for small-size tractors.

4. CONCLUSION

This research designed the parallel swath guidance system based on DGPS for the medium and small size tractors made in China. This system consists of field computer, light bar and DGPS. The field computer is developed using PC/104 CPU module, running on which the embedded real-time operation system was customized by a special development tool. Single chip and CAN controller was selected for guidance light bar design, which use a row of indicator LEDs to guide drivers. Field computer and guidance light bar communicate with each other by using CAN bus. The guidance software running on the field computer was designed in object-oriented manner by integrating embedded GIS development kit. Field test indicated that the guidance system was fit for the small-size tractors. But more tests should be given using RTK-GPS receiver of high precision for reliability and accuracy.

REFERENCES

Buick, R. White, E. 1998. Comparing GPS guidance with foam marker guidance. In: International Conference on Precision Agriculture. Madison: ASA-CSSA-SSSA, 1035-1045.

O'Connor, M., T. Bell, G. Elkaim, and B. Parkinson. 1996. Automatic steering of farm vehicles using GPS. 3rd International Conference on Precision Agriculture, Minneapolis, MN.

CALIBRATION OF MACHINE VISION SYSTEM FOR NONDESTRUCTIVE DETECTION OF PLANTS

Ming Sun[1,2], Dong An[1], Yaoguang Wei[1]

[1] *College of Information and Electrical Engineering, China Agricultural University, Beijing, China, 100083*

[2] *Author for correspondence Address, P.O. Box 63, 17 Tsinghua East Road, Haidian District, Beijing, 100083, P. R. China, Tel: +86-10-62737591, Email: drmingsun@163.com*

Abstract: The detection of plants based on a machine vision system is one of the most important parts of plants simulation. The calibration of the machine vision system is a basic step for nondestructive detection. We discuss common methods such as traditional calibration techniques, active vision based calibration techniques, and self-calibration techniques. The results show that it is feasible to apply the traditional linear calibration method in nondestructive measurement of plants.

Keywords: machine vision, calibration, nondestructive detection

1. INTRODUCTION

Machine vision is a focus of technologies, which is used to simulate the outside or macroscopically vision function of the target by the study of computer's applications. Its chief function is to create or renew the real object by images. Machine vision technique is widely applied to industry, agriculture, spaceflight, remote sensing and medicine, etc. because of its contactless detection, fast speed and high precision. Recently, it also shows great application prospects in botany, agronomy, ecology and forestry (Jain et al., 2003).

Sun, M., An, D. and Wei, Y., 2008, in IFIP International Federation for Information Processing, Volume 259; Computer and Computing Technologies in Agriculture, Vol. 2; Daoliang Li; (Boston: Springer), pp. 1311–1314.

The purpose of the detection of plants based on machine vision system is to study the growing rule of plants quantitatively. The calibration of the machine vision system is a basic step for nondestructive detection, which means that the calibration of the camera in the machine vision system is also necessary. The calibrating techniques of camera can be divided into traditional calibration, self-calibration, and active vision based on calibration. We choose one, which is adapting to the nondestructive detection of plants.

2. METHODS FOR CAMERA CALIBRATION

The methods for camera calibration can be divided into traditional calibration techniques, self-calibration techniques, and active vision based on calibration techniques. The above-mentioned methods will be discussed respectively.

Traditional calibration techniques is to use the frame information of the scenery which usually takes a calibration block whose frame is known and process precision is high as a directional reference, and to simulate the relation between the parameters of the camera model through the corresponding relation of a space spot and a camera spot, and then to get the parameter by optimizing algorithm. The calibration method can use any kind of camera model and has a high precision. But it requests a high-precision calibration block, and we cannot use the block in practical applications sometimes. This method mainly includes direct linear transform camera calibration (Yang et al., 2000), calibration based on the RAC (Tsai, 1987), and planar calibration (Zhang, 1998; Wu and Sun, 2004).

The camera calibration based on active vision is an important branch of the camera calibrations. The so-called camera calibration based on active vision is a method to calibrate cameras in a condition that some movement information of the camera is known, and the movement information includes the quantitative information and the qualitative information. The quantitative information refers to the case that the camera was translated for a given distance to a certain direction in the reference frame of the platform, and the two-translation movement orthogonality of the camera, etc. The qualitative information refers to the case that the camera does absolute linear movement or absolute rotary movement, etc. The main advantage of the camera calibration based on active vision is introduced as follows: The model parameters of the camera can be evaluated by linear equations since the movement information of the camera in the calibration process is known. The current research focuses on evaluating the model parameters through linear equations while reducing the movement restriction of the camera as much as possible. Representative researches include calibration based on three-orthogonal translation movement (Hu and Wu, 2002) and orthogonal movement method based on homographic matrix (Wu and Hu, 2001, 2002).

The self-calibration of camera is a process that calibrating the camera by the corresponding relation of the image points, without any calibration blocks. In the process of self-calibration, it is assumed that the corresponding relation of the image points is determined, and the interior parameters of the camera make no changes when different images are shot (Yu et al., 1999; Wu et al., 2001).

3. RESULTS FOR CAMERA CALIBRATION

Since the machine vision system applied in nondestructive measurement of plants does not require the online calibration, the traditional linear calibration using calibration blocks can easily meet the testing requirements.

Therefore, calibration is done using a one-Yuan coin as a calibration reference. The calibration contains the processes below: collecting images of the calibration reference by CCD camera, shooting images in the condition that the vertical shooting height is fixed and the distance between the lens and the calibration reference and the distance between the lens and the plants are set at the same condition, counting the number of the pixels and calculating the true size that each pixel represents. According to the true size that each pixel represents, the true size of the plant shot in the same condition can be calculated. Figure 1 stands for the three images of two coins collected in different locations when the distance between the lens and the coin is 20 cm and the focal length is 16 cm. Table 1 stands for the corresponding testing results.

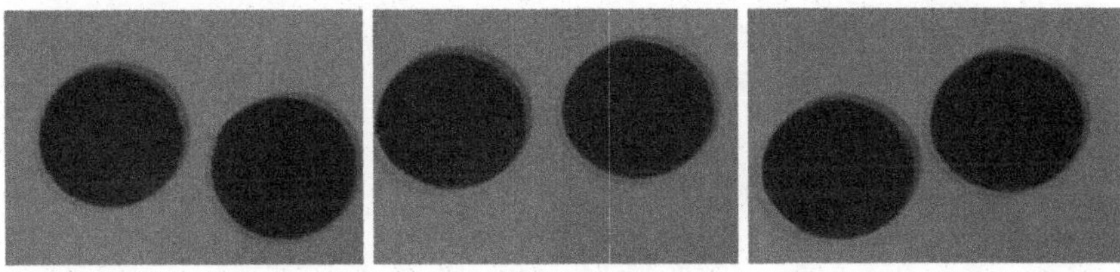

Figure 1. Image of coins

Table 1. Result of measurement

Figure 1	pixels	size per pixel (mm^2)	length per pixel (mm)
Left	61231	0.00566	0.07523
	59802	0.00579	0.07609
Middle	61064	0.00567	0.07530
	59756	0.00579	0.07609
Right	61217	0.00566	0.07523
	60105	0.00576	0.07589
Average	60529	0.00572	0.07564

As plants grow, calibration outcomes in different altitudes can be obtained when adjusting the distance between the lens and the coin, and through this, the values of plants can be calculated in the corresponding distance. The average size per pixel is 0.00572 mm^2.

REFERENCES

Hu Z, Wu F. A review on some active vision based camera calibration techniques. Chinese Journal of Computer, 2002, 25 (11): 1149–1156.

Jain R, Kasturi R, Schunck. Machine vision. China Machine Press, Beijing: 1–20, 2003.

Tsai R Y. A versatile camera calibration technique for high-accuracy 3D machine vision metrology using off-the-shelf TV cameras and lenses. IEEE Journal of Robotics and Automation, 1987, 3 (4): 323–343.

Wu F, Hu Z. Linear determination of the infinite homography and camera self-calibration. Chinese Journal of Automation, 2002, 28 (4): 488–496.

Wu F, Hu Z. A new theory and algorithm of linear camera self calibration. Chinese Journal of Computer, 2001, 24 (11): 1121–1135.

Wu F, Li H, Hu Z. New active vision based camera self-calibration technique. Chinese Journal of Automation, 2001, 27 (6): 736–746.

Wu W, Sun Z. Overview of camera calibration methods for machine vision. Chinese Journal of Computer Applied Research, 2004, (2): 4–6.

Yang N, Yang J, Huang C, et al. Influence of calibration on the accuracy of 3-D reconstruction in direct linear transformation algorithm. Journal of Tsinghua University (Natural Science), 2000, 40 (4): 24–27.

Yu H, Wu F, Yuan B, et al. Camera self calibration technique based on active vision. Chinese Journal of ROBOT, 1999, 21 (1): 1–7.

Zhang Z. Computer vision –Theory and algorithms for calculating. Scientific publishing company, Beijing, 1998.

STUDY ON DETECTION TECHNOLOGY OF MILK POWDER BASED ON SUPPORT VECTOR MACHINES AND NEAR INFRARED SPECTROSCOPY

Jingzhu Wu[1], Shiping Zhu[2], Yun Xu[3], Yiming Wang[3,*]

[1] *College of Information Engineering, Beijing Technology & Business University, Beijing, China, 100037*

[2] *College of Engineering and Technology, Southwest University, Chongqing, China, 400715*

[3] *College of Information and Electrical Engineering, China Agricultural University, Beijing, China, 100083*

* *Corresponding author, Address: P.O. Box 63, China Agricultural University, 17 Tsinghua East Road, Beijing, 100083, P. R. China, Tel: +86-10-62737591, Email: ym_wang@263.net*

Abstract: This paper presents a novel classifier to identify standard and sub-standard milk powder, which is built by support vector machines (SVM) and near infrared spectroscopy (NIR). The training set is composed of 38 samples and the testing set is composed of 12 samples. The correct classification ratio of the training set is up to 100%, while that of the testing set is up to 100%. The result indicates that the combination of SVM and NIR can be used as a fast, convenient, and safe technology to identify standard and sub-standard milk powder.

Keywords: Near Infrared Spectroscopy, Support Vector machines, Milk Powder

1. INTRODUCTION

There appeared many kinds of sub-standard milk powder in recent years in China, which are damaged to people's life. Traditional chemical method to detect milk powder is time-consuming and complex. Near infrared spectrometry (NIR) technology is fast, green and nondestructive. NIR is gradually applied in the field of food and agriculture. Support Vector Machines (SVM) is a new machine-learning algorithm and a new technology

Wu, J., Zhu, S., Xu, Y. and Wang, Y., 2008, in IFIP International Federation for Information Processing, Volume 259; Computer and Computing Technologies in Agriculture, Vol. 2; Daoliang Li; (Boston: Springer), pp. 1315–1316.

to data classification. This paper presents a novel classifier to identify standard and sub-standard milk powder, which is built by SVM and NIR. The study is supported by the foundation of National Advanced Technology Development Project (863 project, No. 2003AA209012) and National Natural Science Foundation of China (project No. 30671198).

2. EXPERIMENT

The sample set is composed of 36unit standard milk powder sample and 14unit substandard milk powder. Near infrared spectrum of sample set were collected by diffuse reflectance from 12500 to 4000 cm-1 in 16 cm-1 on MATRIX-I spectrometer. The software used for this research was implemented in MATLAB v.6.5. The software package about SVM named LIBSVM can be conveniently downloaded from the web, http://www. kernelmachines.

We selected 38 samples, including 30 standard samples and 8 sub-standard samples, as the training set. The testing set is made up of the other 6 standard samples and 6 sub-standard samples. A classifier based on SVM is built to identify the quality of milk powder in this paper. We choose the RBF kernel and the defaulted the penalty parameter C=1.

Table 1. Classifying result by SVM with original input

γ	Support Vectors	Recognition Ratio	Prediction Ratio
0.5	18	89.34% (34/38)	83.33% (10/12)
1.5	19	94.73% (36/38)	100% (12/12)
2.5	19	100% (38/38)	100% (12/12)
3.5	20	100% (38/38)	100% (12/12)
4.5	18	100% (38/38)	83.33% (10/12)
5	19	100% (38/38)	66.67% (8/12)

The result indicates that the SVM classification has good performance to identify the stand and sub-stand milk powder. The theorem states that the less support vectors, the more powerful the model is (Zhang Xuegong, 2000). So the best parameter is γ=2.5 and the number of support vector is 19. It is obvious that while γ is increasing, the Recognition Ratio and the Prediction Ratio is increasing along with γ. After the both ratios arrived at 100%, the Prediction Ratio will drop down and Recognition Ratio will maintain 100% if γ keeps on increasing. So the kernel parameter γ influences the SVM model prediction ability.

REFERENCES

Zhang Xuegong, Introduction to Statistical Learning Theory and Support Vector Machines, *Acta Automation Sinica*26 32-42 (2000)

DESIGN OF CAN-BASED VARIABLE RATE FERTILIZER CONTROL SYSTEM

Wenqian Huang[1,*], Zhijun Meng[1], Liping Chen[1], Chunjiang Zhao[1]

[1] *National Engineering Research Center for Information Technology in Agriculture, China 100097*

[*] *Corresponding author, Address: Room A516, Nongke Building, Shuguang Garden Middle Road, Haidian District, Beijing, 100097, P. R. China, Tel: +86-10-51503425, Fax: +86-10-51503750, Email: huangwq@nercita.org.cn*

Abstract: A novel control system for variable rate fertilizer application based on CAN bus is presented. The system consists of variable rate controller based on DSP, ground speed sampling unit and light-bar guidance unit. These units form a control area network based on CAN bus. The variable rate controller collects signals of GPS position and ground speed, and calculates the conveyor feed shaft rotate speed according to the prescription map stored in local memory. Then the variable rate application can be realized through PID control of the shaft rotate speed. Light-bar guidance unit helps the operator to steer the machine moving by equal spacing parallel passes for less overlap and missing in operation. The effectiveness of the developed system is confirmed by the experiments using the communication protocol special for the application based on CAN bus.

Keywords: Precision Agriculture, variable rate fertilizer, CAN bus, PID control, DSP

1. INTRODUCTION

In recently years, Precision Agriculture (PA) is becoming one of the most important fields in research of agriculture science. Precision Agriculture is the integrated application of GPS, RS, GIS and automatic control technologies. By knowing the different characteristics in each field cell, maximum profit can be obtained by balancing precise amounts of inputs with crop needs, which is determined by weather, soil characteristics and

Huang, W., Meng, Z., Chen, L. and Zhao, C., 2008, in IFIP International Federation for Information Processing, Volume 259; Computer and Computing Technologies in Agriculture, Vol. 2; Daoliang Li; (Boston: Springer), pp. 1317–1320.

historic crop performance. In addition, precision agriculture techniques will more closely meet environmental guidelines because fertilizers, insecticides, and herbicides can be managed to apply the minimum needed for effectiveness (Wang 1995). As an important part of precision agriculture technology, much attention has been paid to the variable rate technology. Variable rate application has the potential to improve fertilizer utilization efficiency, increase economic returns and reduce environmental impacts. Today, developed countries have used variable rate technology widely, but it is still in the starting stage in China. The rapid development of precision agriculture has increased the need for a standardized electronics communications protocol also (Marvin L. Stone et al., 2004).

In this paper, we present a novel system based on CAN bus and provide evaluation experiments for the variable rate fertilizer application. The system structure consisting of the electro-hydraulic proportional valve control system, ground speed sampling ECU and light-bar guidance ECU is presented in this paper. The design of variable rate controller based on DSP and the closed loop control circuit based on proportional valve is discussed in details. Experimental results indicate that the control system can be used for variable rate fertilizer application based on CAN bus. Moreover, the system is open, easy to expand and communicate with other CAN bus based components. It also provides basis for other variable rate applications using CAN bus in precision agriculture.

2. MATIERIALS AND METHOD

2.1 System Construction

The control system links all units together via CAN bus. The variable rate controller based on DSP is the main node in charge of collecting signals from ground speed sampling unit and GPS receiver. The conveyor feed shaft rotate speed can be calculated using these signals and the prescription map stored in local memory. The variable rate fertilizer application is achieved by the PID control of the shaft speed. In the operation progress, the light-bar will help the operator to steer the machine moving along equal spacing parallel passes by lightening different number of LED. With the cooperation of these nodes, the variable rate application can be fulfilled successfully.

In our system, the electro hydraulic proportional valve is adopted as the control component (Figure 1). The developed variable rate controller compares the feedback signal from the optical rotary encoder mounted on the load shaft with an input demand to determine the speed error, and produces a PWM command signal to drive the electro hydraulic proportional valve. The control valve adjusts the flow of pressurized oil to move the hydraulic motor until the desired speed is attained.

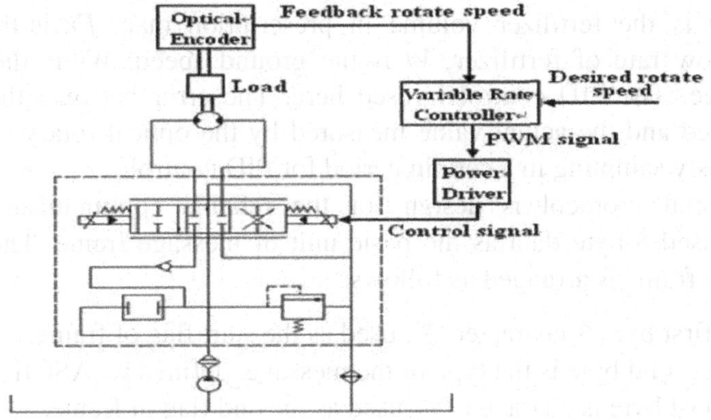

Figure 1. Block diagram of the hydraulic servo control system

2.2 Design of Variable Rate Fertilizer Controller Based on DSP

The variable rate fertilizer controller uses the TI TMS320LF2407A DSP as the main processor. The TMS320LF2407A provides several advanced peripherals such as 10-bit analog-to-digital converter (ADC), serial communications interface (SCI), PWM generation circuit and quadrature encoder pulse (QEP) circuit. The CAN controller has been integrated in the 2407A DSP, so it can realize the CAN communication just using a CAN transceiver 82C250.

In the fertilizer application, the PWM control signal for the electro-hydraulic proportional valve is generated by the DSP PWM circuit. The QEP circuit decodes and counts the quadrature encoded input pulses from the optical rotary encoder as the feedback signal for the PID control. The LCD and keyboard are used for parameters input and system status display. The external FLASH memory is used for the storage of prescription map and guidance parallel lines information.

3. EXPERIMENT

In our system, the control object is the hydraulic motor and the feedback signal is the conveyor feed shaft rotate speed, and the output signal is the PWM pulses with changing duty cycles. If given a certain discharge volume of fertilizer *Ar* in one cell in the field and the machine's breadth *Wi*, and the machine's moving speed from the sampling unit is *Vt*. Then the instantaneous flow rate of fertilizer *Fr* can be calculated by the following formula:

$$Ar = Fr/(Vt * Wi) \tag{1}$$

where *Ar* is the fertilizer volume in prescription map, *Fr* is the instantaneous flow rate of fertilizer, *Vt* is the ground speed, *Wi* is the breadth of machine. The PID control is used here. The error between the desired rotate speed and the actual value measured by the optical rotary encoder is continuously sampling in a certain period for PID control.

A special protocol is design for the reliable communications. The protocol used 8-byte data as the basic unit of message frame. The 8 bytes data in the frame is arranged as follows:

1) The first byte is character '$', used as the start flag of frame.
2) The second byte is the type of the message, defined by ASCII code.
3) The last byte is character '#', used as the end flag of frame.
4) The rest of bytes are used for carrying the parameters.

4. CONCLUSION

In this paper, a control system for variable rate application is presented. The hardware and software design of the control system are described in details. The evaluation experiment shows that the communication in the system using the developed protocol based on CAN bus is fast and reliable. The closed loop hydraulic servo control system is evaluated by the experiment through the PID control, which has a fast response and good performance. The development of this control system will make the variable rate application to implement easily and in a flexible way, and lead to the standardization of the electronic equipment in agricultural machinery.

5. ACKNOWLEDGEMENTS

Our project is supported by the Chinese National 863 Plan (Project No. 2006AA10A306).

REFERENCES

Maohua Wang. 1995. The Development of Precision Agriculture and Innovation of Engineering Technology. Transactions of the CSAE 15(1):1-8.

Marvin L. Stone and Kevin D. McKee. 1999. ISO 11783: An Electronic Communications Protocol for Agricultural Equipment, Agricultural Equipment Technology Conference, Louisville, Kentucky.

A NEURAL NETWORK MODEL FOR PREDICTING COTTON YIELDS

Jun Zhang[1], Yiming Wang[2,*], Jinping Li[1], Ping Yang[1]

[1] *College of Information, Beijing Union University, 100101, Beijing, China*

[2] *College of Information and Electrical Engineering, China Agricultural University, 100083, Beijing, China*

[*] *Corresponding author, Address: P.O. Box 63, College of Information and Electrical Engineering, China Agricultural University, 17 Tsinghua East Road, Beijing, 100083, P. R. China, Tel: +86-10-62737591, Email: ym_wang@263.net*

Abstract: Predicting a realistic target yield is one of the critical problems in precision farming. An artificial neural network was employed to model the nonlinear relationship between cotton yield and the factors influencing yield. Using six-year field data obtained from LuoYang Dry Land Research Center, the neural network model was developed and trained, and the RMSE for test data was 3.70%. The results indicate that the neural network model is a superior methodology for accurately setting cotton yields.

Keywords: artificial neural network, precision farming, cotton yield

1. INTRODUCTION

Factors affecting crop yield are so complex that traditional statistics cannot give accurate results. The artificial neural network can find non-linear relationships by observing a large number of input and output examples to develop a formula that can be used for predictions (Elizondo et al., 1994). Liu Jing et al. developed a neural network model for setting target corn yields and the influencing factors included soil properties, weather and management conditions (Liu et al., 2001). The ANN model resulted in higher r^2 and lower RMSE than the regression models (Kaul et al., 2005). In this research, a feed-forward BP ANN was employed to model the relationship between lint cotton yield and the factors influencing yield.

Zhang, J., Wang, Y., Li, J. and Yang, P., 2008, in IFIP International Federation for Information Processing, Volume 259; Computer and Computing Technologies in Agriculture, Vol. 2; Daoliang Li; (Boston: Springer), pp. 1321–1322.

2. MATERIALS AND METHODS

Historical data from the six-year continuous field experiment (1999-2004) conducted in LuoYang Dry Land Research Center, Chinese Academy of Agricultural Sciences (CAAS) were used to build up the ANN model. Eleven input factors included management factors, soil properties, and weather factors. Eighteen records of eighteen subplots per year and six years of records provided the total set of 108 examples for the ANN model. Trial and error was used to select the parameter values that would give the most accurate predictions. The final network structure was 11-15-1 with the learning rate of 0.05 and the momentum coefficient of 0.6. Initial weights and thresholds were generated randomly, and 11000 epochs were used in training. After 11000 epochs, the network was over-trained.

3. RESULTS AND DISCUSSION

The RMSE for the test examples was 3.70%. Predicted yields were significantly related to the observed yields with the correlation coefficient reaching 0.9239. The calculated yield shows the greatest sensitivity to the rainfall from July to August, the next greatest sensitivity to nitrogen fertilizer, and followed by rainfall from September to October. Growing season GDD and accumulated sunshine hours ranked fourth and fifth in influencing yield. The remaining factors showed less influence on the calculated yield.

4. CONCLUSION

The ANN model has high prediction accuracy and predicted yield trends versus the input factors are realistic. The ANN model has demonstrated here a promising potential in ecology. Further work is planned to test and retrain the ANN for a new field of larger size and with a variety of soil types.

REFERENCES

Elizondo D.A., McClendon R.W., Hoogenboom G. Neural network models for predicting flowering and physiological maturity of soybean. Transactions of the American Society of Agricultural Engineers, 1994, 37: 981-988

Kaul M., Hill R.L., Walthall C. Artificial neural networks for corn and soybean yield prediction. Agricultural Systems, 2005, 85: 1-18

Liu J., Tian L. A neural network for setting target corn yields. Transactions of the American Society of Agricultural Engineers, 2001, 44: 705-713

STUDY ON EXPERT SYSTEM FOR TOWED WATER-SAVING IRRIGATION MECHANIZATION TECHNOLOGY

Na Jia [1,2] , Changle Pang [1,2,*] , Zhuomao E [1,2]
[1] *China Agricultural University, College of Engineering, 17 Tsinghua East Road, Beijing, 100083 China*
[2] *Research and Extension Center of Towed Water-saving Irrigation Mechanization Technology, 17 Tsinghua East Road, Beijing, 100083, China*
[*] *Corresponding author, Tel: +86-10-81682099, Email: pangcl@cau.edu.cn*

Abstract: Expert system techniques have been rapidly disseminating into every scientific domain, with many applications being reported within the last decade. Expert system is bringing a new perspective. This paper introduces the towed water-saving irrigation mechanization technology firstly. Then, an emphasis is made on development of the expert system for towed water-saving irrigation mechanization technology. The system's architecture, components and function are described.

Keywords: artificial intelligence, expert system, water-saving irrigation, towed water-saving irrigation mechanization technology

1. INTRODUCTION

Towed water-saving irrigation mechanization technology is created by Chinese scientists firstly. It is based theoretical principle on no-full irrigation theory and regulated deficit irrigation elements, and taken into account the dry region's circumstances adequately in Northern China. Towed water-saving irrigation mechanization is a kind of new technology that utilizes towing machines which are plentiful in counties, such as various wheeled

Jia, N., Pang, C. and E, Z., 2008, in IFIP International Federation for Information Processing, Volume 259; Computer and Computing Technologies in Agriculture, Vol. 2; Daoliang Li; (Boston: Springer), pp. 1323–1327.

tractors and agricultural vehicles (Kang, S., 1996). Because towed water-saving irrigation mechanization technology is a kind of new technique, the farmers don't know much about it. In addition, a lack of the effective extensive measure induced that the species of water-saving machine which was spread was not much.

Expert system is a kind of artificial intelligence and information system that can be used to inquire and decision-making, it can assist the towed water-saving irrigation mechanization technology to expanse availably (Bruegge, 2000). This paper reports a research effort in developing the expert system for towed water-saving irrigation mechanization technology. One of the strengths of the system is that it can facilitate both query and diagnosis. Feedback collected from the demonstration and evaluation of the system has provided valuable insights into the issues related to the development and implementation of towed water-saving irrigation mechanization technology in China.

2. SYSTEM ARCHITECTURE

Based on users' need analysis, an expert system for towed water-saving irrigation mechanization technology was developed. The system was constructed with Visual C#. Microsoft SQL Server 2000 running on Windows 2000 Advanced Server was used for the back-end databases. The expert system can be reached through Internet, demand was sent from browsers, and the conclusion was made by servers and sent to the browsers (Mehdi Sagheb-Tehrani, 2006). The system can be reached by multi-users simultaneously through Internet. Decision functions can be realized by the system. Users can choose any of them according to their demands and conditions. The system provides guideline and can be used easily.

Fig. 1 shows the architecture for the system (O'Brien, 2004).

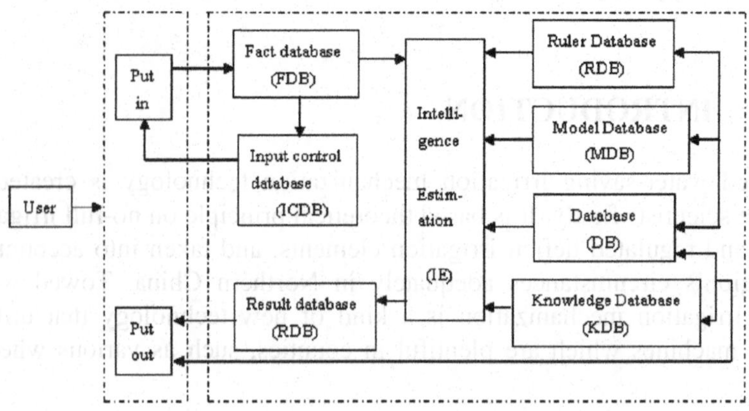

Fig. 1. Structure of the expert system

3. SYSTEM COMPONENTS

To best meet the needs of farmers and experts, four subsystems have been designed and developed in the expert system. The structure of four subsystems can overcome the shortage of each individual system and enhance the overall performance of the expert system. The four subsystems are towed water-saving irrigation machines management subsystem (MMS), water and climate resources management subsystem (WCMS), geography information management subsystem (GMS) and extension system management subsystem (EMS).

The inquiring process is shown in Fig. 2.

More details about each subsystem are discussed as follows:

(1) Towed water-saving irrigation machines management subsystem (MMS):

• To choose the appropriate machine and know about its particular information according to database of machines;

• To ask the users for feedback on the machines.

When a farmer enters into this subsystem, he needs to choose the crop type in first. Then, he needs to choose which period the crop grows to use the irrigation machines. After the farmer submits the demanded informàtion, conclusions such as the best appropriate machine and the machine's detailed information should be given. On the side, farmer can feed back the shortage of machines that he has used.

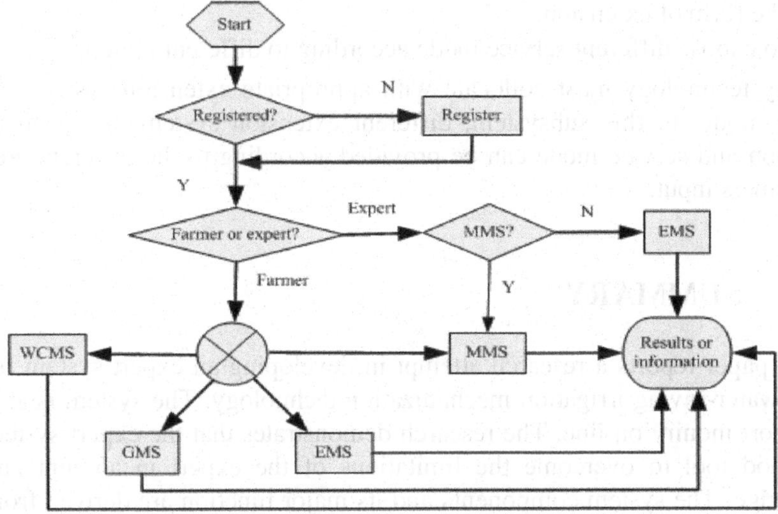

Fig. 2. Working process of the expert system

(2) Water and climate resources management subsystem (WCMS):

- Distributed conditions of well or water cellar;
- The weather forecast;
- Rainfall and capacity of groundwater supply;
- To decide the amount of irrigative water.

In this subsystem, the farmer needs to input regional name firstly. Then, the system searches correlative water and climate information on Internet. At last, according to mathematics model of database, some prediction such as the amount of irrigation and the plan of water-distributed can be given through calculation.

(3) Geography information management subsystem (GMS):

- To analyze soil condition;
- Distributed condition of dry area;
- To account for the degree of drought and define which dry area the region belongs to.

If farmer enters into this subsystem, he needs to input region he belongs to. Then, a map of this area comes forth. Following the cursor's moving on the map, detailed information about each part of land is provided, such as type of the land, type and name of the soil, gradient of landform, amount of water in the field, etc.

(4) Extension system management subsystem (EMS):

- Extension system;
- The form of extension;
- To choose different service mode according to different region.

Every technology must collocate with appropriate extension system and service mode. In this subsystem, different extension system, the form of extension and service mode can be provided according to the different area that farmers input.

4. SUMMARY

This paper reports a research attempt in developing an expert system for towed water-saving irrigation mechanization technology. The system is able to support inquiry on-line. The research demonstrates that the expert system is a good tool to overcome the limitations of the expert in amount and knowledge. The system components and its major function are derived from users' suggestions and needs. Further improvement is still being undertaken alongside the collection of more feedback from users. The research demonstrates the structure of four subsystems can overcome the shortage of each individual system and enhance the overall performance of the expert system. The current system can meet different needs of farmers and experts.

ACKNOWLEDGEMENTS

Funding for this research was provided by the Research and Extension Center of Towed Water-saving Irrigation Technology. We are grateful to Li Tiejun and Yuan Guodong for their contribution to the data collection and the anonymous referees for their careful reading of the manuscript.

REFERENCES

Bruegge, B. and A.H. Dutoit, Object-oriented Software Engineering Conquering Complex and Changing Systems, Upper Saddle River, NJ: Prentice Hall, 2000

Kang, S., Cai, H., Liang, Y., 1996. A Discussion of the Basic Theoretical Problem in Crop Water Management of Water-saving Agriculture. J. Hydraulic Eng. (Chinese) 5, 9±17

Mehdi Sagheb-Tehrani, The Design Process of Expert Systems Development: Some concerns, Expert Systems, vol. 23, issue 2, 2006, pp. 116–125

O'Brien, J.A., Management Information Systems, 6th edn, New York: McGraw-Hill, 2004

DEVELOPMENT OF 3D GIS MODELING TECHNOLOGY

Xiangjian Meng[1], Gang Liu[1,*]

[1] *Key Laboratory of Modern Precision Agriculture System Integration, Ministry of Education, Beijing, China, 10008*

[*] *Corresponding author, Address: P.O. Box 125, Qinghua Donglu 17, Haidian District, Beijing, 100083, P. R. China, Tel: 86-10-62736741, Fax: 86-10-62736746, Email: pac@cau.edu.cn*

Abstract: Recent researches on Three-dimensional Geographic Information System (3D GIS) modeling technology were reviewed. The concept of 3D GIS modeling and its application in various fields were introduced. Essential techniques to realize 3D visualization were studied. A typical 3D GIS modeling usually contains the description and the transmission of GIS spatial data structure. GIS spatial data models were described and species of 3D GIS models were summarized. Current algorithms for transmitting data to 3D models were compared and analyzed.

Keywords: 3D GIS modeling, visualization, review, 3D models

1. INTRODUCTION

Three-dimensional Geographic Information System (3D GIS) modeling technology can provide a more lively and intuitive visual angle. It has been used in many different fields including military affaires, mining, geology, oil exploration and medicine etc. 3D GIS model is a spatial data model, which provides basic information for describing GIS spatial data structure and designing spatial database. It reflects the correlation of spatial entities in the real world. At the same time, it is a theoretical basis for spatial data processing and spatial data mining (Chen Lichao et al., 2004).

Meng, X. and Liu, G., 2008, in IFIP International Federation for Information Processing, Volume 259; Computer and Computing Technologies in Agriculture, Vol. 2; Daoliang Li; (Boston: Springer), pp. 1329–1332.

In 2002, Shinji Araya et al. proposed a flexible and efficient VRML-based terrain model. It can simplify the changes of display range and attach an automatic changeover by placing the given terrain models (tiles) in 3D space (Shinji Araya et al., 2002). In 2003, Elizabeth Burson presented two methods of 3D GIS modeling for professionals to incorporate models such as buildings, caves, viewsheds and many man-made objects into their greater GIS environment (Elizabeth Burson, 2003). Zhang Jing et al. discussed the 3D GIS modeling technology in urban GIS by combining the traditional GIS and virtual reality. Exact information and useful methods for promoting scientific urban management were provided (Zhang Jing et al., 2003). T.H. Kolbe et al. proposed a unified model to represent spatial objects in 3D city and regional models, which describes the real world objects by geometric, topological and thematic properties (Kolbe T.H. et al., 2003). Zhu Qing et al. introduced 3D GIS for the digital documentation of a special T'ang-wooden buildings based on personal computer. They firstly applied 3D GIS modeling technology in digital cultural heritage (Zhu Qing, 2006).

Maps are useful tools to cognize spatial information. In the traditional plat maps, symbols are used to represent spatial objects. As the development of computer techniques and 3D techniques, traditional 2D static plat maps are developed to 3D dynamic virtual environments (Zhu Qing et al., 2003). Based on spatial database, 3D GIS modeling technology adopts Virtual Reality (VR) technique to construct virtual environment. The application of 3D models is accelerated by the development of 3D visualization of spatial information. As Wei Yongjun put forward, 3D models have all the characters and basic functions of the plat maps (Wei Yongjun, 2000). According to the researches on 3D GIS modeling technology, it mainly contains the description and the transmission of GIS spatial data structure.

2. DESCRIPTION OF GIS SPATIAL DATA STRUCTURE

2.1 GIS spatial data models

GIS spatial data models are composed of three basic hierarchies including conceptual data model, logical data model and physical data model. Conceptual data model is an abstract of the relationship between entities. Logical data model is used to express the data in conceptual data model and the correlation between them. Physical data model describes the physical form, storage, path and database structure of data in computer. The spatial data models are expressed by grid model and vector model. Vector model is entity-oriented and considers specific spatial objects as the described objects.

Grid model is space-oriented and regards geographical space as an integer for describing. Usually, grid model is mainly used in digital terrain analysis.

2.2 Species of 3D models

To classify 3D models is a cognitive course to analyze and define spatial objects (Ranzinger M. et al., 1997). According to the efficiency of 3D visualizing expression, 3D models can be classified to three species: (1) has inflexible geometry shape and similar external texture. It also has important characteristics in figure and position. A realistic 3D model can be used repeatedly, such as the model of street lamp. (2) has random geometry shape and similar external texture, which also has important characteristics in figure and position. The texture images used to express these objects include tree, flower and grass etc. (3) has random geometry shape and external texture. One can simulate these objects with a given random function, for example, fountain, waterfall, rainwater, surf and fireworks etc.

3. DATA TRANSMISSION OF 3D MODELS

3.1 3D GIS visualization based on 2.5D GIS

The 2.5D GIS visualizing method is often adopted to realize 3D visualization. It mainly uses high quality Digital Elevation Models (DEM) and high reality 3D vision technology (Wu Xincai, 2002). The 3D visualization is greatly impacted by the quality of DEM which depends on the algorithm for realizing it. Therefore, it's important to select a convenient algorithm with strong practicability, high precision and quick reaction speed. Considering the complex relationship between the points, lines and polygons in the algorithm for creating Triangulated Irregular Network (TIN), a special data structure was adopted to build a preferable topology relationship. In addition, fast index can be established between the points and sides of TIN and triangles, which facilitate the design of the algorithm.

3.2 Euclidean distance converting model and algorithm

Distance calculation is a common algorithm to process spatial data. Euclidean distance converting algorithm is propitious to grid data. Proved by experiments, it is an effective algorithm to calculate distance with its high speed, adjustable precision and simple program. The algorithm is composed of distance conversion and converting algorithm. In grid models, spatial objects are divided into four species including points, lines, polygons and bodies. For any kind of grid data, the position, shape and topological information of objects can be described clearly by using Euclidean distance converting algorithm (Chen Lichao, 2004).

4. CONCLUSION

The recent researches on 3D GIS modeling technology were reviewed. The 3D GIS modeling technology was developed rapidly and applied in many different fields. In recent years, the development of computer techniques and visualizing technologies brought promising advances to 3D GIS modeling technology.

ACKNOWLEDGEMENTS

This paper is supported by Ministry of Education-New century talent program NCET-06-0110.

REFERENCES

Chen Lichao, Zhang Yongmei, Liu Yushu and Zhang Jianhua, 3D Spatial Data Models of GIS Based on Grid. Computer Engineering, Vol. 30 No. 8 (2004).

Shinji Araya, Jin Hyunsuk and Ryo Araya. An Effective Terrain Model for Web-Based 3D Geographic Information Systems. Electronics and Communications in Japan Part 2, Vol. 85 No. 9 (2002).

Elizabeth Burson. Modeling Unordered Point Data for 3D GIS Mapping - approaches for improved usability and accuracy.charlotte.utdallas edu/mgis/prj_mstrs/2003/03_burson/EBurson_Masters.pdf, (2003).

Zhang Jing and Zhang Zhengrong. 3D Visualization about Urban GIS. www.china-simulation.com/esite/preview/0312.htm, Vol. 15 No. 12 (2003).

Kolbe T. H. and Groger G. Towards Unified 3D City Models. http: // www. iuw. uni - vechta, de/personal/geoinf/jochen/papers/28. pdf, (2003).

Zhu Qing, Lu Dandan and Zhang Yeting. Application of 3DGIS in digital cultural heritage. Science of Surveying and Mapping, Vol. 131 No. 11 (2006).

Zhu Qing, Gao Yurong, Wei Yongjun and Huang Duo. Geomatics and Information Science of Wuhan University, Vol. 28 No. 3 (2003).

Wei Yongjun. The Study of 3D Model Theory and Method. Zhengzhou (2000).

Ranzinger M. and Gleixner G. GIS Datasets for 3D Urban Planning. Computer, Environ and Urban Systems, Vol. 21 No. 2 159-173 (1997).

Wu Xincai. Principles and Methods of Geography Information Systems. Beijing, Publishing House of Electronics Industry, 224-226 (2002).

A REGRESSION MODEL OF DRY MATTER ACCUMULATION FOR SOLAR GREENHOUSE CUCUMBER

Weitang Song [1,2,4] , Xiaojun Qiao [3]

[1] *College of Science, China Agricultural University, Beijing, China, 100094*

[2] *Key Lab in Bioenvironmental Engineering of Ministry of Agriculture, China Agricultural University, Beijing, China, 100083*

[3] *National Engineering Research Center for Information Technology in Agriculture, Beijing, China, 100097*

[4] *Corresponding author, Address: Institute of Agricultural Engineering, China Agricultural University, 2 Yuanmingyuan West Road, Beijing, 100094, P. R. China, Email: songchali@cau.edu.cn*

Abstract: The objective of this study is to develop a regression cucumber dry matter production model with a minimum number of parameters. Cucumber (*Cucumis sativus L.*) was cultivated in a soilless system with drip irrigation. The substrate was peat mixed with vermiculite. Five experiments were fulfilled totally in 3 different places in Beijing of China from 2004 to 2005. Cucumber growth data (dry matter weight of leaf, stem, fruit and petiole) were measured and environmental data (temperature, light intensity and day length) were collected. Data collected from 1 experiment in solar greenhouse was used to build the model, which was further validated with the data collected from other 4 experiments in solar greenhouse. A regression model for cucumber dry matter production was established. Based on Logistic curve, the time state variable was expressed as a logistic function about effective temperature accumulation (ETA) and effective light intensity accumulation (ELIA). ETA was defined as the sum of the temperature that was higher than physiological zero point in certain period, and ELIA was defined as the sum of the light intensity that was higher than light compensation point multiplied with time in certain period. Temperature, light intensity and day length were synthetically considered. The model had less state variables, and provided the relationships between the cucumber dry matter accumulation (DMA) per plant and environmental data (temperature, radiation and day length). The result of simulation was satisfied, because RMSE value was less than 6, and the R^2 value of the results was 0.99. It indicated that the regression model for cucumber dry matter production was reasonable and feasible.

Song, W. and Qiao, X., 2008, in IFIP International Federation for Information Processing, Volume 259; Computer and Computing Technologies in Agriculture, Vol. 2; Daoliang Li; (Boston: Springer), pp. 1333–1339.

Keywords: cucumber, dry matter accumulation, regression model, effective temperature accumulation, effective light intensity accumulation

1. INTRODUCTION

Both mechanism models and regression models can be used to describe the course of cucumber dry matter accumulation (DMA) (Challa et al., 1996; Heuvelink et al., 1996). Mechanism models generally have many sub-models for each DMA process, which can describe crop's growing course and have strong explanation function completely (Li et al., 1999). But regression models show directly the relationship between crop development, growth rate, output and environmental factors by functions of multivariate regression equation, index equation, hyperbola equation and "S" curve, etc. It describes the results of observation.

Most present cucumber dry matter accumulation models are explanatory. The application limitations of these models focus on that: (1) they have so many parameters that it is very complicated to be calculated (Challa et al., 1996); (2) they need to input many variables and data, but the output is wavy; and (3) the prediction is not very accurate. Contrarily, regression model parameters are easier to estimate and the calculation process is short, which is practical in real time control for greenhouse.

The accumulation course of dry matter mainly includes photosynthesis and respiration. Temperature and light are two most important factors of cucumber DMA, if nutrition and water are sufficient and plants don't have pests (Chamont et al., 1993). A lot of models describe crop growth with temperature accumulation or effective temperature accumulation (ETA) (Katharine, 1996). Illumination is also important to influence crop growth, even more important than temperature (Wurr et al., 1988). Hand et al. (1992) measured and analyzed the net photosynthesis rate with everyday integrals of solar radiation. A forecasting regression equation of canopy net photosynthesis rate had been gotten. Comparisons of this regression equation with mechanism model of Thornley (1992) showed that light respond curves of the two models were very close.

The objective of this study is to develop a regression cucumber dry matter production model with a minimum number of parameters. Logistic growth equation was selected to build the regression model. Although it is an ideal model for accumulation equation of dry matter (Li, 1996; Liu, 1994), its form is too simple. Because only is time adopted as independent variable, It could not describe the change of crop growth with the change of environmental conditions such as illumination and temperature, etc. Therefore the Logistic growth equation needs to be improved.

2. MATERIALS AND METHODS

Five different tests were carried out for model foundation and verification. Test 1 was for model foundation, and Tests 2-5 for model verification.

2.1 Test 1

Test 1 was achieved in solar greenhouse in Beijing Agricultural Science Academy from April to July of 2003. Cucumber (*Cucumis sativus L.*) was named Beijing New No. 4, and the seedlings were transplanted on April 16 of 2003. Pot soilless culture was applied and substrate was mixture of charcoal and vermiculite with the rate of 2:1 (w/w), and drip irrigation was used to deliver nutrient solution to the plants. In the process of development and growth, all axillaries shoots were removed. Test area was $66m^2$ and density was 5 plants/m^2.

The environmental data, such as light, temperature, and CO_2 concentration were recorded at 10-minute intervals automatically. Six plants were sampled at 10-day (not including roots) intervals, and dry matter weights of leaf, stem, fruit (including flower) and leafstalk were determined. Assumption that root takes 3% of total dry weight was accepted (Marcelis, 1994).

Methods and contents of sampling and measurement in test 2-5 were the same as test 1.

2.2 Test 2

Test 2 was fulfilled in the solar greenhouse in Beijing Agricultural Science Academy from April to July of 2004. The seeds were sown on March 8, 2004, and the seedlings were placed in substrate on April 6, 2004. Cucumber variety and culture type were the same as Test 1, but the difference was that greenhouse was covered with sunshade which transmittance was 35-40% and 37% in average.

2.3 Test 3

Test 3 was completed in the solar greenhouse in China Agricultural University (Beijing) from September to November of 2004. Cucumber was named Jinglv No. 2 and the seedlings were transplanted in charcoal groove built by bricks. Plant density was 7 plants per square meter. The others of water and fertilizer management were the same as Test 1.

2.4 Test 4

Test 4 was gone on in the solar greenhouse in Chinese Academy of Agricultural Sciences (Beijing) from April to July of 2005. The seeds of cucumber, Mini No. 2, were sown on April 3, 2005. The brick grooves were the same as test 3. Substrate was mixture of corn straw and soil in the rate of 3:1 (w/w). Management of water and fertilizer were not difference from Test 1.

2.5 Test 5

Test 5 was accomplished in the solar greenhouse in Chinese Academy of Agricultural Sciences (Beijing) from April to July of 2005. Cucumber variety and culture type were the same as Test 4. The difference was that greenhouse was covered with sunshade which transmittance was 35-45% and 40% in average.

3. MODEL ESTABLISHMENT

The independent variable of time in Logistic model was replaced with a function of effective temperature accumulation (*ETA*) and effective light intensity accumulation (*ELIA*), so the two main factors for crop growth, temperature and light, was considered.

The equation was:

$$DMA = \frac{Dr_{\max}}{1 + e^{\tau(ETA, ELIA)}}$$

where

DMA = dry matter weight per plant, g

$Dr_{\max}$ = the biggest of dry matter weight per plant, g

ETA = effective temperature accumulation, °C d

$ELIA$ = effective light intensity accumulation, μ mol m^{-2}

τ = a function about ETA and $ELIA$

ETA was defined as the sum of the temperature which was higher than physiological zero point in certain period, and was calculated as the following equation:

$$ETA = SUM(T_{mean} - T_b)$$

where

T_{mean} = the average temperature of 2: 00, 8: 00, 14: 00, 20: 00 in a day, °C
T_b = the temperature of cucumber physiological zero point, °C

ELIA was defined as the sum of the light intensity which was higher than light compensation point multiplied with time in certain period. Photosynthetic photon flux density and illumination length were considered together. The calculating equation was:

$$RD = SUM[L_t \times (R - R_b)] \qquad ELIA = SUM(RD)$$

where

RD = the sum of the light intensity which was higher than light compensation point multiplied with time in a day, μ mol m^{-2}

R = photosynthetic photon flux density in certain time, μ mol $m^{-2}s^{-1}$
R_b = light compensation point of cucumber, 51 μ mol $m^{-2}s^{-1}$
L_t = time step, 600s

Finally, the DMA model based on *ETA* and *ELIA* was described as:

$$DMA = \frac{Dr_{\max}}{1 + \exp(\varepsilon + \lambda ETA + \omega ELIA)}$$

4. MODEL VERIFICATION

The above-mentioned model was verified by the Test 2-5 data of dry matter weight as well as temperature and light. $Dr_{\max}$ of Tests 2 and 3 was 70g, but $Dr_{\max}$ of tests 4 and 5 was 160g because of different cucumber variety. The value of regress coefficient ε, λ, ω were calculated from test data.

The simulated values were very near to the actual measurement values, according to their comparison. Generally, root mean square errors (RMSE) value of the model within 15 is excellent. RMSE of the simulated results of four tests were totally smaller than 6. R^2 of the simulated results of four tests were all 0.99. F- statistics results indicated that rejective probability of this equation was below 0.01, i.e. the tenable probability of this regress equation was very notable level. The results of various appraisement indexes, from the all four tests, implied that this model could be applied to simulate accurately the development changes of greenhouse cucumber DMA in certain conditions.

5. CONCLUSIONS

Concept of effective light intensity accumulation (ELIA), the sum of the light intensity that was higher than light compensation point multiplied with time in certain period, has been put forward in this paper. In this way, photosynthetic photon flux density and the length of illumination time were considered and computed as influencing factors of the cucumber DMA. Meanwhile the sensitivity of different crop for light intensity was taken into account.

Logistic equation has been improved. Two environmental factors, temperature and light, have been led into Logistic equation together. A function about ELIA and ETA were applied to replace the time variable. Therefore, the course of cucumber DMA was not only caused by the influence of time change, but also the environmental factors of temperature and illumination.

Compared with mechanism model, regression model parameters were simple and predicting results were more accurate. As tests and data were limited, more test data were necessary in the future to analyze and adjust the coefficients of the model.

ACKNOWLEDGEMENTS

The author wishes to thank the anonymous referees for their careful reading of the manuscript and their fruitful comments and suggestions.

REFERENCES

D. W. Hand, G. Clark, M. A. Hannah, J. H. M. Thornley, J. Warren Wilson. Measuring the canopy net photosythnthesis of glasshouse crops. Journal of Experimental Botany, 1992, 43(248): 375-381.

Heuvelink E. Tomato growth and yield: quantitative analysis and synthesis. Wageningen. 1996.

Hugo Challa, Ep Heuvelink. Photosynthesis driven crop growth models for greenhouse cultivation: advances and bottle-necks. Acta Horticulturae, 1996, 417: 9-22.

Katharine B. Perry, Todd C. Wehner. A heat unit accumulation method for predicting cucumber harvest date. Horticultural Technology. 1996, 6(1): 27-30.

Li Lei, Lou Chunheng, Wen Rujing. Study on the Dry matter Accumulation and Distribution with Various Densities in Cotton. Acta Agriculturae Boreali-Occidentalis Sinica 1996, 5 (2): 10-14.

Li Pingping, Wang Duohui, Deng Qingan. A Study of Dynamic Growth and Potential Production Simulation Model of Salad in Greenhouse. Journal of Biomathematics, 1999, 14(1): 77-81.

Liu Guiru, Zhang Rongzhi, Lu Jianxiang. Study on Grain Filling Characteristics of Wheat Cultivars Tolerant to Drought. Journal of Agricultural University of Hebei. 1994, 17(4): 43-47.

Marcelis L. F. M. A simulation model for dry matter partitioning in cucumber. Ann. Bot 1994. 74: 43-52.

S. Chamont. Modeling dry matter allocation in cucumber crops-competition between fruits and roots. Acta Horticulturae 328, 1993: 195-203.

Thornley J. H. M., Hand D. W., and Waerrn Wilson J. Modelling canopy net photosynthesis of glasshouse row crops and application to cucumber. Journal of Experimental Botany, 1992. 43, 383-391.

Wurr D. C. E., Fellows J. R., Suckling R. F., Crop continuity and prediction of maturity in the crisp lettuce variety Saladin. J. Agric. Sci. 1988. 111, 481-486.

SPATIAL VARIABILITY ANALYSIS OF SOIL PROPERTIES WITHIN A FIELD

Gang Liu[1], Xuehong Yang[2]
[1] *Key Laboratory of Modern Precision Agriculture System Integration, Ministry of Education, China Agricultural University, Beijing, Beijing, China, 10008, Tel: 86-10-62736741, Email: pac@cau.edu.cn*
[2] *China Agricultural University, Beijing, P. R. China*

Abstract: The main efforts and applications up to now are focused on the spatial variability analysis within a field. The spatial variability of soil properties within a field was evaluated using geostatistical analysis and the interpolation methods for mapping soil properties were compared in this paper. This experimentation was conducted in a wheat field of the trail farm in Shunyi County, Beijing. The soil properties tested were N, P, and K. The results suggested that the spatial variability of P were more obvious than N and K, the random sampling error of N was bigger than P and K, and the range of K was larger than N and P. The techniques of Kriging and Inverse Distance Weighting (IDW) can significantly improve estimation precision compared with Spline technique. Comparison of Kriging and IDW estimations revealed that Kriging performed better than IDW.

Keywords: Spatial variability, Geostatistical analysis, Kriging, IDW

1. INTRODUCTION

Most researches on the spatial variability in field are mainly concentrated in precision agriculture for years. However, because of the complexity of crop growing conditions and the difference in crop production within every field, the spatial variability of soil properties that affect especially crop yield should be studied in every field, and the rational sampling interval and sampling density should be also identified. These test data can be finally used to create the spatial variability maps of soil properties with the local

Liu, G. and Yang, X., 2008, in IFIP International Federation for Information Processing, Volume 259; Computer and Computing Technologies in Agriculture, Vol. 2; Daoliang Li; (Boston: Springer), pp. 1341–1344.

estimation method (Wollenhaupt et al., 1997; Ping & Green, 2000). Improvement in estimation quality depends on reliable interpolation techniques that can be used for obtaining soil property values at unsampled locations. Local interpolation techniques commonly used in agricultural applications include IDW, Spline and Kriging (Kravchenko & Bullock, 1998; Vieira, 1996; Johnson et al., 1998). Description of GIS spatial data structure

2. MATERIALS AND METHODS

The experimental field is located at a demo farm in Shunyi County, Beijing. The field surveyed is approximately 11 hectares and the cropping sequences are winter wheat and summer corn. With a DGPS, AgGPS 132, the field was divided into 40*40m grids and 72 soil samples were collected. The soil properties tested are N, P, and K. All soil samples were collected at depth of 0-20 cm.

The spatial variation in the data was explored using geostatistical analysis. The variogram is a central tool. It describes the structure and the spatial scale of variation by measuring the degree of correlation between the sampling points that are a given distance apart. The experimental variogram is fitted with a variogram model and adequacy of the chosen model is tested using a cross-validation technique. In this study we considered spherical, Gaussian, and exponential models for the experimental variogram fitting. The accuracy of the selected variogram model was measured through the error between the measured data and the estimated values. After an appropriate variogram model was selected, Kriging was used to estimate weights based on the variogram model. Thus the variable values at unsampled locations were obtained.

The factors affecting IDW precision is the number of the closest samples used for estimation. In this study, the number of the closest samples was equal to 12 for all studied data sets. The search radius was chosen so that it was large enough to include the required number of the closest samples. Cross-validation technique was used to compare the results obtained with different number of the closest samples. For Spline method, thin-plate smoothing Spline is constructed based on a tradeoff between goodness of fit and smoothness.

3. RESULTS AND DISCUSSIONS

The statistical analysis of the data sets is performanded for N, P and K data. The variance of P content was more evident than the variance of N content and K content, with the coefficient of variation equal to 37 % for P. The values of P contents varied with the maximum of 88.6 mg/kg and the minimum of 16.6 mg/kg. Histograms for N, P and K data were constructed and plotted along with theoretically normal and lognormal probability density functions. P-P probability chart was used to analyze the observed distributions by software SPSS (SPSS Inc., Chicago, USA). The results suggested that N, P and K contents exhibit normal distributions.

The geostatistical analysis of soil properties is accomplished. Experimental variograms for N, P and K data were calculated and fitted with variogram models. The experimental variograms were best fitted with spherical models for N and K data sets, and exponential models were used for P data sets. C_0 is the nugget variance, which means the random spatial variability. $C + C_0$ is the sill, which indicates the amount of the spatial variability of the variable. α is the range and h is the sampling interval. When $h \leq \alpha$, any two sample points have relativity, and the relativity decreases while the value of h increases. When $h>\alpha$, there will be no more relativity between samples. The value of α indicates the degree of the spatial variability of the variable. In other words, α measures the spatial dependence of samples.

The analysis results show that $C_0/(C_0+C)$ of N was 0.769, which indicates that the random spatial variability of N is 76.9% of the total spatial variability. It was the maximum among all the soil parameters. The range of spatial correlation increased from 40.7 to 53.6m, i.e. if we sample soil for these properties, the sampling interval should be about 40m. Of course, as time goes by, the range of each soil parameter keeps changing, but it will not change too much.

The interpolation models and Mean Square Error (MSE) are carried out by comparing measured data with estimates of Kriging, Spline and IDW. The most significant difference was observed in MSE values. Kriging and IDW were overall better than Spline. For the Kriging, spherical model and exponential model were better than gauss model. For IDW, the exponent values of 2 was better than the exponent values of 1 and 3. For the studied data sets, it was the best to create spatial distribution maps using Kriging with spherical model for N and K, and with exponential model for P.

4. CONCLUSION

Experimental variograms showed that the spatial structure of soil properties could be described by spherical, exponential or gaussian models. Variogram model parameters varied markedly according to the soil properties. The extent of spatial variability in this field was given by the range of spatial correlation of the variogram, which varied from 40.7m for P, to 53.6m for K. Above data were also used to compare performances of different interpolation techniques such as Kriging, IDW and Spline. Obtained results indicated that Kriging and IDW could be expected to produce overall better estimations than Spline. For Kriging, furthermore, spherical model was better than exponential model and gauss model for N and K, and exponential model was better than spherical model and gauss model for P. High values of coefficient of variation could be an indication that Kriging with exponential model would be a better choice. Comparing Kriging with IDW revealed that Kriging with the optimal number of neighboring points, carefully selected variogram model produced more accurate estimations than IDW method for such data sets. In a word, proper geostatistical analysis will provide invaluable information for the spatial distribution of soil properties in agricultural fields.

ACKNOWLEDGEMENTS

This paper is supported by Ministry of Education-New century talent program NCET-06-0110.

REFERENCES

Liu Gang, Zhang Man, Wang Maohua. Study on spatial variability of crop yield and soil properties. Proceedings of the 31. International symposium on Agricultural, Opatija, Croatia, 2003, pp. 75–84

Liu Gang, Kuang Jishuang. The study on field soil sampling strategies and interpolation techniques. Information technology of agriculture, ICAST. 2001, pp. 169–178

S.H. Moore, M.C. Wolcott. Spatial Associations between Crop Yield and Soil Characteristics in Corn and Soybean. Proceeding of the 5th International Conference Precision Agriculture, USA, University Minnesota, Oct. 2000, pp. 98-112

John. V. Stafford. Implementing Precision Agriculture in the 21st Century, J. Agric. Engineering Res. 2000 (76), pp. 267–275

Wollenhaupt N.C., Mulla D.J., Gotway Crowford C.A. Soil Sampling and Interpolation Techniques for Mapping Spatial Variability of Soil Properties. The Site-Specific Management for Agricultural Systems. ASA-CSSA-SSSA, 777 S. Segoe Rd., Madison, WI53711, USA, 1997, pp. 19–53

Kravchenko A.N., Bullock D.G. Comparison of Interpolation Methods for Mapping Soil P and K Contents. Proceedings of the 4th International Conference on Precision Agriculture, 19-22 July 1998. St. Paul, MN. pp. 267–280

Jianli Ping, Cary J. Green. Spatial Variability of Yield and Soil Parameters in Two Irrigation Cotton Fields Texas. Proceeding of the 5th International Conference Precision Agriculture, USA, University Minnesota, Oct. 2000, pp. 161–172

Simon Blackmore, Godwin R.J., Taylor J.C. etc. Understanding Variability in Four Fields in the United Kingdom. Proceedings of the 4th International Conference on Precision Agriculture, 19-22 July 1998. St. Paul, MN. pp. 3–18

DESIGN AND DEVELOPMENT OF AUTOMATIC NAVIGATION SOFTWARE FOR FARMING MACHINES

Yunuo Yang[1], Gang Liu[1,*]

[1] *Key Laboratory of Modern Precision Agriculture System Integration Research China Agricultural University*

[*] *Corresponding author, Address: P.O. Box 125, Qinghua Donglu 17, Beijing, 100083, P. R. China, Tel: 86-10-62736741, Fax: 86-10-62736746, Email: pac@cau.edu.cn*

Abstract: The navigation software played an import role in controlling-decision for Automatic Navigation system. It mainly completed data reading, data processing and data outputting. The data reading module accepted date from different sensors via serial ports and extracted useful information. The data processing module included improving positioning accuracy and calculating the value of control parameter. The data outputting module showed the navigation status for users. The software ran well in practice.

Keywords: automatic navigation, map matching, fuzzy logic control, multi-threads

1. INTRODUCTION

Automatic navigation system was the base of precision farming. They could arrive at correct position by tracking predefined path to complete data collecting and seeding. Though the study of automatic navigation system for agriculture had a long history, it is at the starting stage in China (Q. Zhang and H. Qiu, 2004). This paper focused on the research of navigation control technique and positioning method which were implemented in the software.

Yang, Y. and Liu, G., 2008, in IFIP International Federation for Information Processing, Volume 259; Computer and Computing Technologies in Agriculture, Vol. 2; Daoliang Li; (Boston: Springer), pp. 1345–1348.

To show the navigation information, the software provided a friendly interface on which users could detect the position, speed and direction etc. of farming machine. All the information could be saved for further analysis.

The primary goal of this article was to explore a good positioning algorithm and control method and to develop an effective software system for improving accuracy and stability of navigation system.

2. DESCRIPTION OF AUTOMATIC NAVIGATION SYSTEM

The system was constituted by three parts: a main computer, an assisted computer and a farming vehicle. The navigation software running on the main computer accepted GPS data from GPS receiver and posture information from the assisted computer via serial ports. Based on this data, more accurate position coordinates and control decision values would be achieved. Then the main computer communicated with the assisted computer to guide the vehicle.

Different posture sensors were used to get more efficient information. the digital compass was used to evaluate the orientation of the vehicle. The velocity sensor and angle sensor was designed to detect traveling the vehicle' speed and steering angle. The assisted computer read signals from sensors and sent them to the main computer. It got the desired steering angle and sent it to the steering controller which could drive the vehicle.

3. PRINCIPLES AND METHODS

3.1 Navigation positioning algorithm

GPS could accurately measure the absolute position of farming machine in the field, but it might have errors caused by the complex environmental conditions. Though Dead-reckoning (DR) algorithm could calculate the relative position based on information from sensors, errors would be accumulated during a long time. This research put forward Map-Matching (MM) theory to improve the above methods. Fuzzy logic theory was discussed to calculate their weights according to the desired path information (H.W. Griepentrog et al., 2006). The method was helpful for improving the position accuracy in practice.

3.2 Navigation control technique

The key techniques of this part were searching dynamic path and deciding control volume. A dynamic path search was a method to determine the next target point based on both the path curvature and the vehicle speed (Zhou Zhiyan, 2005). The path look-ahead distance determined by fuzzy logic theory was the length of the prospective path and was used to locate the proper target on the path. The target point was used to evaluate the lateral deviation and yaw angle. These data were used to determine the desired steering angle to guide the vehicle accurately along the predefined path.

4. DESIGN AND DEVELOPMENT OF SOFTWARE SYSTEM

4.1 Design of software system structure

To ensure a real-time performance, the system applied the multi-threads method on demand of multi-tasks. As the the body of system, the main thread accomplished displaying tracking position, operating map, MM method and determining desired steering angle. It also would communicate with these modules and co-operated with the assisted threads. The assisted threads were path designing and communication. The former one designed the optimum path based on the database following some principles, and the later one communicated with the assisted computer to accept information and send signal to each other.

4.2 Development of software system

The system was composed of four modules illustrated by the following structure. The data communication module mainly set port parameters, received signals from different sensors and sent decision command to the assisted computer. The implementation of this is applicable applied for CSerialPort class to control three ports (Gong Jianwei, 2005). The method was approved having high-efficiency. The map module included map making, map operating and map displaying. To get an accurate map this research firstly got CAD map, then converted it to a shape map by Arcview 3.2 system and collected 7 base points by GPS receiver to adjust map. The map operating including zooming out, zooming in and panning was developed by GIS ActiveX MapObjects2.2. Path tracking and displaying was also designed in this module. The path designing module could determine the points at the end of the field and set a proper buffer for the farming machine turning. The principles were different based on different field shapes, vehicle structures and farmers needs. Dijkstra algorithm was adopted to fulfill path planning according to several optimum principles. As the most important part the navigation controlling module was constituted of MM, calculating preview point and determining desired steering angle. These modules were associated with the navigation geographic database containing map information and moving vehicle information. Microsoft Access 2003 database was used to save data and Shapefiles method was adopted to manage data (Liu Guang and Liu Xiaodong, 2004).

5. CONCLUSION

Some validation tests were performed on the playground in the east campus of China Agricultural University. According to the results, the software could run well and the navigation system could response in time.

ACKNOWLEDGEMENTS

This paper is supported by the national 863 projects: Control Technique and Product Development of Intelligent Navigation of Farming Machines (2006AA10A304).

REFERENCES

Q. Zhang and H. Qiu, A Dynamic Path Search Algorithm for Tractor Automatic Navigation [J]. American Society of Agricultural Engineers, 2004, 47 (2): 639-646.

H.W. Griepentrog and B.S. Blackmore, Positioning and Navigation, pp. 195-204 of Chapter 4 Mechatronics and Applications, 2006.

Zhou Zhiyan, Study on the Navigation Geographic Information System for the Intelligent Farming Chassis, South China Agricultural University, 2005.

Gong Jianwei, Visual C++/TurboC Serial Ports Communication, Publishing House of Electronics Industry, 2005.

Liu Guang and Liu Xiaodong, GIS Development—VC.NET and MapObjects, Tsinghua University Press, 2004.

ELECTROCHEMICAL SENSORS FOR SOIL NUTRIENT DETECTION: OPPORTUNITY AND CHALLENGE

Jianhan Lin, Maohua Wang [*], Miao Zhang, Yane Zhang, Li Chen
Key Laboratory of Modern Precision Agriculture System Integration, Ministry of Education, Beijing, China, 100083
[1] *Corresponding author, Address: Box 115, China Agricultural University, 17 Tsinghua East Road, Beijing, 100083, China, Tel/Fax: +86-10-82377326, Email: wangmh@cau.edu.cn*

Abstract: Soil testing is the basis for nutrient recommendation and formulated fertilization. This study presented a brief overview of potentiometric electrochemical sensors (ISE and ISFET) for soil NPK detection. The opportunities and challenges for electrochemical sensors in soil testing were discussed.

Keywords: ISE, ISFET, Soil Nutrient Detection, Nitrogen, Phosphorus, Potassium

1. INTRODUCTION

Over-application of fertilizers in China's agricultural production system has caused low fertilizer usage efficiency (~35% in average, NBS, 2006), low agricultural product quality and serious environmental pollution, etc. One measure taken was to test soil for formulated fertilization.

A key in soil testing for formulated fertilization is to determine the amount of soil nutrients, followed by recommendation of nutrient needs and site-specific fertilization. Of the nutrients for crop growth, Nitrogen (nitrate: NO_3^- and ammonium: NH_4^+), Phosphorus (phosphate: PO_4^{3-}, hydrophosphate: HPO_4^{2-} and dihydrophosphate: $H_2PO_4^-$) and Potassium (potash: K^+) are the most important elements. Conventional soil NPK testing methods have been generally performed by three steps: soil sampling, sample pretreatment and

Lin, J., Wang, M., Zhang, M., Zhang, Y. and Chen, Li., 2008, in IFIP International Federation for Information Processing, Volume 259; Computer and Computing Technologies in Agriculture, Vol. 2; Daoliang Li; (Boston: Springer), pp. 1349–1353.

chemical analysis. To date, soil sampling is manually carried out in a field to obtain representative soil samples at a proper depth (~20cm). A vehicle-based hydraulic soil sampler (Wintex 1000, H.M. Agritech Aps, Denmark) was reported to achieve half-automated sampling in fields. Soil sample pretreatment is performed for the purpose of soil extracts through sequential processes: drying, crushing, sieving, extracting and filtering. Most processes can be carried out in batches, but are required to operate and connect by hand. Chemical analysis is handled by trained operators on special instruments to obtain the concentrations of soil nutrients.

So far, soil nutrient detection commonly uses optical measurement. In general, visible/ultraviolet spectrometry is employed for detecting nitrogen and phosphorus and flame spectrometry or atomic absorption spectrometry for potassium. The optical methods are reliable, but time-consuming, complex and high cost per test (~150 Yuan/Sample). This resulted in the limitation of the number of soil samples tested for characterizing the spatial variability of soil nutrients in a field or fields. Therefore, novel soil nutrient detection methods are urgently needed. This study intended to give a brief review of potential electrochemical sensors and a summary of their challenges and opportunities in soil nutrient detection.

2. ELECTROCHEMICAL SENSORS FOR SOIL NUTRIENT DETECTION

An electrochemical sensor consists of an ion selective membrane, which selectively responds to a target ion, and a transducer, which transforms the reactions into detectable electrical signals. Ion Selective Electrode (ISE) and Ion Selective Field Effect Transistor (ISFET) are two types of commonly-used potentiometric electrochemical sensors for soil nutrient detection.

2.1 Ion Selective Electrode

The response mechanism of ISE method can briefly described by the Nernst equation as a change of an ISE's potential, compared with a reference electrode, is linear to the change of the ionic activity (in logarithmic units) of the target ion.

ISEs were reported to detect soil nitrate (Dahnke, 1971; Hansen, 1977), ammonium (Banwart, 1972; Simeonov, 1976) and potassium (Mei, 1982; Wang, 1992). To date, no promising ISE for phosphorus detection was reported, but several literatures presented that the PVC-based membrane ISEs could be used to measure phosphate content in biological samples (Glazier, 1988; Carey, 1994; Liu, 1997, Fibbioli, 2000; Wroblewski, 2001). ISEs were used for soil nutrient detection in two directions: (1) Flow

Injection Analysis (FIA) systems (Ruzicka, 1977; Hongbo, 1985; Ferreira, 1996), and (2) vehicle-based soil sensing systems (Adsett, 1991; V.I. Adamchuck, 2004). However, ISEs might not have been ready for real-time sensing applications because of their response delay (several minutes).

2.2 Ion Selective Field Effect Transistor

ISFET is the integration of an ISE and a field effect transistor (FET). The ion selective membrane is placed on the top of the insulator layer of the FET structure, so the threshold voltage of the ISFET can be chemically modulated and the measured voltage is related with the concentrations of a target ion. ISFETs have several advantages over ISEs, such as small dimensions, low output impedance, high signal-to-noise ratio, fast response and the ability to integrate mulit-ISFETs on one chip.

ISFETs were reported to detect soil ammonium (Oesch, 1981), nitrate (Van, 1994; J. Artigas, 2001) and potassium (Moss, 1975; 1978; Van, 1994; J. Artigas, 2001). Also, ISFETs were used in FIA systems and vehicle-based real-time soil sensing systems by researchers (A.U. Ramsing, 1980; Loreto, 1996). A successful automated system for soil pH mapping was reported to be tested under field conditions by Adamchuck et al. (2002). However, ISFET's high cost and inconsistent repeatability limited their wide extension use in practical systems.

3. OPPORTUNITIES AND CHALLENGES

Some emerging advanced techniques, such as MicroElectroMechanical System (MEMS), Microfluidics and Lab on valve (LOV), have been attempted for electrochemical detection in biological, chemical and medical fields and provide new ideas and good opportunities for electrochemical sensors from fabrication to measurement.

Multi-targets Simultaneous Detection. So far, most soil nutrient detection methods can only measure one target ion by using conventional electrochemical measurement due to an ion selective membrane, in theory, only selectively responding to one target ion. Electrochemical sensors can be integrated onto one chip as a sensor array to provide a feasible approach of multi-targets simultaneous detection. However, all the membranes developed for soil NPK detection did not respond to only one specific ion, but also to other interfering ions present in the analyte. One main challenge faced is the reliability of sensor array, that is, to avoid or diminish the interferences from other ions while using a sensor array for simultaneous detection.

Automation. Both ISEs and ISFETs have successfully been used in FIA systems. At present, micropumps and microvalves suitable for the development of accurate FIA systems are commercially available now. It has been possible to integrate one sensor or sensors with a FIA system to develop an automated ISE/ISFET-FIA system, even a microfluidic detection system on one chip. This kind of microsensors may be of automatic control and easily used even for non-trained users. However, the blocking of the microchannels by small particles should be considered.

Lower cost. Electrochemical sensors have the potential to be produced in batches to very small sizes by using MEMS-based microfabrication technology at a very low cost. The sensors can even be designed for one-off or multiple uses. Besides, small-sized sensors require small volume of reagents and samples, which can also reduce the cost in soil testing.

Rapidness. Time is a critical factor for soil nutrient detection since the variability of soil nutrient levels may be quite high over time (Sudduth et al., 1997). Due to complex soil pretreatment and chemical analysis, standard testing methods for soil NPK are time-consuming (in hours). Electrochemical sensors can rapidly respond to the target ions in minutes, suitable for in-field rapid detection.

4. CONCLUSION

The advantages of potentiometric electrochemical sensors are stimulating the interest of their applications in soil nutrient detection. They have potentials for automated multi-target rapid detection of soil nutrients. As such, they are also faced with the challenge from their reliability. Advanced engineering technologies have opened our mind and provided new approaches for soil testing to follow the KISS (Keep It Simple and Stupid) principle to treat the complex soil testing procedures with simpler methodology at a lower cost.

ACKNOWLEDGEMENTS

This research was supported by National 863 Project (2006AA10Z216).

REFERENCES

Ramsing A.U. et al. Anal. Chim. Acta, 1980, 118: 45

Adamchuk V.I. Feasibility of on-the-go mapping of soil nitrate and potassium using ionselective electrodes. Transaction of ASAE., 2002, 02, 1183

Adsett J.F., G.C. Zoerb. Automated field monitoring of soil nitrate levels. In: Automated agriculture for the 21st century, Mich.: ASAE Publ., 1991, 326-335

Banwart W.J., et al. Commun. Soil Sci. Plant Anal., 1972, 3: 443-458

Carey C.M., W.B. Riggan. Cyclic polyamine ionophore for use in a dibasic-phosphate selective electrode. Anal. Chem., 1994, 66: 3587-3591

Dahnke W.C. Use of the nitrate specific ion electrode in soil testing. Soil Sci. and Plant Anal., 1971, 2(2): 73-84

Ferreira A.M.R., Lima J.L.C., Rangel A.O.S.S. Potentiometric determination of total nitrogen in soils by flow injection analysis with a gas-diffusion unit. Aust. J. Soil Res., 1996, 34: 503

Fibbioli M., M. Berger, F.P. Schmidtchen, et al. Polymeric membrane electrodes for monohydrogen phosphate and sulfate. Anal. Chem., 2000, 72(1): 156-160

Glazier S.A., M.A. Arnold. Phosphate-selective polymer membrane electrode. Anal. Chem., 1988, 60: 2540-2542

Hansen E.H., A.K. Ghose, J. Ruzicka. Flow injection analysis of environmental samples for nitrate using an ion-selective electrode. Analyst, 1977, 102: 705-713

Hongbo C., Hansen E.H., Ruzicha J., Evaluation of critical parameters for measurement of pH by flow injection analysis. Anal. Chim. Acta 169, 209 (1985) J. Artigas, A. Beltran, C. Jimenez, et al. Application of ion sensitive field effect transistor based sensors to soil analysis. Computers and Electronics in Agriculture, 2001, 31: 281-293

Liu D., W.C. Chen, R.H. Yang, et al. Polymeric membrane phosphate sensitive electrode based on binuclear organotin compound. Analytica Chimica Acta, 1997, 338: 209-214

Loreto A.B., Morgan M.T. Development of an Automated System for Field Measurement of Soil Nitrate. Transaction of ASAE, 1996

Mei Shourong. Soil Total Potassium Detection Using Potassium Electrode. Shanghai Agriculture Science and Technology, 1981, (01) (in Chinese)

Moss S.D., Janata J., Johnson C.C. Potassium, ion-selective field effect transistor. Anal. Chem., 1975, 47 (13): 2238

Moss S.D., Johnson C.C., Janata J. Hydrogen, calcium and potassium ion-sensitive FET transducers: a preliminary report. IEEE Trans. Biomed. Eng. 1978, BME-25, 49

National Bureau of Statistics of China. China Statistical Yearbook 2006. Beijing: China Statistics Press, 2006 (in Chinese)

Oesch U., Caras S. Field effect transistor sensitive to sodium and ammonium ions, Anal. Chem., 1981, 53: 1983

Ruzicka J., Hansen E.H., Zagatto E.A. Flow injection analysis. Part II: use of ion-selective electrodes for rapid analysis of soil extracts and blood serum. Determination of potassium sodium and nitrate. Anal. Chim. Acta, 1977, 88(1)

Simeonov V., et al. Anal. Letter, 1976, 9: 1025-1029

Sudduth K.A., J.W. Hummel, S.J. Birrell. 1997. Sensors for site-specific management, 183-210, In: Pierce F.J. and E.J. Sadler (ed). The state of site-specific management for agriculture. Madison, Wisc. ASA-CSA-SSSA

Adamchuk V.I., J.W. Hummel, M.T. Morgan, et al. On-the-go soil sensors for precision agriculture. Computers and Electronics in Agriculture, 2004, 44:71-91

Van der Wal P.D., Van der Berg A., De Rooij N.F. Universal approach for the fabrication of Ca_2-, K+ and NO_3- sensitive membranes ISFETs. Sensors and Actuators, 1994, B18-19, 200

Wang Baoli, Qu Dong, Tian Hua, Simultaneous Determination of Sodium and Potassium in Soil Extract by Ion Selective Electrode, Journal of Northwest Sci-Tech University of Agriculture and Forestry (Natural Science Edition), 1992, 20(3): 130-134 (in Chinese)

Wroblewski W., K. Wojciechowski, A. Dybko, et al. Durable phophate-selective electrodes based on uranyl salophenes. Anal. Chim. Acta, 2001, 432: 79-88.

RELATION EXTRACTION BASED ON AGROVOC

Qian Wang
Computer and Information Management Center, Tsinghua University, Beijing, China, 100084
Address: Computer and Information Management Center, Tsinghua University, Beijing, 100084, P. R. China, Tel: +86-10-62850228, Email: cauwq@163.com

Abstract: Relation extraction is an important step in Ontology construction. This paper provides a method to extract relations among conceptions with AGROVOC.

Keywords: relation extraction, AGROVOC, thesaurus

1. INTRODUCTION

Ontology is a conceptualization of a domain into a human understandable, but machine-readable format consisting of entities, attributes, relationships and axioms (Guarino et al., 1995). How to extract relationships among entities is a difficult task in ontology construction. AGROVOC (Fao, 2007) is a multilingual thesaurus in agriculture which has indexed lots of literatures. We can obtain the relationships from AGROVOC and its text resource.

2. RELATION EXTRACTION

Description logic is used to organize the ontology knowledge system. Description logic can be divided into two parts, and one is Tbox which contains the limited aggregate of terminological, the other is Abox which contains the limited aggregate of assertion. In this paper, the concept schema transformed by AGROVOC can be considered as Tbox. We can obtain Abox

Wang, Q., 2008, in IFIP International Federation for Information Processing, Volume 259; Computer and Computing Technologies in Agriculture, Vol. 2; Daoliang Li; (Boston: Springer), pp. 1355–1357.

from literature database indexed by thesaurus. We can extract the further relations and properties from literature database. For instance, if a property, P, is tagged as a SymmetricProperty property, we can conclude:

$$P(x,y) => P(y,x)$$

Contrarily, if the rule above is given, the schema can be built as follows:

```
<owl:ObjectProperty rdf:about="P">
  <rdf:type rdf:resource="http://www.w3.org/2002/07/owl#SymmetricProperty"/>
</owl:ObjectProperty>
```

Each literature can be converted into an instance of the tree by distributing the tags (Muhammad et al., 2005). To obtain the basic relationship, we should input those verbs surrounded by agricultural concepts. Some of the key behavioral features of concepts can be extracted by algorithm Morphological variants (Muhammad et al., 2005). Thus the system can obtains the frequently occurring triplets of the form < *s*, *p*, *o*,> efficiently. But few property rules defined in OWL (W3C 2004) are identified.

We will give an algorithm for the further Relation Extraction (RE). In this algorithm, we use some functions and sets explained as follows:

- *I* is a set of instance statements obtained from literature database. Every statement is represented as *<s,p,o>*. *s* is subject, *p* is predicate and *o* is object;
- $P=\{<s,p,o>\mid <s,p,o>\in I\}$, where $<s,p,o> \in I$, return all the instances which contain property *p*;
- *R* is a set of property rules defined in OWL;
- *Support* (p_r) is the support degree where *p* satisfies *r*, which has specific formula according to *r*;
- *Confidence* (p_r) is the confident degree where *p* satisfies *r*, which also has specific formula according to *r*;

Algorithms: Relation Extraction (RE)

Input: Literature database indexed by thesaurus;

Threshold *S* of *Support* (p_r);
Threshold *C* of *Confidence* (p_r)

Output: Property definition of OWL

1. First, find instances which can be represented with ID composed of literature and indexing order from literature database.
2. Prepare *I* from instances database.
3. Prepare *P* by classifying *I* by property.
4. for each $p \in P$ {

```
        for each r ∈R {
            compute Support (p_r) according to r;
            compute Confidence (p_r) according to r;
            If ( Support (p_r) < S || Confidence (p_r) < C ){
      continue ;
            }
            p_r is true;
        }
    output p_r definition of OWL;
    }
```

3. CONCLUSION

In this paper we propose a new method by which we can extract the OWL relationships from literature database indexed by thesaurus. The proposed algorithm is used to construct domain ontology in Agriculture. The extraction process can be completed automatically or semi-automatically by computer so that it depends less upon experts. However, it is not perfect in multiple-level relation mining and other axioms. As for future research, we intend to adjust the extraction arithmetic to support multiple-level relations and mine other axioms.

REFERENCES

Guarino N and Giaretta P. Ontologies and Knowledge Bases: Towards a Terminological Clarification. Toward Very Large Knowledge Bases: Knowledge Building and Knowledge Sharing, IOS Press, Amsterdam, 1995.

Fao. AGROVOC Thesaurus. http://www.fao.org/aims/ag_intro.htm, 2005.

Muhammad Abulaish, Lipika Dey: Biological Ontology Enhancement with Fuzzy Relations: A Text-Mining Framework. Web Intelleigence 2005: 379-385.

W3C. OWL Web Ontology Language Guide. http://www.w3.org/TR/2004/REC-owl-guide-20040210/, 2004.

LOCUST IMAGE SEGMENTATION USING PULSE-COUPLED NEURAL NETWORK

Xuemei Xiong, Yiming Wang
College of Information and Electrical Engineering, China Agricultural University, 100083 Beijing, China

Abstract: The main objective of this study was to evaluate the feasibility of identifying locusts by the image segmentation method based on pulse-coupled neural network (PCNN). Segmentation results by simplified PCNN and traditional morphology open operation method were given. The simulation results demonstrated that the PCNN was powerful enough to perform the detecting.

Keywords: Pulse-coupled neural network, Image segmentation, Locusts detection

1. PCNN MODEL

In this work, a PCNN with simplified parameters was applied to locust image segmentation. The area recognition rate (ARR) was used to measure the performance of this method.

2. EXPERIMENTS ON LOCUST IMAGES

The performance of PCNN on locust images is shown in Table 1. The best quality of image segmentation for tiny objects (Fig. 1) is obtained by a relatively larger linking strength β. The time response is different depending on the coupling strength between neighboring neurons and strong coupling neurons have a tendency to fire together for similar stimulus.

Xiong, X. and Wang, Y., 2008, in IFIP International Federation for Information Processing, Volume 259; Computer and Computing Technologies in Agriculture, Vol. 2; Daoliang Li; (Boston: Springer), pp. 1359–1360.

Table 1. Performance of PCNN on locust images

Images in	Parameters	Area Recognition Rate	Average Processing Time
Fig. 1	$\Delta T = 0.02, V_T = 100$, $\beta = 0.8, \alpha = 0.2$	92%	1s
Fig. 2	$\Delta T = 0.02, V_T = 100$, $\beta = 0.8, \alpha = 0.2$	81%	1s

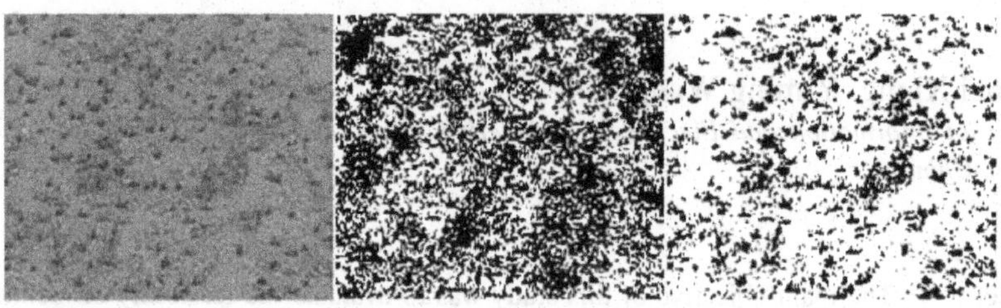

Fig. 1. Locusts from a long distance (left, converted as a gray-scale image) and the segmentation results (middle: result of open operation; right: result of PCNN at n=2)

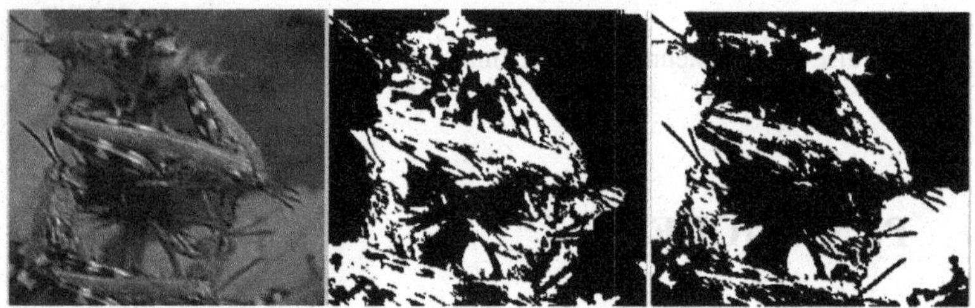

Fig. 2. Locusts from a short distance (left, converted as a gray-scale image) and the segmentation results (middle: result of open operation; right: result of PCNN at n=2)

3. CONCLUSION

We applied the simplified PCNN algorithm to locust image segmentation. The performance of PCNN is compared with the traditional morphology open operation method. Experimental results of PCNN in terms of visual effects are better than the open operation of mathematic morphology.

LIGHT SIMULATION INSIDE TOMATO CANOPY BASED ON A FUNCTIONAL-STRUCTURAL GROWTH MODEL

Qiaoxue Dong[1], Weizhong Yang[1], Yiming Wang[1,*]
[1] *College of Information and Electrical Engineering, China Agricultural University, Beijing, China, 100083*
[*] *Corresponding author, Address: P.O. Box 63, China Agricultural University, 17 Tsinghua East Road, Beijing, 100083, P. R. China, Tel: 86-10-62737591, Email: ym_wang@263.net*

Abstract: This paper present how the light transfer and interception inside the canopy was computed by combining a tomato dynamic growth model. The tomato functional structural growth model was applied to produce the topology and geometry of individual plant, and through transformation, the canopy could be composed of multi-individual plants. In this simulation system, a radioactive transfer model named hierarchical instantiation for radiosity, was coupled effectively into this tomato growth model to compute the repartition of light energy throughout the canopy, so the key interface technology when integration were explained in the paper. Finally, the visualization results were presented for the light simulation inside canopy.

Keywords: Tomato, Light simulation, Functional structural growth model, Visualization

1. INTRODUCTION

Computation of the distribution of light energy in vegetation mainly depends on the illumination model and representation of vegetation. Up to now, in the field of agronomic research, most approaches consider the vegetation as a turbid medium (Ross 1981), and attenuation factors was used to estimate direct illumination in the plant without accounting for light scattering inside the vegetation. Recent researches on virtual plants (Goel et al., 1990) enable detailed three dimensional canopy structure to be computer-

Dong, Q., Yang, W. and Wang, Y., 2008, in IFIP International Federation for Information Processing, Volume 259; Computer and Computing Technologies in Agriculture, Vol. 2; Daoliang Li; (Boston: Springer), pp. 1361–1364.

generated (Fournier et al., 2003). This kind of representation makes it possible to properly simulate light transfer and distribution within virtual plant scenes. Furthermore, by combining illumination model (Soler et al., 2003) with plant growth models, structural-functional mode (Yan et al., 2004) of crop growth not only provide an accurate tool to compute light interception and photosynthesis, but also enable to investigate plant growth response to external light signal or to quantify mutual influence of plant morphogenesis and plant physiological functions (de Reffye et al., 1997).

2. OVERVIEW OF SIMULATION

Figure 1 summarizes the architecture of this simulation system. The tomato growth simulator kernel on the left-hand side is in charge of three tasks: computing the plant topology, computing the volume of organs, and computing the geometry of the plant. The light simulation module depends on the solution of some key interface technology such as creation of instances, light sources and leaf optical property.

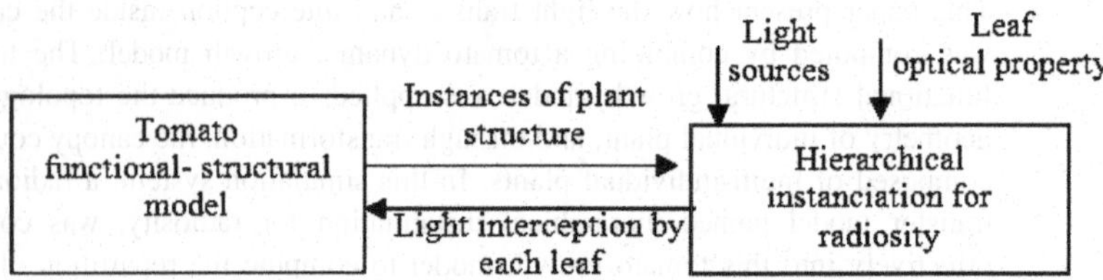

Fig. 1. Architecture overview of tomato growth simulation system

2.1 Tomato functional-structural model

Tomato functional-structural model combines a process-based model (Heuvelink, 1999) with a three-dimensional description of plant in one modeling framework. The process-based model deals with the growth and development of individual organs, so in such model, the organs not only play a functional role, but provide reliable information about biomass production and allocation as well as construction of organ shape, which depends on the cumulated biomass inside it. So the model is driven to simulate the growth of tomato plant by their interaction among above modules (Dong et al., 2003).

2.2 Light simulator

Here, hierarchical instantiation for radiosity (Soler et al., 2000) was selected as radiation transfer model, where a new radiosity algorithm based on the concept of hierarchical instantiation was proposed. So one of the key interface technologies is how to identify similar geometry, thus insatiable structures in the plantation scene. The tomato structural model provides this possibility by producing topological and geometrical information. Plants are defined as hierarchies of botanical structures, each one attributed with a collection of botanical parameters, such as physiological age, the kind of axis, branching order, and so on. This useful information can be used as a criterion of similar structures (Fig. 2).

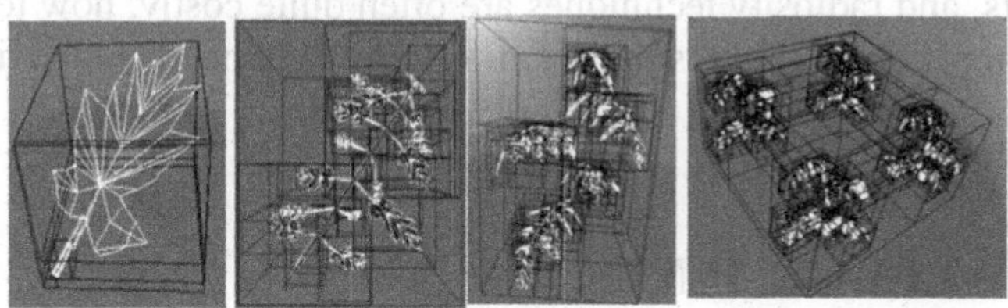

Fig. 2. Hierarchical insatiable structures of tomato growth scene

When simulating light distribution inside canopy, the greenhouse light environment needs to be provided to hierarchical instantiation algorithm. An advanced technology named image-based lighting technique (Debevec, 1998) was used to capture real-world illumination inside greenhouse and then to generate directional light sources feed into light simulator.

3. RESULTS AND DISCUSSIONS

The parameters of tomato growth model were obtained by field measurement and optimization procedure (Dong et al., 2003). Based on these parameters, tomato growth engine can produce the geometrical vegetation for each growth stage, and then light simulator was applied to compute the light distribution at the organ level. (Fig. 3 (a)). And a visualization result was also shown in Fig. 3(b).

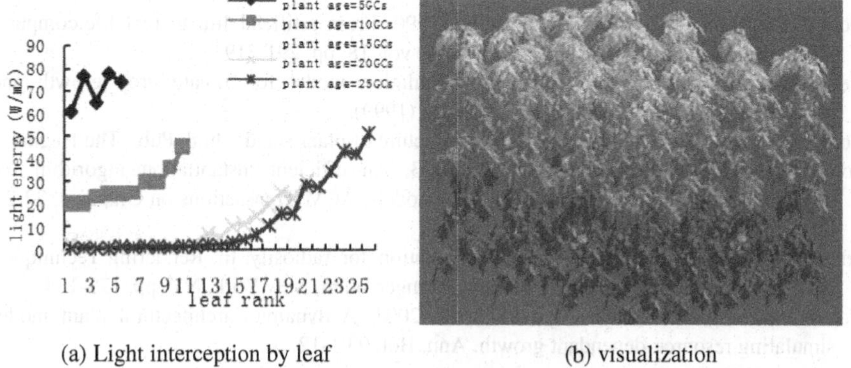

(a) Light interception by leaf (b) visualization

Fig. 3. Light distribution inside plantation

It have showed that the tomato functional-structural model can output the geometrical and topological information dynamically, which enables an efficient combination with the radioactive transfer model to compute the light absorption by each individual organ. This will allow some tests of hypotheses by model experiments, concerning the question how the local radiation interception influences the control of shoot growth and plant architecture. But because canopy geometry are based on the very geometry of the plants, and radiosity techniques are often quite costly, how to improve the efficiency of this light simulation system will be included in further study.

ACKNOWLEDGEMENTS

This study was supported by the National Natural Science Foundation of China (#60073007), and the LIAMA laboratory in Beijing, China.

REFERENCES

de Reffye P, Houllier F. Modeling plant growth and architecture: some recent advances and application to agronomy and forestry. Current Science, 73(11):984-992 (1997).

Debevec P. 1998. Rendering synthetic objects into real scenes: bridging traditional and image-based graphics with global illumination and high dynamic range photography. In: Proc. Siggraph 98, Computer Graphics, ACM Press, New York,, pp. 189-198.

Dong Q.X., Wang Y.M., Barczi J.F., et al. 2003. Tomato growth modeling based on interaction of its structure-function. In: Hu BG, Jaeger M (eds.). Proc. Plant Growth Modeling and Applications (PMA'03), Beijing, China. Tsinghua University Press and Springer, pp. 250-262.

Fournier C., Andrieu B., Ljutovac S., et al 2003. ADEL-wheat: A 3D architectural model of wheat development. In: Hu BG, Jaeger M (eds.). Proc. Plant Growth Modeling and Applications (PMA'03), Beijing, China. Tsinghua University Press and Springer, pp. 55-63.

Goel N.S., Knox L.B., and Norman M.N., 1990. From artificial life to real life:computer simulation of plant growth. Int. J. Gen. Syst., vol. 18, pp. 291-319.

Heuvelink E. Evaluation of a dynamic simulation model for tomato crop growth and development. Annals of Botany, 83:413-422 (1999).

Ross, J. 1981. The radiation regime and architecture of plant stands. Junk Pub., The Hague.

Soler C., Sillion F., Blaise F., et al. 2003. An efficient instantiation algorithm for simulating radiant energy transfer in plant models. ACM Transactions on Graphics, 22(2): 204-233.

Soler C., Sillion F., 2000. Hierarchical instantiation for radiosity. In: Rendering Techniques' 00, B. Peroche and H. Rushmeier (Eds). Springer Wien, New York, NY, pp. 173-184.

Yan H.P., Kang, M.Z., de Reffye P., et al. 2004. A dynamic, architectural plant model simulating resource-dependant growth. Ann. Bot, 93:1-12.

THE THERMAL MEASUREMENT SYSTEM

Yanzheng Liu[1,2], Guanghui Teng[1,*], Chengwei Ma[1], Shirong Liu[3]
[1] *College of Water Conservancy and Civil Engineering China Agricultural University, Beijing, China, 100083*
[2] *Beijing Vocational College of Agriculture, Beijing, China, 102442*
[3] *Yantai Research Institute China Agricultural University, Yantai, 264670*
[*] *Corresponding author, Address: P.O. Box 195, College of Water Conservancy and Civil Engineering, China Agricultural University, 17 Tsinghua East Road, Beijing, 100083, P. R. China, Tel: +86-10-62737583, Fax: +86-10-62736413, Email: futong@cau.edu.cn*

Abstract: The K-value (the overall heat transfer coefficient) is an important factor to reflect thermal performance of the covering material in a green house. The measuring process of K-value was burdensome in labors, low accuracy, complicated in dealing with data that is difficult to meet the development of modern agriculture. New testing methods need to be developed, and specially designed for greenhouse covering materials.

Keywords: greenhouse, virtual instrumentation, thermal measurement system

1. INTRODUCTION

The K-value (the overall heat transfer coefficient) is an important factor to reflect thermal performance of the covering material in a green house. In general, we test the K-value in the nature environment condition or simulated real environment in lab. Mihara and Hayashi kept the constant temperature inside of the room using the control instrument, and calculated the electricity quantity to get the K-value (Mihara and Hayashi, 1970). P. Feuilloley and G. Issanchou compared thermal performance between the glass and the plastic covering materials (P. Feuilloley and G. Issanchou, 1996). Xinqun Zhou tested the thermal insulation of honeycomb covering materials with plastic sheets, and he got the K-value using multi-conversion

Liu, Y., Teng, G., Ma, C. and Liu, S., 2008, in IFIP International Federation for Information Processing, Volume 259; Computer and Computing Technologies in Agriculture, Vol. 2; Daoliang Li; (Boston: Springer), pp. 1365–1369.

switch and the direct current potentiometer (Xinqun Zhou, 1998). The measuring process of K- value was burdensome in labors, low accuracy, complicated in dealing with data that is difficult to meet the development of modern agriculture. New testing methods need to be developed, and specially designed for greenhouse covering materials.

2. MATIERIALS AND METHOD

2.1 Geography and meteorology

In our study, we measured the cover material of a Static Hot Box to simulate the real condition, and we used the testing platform for protected agriculture cover material (Fig. 1).

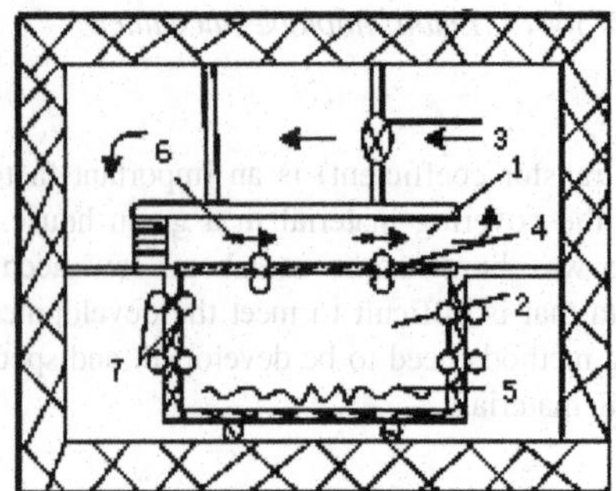

Fig. 1. The cross-section diagram of the testing platform

The box is airproof and is not subject to the sun radiation, $Q_t = Q_s = 0$; there are neither soil nor plants, $Q_r = Q_e = Q_p = Q_d = 0$; there are no lights on during the test, $Q_a = 0$; the temperature is kept constant during the test, $\Delta Q = 0$. With these conditions, the equation (1) is simplified as:

$$Q_w = KA_w(t_i - t_0) = Q_g - \sum_{j=1}^{5} Q_j \tag{1}$$

$$K = \frac{Q_c}{A_c(t_i - t_o)} = \frac{Q - \sum_{j=1}^{5} Q_j}{A_c(t_i - t_o)} = \frac{Q - \sum_{j=1}^{5} A_j(t_{wij} - t_{woj})/R_{wj}}{A_c(t_i - t_o)} \tag{2}$$

where Q_w is the heat conducted through the cover material per second; Q_j is the heat through the other five walls of the heating box; Q_g is the overall heat given by the box (power of heating when the system is at balance); K is the heat transfer coefficient; A_w is the area of the cover; A_j

is the area of the other five walls; t_i, t_0 is the inside and outside temperatures of the cover respectively; R_{wj} is the thermal resistance of the other five walls. Based on the studied conditions (K.V. Garzoli and J. Blackwel, 1981), we can calculate the K-value of the cover in the testing platform.

2.2 The hardware components of the virtual measurement system

The virtual instrument consists of computer hardware, an instrument module, and a measurement and control unit with software for data analysis, process communication, and graphic interface. The flow diagram of this system is shown in Fig. 2.

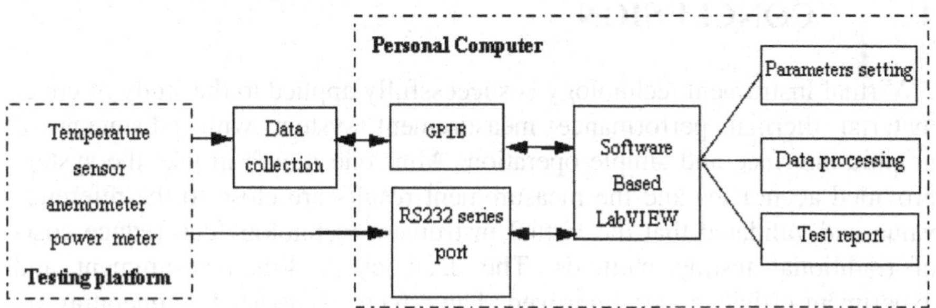

Fig. 2. System structure block diagram

3. RESULT AND DISCUSSION

We tested the K-value of a 5mm thick flat glass in the TMS. The power of heating is 55±1 W, and the airspeed is 3.0±0.125 m/s and the temperature of sky radiation board is -30±2°C, (These values are based in the weather information of Beijing), so the testing platform simulates the real condition. The measurement results are shown in Table 1. Form the Table 1 we can see that the testing K-value of glass is $7.59\,W \cdot m^{-2} \cdot {}^\circ C^{-1}$, and the deviation compared with the reference value (Qin Midao, 2004) is between 0.7% and 1.1%, that shows that the precision of the measurement system is accurate. There are only a few standard testing methods used for thermal performance of covering materials for greenhouses. The values presented are found to vary considerably among the materials considered. The heat transfer coefficient of different covering materials are not the same, for example, the K-value of 3mm thick of flat glass is from 6.29 to 6.86 (Zhou

Changji, 2003). These variations are attributed mainly to the testing method employed and also to the type of materials.

Table 1. Result of measurement

Testing cycle	Heat loss/W	Heat flux of glass/W	K testing/$W \cdot m^{-2} \cdot {}^{\circ}C^{-1}$	K reference/$W \cdot m^{-2} \cdot {}^{\circ}C^{-1}$
1	38.4	232.1	7.60	7.54
2	38.6	231.4	7.63	7.56
3	38.7	231.3	7.59	7.51
4	38.6	230.7	7.53	7.49
5	38.5	231.5	7.60	7.58
Average	38.6	231.36	7.59	7.54

4. CONCLUSION

Virtual Instrument technology is successfully applied to the study of cover material thermal performance measurement system with advantage of graphic interface and simple operation. After one month in use, the system provided accuracies and the measurement results are close to the reference value and validated that the virtual instrument technology can reduce costs of traditional testing methods. The accuracy of this measurement and experiment efficiency are improved. The effects of airspeed, temperature on both sides of the material and etc on the K-value need more experiments and studies.

REFERENCES

Dai Guosheng. Heat Transfer (second edition), Higher Education Press, Beijing, 1999

K.V. Garzoli, J. Blackwell. An analysis of the nocturnal heat loss from a single skin plastic greenhouse. J. Agric. Eng. Res. 1981, 26: 203-214

Mihara and Hayashi. Studies on the Insulation of Greenhouse (1) Overall Heat Transfer Coefficient of Greenhouse with single and double covering using several material curtains. J. Agr. Met. 1979, 35: 1 13-19

P. Feuilloley, G. Issanchou. Greenhouse covering materials measurement and modeling of thermal properties using the hot box method, and condensation effects. J. Agric. Eng. Res. 1996, 65: 129-142

Qin Midao. Research on the Test Apparatus and Technology for Measuring the Thermal Transmittance of Greenhouse Covering Materials. Master Dissertation, Beijing: China Agricultural University, College of Water Conservancy & Civil Engineering. 2004.

The State Bureau of Technical Supervision P. R. China. Adiabatic material and correlation term. 1996. GB/T 4132-1996. Beijing: Standards Press of China. 1997

Zhou Changji. Modern Greenhouse Engineering. China Beijing: Chemical Industry Press. 2003

Zhou Xinqun, Dong Renjie, Zhang Shumin. Study on Thermal Insulation Covering Materials of Honeycomb Plastic Sheet. Transactions of the Chinese Society of Agricultural Engineering. 1998, 12: 159-163

EXPERT SYSTEM OF NON-POLLUTION FEICHENG PEACH PRODUCTION IN CHINA

Ming Li, Yan'an Wang*, Lihui Wang, Zhiqiang Yan, Qingyan Yu
College of Life Science, Shandong Agricultural University, Taian, Shandong, China, 271018
* *Corresponding author, Address: College of Life Science, Shandong Agricultural University, 61 Daizong Street, Tai'an, 271018, P. R. China, Tel: +86-538-8249144, Email: wyasdau@126.com*

Abstract: In order to promote standard Feicheng peaches production, based on long-term research and integrated achievements of many disciplines on Feicheng peach, following the standard of localization, characteristic and non-pollution, adopting the knowledge engineering, and using PAID (Platform for Agricultural Intelligence-system Development), the expert system of non-pollution Feicheng peach production has been developed. As the samples of knowledge representation, the model of water-saving irrigation and pattern of search list for differentiation between diseases and insect pests are discussed. Through application of the system, the users have acquired significant social and economic benefit.

Keywords: Feicheng peach (*Prunus persica* cv. Feicheng), non-pollution, expert system, knowledge representation

1. INTRODUCTION

Feicheng peach (*Prunus persica* cv. Feicheng), a kind of fruit mainly produced in Feicheng, Shandong province, China, is famous at home and abroad for big size, excellent taste and abundant nutrition, and praised as "the champion of peaches". Up to January, 2005, the city has 4,800 ha of Feicheng peaches as green food for grade A. In future, the city is going to enhance the informatization of Feicheng peach industry in order to drive marketization and promote industrialization. Agricultural expert system is a

Li, M., Wang, Y., Wang, L., Yan, Z. and Yu, Q., 2008, in IFIP International Federation for Information Processing, Volume 259; Computer and Computing Technologies in Agriculture, Vol. 2; Daoliang Li; (Boston: Springer), pp. 1371–1374.

useful decision tool for growers and county extension faculty with standard cultivation.

Since the first agricultural expert system (PLANT/ds) was developed (Michalski et al., 1982), a lot of expert systems (ES) or decision support systems (DSS) on peach have been readily available and affordable, such as an expert system for the diagnosis of peach and nectarine disorders (Plant et al., 1989), a user friendly peach tree growth and yield simulation model for research and education (Allen and Dejong 2002), peach expert system (Wang et al., 2001). A literature search reveals no attempt in developing localized, characteristic and non-pollution expert system for peach. Also the existing diagnosis expert systems are lack of differentiation of diseases and insect pests. In order to promote the non-pollution Feicheng peaches production, from 2001 to 2005, the expert system of non-pollution Feicheng peach production was developed and applied in Feicheng city.

2. SYSTEM DESIGN

2.1 Architecture

The system architecture is designed according to existing conditions of demonstration districts and framework of PAID. The decision-making system of Feicheng peach production includes overall agronomic practices, such as decision-making for orchard selection, orchard plan, soil management methods, soil fertilizing methods, fertilizing dose rate, water-saving irrigation, methods of training and pruning, course of training and pruning, flower and fruit thinning, manual pollination and bagging technique, evaluation of quality traits of fresh fruits, differentiation between diseases and insect pests, diagnosis and control of infectious diseases, diagnosis and control of non-infectious diseases, morphological diagnosis of malnutrition, diagnosis and control of foliar nutrition, diagnosis and control of insect pests and so on.

2.2 Knowledge representation

Agronomic practice is not only calculating crop water requirements but it includes other issues which needs human judgment. Expert system technology can help in two ways: first, the mathematical models can be implemented using expert systems technique with the advantage of explaining to the user how the model calculates a certain quantity; second, the heuristic and symbolic knowledge which cannot be mathematically modeled is to be included easily. So, the mathematic model and production rule is applied in knowledge representation. The examples are following.

2.2.1 Mathematic model

The model of water-saving irrigation is expressed as the formula (1) reflecting the relationship of each parameter in the model (Xu, 1998):

$$X = Y \times Q \times A \times P \times (B - R) \tag{1}$$

where X is net irrigation water(kg/ha), Y is irrigation area(m^2), Q is cover rate of tree canopy(%), A is irrigation depth(m), P is soil bulk density, B is expected soil water content(%), R is practical soil water content(%).

2.2.2 Production rule

Facing a kind of abnormal phenomenon, it is important to judge whether it is caused by diseases or insect pests scientifically and correctly; If it is a disease, then whether is it infectious or non-infectious one? If it is infectious disease, which kind of factor causes it? When it is non-infectious disease, which kind of pathogen causes it? In order to solve the problems, the model of diagnosis & control of diseases and insect pests offers three diagnosis paths: in the first path, the users can carry out differentiation between diseases and insect pests at first, if it is a disease, then decide whether it is infectious or non-infectious, later enter related decision-making module to diagnosis; As the second path, the users can enter directly related decision-making for diagnosis & control of infectious diseases, decision-making for diagnosis & control of non-infectious diseases, decision-making for morphological diagnosis & control of malnutrition, decision-making for diagnosis & control of foliar nutrition, or decision-making for diagnosis & control of insect pests to diagnose; In the third path, the users optionally can consult directly the integrated information querying system to query related content of diseases and insect pests if they know the causes of problem (Wang et al., 2005). In summary, the search list of differentiation between diseases and insect pests is made.

3. SYSTEM IMPLEMENTATION

Coming to editing of knowledge rules, referred to the CommonKADS methodology (Schreiber, Shi et al., translated, 2003), the knowledge rules are edited by template and formalized according to the request of PAID, which improves the efficiency of development.

4. SYSTEM APPLICATION

This system has demonstrated and extended for more than 670 ha totally, the yield of Feicheng peaches increases by 8%–10%, cost decreases by 10%–12%, the rate of high-quality fruit increases by 15%–20%, and 95% of fruit has reached the standard of non-pollution food. The ordinary users have been trained for about 5300 person-times, which acquires significant social and economic benefit.

ACKNOWLEDGEMENTS

The research was funded by Taian "The Tenth Five-year Plan" key project (2001003) and Shandong provincial key extention project in 2003 (2003129).

REFERENCES

Allen M.T., Dejong M.T. Using L-system to model carbon transport and partitioning in developing peach trees, Acta Horticulturae, 2002, 584: 29-34.

Michalski R.S., Davis J.H., Bisht V.S. and Sinclair J.B. PLANT/ds: an expert consulting system for the diagnosis of soybean diseases, Proc. European Conference on Artificial Intelligence, July, 1982, Orsay, France: 133-138.

Plant R.E., Zalom F.G., Young J.A. and Rice R.N. An expert system for the diagnosis of peach and nectarine disorders. HortScience, 1989, 24: 700.

Schreiber G. Shi Z.Z., Liang Y.Q., Wu B. et al (translated). Knowledge Engineering and Knowledge Management. China Machine Press, Beijing, 2003: 79-106 (in Chinese).

Wang J.H., Yang X.T., Wang B.H. The construction of the agricultural expert system knowledge database based on the HPC develop platform. Computer and agriculture, 2001 (4):13-16 (in Chinese).

Wang Y.A., Li M., Wang L.L. et al. The knowledge database construction of fruit trees expert system for diagnosis, prevention and control of disease and pest, Journal of Shandong agricultural university. 2005, 36: 475-480 (in Chinese).

Xu M.X. The Cultivation Technology of Fruit Trees in Drought Area. Jindun Publishing House, Beijing: 1998 (in Chinese).

TOWARDS DEVELOPING AN EARLY WARNING SYSTEM FOR CUCUMBER DISEASES FOR GREENHOUSE IN CHINA

Ming Li[1,2], Chunjiang Zhao[1,*], Daoliang Li[2], Xinting Yang[1], Chuanheng Sun[1], Yan'an Wang[3]

[1] *National Engineering Research Center for Information Technology in Agriculture, Key Laboratory for Information Technologies in Agriculture, the Ministry of Agriculture, Beijing, China,100097*

[2] *College of Information and Electrical Engineering, China Agricultural University, Beijing, China,100083*

[3] *College of Life Science, Shandong Agricultural University, Taian, Shandong Province, China, 271018*

[*] *Corresponding author. Address: Room 320, Beijing Agricultural Science Mansion, Building A, Banjing, Haidian district, Beijing, 100097, P. R. China, Tel: +86-10-51503411, Fax: +86-10-51503750, Email: zhaocj@nercita.org.cn (Chunjiang Zhao)*

Abstract: The integrated management of cucumber (*Cucumis sativus L.*) diseases play a key role in guaranteeing the high quality and security of cucumber production in greenhouse, moreover, the early warning of cucumber diseases is the chief precondition for IPM (Integrated Pest Management). This paper describes an attempt to develop an early warning system for cucumber diseases in greenhouse. By analysing plant disease epidemiology and early warning theory, the conceptual model of early warning on cucumber disease of greenhouse is developed. The data collection, data transfer system, database system, forecast system, warning system, and so on are integrated and an early warning system for cucumber diseases in greenhouse has been designed.

Keywords: cucumber (*Cucumis sativus L.*), greenhouse, disease, early warning system

Li, M., Zhao, C., Li, D., Yang, X., Sun, C. and Wang, Y., 2008, in IFIP International Federation for Information Processing, Volume 259; Computer and Computing Technologies in Agriculture, Vol. 2; Daoliang Li; (Boston: Springer), pp. 1375–1378.

1. INTRODUCTION

Concerns about food safety, environmental quality and pesticide resistance have dictated the need not only for diagnosis and controlling the diseases, but also for forecasting them correctly and providing preventive measures for planters, namely providing an early warning. In order to develop a reliable and effective early warning system, it is necessary to understand related information concerning epidemiology of pathogens, greenhouse microclimate, cucumber growth and agronomic practices (Jeger, 2004).

The early warning theory has been originated from military affairs, and extended to economy, earth quake, flood, biological disaster early warning and so on. There have been many pest-warning system, such as for processing tomato (Gleason et al., 1995), pear blossom blast (Latorre et al., 2002) and cotton bollworm (Bai et al., 2002), but few studies of cucumber diseases early warning/forecast system in greenhouse have been conducted. Xu et al. (2003) developed a forecasting and management system of cucumber powdery mildew and downy mildew in plastic greenhouse tunnel, which is not based on real-time environmental information. A greenhouse intelligent ecohealth wardship system for tomato (Lu et al., 2004) and ecosystem monitoring and the decision system for diseases and insect pest management of cucumber in non-heated greenhouse (Ju, 2006) are reported, but both need more support of accurate predictable warning models. The lack of such information represents a gap in our knowledge of disease warning systems, thus combining disease early warning model with greenhouse environment data logger, an early warning system for cucumber diseases in greenhouse has been designed.

2. CONCEPTUAL MODEL OF EARLY WARNING

According to plant epidemiology and early warning theory, the conceptual model of early warning on cucumber disease in greenhouse is designed, which includes five following steps. The first one is warning index confirmation. The early warning model predicts outbreak time and grade of diseases, which are also warning index. Outbreak time is the date when diseases become visible, which is calculated from the transplanting date. Coming to the grade of diseases, disease index (DI) is the integration of disease incidence and disease severity as an acceptable sign. The second one is warning indicators analysis. As to a specific disease, such as cucumber downy mildew or powdery mildew, there will be some domain factors for plant disease epidemic. Taking cucumber downy mildew for example, relative humidity (RH), duration of leaf wetness and air temperature are main environmental factors for infection and epidemic. Then, warning source search is the third one. It will present the warning source, such as the times of irrigation is excessive, the greenhouse ventilation effect is poor or the daily temperature range is much and so on. Later one is warning in time. After warning source search, the warnings are issued through sound alarm and caution light. The caution light uses a key with 5 disease classes (0, 1, 2, 3, 4; from none to severity) and defines green, blue, yellow, orange and red warning signal respectively. Then, the users can judge the warning based on integrated greenhouse circumstances and their experiences. The last one is warning obviation. During the process of warning obviation, the users can input the real-time information and acquire change of disease-warning, which is helpful for decision-making. If the warning management is over, the warning will be relieved by users.

3. SYSTEM ARCHITECTURE

The early warning system uses techniques of inference machine from artificial intelligence and expert systems to provide smarter support for the decision-maker, which include four parts mainly. The core of the system is model base. The development of models is embedded in an object-oriented approach that emphasizes the separation of models from data and promotes model and data integration and reuse (A.E. Rizzoli et al., 1998). So the model management system allows the user to find a model corresponding to historical data, to modify a selected model and to compose a new model, possibly assembled from existing ones. Then the forecast (inference) machine integrates the method of the mathematical model and rule-based model. The rule-based model is used for predicting the disease outbreak day and the mathematical model is employed for calculating the epidemic degree of disease in a few days later. Followed, the warning machine is started up. Warning machine in office is dependent on the main computer and work in ways of sound alarm and caution light, which consists of components for user login, data import, query and statistics, and system maintenance.

4. CONCLUSION

To avoid known barriers hindering the broader acceptance of early warning system, the development team identifies a prerequisite which has guided development and applications. In developing an early warning model, pathogen, agronomic, environmental and host factors should be taken into account (Shtienberg, 2000). An important aspect comprises the possible interactions among parameters, which sometimes may be complex. For example, the amount of fertilization may directly affect the response of the host plant to the pathogens, while excessive fertilization might result in excessive growth of the foliage, which, in turn, would affect the micro-climatic conditions within the canopy. Consideration of all parameters, and their interactions, may enable the grower to adequately predict the likelihood of a disease outbreak. So integrated above knowledge and models, the early warning system for greenhouse cucumber diseases will be more practical.

ACKNOWLEDGEMENTS

It is an outcome of National Key Technology R&D Program (No: 2006BAD10A08) and Huo Yingdong Young Teacher's Foundation (Contract No: 94032).

REFERENCES

Bai L.X., Sun Y.W., Jin Z.Q. The Development of the Early Warning to the Cotton Bollworm Calamity and Its Auxiliary Control Decision-making System (MLCYJJC-CDROM), Cotton Science, 2002, 14: 166-170 (in Chinese).

Gleason M.L., MacNab A.A., Pitblado R.E., et al. Disease-Warning System for Processing Tomatoes in Eastern North America: Are We There Yet? Plant Disease, 1995, 79: 113-121.

Jeger M.J. Analysis of disease progress as a basis for evaluating disease management practices. Annual Review of Phytopathology, 2004, 42: 61-82.

Ju R.H. Ecosystem monitoring and the decision system for diseases and insect pest management of cucumber in SES greenhouse. PhD thesis, China Agricultural University, Beijing, China (in Chinese), 2006.

Latorre B.A., Rioja M.E., Lillo C. The effect of temperature on infection and a warning system for pear blossom blast caused by Pseudomonas syringae pv. Syringae. Crop Protection, 2002, 21: 33-39.

Lu J. Research on Greenhouse Intelligent Ecohealth Wardship System (GH-Healthex) Base on Tomato. Masteral Dissertation, China Agricultural University, China, 2004 (in Chinese).

Rizzoli A.E., Davis J.R., Abel D.J. Model and data integration and re-use in environmental decision support systems. Decision Support Systems, 1998, 24: 127-144.

Shtienberg D. Modelling: the basis for rational disease management. Crop Protection, 2000, 19: 747-752.

Xu N. Forecasting and management system of cucumber powdery mildew and downy mildew in plastic greenhouse tunnel. Masteral Dissertation, Nanjing Agricultural University, China, 2003 (in Chinese).

HEAT LOSS THROUGH COVERING LAYER

Yi Zhang, Chengwei Ma*, Junfang Zhang, Midao Qin, Ruichun Liu

* *Corresponding author. Director of Key Lab of Agricultural Bio-environment Engineering, Ministry of Agriculture, P. R. China, 17 Tsinghua East Road, Beijing, 100083, P. R. China, Email: macwbs@cau.edu.cn*

Abstract: A model of heat transfer through greenhouse covering materials was developed in this study. The influences of the thermal properties of covering materials, the construct parameters of greenhouse and the environment inner and outer of greenhouse were completely considered in the model. Heat transfer flux density through covering layer can be simulated and forecasted by using this model. The simulated values of heat transfer flux density through covering layer were in accordance with the values got from some experiments. This model can be used for many purposes, such as greenhouse design, analysis and control for thermal environment in greenhouse.

Keywords: greenhouse; covering material; heat transfer; heat radiation; theoretical model

1. INTRODUCTION

More than 70% of the energy consumption in a greenhouse in winter is owing to the heat loss through covering layer. For this reason, an accurate analytical prediction of the heat loss through covering material is the key point for predicting energy consumption in a greenhouse.

In the past, some methods applied to analyzing the heat transfer through ordinary building materials were used to calculate the heat transfer through greenhouse covering material (Papadakis et al., 2000). But heat radiation through covering material is different from ordinary building materials. Some models have been established and developed, which considered the transmission of infrared radiation through covering layer and the direct

Zhang, Y., Ma, C., Zhang, J., Qin, M. and Liu, R., 2008, in IFIP International Federation for Information Processing, Volume 259; Computer and Computing Technologies in Agriculture, Vol. 2; Daoliang Li; (Boston: Springer), pp. 1379–1382.

radiant heat exchange between inner and outer of a greenhouse (Feuilloley et al., 1996; Li et al., 2004).

The objective of this study was to develop a heat transfer model which can exactly reflect thermal properties of covering materials, greenhouse structures and working conditions.

notation			
t, T	air temperature, □, K	ρ	reflectance of a surface for thermal radiation
F	area, m^2	ε	emittance of a surface for thermal radiation
J	effective radiation heat flux density, W/m^2	σ_b	black radiation constant($=5.67\times10^{-8}$), $W/(m^2\cdot K^4)$
G	radiation heat flux density, W/m^2	α	convective heat transfer coefficient, $W/(m^2\cdot$□)
Q, q	convective heat flux density, W, W/m^2	ϕ	radiation shape factor
v	wind speed, m/s	τ	transmittance of a surface for thermal radiation
β	ratio of heat prevention($= F_s/ F_1$)	K	overall coefficient of heat transfer of covering layer, $W/(m^2\cdot K)$
subscripts			
v	sky	i	inside
s	ground	1u	outside surface of covering
1	covering	1d	inside surface of covering
o	outside		

2. HEAT TRANSFER MODEL OF GREENHOUSE COVERING LAYER

A heat transfer model of greenhouse covering layer is illustrated schematically in Figure 1.

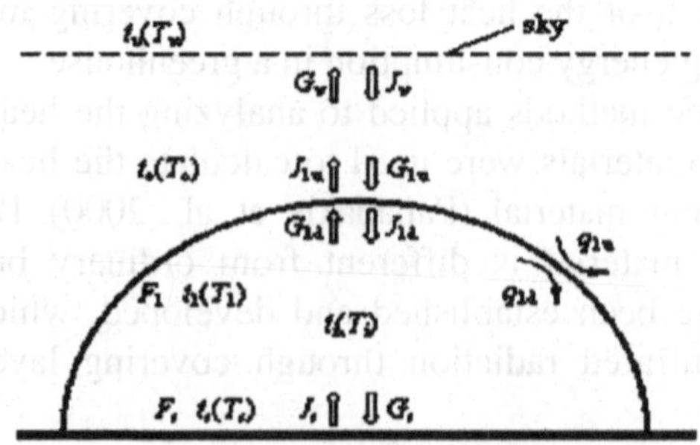

Figure 1. Sketch of heat transfer model of greenhouse covering layer

The method of analyzing effective radiation was used in this study. For each surface related to radiant heat transfer, effective radiation heat flux and radiation heat flux toward the surface could be calculated as follows.

$$J_v = \varepsilon_v \sigma_b T_v^4 + G_v \rho_v \quad (1)$$

$$G_v = J_{1u} \phi_{1u,v} F_1 / F_v \quad (2)$$

$$J_{1u} = \varepsilon_{1u} \sigma_b T_1^4 + G_{1u} \rho_{1u} + G_{1d} \tau_1 \quad (3)$$

$$G_{1u} = J_v \phi_{v,1u} F_v / F_1 + J_{1u} \phi_{1u,1u} \quad (4)$$

$$J_{1d} = \varepsilon_{1d} \sigma_b T_1^4 + G_{1d} \rho_{1d} + G_{1u} \tau_1 \quad (5)$$

$$G_{1d} = J_s \phi_{s,1d} F_s / F_1 + J_{1d} \phi_{1d,1d} \quad (6)$$

$$J_s = \varepsilon_s \sigma_b T_s^4 + G_s \rho_s \quad (7)$$

$$G_s = J_{1d} \phi_{1d,s} F_1 / F_s \quad (8)$$

The sky temperature has relationships with many factors such as air temperature, so there are lots of methods to calculate it, for example, $T_v = 0.0552 T_o^{1.5}$.

Where there are 9 unknown quantities that are 8 radiation heat fluxes and covering temperature $T_1(t_1)$ in above equations. In order to calculate $T_1(t_1)$, energy-balance equation of the covering layer and convective heat flux equations must be researched.

Convective heat flux between air and surface covering is given by Eqn. (9-12) (Roy et al., 2002):

$$Q_{1u} = F_1 q_{1u} = F_1 \alpha_{1u} (T_1 - T_o) \quad (9)$$

$$\alpha_{1u} = 7.2 + 3.84 v_o \quad (10)$$

$$Q_{1d} = F_1 q_{1d} = F_1 \alpha_{1d} (T_i - T_1) \quad (11)$$

$$\alpha_{1d} = 2.21 (T_i - T_1)^{0.33} \quad (12)$$

According to the principle of energy conservation, covering temperature is calculated from

$$(G_{1u} + G_{1d} + q_{1d}) - (J_{1u} + J_{1d} + q_{1u}) = 0 \quad (13)$$

3. CONCLUSIONS (SIMULATED CALCULATION AND EXPERIMENTAL VERIFICATION)

The author worked out the computer program of covering heat transfer which was based on the heat transfer model. Simultaneously, some heat transfer experiments of a few kinds of covering materials were tried out in a laboratory using test apparatus (Zhang et al., 2005). Each group of simulation and experiment was at the same environmental conditions, so the results represented in table1 could be compared.

Table 11. Simulated values and experimental results

Covering material	Environmental conditions					Heat flux density through covering	
						Experimental	Simulated
	t_o/□	t_i/□	v_o/m·s^{-1}	t_v/□	t_s/□	q_w/W·m^{-2}	
PE	-1.3	27.8	1.0	-27.8	27.2	253.2	235.9
PE	-2.5	27.0	2.1	-32.8	26.2	273.2	255.7
PVC	-2.2	25.0	4.0	-27.2	23.6	231.5	226.0
Glass	-1.3	27. 5	1.0	-28.3	26.7	190.1	192.8
Glass	-2.3	28.3	3.0	-30.3	27.1	232.3	226.0

For the heat flux density through covering, simulated predictive values are in accordance with the results got from the experiments, with the max relative error below 10%.

A heat transfer model of greenhouse covering layer is developed in this paper. The influence of the greenhouse structure on the covering heat transfer is reflected in the model. Radiant and convective heat flux density inside and outside covering could be simulated and calculated by this model, then overall energy consumption could be predicted.

ACKNOWLEDGEMENTS

The author is indebted to Beijing Municipal Commission of Education for the financial support by the cooperation program (XK100190550).

REFERENCES

G. Papadakis, D. Briassoulis, G. Scarascia Mugnozza, et al. Radiometric and Thermal Properties of and Testing Methods for Greenhouse Covering Materials. *J. agric. Engng. Res.*, 2000, 77(1): 7-38

P. Feuilloley, G. Issanchou. Greenhouse Covering Materials Measurement and Modeling of Thermal Properties Using the Hot Box Method, and Condensation Effects. *J. agric. Engng. Res.*, 1996, 65: 129-142

Shuhai Li, Chengwei Ma, Junfang Zhang, et al. Thermal model of multi-span greenhouses with multi-layer covers. *Transactions of the CSAE*, 2004, 20(3): 217-221

J.C. Roy, T. Boulard, C. Kittas, et al. Convective and Ventilation Transfers in Greenhouses, Part 1: the Greenhouse considered as a perfectly stirred tank. *Biosystem Engineering*. 2002, 83(1): 1-20

Junfang Zhang, Chengwei Ma, Midao Qin, et al. Research and development of the test apparatus for measuring the overall heat transfer coefficient of greenhouse covering materials. *Transactions of the CSAE*, 2005, 21(11): 141-145

OPTIMIZATION OF POSITION OF REFLECTIVE BOARDS FOR INCREASING LIGHT INTENSITY INSIDE CHINESE LEAN-TO GREENHOUSES

Yun Kong[1,2], Shaohui Wang[2], Yuncong Yao[2], Chengwei Ma[1,*]

[1] *Key Laboratory of Bioenvironmental Engineering of Ministry of Agriculture, College of Hydraulic and Civil Engineering, China Agricultural University, Beijing, 100083, China*

[2] *Department of Plant Science & Technology, Beijing Agricultural College, Beijing, 102206, China.*

[*] *Corresponding author, Address: College of Hydraulic and Civil Engineering, China Agricultural University, Beijing, 100083, China, Tel: +86-10-62736413, Email: macwbs@cau.edu.cn*

Abstract: Equations were derived based on the principles of light reflection to optimize the height and angle parameters of the reflective board in real time and location. The optimized parameters were applied to enhance the light intensity inside the lean-to greenhouses in China. The calculated results were significantly correlated to the actual measurements. Therefore, the equations developed in this study can be used to optimize the design parameters for the reflective board in practice. Additionally, based on the derived calculation equations, the optimized ranges of height and angle parameters were calculated for setting reflective board from autumn to winter and spring at latitude 32-42° in China.

Keywords: position optimization, reflective board, sunlight intensity, Chinese lean-to greenhouse

Kong, Y., Wang, S., Yao, Y. and Ma, C., 2008, in IFIP International Federation for Information Processing, Volume 259; Computer and Computing Technologies in Agriculture, Vol. 2; Daoliang Li; (Boston: Springer), pp. 1383–1389.

1. INTRODUCTION

Installing reflective materials inside a greenhouse, especially a lean-to greenhouse, is one of the most effective and economical solutions to improve light intensity (Li et al., 1998; Cai, 1994; Kong, 2004). Previous research reported that hanging reflective screen under the ridge of lean-to greenhouses has more light enhancement than that on the back wall (Li et al., 1998). But the technical challenge in adopting this measure is to determine the optimal installation position of the reflective board with a width of about 1 m, the popular size of commercial products. If the position is too low, it is very inconvenient for operational and management activities. Tall plants may also block some of the light from reaching the reflective surface. On the other hand, if the board is installed too high, its light enhancing effect will be reduced (unpublished data). As to the angle between the reflective board and ground level, some researchers reported that it should be in the range of 75-85° (Jia, 2000). In fact, the optimal height and angle of reflective board should change with the time and location of the specific greenhouse (Pucar, 2002). In this study, mathematical equations were developed and tested to enhance light intensity by optimizing the parameters of height and angle of reflective board inside lean-to greenhouses in various time and locations.

2. OPTIMIZATION THEORY

2.1 Optimization Consideration

A few assumptions were made as follows to simplify the calculations. (1) Light enhancement of the reflective board depends mainly on the reflection of direct light. (2) Sunbeam light does not change its initial direction when passing through the greenhouse cover until it reaches the reflective board. (3) When the direct light travels inside greenhouse, light intensity attenuation effect is negligible.

With the above assumptions, the optimal objectives of reflective board location were described as follows: (1) To adjust height of the reflective board to make its bottom edge as far away from the top of crop canopy as possible. (2) At the optimal height, to determine the inclination of the reflective board to cast a minimal shading area on the back-wall and to produce a maximal light enhancement in the cultivation area from 1 m away from reflective board to northward, where is normally associated with the greatest light enhancement by reflective board (Cai, 1994; Kong, 2004).

2.2 Optimization Calculation

Suppose that there is a point A, from which the vertical distance to the vertically installed reflective board is R (m), the azimuth angle of the greenhouse is θ^*, the vertical distance from the top of the crop canopy to the bottom edge of the reflective board is H (m) (Fig. 1), and the height of the reflective board is L (m).

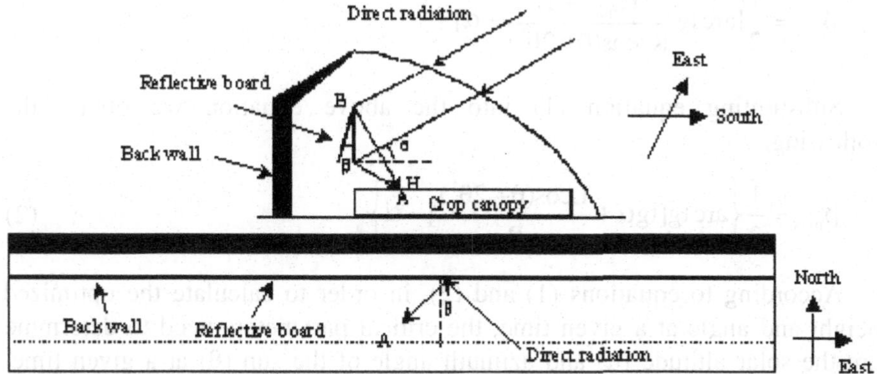

Figure 1. Schematic diagrams of cross-section (upper) and planform (lower) of the greenhouse with reflective board

In order to achieve a maximal light enhancement in the area from point A to the northward aspect, the area must always be in the range of the reflected direct sunlight. For objective (1), the optimized height of the reflective board at a given time should be the maximum value of H. H_{max} at a given time can be calculated using the following equation:

$$H_{max} = \frac{R\,tg\alpha}{\cos(\theta + 2\theta^*)} \tag{1}$$

where, α: solar altitude (degree); θ: azimuth angle of the sun (degree); θ^*: azimuth angle of the greenhouse, when the south roof faces to south, $\theta^*= 0$, negative values for east aspect, and positive values for west aspect; and, R: vertical distance of point A to the vertically installed reflective board.

Let the bottom edge of the reflective board at its optimized height move northward around its top edge, point B (Fig. 1). Suppose that the rotation angle of reflective board is β (Fig. 1). In order to make the point A and its northward area be in the range of the reflected sunlight, the reflective board can be rotated at a angle ranging from 0 to $\beta_{max.}$ For objective (2), β_{max}, can be regarded as the optimized angle. From the rules of geometry and the laws of light reflection, we can derive

$$tg(\alpha+2\beta_{max})=\frac{H_{max}+L}{R/\cos(\theta+2\theta^*)}$$

i.e.

$$\beta_{max}=\frac{1}{2}[arc\,tg\frac{H_{max}+L}{R/\cos(\theta+2\theta^*)}-\alpha]$$

Substituting equation (1) into the above equation we obtain the following:

$$\beta_{max}=\frac{1}{2}\left\{arc\,tg[tg\alpha+\frac{L\cos(\theta+2\theta^*)}{R}]-\alpha\right\} \tag{2}$$

According to equations (1) and (2), in order to calculate the optimized height and angle at a given time, the critical parameters need to determine are the solar altitude (α) and azimuth angle of the sun (θ) at a given time, with L and R are known. The two parameters at a given time can be calculated from several well known equations, with the local longitude and latitude are known (Zhou, 2003).

3. EXPERIMENTAL VALIDATION

A compact experiment was conducted inside a Chinese type lean-to greenhouse with an azimuth angle of -4° on December 12th, 2004, when the sky was clear. The reflective board with a width of 0.98 m was hung vertically under the ridge of the greenhouse with different height (H) of 0 m, 0.2 m, 0.4 m, 0.6 m, 0.8 m and 1 m. When the reflective board was fixed at the height of 0.5 m, its bottom edge rotated northward around its top edge by angles of 0°, 15°, 30°, 45° and 60°. A calibrated light sensor (Li-COR, Nebraska, USA), which was positioned on the top of crop canopy at a 1-m vertical distance from the reflective board southward, was used to measure the PPFD (Photosynthetic Photon Flux Density) at different time of that day from 8:00 to 16:00 at an interval of one hour.

As shown in Fig. 2, the PPFD decrease slowly from 0 m to a certain height and rapidly at higher positions. The critical height varied with time, with about 0.3 m to 0.5 m during 9:00-14:00. The critical heights from experiments and from the calculation of optimized height of the reflective board matched very well during 9:00-14:00 as shown in Table 1. As shown in Fig. 3, like the critical height described previously, there was also a critical angle. The PPFD decreased rapidly when the rotation angle increased beyond the critical angle. As can be observed from Fig. 3, the critical angle ranged from 15° to 30°, which was consistent with the calculated value of the optimal angle shown in Table 1.

Table 1. Calculated results for optimizing the installing height and rotation angle of reflective board in the experimental greenhouse on December 12, 2004

Time (hh:mm)		8:00	9:00	10:00	11:00	12:00	13:00	14:00	15:00	16:00
Calculated parameters	α (°)	6.5	14.9	21.4	25.6	26.8	25.0	20.4	13.5	4.9
	θ (°)	-52.2	-41.1	-28.3	-13.9	1.4	16.6	30.7	43.1	54.1
Calculated results	Hmax (m)	0.23	0.41	0.49	0.52	0.51	0.47	0.40	0.29	0.1
	βmax (°)	11.6	13.3	13.9	14.2	14.5	15.1	15.9	16.6	16.7
	β'max (°)	18.6	16.6	15.2	14.4	14.5	15.4	17.2	19.7	22.7

Note: β'_{max} is the optimal rotation angle when the installation height of reflective board was 0.5 m throughout the whole day.

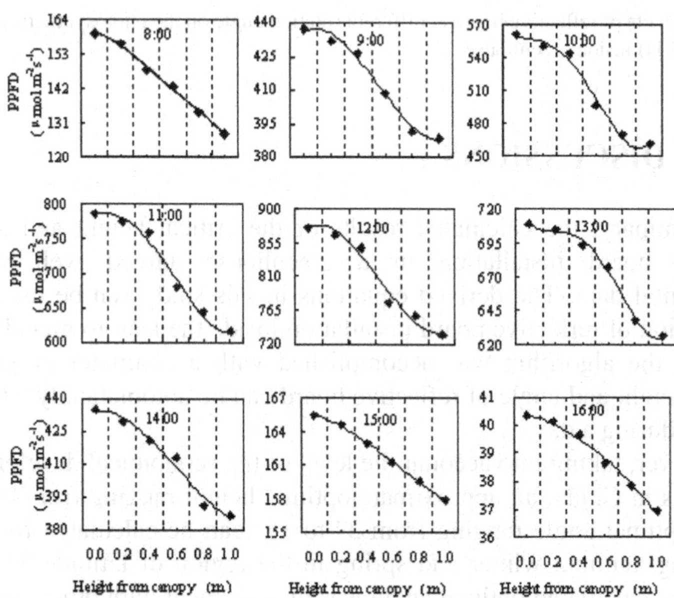

Figure 2. Effects of reflective board at different height on the photosynthetic photon flux density (PPFD) inside greenhouse

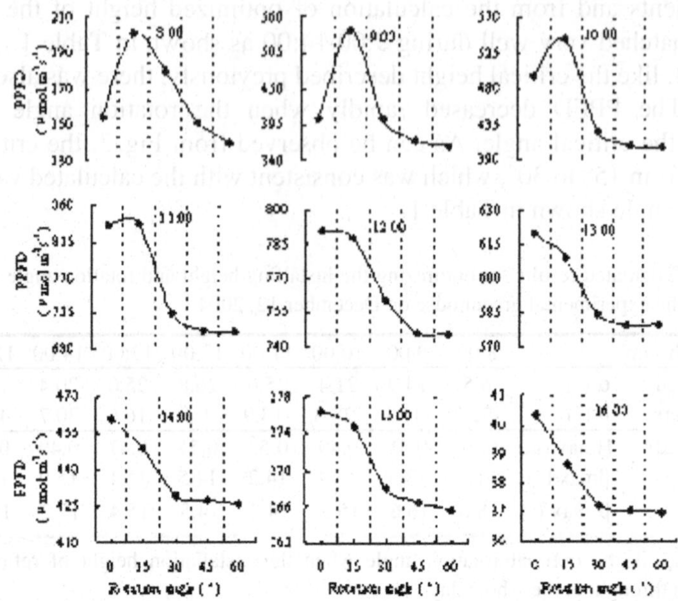

Figure 3. Effects of reflective board at different rotation angle on the photosynthetic photon flux density (PPFD) inside greenhouse

4. DISCUSSIONS

In summary, the calculated results of the critical height and angle for reflective board installation in a greenhouse agreed well with the experimental data. The derived equations in this study can be used for the optimization of reflective board installation inside the lean-to greenhouses in China. If the algorithm was accomplished with a computer program, the optimal height and angle of reflective board can be automatically adjusted in real time during a day.

However, taking into account the level of the economical development in rural areas in China, an approximate optimal height ranging from 0.5 to 2.4 m, and optimal angle ranging from 3° to 15° can be calculated for a given day during autumn, winter and spring in the region of latitude 31°–42° in China. The calculated optimal angle was in agreement with other studies that indicated that the inclination of reflective board should be in the range of from 75° to 85° (Jia, 2000).

ACKNOWLEDGEMENTS

The authors are indebted to the cooperation program between Beijing Education Committee and China Agricultural University (XK100190550) for the financial support. We wish to thank Dr. Qiangxian Wu and Dr. Shaolin Yang for their careful reading and revision of the manuscript.

REFERENCES

Cai D. Experiment of application of reflective film to improving light intensity inside greenhouse. China Agricultural Meteorology, 1994, 15(1): 45, 41 (in Chinese).

Jia T. Light supplement technique for protected cultivation. West-southern Horticulture, 2000, 28(4): 50 (in Chinese).

Kong Y, Meng L. Effect of reflective screen on sunlight distribution and morphological characters of tomato inside solar greenhouse. Northern Horticulture, 2004, 5: 10-11 (in Chinese).

Li S, Kurata K, Takakura T. Solar radiation enhancement in a lean-to greenhouse by use of reflection. Journal of Agricultural Engineering Research, 1998, 71: 157-165

Pucar M Dj. Enhancement of ground radiation in greenhouse by reflection of direct sunlight, Renewable Energy, 2002, 26: 561-586

Zhou C J. Advanced Greenhouse Engineering. Chemistry Industry Press, 2003, Beijing (in Chinese).

BOUNDARY SETTING IN SIMULATING GREENHOUSE VENTILATION BY FLUENT SOFTWARE

Cuiping Hou[1,2], Chengwei Ma[1,2,*]

[1] *College of Water Conservancy and Civil Engineering, China Agricultural University, Beijing, Beijing, China, 100083*

[2] *Bia-Environmental Engineering Lab. Key Lab of Mart., 100083, China*

* *Corresponding author, Address: P.O. Box 195, China Agricultural University, 17 Tsinghua East Road, Beijing, 100083, P. R. China, Tel: +86-10-62736413, Fax: +86-10-62736413, Email: macwbs@cau.edu.cn*

Abstract: Fluent, as one of the commercial CFD packages, the author discussed its application in greenhouse ventilation research. In the paper, the under problems would be expounded: in fluent what models will be used in greenhouse studying, and the details during in these models used, and the comparing of the same style models. The steps to solve the greenhouse problem also was expatiate in this paper, mainly including: how to abstract its physical models; how to define its compute domains; the factor should be considered in creating and meshing the geometry; the types of boundary condition, and how to give the boundary conditions, in this section, the author will take some instances to emphasize discuss; the selection of parameter in simulation, and a few of methods of post processing. At the last, the author introduced some even existing problem in simulating.

Keywords: fluent, simulation, boundary condition

Hou, C. and Ma, C., 2008, in IFIP International Federation for Information Processing, Volume 259; Computer and Computing Technologies in Agriculture, Vol. 2; Daoliang Li; (Boston: Springer), pp. 1391–1395.

1. THE PROBLEM OF GREENHOUSE VENTILATION CAN BE SOLVED BY FLUENT

1.1 Abstract of the Physical Models

To a problem of ventilation in greenhouse, at first we should make some assumption to determine a physical model. Subsequent factors should be think about in this process: two-dimensional or three-dimensional; steady-state or unsteady-state; laminar flow or turbulence flow; physical parameters, change or unchanged, compressible or uncompressible, uniform or not uniform in speed; the style of the boundary condition. All computations were performed assuming conditions.

1.2 Computational Domain of the Simulation

The size of the greenhouse model is significantly limited by the available computer memory and processor speed. The solution domain is subdivided into a finite number of contiguous control volumes. Usually we selected a large domain including the greenhouse, and the determine of the size of the computational domain need our try, the result of the simulation should be no more than significant improvement (Kacira et al., 1998; Mistriotis et al., 1997; Short, 1996).

1.3 The Grid

The CFD model calculating time was strongly dependent on the size of the grid domain and a node adjustment factor was used to minimize grid numbers. The grid size should be dense enough to describe accurate flows around complex shapes and sparse enough to minimize calculations in uniform flow areas. The relatively small accuracy increase between the different grids indicates that the grid dependency on the solution has become minimal. It results from an empirical compromise between a dense grid, associated with a long computational time, and a less dense one, associated with a small deterioration of the simulated results. Maximum level allowed by the software; In the case of the greenhouse with crop the necessary grid refinement and the problems of convergence did not allow a reduction in the number of elements; In the case of the empty greenhouse a diminution in the number of elements did not cause a significant variation in the accuracy of simulations, neither it produce convergence problems. From the general grid specifications, a body-fitted coordinate (BFC) grid method was used with fluent to acquire the most accurate flow patterns and computations at the boundaries of the greenhouse. Smooth and uniform enough to guarantee the best results. The discretisation is performed with triangular and quadrilateral elements. The grid was an unstructured grid with higher density at the vent openings, where the flow was subject to strong gradients.

The scale of detail which should be applied must be investigated to achieve a good balance between increased complexity, demanding modeling skills and computational effort, and the required accuracy of the model output.

2. BOUNDARY CONDITIONS

The choice of suitable numerical parameters and sufficiently accurate boundary conditions required for accurate results. The velocities are obtained from the conservation principle, the pressure is obtained from the conservation of mass principle and the temperature is obtained from the law of conservation of energy. Crop canopy, the flow would take into account the resistance presented by the canopy, the heat balance would include effects of transpiration, convection and radiation and mass balance would include the effect of photosynthesis on carbon dioxide levels, transpiration and convectional. Boundary conditions contain the effects of external factors on the flow and temperature.

FLUENT required input variables associated with the properties of the air and boundary conditions. The dynamic boundary conditions prescribed a null pressure gradient in the air, at the limits of the computational domain, and wall-type boundary conditions along the floor, the walls and the roof and vents surfaces of the scale greenhouse model. The thermal boundary conditions imposed fixed temperatures at the limits of the computational domain and at roof and floor levels, and adiabatic conditions along the side walls of the scale greenhouse model. The dynamic boundary conditions prescribed a null pressure gradient in the air, at the limits of the computational domain, and wall-type boundary conditions along the floor, the walls and the roof and vents surfaces of the scale greenhouse model. The thermal boundary conditions imposed fixed temperatures at the limits of the computational domain and at roof and floor levels, and adiabatic conditions along the side walls of the scale greenhouse model.

The values of those inputs were chosen from the fluent manual and from the experimental greenhouse. A list of the input values usually is: momentum factors, inlet velocity or inlet profile, model constants, inlet kinetic energy, outer boundary conditions (usually be setting as fixed pressure), if think about the plant, crop porosity should be taken account; energy factors, side wall temperature (usually be setting as adiabatic), cover/soil temperature, air temperature, heat flux from floor/roof, sky temperature, radioactive heat transfer; mass factors should setting air properties, include, specific heat, thermal conductivity, dynamic viscosity, atmospheric pressure, also include the aid density, which can be set as ideal gas, also can be set as boussinesq, in this case the density becomes constant in all equations except for the buoyancy term in the momentum equations. Fluid flow, heat and mass transfer processes, once all inputs were defined in fluent, a case file was generated. Fluent utilized the case file to solve conservation equations using the SIMPLE algorithm with an iterative line-by-line solver and a multi-grid acceleration. When the numerical solutions were converged, a numerical and graphical output of the results is obtained.

If only ventilation due to the wind is studied, the gravity is considered zero, the heat input for this situation can be set to any value not influencing the ventilation rate. If the buoyancy effect is included, the heat input was set to a value equal to the heat supplied by solar radiation and gravity is reset.

Body-fitted coordinates, which is a coordinate system with grid lines that coincide with the boundaries of the geometry. Body-fitted coordinates to acquire the most accurate flow patterns and computations at the boundaries (Fluent, 1993). Body-fitted coordinates were also applied to exactly conform the grid to the contours of the boundary conditions.

To limit the control volume of the large hall containing the reduced-scale greenhouse, pressure-type boundary conditions were selected for the ambient air. These boundary conditions prescribe fixed pressure and temperature conditions at the limits of the computational domain and the inlet and outlet air velocities are automatically computed to satisfy the continuity conditions.

3. DISCUSSION

Until now, because of the great variability of the climatic parameters involved nobody has found a model of greenhouse ventilation for general application (Bailey et al., 1999). CFD can be a useful tool in the study of in the internal greenhouse climate. With a CFD model a 'virtual reality' simulation may be created that is very versatile and relatively cheap. But simulation result is not so believable, in many times, its result even need validate, present methods include, sonic anemometers and smoke tracing techniques, which can measure and visualize the internal airflow, these and other tools like PIV, hot-wire anemometers, laser-Doppler anemometers, thermo-couples, carbon dioxide monitors and other specialized equipment. Future applications of fluent even lie in conducting parametric studies.

REFERENCES

Bailey, B.J. 1999. Constraints limitations and achievements in greenhouse natural ventilation. Acta Hortic, 534, 21–30.

Kacira, M., Short, T.H., Stowell, R.R. 1998. A CFD evaluation of naturally ventilated multi-span, sawtooth greenhouses. Trans. ASAE, 41 (3), 833–836.

Mistriotis, A., De Jong I. 1997. Computational fluid dynamics study as a tool for the analysis of ventilation and indoor microclimate in agricultural building. Neth. J. Agric. Sci. 45, 81–96.

Short I.H. 1996. Dynamic model of naturally ventilated poly houses, Grower Talks Magazine, summer.

MONITORING CROP GROWTH STATUS BASED ON OPTICAL SENSOR

Di Cui[1], Minzan Li[1], Yan Zhu[2], Weixing Cao[2], Xijie Zhang[1]
[1] *Key Laboratory of Modern Precision Agriculture System Integration Research, Ministry of Education, China Agricultural University, Beijing 100083, China Corresponding author, Email: limz@cau.edu.cn*
[2] *Key Laboratory of Crop Growth Regulation; Ministry of Agriculture, Nanjing Agricultural University, Nanjing 210095, China*

Abstract: In order to detect the growth status and predict the yield of the crop, crop growth monitor measuring nitrogen content in the plant is developed based on optical principle. The monitor measures the spectral reflectance of the plant canopy at the 610 nm and 1220 nm wavebands, and then calculates the nitrogen content in the plant with the measured data. The field test was carried out to evaluate performance of the monitor. A portable multi-spectral radiometer named Crop Scan was used to measure the reflectance as a reference instrument. The result shows that the leaf reflectance measured by the monitor has a close linear correlation with that measured by Crop Scan at the 610 nm waveband (R2 = 0.7604), but the correlation between them is needed to be improved at the 1220 nm waveband. The hardware and the software of the monitor are also explained in detail. It is still need to be improved to satisfy the demand of ground-based remote sensing in precision farming.

Keywords: nitrogen content; crop growth; ground-based remote sensing; precision agriculture

1. INTRODUCTION

Nitrogen content of the crop is an important index, which can be used to evaluate the growth status and predict the yield and the quality of the crop. So it is necessary to real time detect and diagnose the nitrogen status of the

Cui, D., Li, M., Zhu, Y., Cao, W. and Zhang, X., 2008, in IFIP International Federation for Information Processing, Volume 259; Computer and Computing Technologies in Agriculture, Vol. 2; Daoliang Li; (Boston: Springer), pp. 1397–1401.

crop. Nitrogen fertilizer has a great effect on the reflectance of the light at visible band since the nitrogen fertilizer and the chlorophyll content are correlated (Stone, M. L. 1996). So the leaf nitrogen content can be estimated by analyzing the useful spectral features of crop canopy. Our object is to design an optical sensor to measure spectral reflectance of the plant canopy for determining N status in the plants.

2. DEVELOPMENT OF A PORTABLE MONITOR FOR CROP GROWTH CONDITION

2.1 Structure of the portable Monitor

Figure 1 shows the structure of the monitor. It was composed of the sensor and the controller. The sensor was designed with four channels representing different spectral bands. The channel 3 and channel 4 measured the sunlight, while the channel 1 and channel 2 measured the reflected light from the plant canopy. The angle of solar incidence can make the output of photodiode changing. In order to avoid this influence, the milky diffuse glass was used as the optical window of the channel 3 and channel 4. The optical mirror was chose as the window of the channel 1 and channel 2. When using this detector to measure reflectance ratios, the sensor was lift on the top of the canopy and the output of the sensor was transmitted to the controller. Then the controller dealt with the output and saved the data.

Each channel consisted of an optical window, a washer, a filter, and a sensor. The channel 1 and the channel 3 had the same filters and sensors. The wavelength of the filter was 610 nm and the electric eye was made up of Si. And the wavelength of the filter in the channel 2 and channel 4 was 1220 nm. The photodiode in those was made up of InGaAs. The wavebands were determined based on the research report (Xue Lihong et al., 2004).

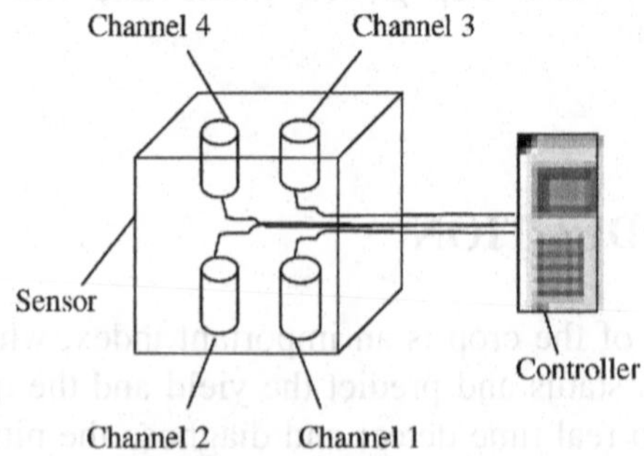

Figure 1. The structure of the portable crop growth monitor

2.2 Hardware Design

Figure 2 shows the structure of electric circuits of the hardware. The photodiodes converted the optical signal into the electronic signal. The electronic signal went through an amplifier and an A/D transducer to be digitalized. The digital signal was calculated into chlorophyll content in the plants by the microcontroller, and the result was displayed on the LCD and stored in the Flash Disk through the USB port. The keyboard gave the indications to the microcontroller.

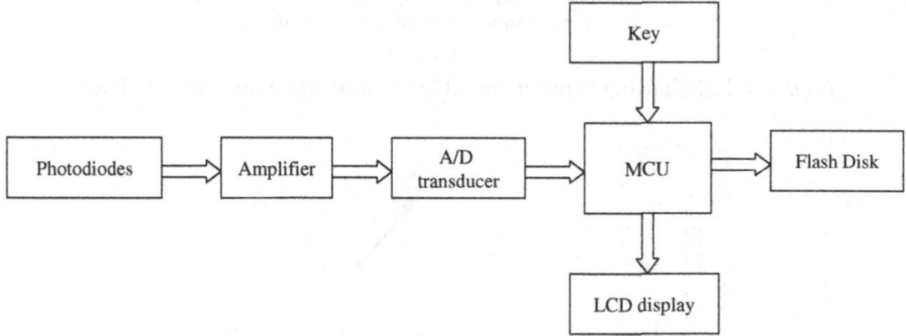

Figure 2. The Structure of electric circuits of the hardware

3. PERFORMANCE TESTS AND RESULTS

To evaluate performance of the monitor, we did field text in the wheat field of Nanjing Agricultural University on March 24. The weather was fine. There were 24 grids in the experiment field and the nitrogen content of each plat was different. The area of each grid was 5 m×5.5 m. The wheat was 10 cm in height and didn't cover the ground. The soil between the two rows could be observed clearly. We used the monitor and Crop Scan to measure the reflectance of the wheat canopy. Each grid was measured five times along the two diagonals of each grid and the average was taken as the final result. At every sampling point, the samples were collected to measure the content nitrogen in the laboratory.

Figure 3 shows that the leaf reflectance measured by portable monitor has a close linear correlation with that measured by Crop Scan (R2 = 0.7604) at the 610 nm wavelength. However, the correlation between those is needed to be improved at the 1220 nm wavelength (Figure 4).

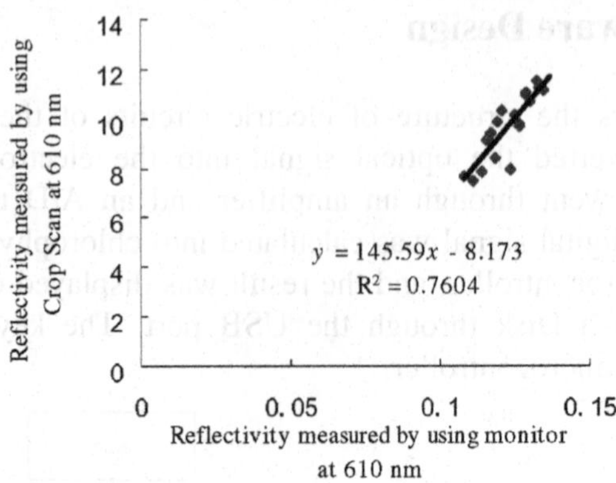

Figure 3. Relationship between portable monitor and crop scan at 610 nm

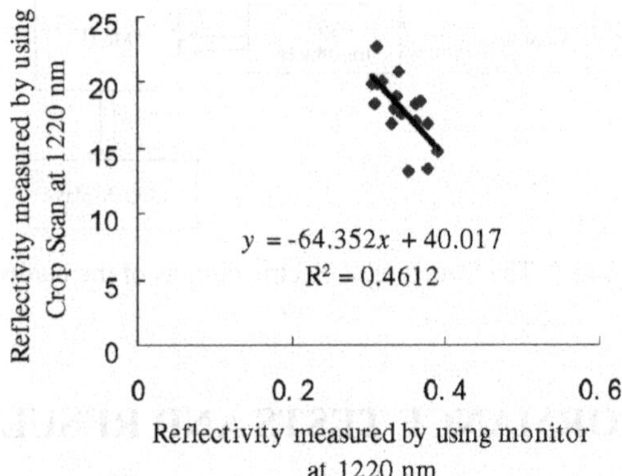

Figure 4. Relationships between portable monitor and crop scan at 1220 nm

4. CONCLUSION

The portable crop growth monitor based on optical principle was developed and tested in the field. The Spectral reflectance was calculated using the monitor's outputs and showed a close correlation with the result measured by Crop Scan at 610nm wavelength (R2 = 0.7604), but not good at 1220 nm wavelength (R2 = 0.4612). It may be caused by the height of wheat. In the future, more field tests will be performed to further test and to improve the device in order to make it satisfy the demand of precision farming.

ACKNOWLEDGEMENTS

This study was supported by Key Program of Science and Technology, Ministry of Education (105013).

REFERENCES

Stone, M. L., J. B. Solie, W. R. Raun, R. W. Whitney, S. L. Taylor, and J. D. Ringer. Use of spectral radiance for correcting in–season fertilizer nitrogen deficiencies in winter wheat. Transactions of the ASAE, 1996, 39(5): 1623-1631

Xue Lihong, Cao Weixing, Luo Weihong, Zhang Xian. Correlation between leaf nitrogen status and canopy spectral characteristics in wheat. Acta Phytoecologica Sinica., 2004, 28 (2): 172-177

REAL-TIME SOIL SENSING WITH NIR SPECTROSCOPY

Jianying Sun, Minzan Li, Lihua Zheng, Ning Tang
Key Laboratory of Modern Precision Agriculture System Integration Research, China Agricultural University, Ministry of Education, Beijing 100083, China
Corresponding author, Email: limz@cau.edu.cn

Abstract: The grey-brown alluvial soil is a typical soil in the Northern China. It was selected as research object to reveal feasibility and possibility of real-time analyzing soil parameter with NIR spectroscopic techniques. 150 samples were collected from a winter wheat farm. And then the NIR absorbance spectra were rapidly measured under the original conditions by a Nicolet Antaris FT-NIR analyzer. Three soil parameters, soil moisture, SOM, TN content, were analyzed. For soil moisture content, a linear regression model was available, using 1920 nm of wavelength with correlation coefficient of 0.937, So that the results obtained could be directly used to real time evaluate soil moisture. SOM content and TN content were estimated with a multiple linear regression model, 1870 nm and 1378 nm wavelengths were selected in the SOM estimated model, while 2262 nm and 1888 nm wavelengths were selected in the TN estimated model. The results showed that soil SOM and TN content could be evaluated by using NIR absorbance spectra of soil samples.

Keywords: NIR spectroscopy, chemometrics, soil moisture, soil organic matter, soil total nitrogen

1. INTRODUCTION

Describing the variability of soil parameters is an important step for promoting precision agriculture. So it is necessary to look for a real-time evaluating method of soil parameters.

NIR spectroscopy appears as a rapid, convenient and simple nondestructive technique that can be used to quantify several soil properties

Sun, J., Li, M., Zheng, L. and Tang, N., 2008, in IFIP International Federation for Information Processing, Volume 259; Computer and Computing Technologies in Agriculture, Vol. 2; Daoliang Li; (Boston: Springer), pp. 1403–1406.

in many researches (Shimano T.). Li Minzan, A. Sasao, and S. Shibusawa have taken Kanto Loam as the research object and developed some estimation models of soil parameters based on spectral reflectance of raw soil samples (Li M.Z. 2000). In addition, research showed that NIR spectral of soil was largely affected by soil type. Thus, the grey-brown alluvial soil in the Northern China was selected as research object, and the estimation models of soil moisture content, SOM, and TN were developed with NIR spectra of raw soil samples.

2. MATERIALS AND METHODS

Soil samples were collected from an experimental farm of winter wheat in China Agricultural University. After the soil samples were taken into the lab, NIR absorbance spectra and soil moisture were rapidly measured under their original conditions, The NIR absorbance spectra were measured by a Nicolet Antaris FT-NIR analyzer. Then each soil sample was divided into two parts. One was used to directly analyze soil moisture. The other was used to analyze soil parameters after air-dried. Moisture was measured by 1050C-24h method with an electric fan heater. Air-dried soil samples were analyzed using chemistry method. The soil parameters analyzed were soil organic matter (SOM), soil total nitrogen (TN), soil total phosphorus, soil total potassium and soil pH.

3. RESULT AND DISCUSSION

3.1 Moisture

The soil moisture content has high correlation with original absorbance at every waveband. 90 data sets were used as calibration group to build an estimation model, and the other 60 data sets were used as validation group to check the model. Using absorbance at 1920 nm, the wavelength with highest absorbance peak, a single linear regression model was established.

$$y = 17.706 - 2.167x_{1920} \tag{1}$$

where: x_i is the original absorbance at wavelength of i;
y is the soil moisture content

The correlation coefficient of the regression model is (Rc2)0.936, F and t test showed that the characteristics of the model and the regression parameters were significant. The results of the calibration and validation of the model is shown in Figure 1:

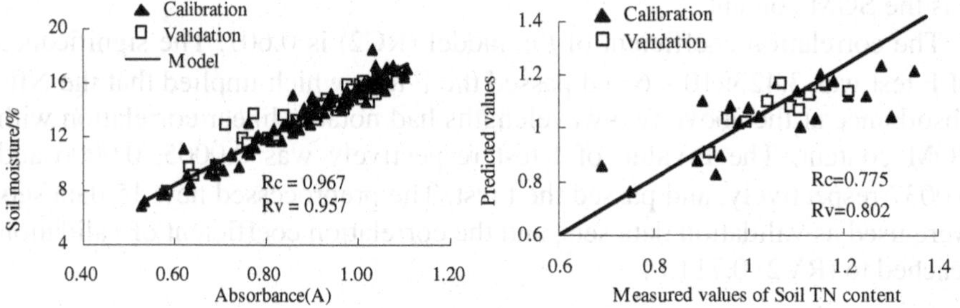

Figure 1. Calibration and validation of soil moisture SLR model

Figure 2. Calibration and validation of TN PLSR Model

3.2 TN

Four wavelengths were selected form the wavebands with higher correlation coefficient, which were 2240 nm, 2090 nm, 1901 nm and 1827 nm. Using all four data sets a multiple linear regression was executed. In order to eliminate the effect of multicollinearity, PLSR was carried out. Using Leave One Out – Cross Validation, the PLS regression was conducted. The model was as follows:

$$y = 0.237 - 2.492x_{2240} + 0.263x_{2090} + 6.138x_{1901} - 4.285x_{1827} \quad (2)$$

where: x_i is the NIR absorbance at wavelength of i;
y is the soil TN content

The result of calibration and validation is shown in Figure 2. The correlation coefficient of the model (RC2) was 0.601, and the correlation coefficient of validation (RV2) was 0.643. In addition, it would be observed that there were non-linear factors in the model from Figure 2.

3.3 SOM

Difference of soil SOM content is also very little, and the correlation coefficient between soil SOM content and the original absorbance spectra or its first spectra derivation was very low. The maximum is lower than 0.4. Regression model was established using 30 new data preprocessed from original NIR absorbance spectra. The spectral variable used was the first derivative of NIR absorbance. Two wavelengths were selected based on correlation analysis, which are 1870 nm and 1038 nm. Using the above two wavelengths a multiple linear regression model was established as follows:

$$y = 1.195 - 515.471x_{1870} + 159.268x_{1038} \quad (3)$$

where: x_i is the first spectral derivation at wavelength of i;
y is the SOM content

The correlation coefficient of the model (RC2) is 0.607. The significance of F test was 3.423×10 – 6 and passed the F test, which implied that the NIR absorbance at the above two wavelengths had notable linear correlation with SOM content. The P-value of t test respectively was 0.0005, 0.0006 and 0.0037 respectively, and passed the t test. The preprocessed new 15 data sets were used as validation data sets, and the correlation coefficient of validation reached to (RV2) 0.711.

4. CONCLUSION

The soil moisture content has high correlation with NIR original absorbance. It was feasible evaluating soil TN and SOM content with NIR absorbance spectra of raw soil samples, and the regression model obtained from the first deviation data was more feasible since the deviation spectra could eliminate noises. And the results obtained could be taken as theoretic basis for the analysis of real-time soil parameters. In order to further improve precision of model, it is necessary to intentionally implement soil nutrition-stress in next experiment, and developed a more feasible evaluation model.

ACKNOWLEDGEMENTS

This study was supported by National 863 Program (2006AA10A301).

REFERENCES

Li M.Z., Sasao A., Shibusawa S., et al. Soil parameters estimation with NIR spectroscopy. Journal of the Japanese Society of Agricultural Machinery, 2000, Vol. 62(3), pp. 111-120

Li N., Min S.G., Tan F.L., Zhang M.X., Ye S.F. Nondestructive Analysis of Protein and Fat in Whole-kernel Soybean by NIR, Spectroscopy and Spectral Analysis, 2004, Vol. 24(1), pp. 45-49

Shimano T. New design of geodesic lenses. MOC/GRIN'89, Tokyo: 130-135

YIELD MAPPING IN PRECISION FARMING

M. Zhang, M.Z. Li , G. Liu, M.H. Wang
Key Laboratory of Modern Precision Agriculture System Integration Research, China Agricultural University, Ministry of Education, Beijing 100083, China

Abstract: Spatial variability of yield is very important in precision farming. With a yield monitor system installed on combine harvester, the yield data can be collected automatically while harvesting. In this paper, the component and the working theory of the yield monitor system were introduced. Performance of main sensors and the method improving GPS position accuracy were analyzed. Based on the analysis, a new grain yield monitor is developed and performance tested. The crop harvested was wheat, and the harvesting combine used to equip the monitor was JL1603, a typical machine in northern China with 4 m of header width. And. The yield map was created to support decision making in precision agriculture.

Keywords: precision agriculture, yield monitor system, grain flow sensor, yield map

1. INTRODUCTION

In practice of precision agriculture, most important thing is to realize the spatial and temporal variability of the field conditions, yield, soil fertilizer, crop growing status, and so on. Then, it is needed to analyze the reason influence crop growing according to above information. And finally, it is necessary to input fertilizer, pesticide etc based on the crop demand.

A yield map is the basis for understanding the yield variability within a field, analyzing reasons behind the yield variability, and improving management according to the increase of the profit (M.H. Wang, 1999). So a yield monitors system need to be installed on combine harvester, and the yield data can be recorded automatically while harvesting (L. Thylen, 1996).

Zhang, M., Li, M.Z., Liu, G. and Wang, M.H., 2008, in IFIP International Federation for Information Processing, Volume 259; Computer and Computing Technologies in Agriculture, Vol. 2; Daoliang Li; (Boston: Springer), pp. 1407–1410.

2. THE DEVELOPMENT OF INTELLIGENT CONTROLLER

2.1 Development of Hardware

The intelligent controller collected four analog signals (grain flow, header height, grain moisture, and grain temperature), two pulse signals (ground speed and elevator speed), and DGPS data. After a series of process, the system stored all data on a CompactFlash® (CF) card once per second. The yield monitor also included a liquid crystal display (LCD) and a touch screen as the Man-Machine interface. Figure 1 shows the diagram of the intelligent controller developed. It uses a P80C592 manufactured by Philips Semiconductors as the microcontroller.

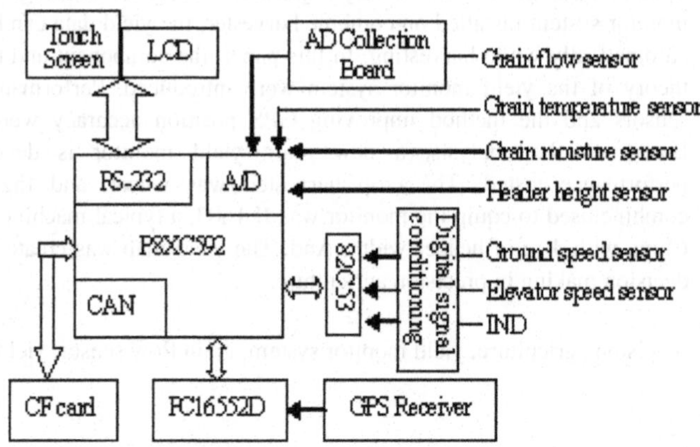

Figure 1. Diagram of the intelligent controller developed

Pretests showed that a sampling frequency of at least 200Hz was needed for the system to properly record the impaction on the flow sensor. However, it was hard to increase the sampling frequency on the main microcontroller since its load had been heavy enough. Therefore, a new signal acquisition circuit board was specially designed to collect grain flow signals with a frequency of 250Hz. After sampling, the board calculated the mean of 250 data obtained in one second, and then sent the result to the main microcontroller.

2.2 Programming of Software

There are five tasks to be dealt with by the program; they are data acquisition, data processing, LCD displaying, touch screen input, and data saving. The details of them are shown in Figure 2.

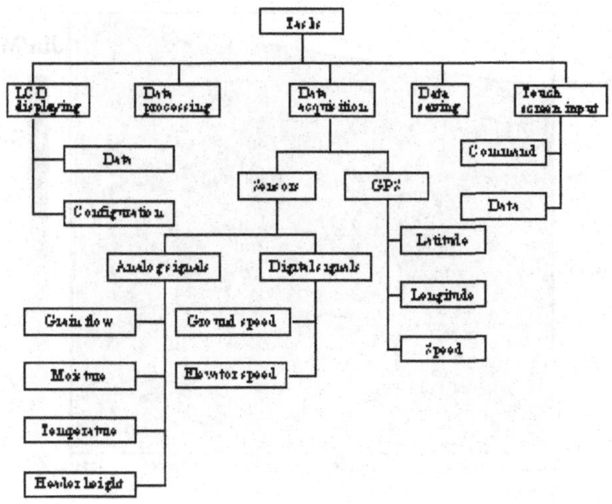

Figure 2. Tasks of the program

These tasks make up a multitasking real time system. However, it may results in some conflicts of the tasks in time sequence. So four interrupt sources of P80C592 are used in the system, which are the timer interrupt source, the external interrupt source, the ADC end-of-conversion interrupt source and the UART serial I/O port interrupt source.

3. FIELD TRIAL

The system was installed on the combine harvester and began to harvest wheat on June 11, 2004. And the harvesting combine used to equip the monitor was JL1603, with 4 m of header width. Before harvest, the yield monitor system was setup. In order to reduce the systematic error, before harvesting a whole field, the system was calibrated in a smaller area in the similar condition with the whole field. After harvest, the data that collected by the yield monitor system were output and be analyzed by some software. The winter wheat yield map of the field F1 generated after the error yield data processing is shown in Figure 3. The maps were generated using Surfer6 via IDW (Inverse Distance Weighted) interpolation. The output grid size is 20 m, and search radius is 20 m. In this map, yield decreased with color. The yield around the field was lower than the center. According to this yield map, understanding the yield variability within a field, analyzing reasons behind the yield variability, improving management, then the profit can be increased.

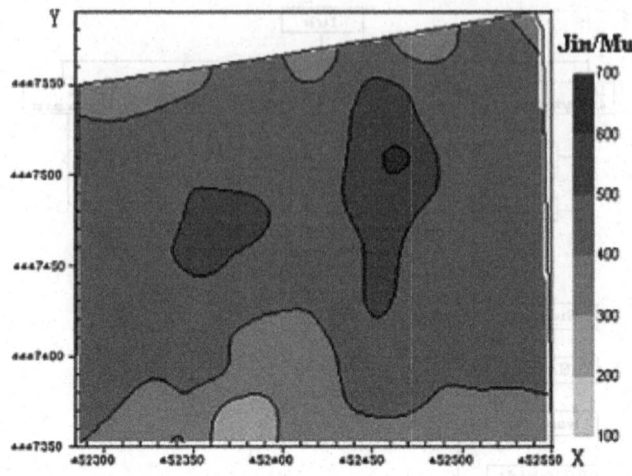

Figure 3. Yield map of field F1

4. CONCLUSION

This paper introduced components and the working theory of yield monitor system; compared the performance of main sensors and the method of improving GPS position accuracy. Then, a grain intelligent controller based on an 8-bit microcontroller was developed, and the programming of the software used for the monitor was conducted. A long time of laboratory and field tests showed that the whole worked steadily. However, the capability of handling multitask should be optimized. Finally, grain yield map was created. According to this map, to analyze the reason that caused the yield variability within a field, further to improve crop management and decrease input.

ACKNOWLEDGEMENTS

This study was supported by National 863 Program (2006AA10A305).

REFERENCES

L. Thylen and Donal P.L. Murphy, The Control of Errors in Momentary Yield Data from Combine Harvesters, Journal of Agriculture Engineering Research, 1996, 64: pp. 271-278

M.H. Wang, Field Information Collection and Process Technology. Agriculture Mechanization [J], 1999, Vol. 7, pp. 22-24

STUDY ON EXTRACTING EDGE OF CROPLAND SCENERY

Mingxia Shen, Zhiye Yan, Changying Ji
Nanjing Agricultural University, Nanjing Agricultural University, 40# Dianjiangtait Road, Pukou Nanjiang, 210031, P. R. China, Tel: +86-025-58606585, Fax: +86-025-58606585

Abstract: Vision navigation and location is a main function in the vision system of intelligent agricultural mobile robot, and the edge feature of images is an important feature for vision navigation and location. According to the characteristic of cropland scenery, a compactly supported dyadic antisymmetric wavelet with respect to origin is brought forward to detect edges of cropland image. A set of filters were given that can be used to construct the edge detecting wavelet. The edge features are provided by determining the local maxim of wavelet coefficient at dyadic scale of the image. After the computer simulation was carried out on the cropland image, the continuous and smooth edge image can be got. The edges of cropland scenery are extracted accurately. The experiment result reveals that the method is efficient and practicable.

Keywords: Farm Crops, Marginal information, Wavelet transforms, Computer vision

1. INTRODUCTION

The detection of the edges in cropland scenery is the fundamental content of the wireless navigation system in mobile agricultural robot and it still remains an issue which has not been settled satisfactorily. In the common method of edge detection, one thing still stands in the way: the noises of the image are increased when the edges are maintained and enhanced, whereas

Shen, M., Yan, Z. and Ji, C., 2008, in IFIP International Federation for Information Processing, Volume 259; Computer and Computing Technologies in Agriculture, Vol. 2; Daoliang Li; (Boston: Springer), pp. 1411–1416.

faintness was caused when the random noises are eliminated. We need to look for a kind of detection means which is insensitive to noise and can locate accurately, neither neglecting the real edges nor bringing about fake edges.

The edges points are usually located at where the grayness of the image mutates. Because wavelet transform is denoted by multi-resolution that can provide signals, it has good analytic characteristics of time-frequency localization and can meanwhile provide the time domain and frequency domain of the image signals. When detecting and locating the image edges, we can make use of the transmitting information between signals and resolution under every resolution, which features dual selectivity of space and frequency and can quantitatively describe the features of different structure boarders in the image. Therefore, this method can effectively detect and locate the edges of images and has been widely used in image processing and pattern recognition.

As far as the cropland scenery image is concerned, the boarders of some important structures in the image are usually some regular curves, which have strong internal geometrical characteristics. The grayness transformation in the curve direction is smooth while that in the vertical direction is singular. So the modular amplitude value, angle and position of the wavelet transformation in curve direction will not change notably. The random distributed boundary points in the image caused by noise will not bring about smooth singular curve; the maximal value point caused by noise is irregular at amplitude, angle and position. The various features of cropland scenery and noise provide the major reference for edge detection by using wavelet transform. Because the quality of wavelet edge detection is closely related with the features of wavelet function, this thesis brings forward a compactly supported dyadic antisymmetric wavelet with respect to origin to detect the edges in the image.

2. MATIERIALS AND METHOD

2.1 The Principle of Wavelet Edge Detection

Suppose $\theta(x, y)$ is an adequately smooth duality function, which satisfies the following conditions (Mallat S, 1992)

$$\int_{-\infty}^{\infty}\int_{-\infty}^{\infty}\theta(x,y)\,\mathrm{d}x\,\mathrm{d}y=1 \quad \lim_{x^2+y^2\to\infty}\theta(x,y)\to 0 \tag{1}$$

Introduce signal $\theta_s(x,y)=\frac{1}{s^2}\theta(\frac{x}{s},\frac{y}{s})$ (2)

Define $\psi_s^1(x,y)=\frac{\partial\theta(x,y)}{\partial x}$ $\psi_s^2(x,y)=\frac{\partial\theta(x,y)}{\partial y}$ (3)

Then function $\psi^1(x,y)$ and $\psi^2(x,y)$ are two-dimension wavelets.

Transforming the image by using dyadic wavelet and then detecting in the gradient direction the maximum point of the modular value and its corresponding revulsion point (singular point) of grayness in the image, we can get edge points, the amplitude of modular value and the strength of the edges. Thereafter, connecting the neighbor edge points that are of similar direction and modular value into one edge chain and eliminating the chain whose length is shorter than the threshold, we can get the edge of certain scale.

2.2 The Structure of Compactly Supported Dyadic Antisymmetric Wavelet with Respect to Origin

The dyadic expansion of $\hat{\psi}^1(\omega_x,\omega_y)$, $\hat{\psi}^2(\omega_x,\omega_y)$ to cover the whole Fourier surface.

By further study we can know that in order to eliminate the influence of the edges and increase the operation speed, we can choose two wavelets (Yang Fusheng, 2000) $\psi^1(x,y),\psi^2(x,y)$, which can be expressed as the product of a univariate function with x as the independent variable and that with y as the independent variable, i.e., $\psi^1(x,y)=\psi(x)\xi(y)$, $\psi^2(x,y)=\xi(x)\psi(y)$, in which $\hat{\psi}(2\omega)=G(\omega)\hat{\varphi}(\omega)$, $\hat{\xi}(2\omega)=L(\omega)\hat{\varphi}(\omega)$, $\psi(x)$ and $\xi(x)$ are the anti-symmetrical and symmetrical functions respectively at $x=0$. The symmetrical dyadic wavelet is the tool for detecting roof-style edges (Xu Peixia, 1996), whereas anti-symmetrical dyadic wavelet is that for step-lifting edges. By using both of them simultaneously, we can detect and classify all the edges in the cropland scenery. $\hat{\varphi}(\omega)$ is the Fourier transform of univariate scale function $\varphi(x)$, which can be expressed as $\hat{\varphi}(\omega)=\prod_{j=1}^{\infty}H(2^{-j}\omega)$, $H(\omega)\in L^2(-\Pi,\Pi)$, $|H(\omega)|\le 1$. $G(\omega),L(\omega)$ respectively refer to the Fourier transforms of digital filters $g(n),l(n)$. Then make $G(\omega),L(\omega)$ satisfy

$$\begin{cases} G(\omega) = j\,\mathrm{sgn}(\omega)(1+\sum_{k=1}^{N} a_k \cos k\omega)\sqrt{1-|H(\omega)|^2} \in L^2(-\pi,\pi) \\ L(\omega) = \frac{1}{\sqrt{2}}(1+\sum_{k=1}^{T} b_k \cos k\omega)\sqrt{1+|H(\omega)|^2} \in L^2(-\pi,\pi) \end{cases} \tag{11}$$

in which $(a_1, a_2 ....a_N)$ is an optional N-dimensional real-number group that satisfies $\sum_{k=1}^{N}|a_k| < 1$, while $(b_1, b_2 ....b_T)$ is an optional P-dimensional real-number group that satisfies $\sum_{k=1}^{T}|b_k| < 1$. Then we have

$$\sum_{j=\infty}^{\infty}\left[\left|\hat{\psi}^1(2^j\omega_x, 2^j\omega_y)\right|^2 + \left|\hat{\psi}^2(2^j\omega_x, 2^j\omega_y)\right|^2\right] = 1.$$

That means ψ^1, ψ^2 make up the basic dyadic wavelet group that detects the edges in the two-dimensional image of the cropland scenery. Set dual scale function $\varphi(x,y) = \varphi(x)\varphi(y)$, and we can get the wavelet transform of the image in the above dyadic wavelet group:

$$S_2^j f(x,y) = \sum_{m\in Z}\sum_{n\in Z} h_{(m)} h_{(n)} S_2^{j-1} f(x-2^{j-1}n, y-2^{j-1}m) \tag{12}$$

$$W_{2^j}^j f(x,y) = \sum_{m\in Z}\sum_{n\in Z} l_{(m)} g_{(n)} S_2^{j-1} f(x-2^{j-1}n, y-2^{j-1}m) \tag{13}$$

$$W_{2^j}^j f(x,y) = \sum_{m\in Z}\sum_{n\in Z} g_{(m)} l_{(n)} S_2^{j-1} f(x-2^{j-1}n, y-2^{j-1}m) \tag{14}$$

其中:

in which

$$S_2^j f(x,y) = f(x,y) * \varphi_2^j(x,y), \quad W_{2^j}^1 f(x,y) = f(x,y) * \psi_{2^j}^1(x,y)$$

$$W_{2^j}^2 f(x,y) = f(x,y) * \psi_{2^j}^2(x,y).$$

2.3 The Choosing of Compactly Supported Dyadic Antisymmetric Wavelet with Respect to Origin and the Analysis of the Characteristics of the Filter

From the structure of the above edge detection wavelet, we can see that choosing the edge detection wavelet equals to choosing digital filters $g(n)$

and $l(n)$. Because the edge detection of the cropland scenery demands that the detected edges have good locality (i.e., correct edge location) and no position excursion, digital filter group must have linear phase and should have symmetry about zero point (Zhang Guobao, 1998).

In order to detect the local maximum modular value of the image and evaluate the gradient value of the grayness variation, we choose high pass filter $g(n)$ as odd symmetry, i.e., $g_{(n)} = -g_{(-n)}$ and $g_0 = 0$, then the Fourier transform is $G(\omega) = \sum_{n \in Z} g_{(n)} e^{-jn\omega}$, in which $G(\omega)$ is real function and $G(-\omega) = -G(\omega)$.

Because low-pass filter $l(n)$ has smoothening function, it can't have odd symmetry and can only have even symmetry, i.e., $l_{(n)} = l_{(-n)}$, and its Fourier transform is $L(\omega) = \sum_{n \in z} l_{(n)} e^{-jn\omega}$, in which $L(\omega)$ is real function and $L(\omega) = L(-\omega)$.

Calculate the filtering coefficient g_n, l_n and we can get the following table:

3. THE RESULT AND ANALYSIS OF THE EXPERIMENT

Adopt the filter group proposed in this thesis, wavelet transform the tested image by using the filtering coefficients listed in Table 1; adopt the method put forward in this thesis to detect the local maximum modular value;

Table 1. Dyadic wavelet transform filter coefficient

	0	1	2	3	4
g_n	0	0.0506770	0.180425	0.025216	-0.003602
l_n	0.190682	0.066190	0.031165	0.007269	0.000035
	0	-1	-2	-3	-4
g_n		-0.506770	-0.180425	-0.025216	0.003602
l_n		0.066190	0.031165	0.007269	0.000035

assume the points of the local maximum modular values that are greater than the wavelet transform modular value mean as the edge points; repeat the above operations on the given images, and connect the edge points in the 8-neighborhood according to the 8 connection rule; assume the threshold value and eliminate the chains who are less than the threshold value, because they might be caused by noises, and then we obtain the edge image.

The thesis is based on the compactly supported dyadic antisymmetric wavelet with respect to origin to detect the edges in the cropland images. We can see that the detected edge in the image is of fine locality and is abundant in details—every edge has been extracted. This shows that the wavelet put forward in the thesis is of the same quality in each direction and all the edges are continuous and of mono picture elements. Because the selected high-pass filter is symmetric about the zero point, the detected edges in the image have no translation of picture elements and are of accurate localization.

The experiment proves: if we use a wavelet transform of a certain scale to detect the fine edges in the cropland image, we should choose small neighborhood; if to detect the main edges, we should choose appropriate neighborhood. The experiment has also provided a thought for detecting various edges caused by different revulsion points, such as shadows, oscillations, peaks and grains.

REFERENCES

Mallat S, Whwang W. Singularity Detection and Processing with Wavelets, IEEE Trans. Information Theory, 1992, 38(20):617-643

Xu Peixia, Sun Gongxian, *Wavelet analysis and applicable examples*, China Science and Technology University Press, 1996, 21-38

Yang Fusheng. *The engineering analysis and application of wavelet transform*. Science Press, 2000, 112-176

Zhang Guobao, Chen Weinan. *Multi-scale edge extraction based on orthogonal wavelet transform*, China Image and Graph Learned Journal, 1998, (8):651-654, 38(20):617-643

STUDY ON GREENHOUSE TEMPERATURE VARIABLE UNIVERSAL ADAPTIVE FUZZY CONTROL

Weizhong Yang, Qiaoxue Dong, Yiming Wang*

College of Information and Electronic Engineering, China Agricultural University, 10083 Beijing, China

** Corresponding author, Address: P.O. Box 63, China Agricultural University, 17 Tsinghua East Road, Beijing, 100083, P. R. China, Tel: +86-10-62737824, Email: ym_wang@263.net*

Abstract: Basic fuzzy controller (BFC) has been successfully adopted to control greenhouse environment temperature. Variable universal adaptive fuzzy controller (VUAFC) which can contracts or expands the universal of the fuzzy variables of the BFC by the value of response error has been developed to get better control performance. By analyzing the result of experiments, VUAFC improved the performance of the fuzzy controller greatly.

Keywords: fuzzy control, greenhouse temperature, variable universal

1. INTRODUCTION

Suitable environment temperature is very important for the greenhouse-crop. Fuzzy control is successful to use on control of the greenhouse environment temperature (Yang et al., 1999). As the nature of greenhouse temperature dynamic process is time-variance, the fuzzy controller must have the capability of adaptivity. VUAFC can adjust the universes of some fuzzy variables by the change of error. It will be higher on the control precise.

In the paper, firstly, the design of VUAFC of greenhouse environment temperature is introduced. Then, some experiments are carried out, and the control effects of BFC and VUAFC are compared.

Yang, W., Dong, Q. and Wang, Y., 2008, in IFIP International Federation for Information Processing, Volume 259; Computer and Computing Technologies in Agriculture, Vol. 2; Daoliang Li; (Boston: Springer), pp. 1417–1420.

2. DESIGN OF VUAFC OF GREENHOUSE ENVIRONMENT TEMPERATURE

Variable universe (Li et al., 2002) means that some universes such as X, can change with changing variables x, denoted by

$$X(x) = \left[-\alpha(x)E,\ \alpha(x)E\right] \tag{1}$$

where $\alpha(x)$ are contraction-expansion factors of the universe X. Being relative to variable universes, the original universes $X = [-E, E]$ are called initial universes. The situation of variable universes changing is shown as Figure 1.

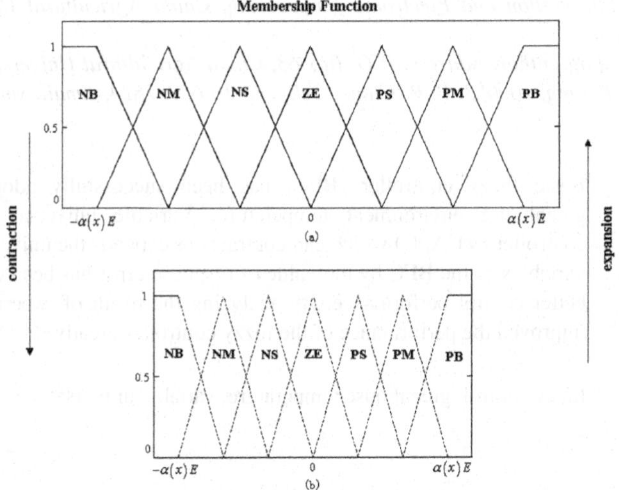

Figure 1. The schematic of contracting/expanding universe

In the paper, $\alpha(x)$ is as follow:

$$\alpha(x) = \frac{1-\varepsilon e^{-kx^2}}{1-\varepsilon e^{-kE^2}} \quad \text{and} \quad k>0 \tag{2}$$

where k is a parameter that can be chosen by designers, and ε is a very small constant. Set $\varepsilon = 0.8$, $k = 1$.

3. EXPERIMENTS AND ANALYSIS

In order to evaluate the system capability roundly, three kinds of characteristics were applied. They are dynamic performance characteristics, stable-state performance characteristics and synthetic performance characteristics. The characteristics adopted in this paper are: overshoot, denoted by σ; 5% rise time, denoted by 5% t_s; steady-state error, denoted by e_{ss} and MITAE (Integral of product of Time and Absolute-value of Error) criterion (Xiang, 1986), denoted by J_{MITAE}.

MITAE criterion is denoted by

$$J_{MITAE} = \frac{1}{t_1 - t_0} \int_{t_0}^{t_1} t\left|e(t)\right| dt \tag{3}$$

In the formula above, the time interval from t_0 to t_1 is the transient time, And $e(t)$ is the static error of the system response.

From (3), J_{MITAE} is the time weighting sum of system error. Since error of steady system reduces gradually as time elapses, it can be concluded that J_{MITAE} can reflect the dynamic and static performance of system. The smaller value of J_{MITAE} can indicate quick response speed, short regulation time, small over-regulation and small steady-state error. When J_{MITAE} reaches minimum, dynamic process of system response is optimized. It is very suitable to use this index to evaluate the dynamic response performance of greenhouse environmental temperature.

The chamber temperature control experiments were carried out in the greenhouse located in China Agriculture University in Dec. 2003. During the experiment period some tomato plants were planted in the greenhouse, and began to be harvested. Because air temperature is low in the winter night, heater is needed to maintain the growth of tomato; the experiment is executed during the night. The chamber temperature set point is 13 degrees Celsius. The chamber temperature depends on the hot water flow rate of the heating system, and the flow rate is related to the valve travel. The supervisory computer measures the chamber temperature and calculates e_1 and e_2, the inputs of the controller then, infer u, the output of the controller. u is the time that the valve is powered, if $u > 0$ the valve position augments and vice versa.

The course of the experiment was that the BFC was adopted firstly, the parameters was k_1=2, k_2=20 and k_3=30, the response curve was logged. Then the VUAFC was applied, and the response performances were measured. Table 1 lists the performances of the 2 fuzzy controllers; they are over-regulation (σ), 5% rise time (5% ts), steady-state error (e_{ss}), and J_{MITAE}. The Figure 2 shows the response curves of them.

Table 1. Performances of BFC and VUAFC

	σ (%)	5% t_s (min)	e_{ss}	J_{MITAE}
BFC	6.03	252	0.27	117.72
VUAFC	2.75	55	0.04	31.05

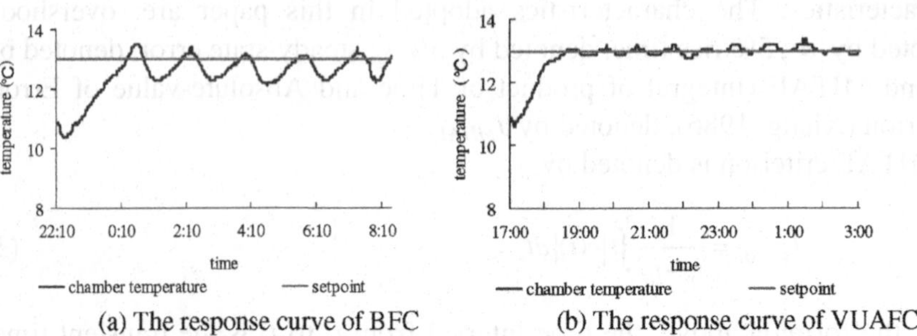

(a) The response curve of BFC (b) The response curve of VUAFC

Figure 2. The response curves of 3 fuzzy controllers

4. CONCLUSION

VUAFC can improve the performance of environmental temperature control efficiently. Since it doesn't occupy too much resource, it can not only be used on normal computer, but also be used on single-chip microcomputer system as well.

ACKNOWLEDGEMENTS

This study was supported by National Tenth-Five year-Plan of China (Grant (Grant No. 2001BA503B01).

REFERENCES

Li Hongxing, et al. Variable universe stable adaptive fuzzy control of nonlinear system. *Science in China. Ser. E.*, 2002, 45: 225-240.

Xiang Guobo, Optimal control based on ITAE. Mechanical Industry Press, Beijing, 1986.

Yang Weizhong, et al., Experiment and research of greenhouse automatic control. *Trans. of the CSAE*, 15 259-261 (1999).

MACHINE VISION BASED COTTON RECOGNITION FOR COTTON HARVESTING ROBOT

Yong Wang [1,2], Xiaorong Zhu [3], Changying Ji [1,*]
[1] *College of Engineering, Nanjing Agricultural University, Nanjing, China, 210031*
[2] *College of Sciences, PLA University of Science and Technology, Nanjing, China, 210007*
[3] *Department of Radio Engineering of South East University, Nanjing, China, 210049*
[*] *Corresponding author, Address: College of Engineering, Nanjing Agricultural University, Nanjing, 210031, P. R. China, Tel: +86-10-58766570, Email: chyji@njau.edu.cn*

Abstract: A new cotton recognition method is proposed in this paper. It provides parameters for motion of the manipulator so that it can acquire precise location information of cotton, identify cotton from surroundings correctly, and accordingly pick up them automatically. This method is based on color subtraction information of different parts of cotton. Furthermore, in order to increase accuracy rate of cotton recognition, dynamic Freeman chain coding is used to remove noise. Experimental results show that the proposed method has good performance for cotton identification.

Keywords: cotton recognition, image processing, Freeman chain coding

1. INTRODUCTION

Agricultural automation and intelligence have become a major issue in recent years (Shiqchil et al., 2001; Alessio et al., 2001; Slaughter et al., 1999). However, autonomous harvesting robots have not yet been commercially applied in horticulture practices. Obviously, the price and performance of harvesting robots have not yet met the requirements for their successful introduction in harvesting practices (Yud-Ren et al., 2002; Sander, 2005; VanHenten et al., 2003; Elias et al., 2003; Tadhg et al., 2004). In this

Wang, Y., Zhu, X. and Ji, C., 2008, in IFIP International Federation for Information Processing, Volume 259; Computer and Computing Technologies in Agriculture, Vol. 2; Daoliang Li; (Boston: Springer), pp. 1421–1425.

paper, a new cotton recognition method is introduced, which is based on color subtraction information. Besides, the modified dynamic Freeman chain coding is used to further increase recognition accuracy.

2. COLOR ANALYSIS FOR COTTON

In image processing, there are six color spaces widely used, namely, RGB, normalized rgb, HIS, YCrCb, $L^*a^*b^*$ and $I_1I_2I_3$. The objects investigated are four hundred images of cotton taken in Jiangpu Farm of Nanjing Agricultural University from ten to twelve in the morning and from two to four in the afternoon from October to November in sunshiny weather. Each image is 24-bit true-color. Color data of cotton fruits, leaves and stems are extracted from these images. According to transition formulas of different color space (Stephen et al., 2003), color mean distributions of them in six color space models are respectively drawn as shown in Figure 1.

Although $I_1I_2I_3$ and $L^*a^*b^*$color space can be used to identify cotton fruits from background, they need much time to convert from RGB module to other spaces. In general, none of the six color space can precisely identify cotton fruit from the background in real-time.

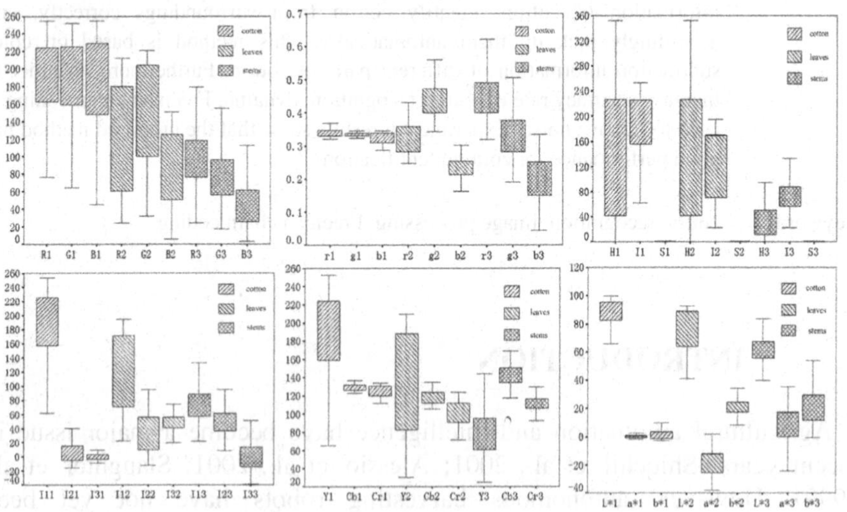

Fig. 1. Mean distribution of color of cottons, leaves and stems in six color models

3. IMAGE PROCESSING ALGORITHM FOR COTTON RECOGNITION

3.1 Color Subtraction Module

In order to correctly identify cotton fruits from the background in real-time, a new color module was developed. According to above analysis, it can be seen that the color values of red, green and blue of cotton fruits are quite the same, whereas other two are quite different. Hence, this character can be used to differentiate cotton fruits from the other two. The distribution of red-blue (*R-B*), red-green (*R-G*) and green-blue (*G-B*) of cotton fruits, leaves and stems are shown in Figure 2.

In this paper the subtraction value of red and blue is selected. And the algorithm is as follows:

$$g(x,y)=\begin{cases}0 & ,if\ f(x,y)<T \\ 255, & otherwise\end{cases} \tag{1}$$

where f(x, y) is the gray value of pixel (x, y), and T is the threshold. As example, Figure 3(b) is recognition result by color subtraction module.

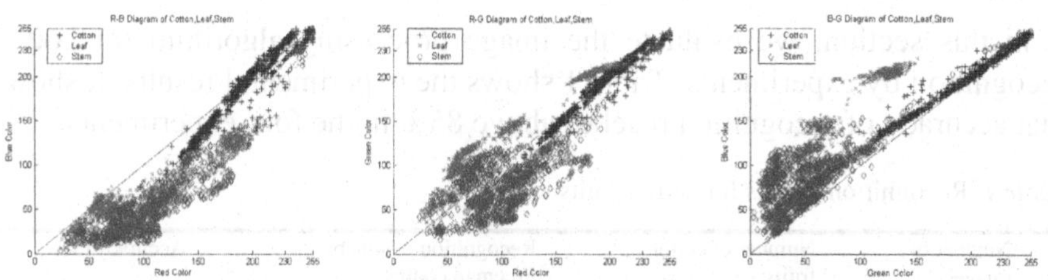

Fig. 2. Distribution of R-B, R-G and G-B

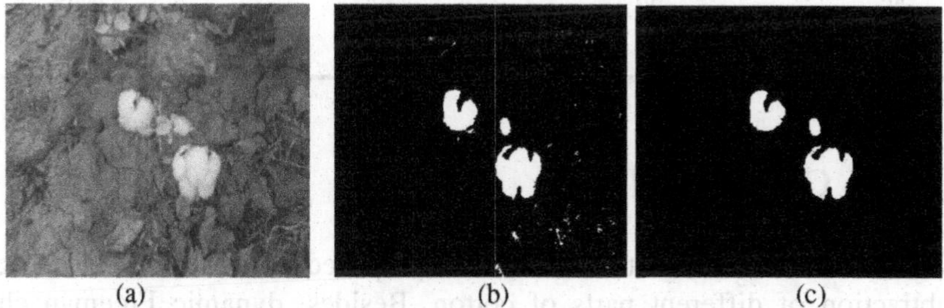

Fig. 3. Original image and recognition results

3.2 Dynamic Freeman Chain Coding

In order to solve the white points in Figure 3(b), a new method of dynamic Freeman chain coding is proposed.

The algorithm of dynamic Freeman chain coding is as following:

(1) Scan the binary image, if the pixel value is 255, then labeled, and use the rule of Freeman to trace anticlockwise.

(2) Compute the perimeters of different regions.

(3) Remove the region which perimeter is less than T which is calculated by the formula:

$$T = \mu_1 P_1 + \mu_2 P_2 \quad (2)$$

$$\mu_1 = \frac{M1}{\sum_{i=1}^{n} perimeter}, \quad \mu_2 = 1 - \mu_1 \quad (3)$$

where P_1 is the median value of perimeter, P_2 is the maximum value of perimeter, μ_1 and μ_2 are weight indexes, M_1 is the number which perimeter is less than the median value. If perimeter is less than T, the value of the pixel in the region is 0, otherwise 255.

Figure 3(c) shows the recognition result by dynamic Freeman coding.

4. PERFORMANCE EVALUATION

In this section, we evaluate the image processing algorithm for cotton recognition by experiments. Table 1 shows the experimental results. It shows that accuracy of recognition reaches above 85% in the four experiments.

Table 1. Recognition results for cotton fruits

Number of images	Number of cotton fruits	Recognition results by Freeman coding	Accuracy (%)
20	42	37	88.09
50	101	87	86.14
100	182	158	86.81
150	273	237	86.81

5. CONCLUSION

A new cotton recognition method is proposed, which is based on color subtraction of different parts of cotton. Besides, dynamic Freeman chain coding is used to increase recognition accuracy. Experimental results show that the method has high performance for no-overlapping individual cotton recognition.

REFERENCES

Alessio Plebe, Giorgio Grasso. Localization of spherical fruits for robotic harvesting. Machine Vision and Application, (2001) 13:70-79

E.J. Van Henten, B.A.J. Van Tuijl, J. Hemming, et al. Field Test of an Autonomous cucumber picking robot. Biosystem Engineering (2003) 86(3):305-313

Elias N. Malamas, Euripides G.M. Petrakis, Michalis Zervakis, et al. A survy on industrial vision systems, applications and tools. Image and Vision Computing 21 (2003): 171-188

K.F. Sanders. Orange Harvesting System Review. Biosystems Engineering, 2005, 90(2): 115-125

Shiqchik Hayashi, Katsunobu Ganno, Yukitsugu Ishii, et al. Robotic Harvesting System for Eggplants. JARQ, 2000, 36(3): 163-168

Stephen Westland and Caterina Ripamont. Computational Colour Science using Matlab. John Wiley & Sons. Ltd, 2003, 2

Tadhg Brosnan, Da-Wen Sun. Improving quality inspection of food products by computer vision-a review. Journal of Food Engineering, 61 (2004):3-16

W.S. Lee, D.C. Slaughter, and D.K. Giles. Robotic weed Control System for Tomatoes. Precision Agricultural, 1 (1999): 95-133

Yud-Ren Chen, Kuaiglin Chao, Moon S. Kim. Machine Vision technology for agricultural application. Computer and Electronics in Agriculture, 36 (2002):173-191

LEAF AREA CALCULATING BASED ON DIGITAL IMAGE

Zhichen Li, Changying Ji [*], Jicheng Liu
[*] *Corresponding author: College of Engineering, Nanjing Agricultural University, Nanjing, Jiangsu, 210031, China, Email: chyji@njau.edu.cn*

Abstract: A new and simple method for plant leaf area measurement using image processing technique was provided. Sixty leaves with different shape and size of six plant species were selected for study. The software using for image processing and leaf area calculation was matlab6.0. Experiments were performed to test the performance of the estimating system by comparing the estimated leaf area with the grid square leaf area. The range of relative errors of estimation were 4.42-7.81%, 3.56-8.01%, 4.12-6.97%, 3.89-7.28%, 4.23-8.12% and 3.56-7.03% for Eucommia Bark, Paulownia, Maidenhair tree, Bamboo, Cycad, Weed respectively. The results illuminate the more close to rectangle area the more accurate of the estimating leaf area. So it can be concluded that the method developed in this study obtained the sufficient accuracy for the possible non-destructive leaf area measurement.

Keywords: leaf area, image processing, grid square

1. INTRODUCTION

The leaf is the most important photosynthesis organ of plant, and it affects crop growth and bio-productivity. Leaf area is a valuable parameter in studies of plant nutrition, plant protection measure, plant soil-water relations, crop ecosystem etc. (Mohsenin, 1986). Accurate and rapid non-destructive leaf area measurement is important in plant studies of understanding and modeling ecosystem function. The methods for leaf area measurement include:

Li, Z., Ji, C. and Liu, J., 2008, in IFIP International Federation for Information Processing, Volume 259; Computer and Computing Technologies in Agriculture, Vol. 2; Daoliang Li; (Boston: Springer), pp. 1427–1433.

1. Weighing. Copying the shape of the leaf on a piece of paper and weighing the copy. So leaf area is estimated by the following equation: $LA = W/c$ LA is the leaf area, W is the weight of the paper and c is the coefficient of the paper (weight of unit area).

2. Counting grid. Projecting leaf on a piece of grid paper and counting the number of panes. So Leaf Area is estimated by the following equation: $LA = GN \times GA$ LA is the leaf area, GN is the number of grid and GA is area of one grid.

3. Regression equation. This method includes two steps. Firstly one specific species of plant leaf of different size is selected for study. The length and width of the leaf is measured and the leaf area is also measured using method 1 or 2. the leaf area prediction model was developed as the following: LA=f (L, W) LA is the leaf area, L is leaf length and W is leaf width. Second, when you want to know the specific plant leaf area, you can calculate the leaf area using the equation only by measuring the length and width of the leaf.

4. Photoelectric scanning. Scanning the reference object e.g. 50×50mm by specific dot per inch (dpi). Calculating the area of one pixel. Then scanning the leaf by the same dpi. And calculating the leaf area by the following equation: $LA = PN \times RAP$ LA is the leaf area, PN is the number of leaf scanning image pixels and RAP is area of one reference object scanning image pixel.

5. Utilizing leaf area instrument. It is reliable and convenient to estimate leaf area using mechanical, digital or portable scanning planimeters (Daughtry, 1990).

Every method of the above is reliable, but method 1 and method 2 take much time; method 3 requires many different equations and one equation is only applicable for one specific species of plant; method 5 is expensive and method 1-5 are destructive.

Due to the developments of computer, PC-based image processing has become a feasible tool in plant studies. Digital image processing has been used for leaf area measurement. Meyer and Davidson (1987) used the stereoscopic system and measured the leaf area from the three-dimensional coordinates of edge points identified by the user-interactive program. Nyakwende et al. (1997) determined the leaf area by the regression of the project leaf areas in three viewpoints, from the top, from the side and from the oblique angle. In a series of previous research activities, a non-destructive image processing technique to measure time-dependent growth curves of selected vegetable seedlings has been developed (Chien and Lin, 2000a, 2000b, 2002).

The leaf area measurement method introduced in this paper is rapid and simple, the metrical result is reliable. The system is not expensive and it is possible to realize non-destructive measurement.

2. MATERIALS AND METHOD

2.1 Material

The image acquisition system is depicted in figure 1. A hole was drilled in the centre of the top face of the box and the camera was placed on the hole. The diameter of the hole is a little bigger of the lens of the camera. The distance between top and bottom face is 450mm. It is easy for adjusting the focus of the camera because the axis of the lens is vertical to the flat. The left, right, front and back face of the box is hollow, so it is easy for the light projecting in and it is difficult for forming shadow. It is also easy for the leaf to be placed under the camera. The source of the light is the natural light. A piece of white paper with a rectangle was placed on the platbed and the side of the rectangle is black. A lot of paper with varied rectangle was prepared for the leaf of different size. The leaf on the rectangle was covered a transparent glass plate to obtain a high-contrast image of the leaf. So it is easy for getting reliable threshold for segmentation. Another function of the glass is to flat the leaf. The nut is used to tighten the glass.

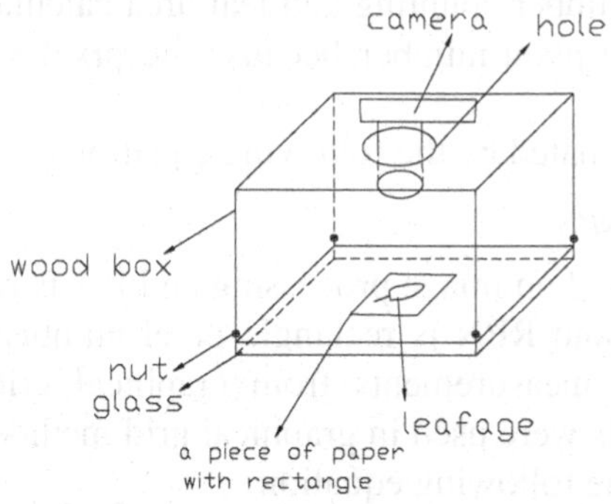

Fig. 1. Image acquisition system structure

The model of the CCD camera is Olympus N438. The image was acquired outdoors and was digitized into 24-bit (RGB) images with a resolution 640*480. The image is processed by computer (Pentium III 700 MHz CPU). The images were recorded on a U-disc of the camera and then downloaded into the personal computer. An image processing and determining leaf area program was developed using Matlab 6.0 under the Microsoft Windows operating system.

To compare the estimated leaf area with the actual leaf area, total 60 leaves of 6 species were sampled for image acquisition and measurement. The actual leaf areas were determined by method 2 (counting grid). Leaves selected for area measurement were Cycad, Paulownia, Bamboo, Weed, Maidebhair tree, and Eucommia Bark representing varied shapes (Fig. 2). Cycad leaves are heart-shaped and Paulownia leaves are star shaped. Bamboo leaves are fish-shaped. Weed leaves are long like a sword. Maidebhair tree leaves are fan shaped and Eucommia Bark leaves are elliptical.

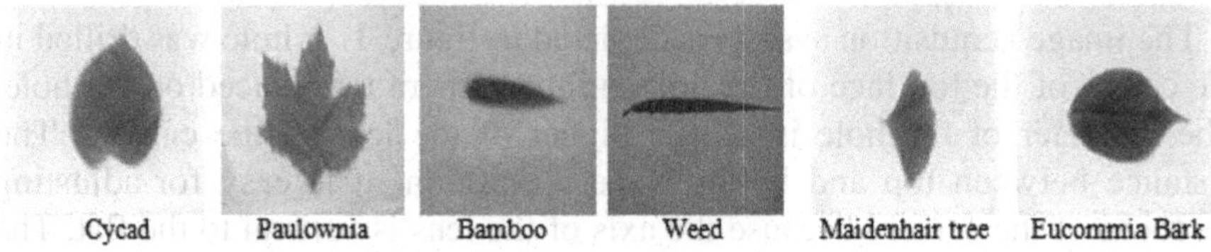

Fig. 2. Leaf image of plants

2.2 Algorithms and Image processing

The software algorithm for the system was written in Matlab6.0. the algorithm execute the operations of image processing including setting thresholds and binary image, the estimated leaf is segmented from surroundings by set threshold, the pixel value is set to 1 and 0 by the threshold; of pixels number counting and leaf area calculating. It is easy and rapid for counting the pixel number because the pixel value has only two, 1 and 0.

The leaf area is estimated by the following equation.

$$ILA = RA \times LPN / RPN \tag{1}$$

ILA is leaf area based on image processing and RA is rectangle area. LPN is leaf pixel number and RPN is rectangle pixel number. Leaf area results were compared with measurements from graphical grid methods. Graph sheets with 1mm2grids were used in graphical grid methods. So the total leaf area is estimated by the following equation

$$GLA = ISN + PSN / 2 \tag{2}$$

GLA is the estimated leaf area by grid method, ISN is the number of enclosed squares by the profile and PSN is the number of passed through squares by the profile.The error rate is determined by the following equation: ER is the error rate, GLA is the estimated leaf area by grid method and ILA is leaf area based on image processing.

$$ER = ((GLA - ILA) / GLA) \times 100\ \% \tag{3}$$

3. RESULTS AND DISCUSSIONS

Ten leaves of every species of plants were selected for leaf area estimation. Five leaves are normal and the other five leaves are abnormal, some leaves are incomplete and some leaves are blurred, they are shown in fig. 3. One leaf area was calculated three times. The leaf area value equals the average of the three values. The estimated leaf area was compared with the results of grid square method.

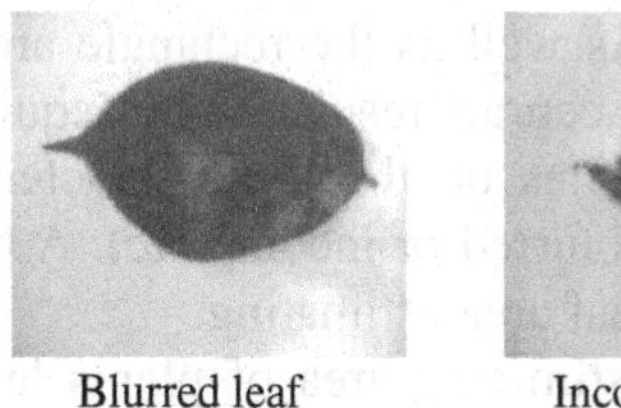

Blurred leaf Incomplete leaf

Fig. 3. Abnormal leaf

The analyses of relative errors are included in table 1. So it can be concluded that the result is reliable. It has the maximum error rate for the abnormal leaf area calculation and minimum error rate for the normal leaf area. The result is showed in table 2. The error rate of the abnormal leaf area is bigger than the normal leaf area. The cause is that the pixels of the blurred or incomplete leaf were set zero in the process of binary, the number of the LPN is little and then the error is big.

Table 1. The error rate of estimated leaf area

Species	Eucommia bark	Paulownia	Maidenhair tree	Bamboo	Cycad	Weed
Minimum Error rate (%)	4.42	3.56	4.12	3.89	4.23	3.56
Maximum Error rate (%)	7.81	8.01	6.97	7.28	8.12	7.03
Average error (%)	5.23	5.02	4.83	4.96	5.35	4.92

Table 2. The error rate of normal and abnormal estimated leaf area

Species	Normal Leaf		Abnormal Leaf	
	MinER	MaxER	MinER	MaxER
Eucommia bark	4.42	4.48	4.62	7.81
Paulownia	3.56	4.12	4..89	8.01
Maidenhair tree	4.12	4.97	4.38	6.97
Bamboo	3.89	6.28	7.22	7.28
Cycad	4.23	5.68	4.23	8.12
Weed	3.56	7.03	4.56	7.03

MinER is minimum error rate and MaxER is maximum error rate (%)

The influence of rectangle area is also studied. Five pieces of paper with different rectangle were used for the background and reference object of the image. Three leaves for every species of plant were selected for studying. The relative error equals the average of the estimated three leaves area error value. The relations between error and rectangle area were showed in table 3

Table 3. The relation of relative error and rectangle area

Rectangle Area (mm^2)	12000	15000	18000	21000	24000
Relative Error (%)	4.36	5.36	6.08	6.25	6.48

4. CONCLUSIONS

This study has shown that the accuracy of estimated leaf area is correlated to the area of the rectangle. As well as the rectangle area is a little bigger than the leaf area, a very high accurate result can be acquired. This study has also shown that the accuracy lies on the leaf characteristic, for example: normal leaf or abnormal leaf (blurred or incomplete). A high accurate result can be getting for the normal leaf area estimating.

The system we made for estimating area of plants leaf having different shapes is reliable and simple. Experiments were performed to test the performance of the estimating system by comparing the image estimated leaf area with grid square method. The method developed in this study obtained very high accuracy.

Because the study of understanding plant growth and development has been increasing, this system will be very useful tool for estimating leaf area for plants without using additive expensive devices. It is not expensive and it can utilize the existing computer and camera, therefore it does not require additional hardware. The device is very convenient for taking to field. So it is possible to enable researchers to make non-destructive and repeated measurement.

REFERENCES

Chien C.F., and T.T. Lin. 2000a. Non-destructive measurement of vegetable seedling leaf area using elliptical Hough transform. ASAE Parpe No. 003023. St. Joseph, Mich: ASAE.

Chien C.F., and T.T. Lin. 2000b. An image processing method to measure overlapped leaf area using elliptical Hough transform. J. Agric. Machinery 9 (4):47-64 (in Chinese), English summary.

Cirak C., Odabas M.S., Ayan A.K., 2004. Leaf area prediction model for summer snowflake (Leucojum aestivum) International Journal of botany.

Daughtry. 1990. Direct measurements of canopy structure. Remote Sensing reviews. 5, 45-60

Eschenbach, C., Kappen, L., Leaf area index determination in an alder forest: a comparison of three methods. J. Exp. Bot. 47, 1457-1462.

Gamiei. Y.S., Randle W.M., Milisha, Smittleda, 1991. A rapid and non-destructive method for estimating leaf area of onions. Horticultural Science, 26: 206.

Gholz, H.L., Ewel, K., Tesky, R.O., 1990. Water and forest productivity. For. Ecol. Manage. 30, 1-18.

Iakahiro Kanuma, K. Ganno, S. Hayashi etc. Leaf Area Measurement Using Stereo vision. Artificial Intelligence in Agricultural. 1998.

Lieih J.H., James F.R., Hugo H.R., 1986. Estimation of leaf area of soybeans grown under elevated carbon dioxide levels. Field Crop Research, 13: 193-203.

Meyer G.E., and D.A., Davidson, 1987. An electronic image plant growth measurement system. Trans. of the ASAE, 30(1), 242-248.

Mohsenin N.N., 1986. Physical Properties of Plant and Animal Materials. Gordon and Breach Science Publisher. New York, pp. 107-110.

Nyakwende E.C., J. Paul and J.G. Atherton, 1997, Non-destructive determination of leaf area in tomato plants using image processing. Journal of horticultural science, 72 (2), 255-262.

RESEARCH ON SPRAY PRECISELY TOWARD CROP-ROWS BASED ON MACHINE VISION

Honghui Rao [1], Changying Ji [2,*]

[1] *College of Engineering, Jiangxi Agricultural University, Nanchang, China, 330045*

[2] *College of Engineering, Nanjing Agricultural University, Nanjing, China, 210031*

[*] *Corresponding author, Address: Department of Agriculture machinery, Nanjing Agricultural University, 40 Dianjiangtai Road, Pu kou, Nanjing, 210031, P. R. China, Tel: +86-25-58606670, Email: chyji@njau.edu.cn*

Abstract: A method to aim toward crop-rows was put forward for spray control in this article. At first, the image of crops captured by a CCD camera was passed on computer, then crops were segmented from background by obtaining H value and its binary image was obtained by means of OSTU. Finally, the center line of crop-row was regressed by Hough transform after the binary image was morphologically eroded. A spray control system was designed to move the spray nozzle accurately on crop row.

Keywords: machine vision; Hough transform; center line of crop-row

1. INTRODUCTION

The characters of modern agriculture are precise, environmental and sustainable. With the development of agriculture, more and more pesticide is used to protect crops, which leads to soil pollution and reduction of pesticide. Both soil pollution and reduction of pesticide give much trouble to human and environment. It is necessary to use precision technique for various types of agriculture operations, so that the pesticides can be placed where they have an optimal effect with minimum quantity. Machine vision,

Rao, H. and Ji, C., 2008, in IFIP International Federation for Information Processing, Volume 259; Computer and Computing Technologies in Agriculture, Vol. 2; Daoliang Li; (Boston: Springer), pp. 1435–1439.

a classic technique used in precision agriculture, has been applied widely in agricultural production. Many researchers have investigated strategies to control weeds with less herbicide to reduce production costs and to protect the environment. For example, Lei Tian has developed an automated equipment of weed control in tomato fields based on machine vision and spraying system based on differential GPS (Tian et al., 1997; Tian et al., 1999). A machine-vision guided precision band sprayer for small-plant foliar spraying demonstrated a target deposition efficiency of 2.6 to 3.6 times than that of a conventional sprayer, and the non-target deposition was reduced by 72% to 99% (Giles et al., 1997). For high-value crops, high-accuracy machine vision and control systems have been studied for outdoor field applications in California (Lee et al., 1999). To detect line structures, a method to determinate crop rows by image analysis without segmentation was put forward (Søgaard et al., 2003). What's more, a vision based row-following system for agricultural field machinery has been designed in Halmstad University (Björn et al., 2005). It was showed that mobile robots could be most probably navigated through the fields autonomously to perform different kind of agricultural operations. In China, there are also some researchers to study smart spraying based on machine vision. A method based on vertical projection was described to detect orientation of crop rows (Yuan et al., 2005). However, with the exception of a few recent studies, much vision recognition of crop-rows is still at the stage of studying artificial crops in the laboratory, or studying the target crops under a controlled indoor lighting environment. So there still is a long way to make autonomous mobile robots based on machine vision practical in China.

In the research on the autonomous variable spraying systems based on machine vision, there are three problems need to be solved. At first, recognizing the target autonomously under complicated lighting environment is a dilemma, the algorithm on segmenting crops from background and segmenting crops from weed mostly depend on the lighting conditions. This is an area needs to deep research in spite of many scientists putting forward many efficient ways. Secondly, the form of spraying control toward the crops have two ways: one way the nozzles are fixed to spray where there are crops or weed; the other way is moving the nozzles toward the crops. Compared with the former, moving the nozzles is obviously more difficult and practical, which is becoming a hotspot all over the world. Thirdly, the variable technology of spraying is still explored in recent years, some researchers put forward to using PWM technology on changing the flux of nozzles (Cheng et al., 2003).

2. MATIERIALS AND METHOD

As many field crops are planted by rows and each crop grows like blob in early state, in order to make study conveniently, it is supposed that all row pitches are equal, the crops in different rows have not been interleaved and there is no weed in the field. Under these conditions, a horsebean image of 640×480 pixels was captured by a CCD camera, then the image was passed to computer for processed and later the center line of crop-row was obtained by Hough transform.

2.1 Image preprocessed

The true color of horsebean image (figure a) was transformed from the RGB model to HSI model, which can be segmented from its background by obtaining its H value of HSI model. Before the horsebean image is processed with morphological erode method, its binary image should be obtained. Through many experiments it was fund that using the OSTU algorithm can make perfect performance (He et al., 2001).

After obtaining its binary image by the algorithm of OSTU, we used the structure 3×3 to process the image with morphological erode. Through which the binary image was processed by 12 times, the result showed as figure b. Then the center line of horsebean (figure c) was regressed by Hough transform. Finally, the subtract image (figure d) is to test its efficiency. Obviously the method is available to detect the center line of crop-row.

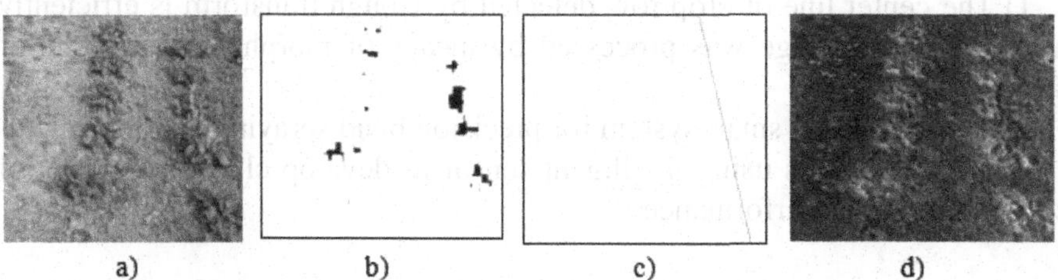

a) b) c) d)

Fig. 1. Original image (a), erode image (b), the center line of crops by Hough transform (c) and its subtract image (d)

2.2 Experiment and procedures

A spray control system based on machine vision was designed to move the nozzle toward the centerline of the crop row. The basic structure and working principle are shown as follows:

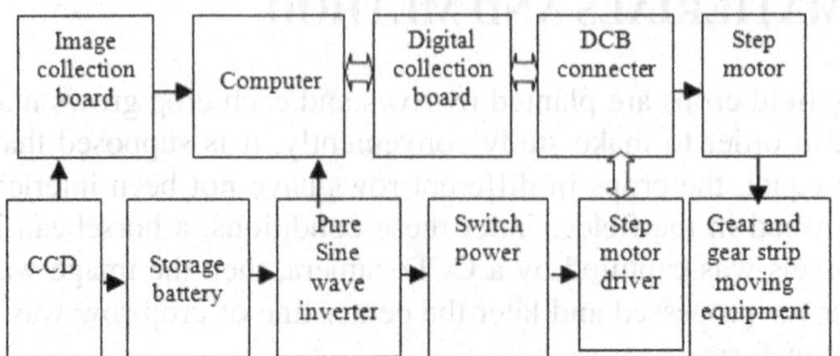

Fig. 2. Spray control system for field crop-rows based on machine vision

When the system works, a crop image is obtained by CCD camera and passed to computer through an image collection board firstly. After the centerline of the crop row is obtained with the method as mentioned above. The distance from the nozzle mark to real crop row can be computed. At the same time, the corresponding pulses are sending through a counter of digital collection board to move the equipment toward the right crop row.

The character of the equipment lies in its adaptation. Before sending pulses, the direction and quantities of the pulse should be computed firstly. Experiment was implemented with a computer of Pentium Π and the results showed that the method is available to aim at the crop row.

3. CONCLUSIONS AND FUTURE WORKS

1) The center line of crop row detected by Hough transform is efficiently after its binary image was processed by means of morphological erode of structure 3×3.

2) A vision based smart system for precision band spraying is established.

3) Future work is using intelligent sensor to develop closed-loop control system for higher performance.

ACKNOWLEDGEMENTS

This research has been supported by Ministry of Education of People's Republic of China (contract number: 03091).

REFERENCES

Björn Åstrand, Albert-Jan Baerveldt. A vision based row-following system for agricultural field machinery [J]. Mechatronics, 2005 (15):251-269

Cheng Yong, Zhen Jiaqiang, et al. R&D of variable rate technology in precision agriculture management system [J]. Transactions of the Chinese Society of Agricultural Machinery, 2003, 6 (34): 156-159

D.K. Giles, D.C. Slaughter. Precision band spraying with machine-vision guidance and adjustable yaw nozzles [J]. Transaction of the ASAE, 1997, 40(1):29-36

H.T. Søgaard, H.J. Olsen. Determination of crop rows by image analysis without segmentation [J]. Computers and Electronics in Agriculture, 2003 (38):141-158

He Bing, Ma Tianyu, et al. Visual C++ Digital Image Process [M]. Beijing, 2001

L. Tian, D.C. Slaughter, R.F. Norris. Outdoor field machine vision identification of tomato seedlings for automated weed control [J]. Transaction of the ASAE, 1997, 40(6):1761-1768

L. Tian, J.R. Reid, J.W. Hummel. Development of precision sprayer for site-specific weed management [J]. Transaction of the ASAE, 1999, 42(4):893-900

W.S. Lee, D.C. Slaughteer, D.K. Giles. Robotic weed control system for tomatoes [J]. Precision Agriculture, 1999(1):95-113

Yuan Zuoyun, Mao Zhihuai, et al. Orientation technique of crops rows based on computer vision [J]. Journal of China Agricultural University, 2005 (10): 69-72

A WEB BASED EXPERT SYSTEM FOR MILCH COW DISEASE DIAGNOSIS SYSTEM IN CHINA

Libin Rong [1,2] Daoliang Li [1,2,*]
[1] *China Agricultural University, College of Information and Electrical Engineering, Beijing 100083, China*
[2] *Key Laboratory of Modern Agriculture System Integration, China Agricultural University, Beijing 100083, China*
[*] *Corresponding Author, Tel: +86-10-62736717, Fax: +86-10-62737679, Email: dliangl@cau.edu.cn or*

Abstract: A web-based multi-models expert system called DCDDS is presented in this paper, which developed for diagnosis of dairy cow diseases through the symptoms submitted by users on web. As it is accepted that the inference engine and the relevant knowledge representation are the crucial part of diagnosis expert system, which limits its application and popularization in animal disease diagnosis. To break the limit and raise accuracy, this paper compares and appraises the existed systems and presents a solution that contains three models—Case-based reasoning (CBR), Subjective Bayesian theory and D-S evidential theory. Accordingly a knowledge representation method which can support the three different models is also designed. Up to the complicacy of the group of symptoms users acquired, they can choose which of the three models should be adopted to meet the best resolve. The performance of the proposed system was evaluated by an application to the field of dairy cow disease diagnosis using a real example of dairy cow diseases. The result indicates that the new methods have improved the inference procedures of the expert systems, and have showed that the new architecture has some advantage over the conventional architectures of expert systems on both efficiency and accuracy.

Keywords: Dairy cow disease diagnosis; Web-based expert system; Case-based reasoning; Subjective Bayesian theory; D-S evidential theory; China

Rong, L. and Li, D., 2008, in IFIP International Federation for Information Processing, Volume 259; Computer and Computing Technologies in Agriculture, Vol. 2; Daoliang Li; (Boston: Springer), pp. 1441–1445.

1. INTRODUCTION

Dairy cow disease diagnosis is rather a complicated process in stockbreeding production activity because the disease commonly resulted from many conditions difficult to detect contains nutritional and environmental problems, microbe conditions and incorrect method of milking, etc. The diseases diagnosis is rather a difficult task for dairy farmers, which can take a cow breeding worker or a dairyman more than three years to be able to correctly identify diseases by self-learning and practicing. So it needs the specialist to deal with. But the experts do not often on the port for most of the culture sites are remote from expert centers which are often seated in big cities. The serious confliction of the need of dairy cow disease expertise and the fact that the supply fall short of demand present a practical need to develop a computer-based system with the ability of providing accurate, timely and remote service of disease diagnosis.

A web-based expert system for dairy cow disease diagnosis has been developed by Key Laboratory of Modern Agriculture System Integration at China Agricultural University. It was a major outcome of Huo Yingdong Young Teacher's Foundation (Contract No. 94032). This web-based intelligent system can mimic human cow disease expertise and diagnose 67 types of common dairy cow diseases with a user-friendly interface. It contains a large amount of cow disease data and images, which are used to conduct online disease diagnosis.

Experiences with developing web-based expert systems, lessons learnt and test evaluation are discussed and conclusions are provided at the end of the paper.

2. DOMAIN DESCRIPTION

2.1 The knowledge of dairy cow disease

Profitable milk production relies upon a flock of healthy, productive cows. But failures of cattle diseases diagnosis are one of the important reasons for considerable losses of income in dairy farming. Many factors can be related to dairy cow diseases and the main impact elements contain the follows:

✧ The microbe factor which contains infective pathogen and environmental pathogen.

✧ The environmental factor that indicates improper management and bad sanitation, not to clean or disinfect the cowshed for example.

✧ The unscientific milk technology which is the most important reason for mastitis.

✧ The management of feeding which is to say the cow should be fed timely and the feedstuff should be nutritious.

✧ The individual factor.

2.2 User's needs for expert system

China is a large country and most dairy farms are spread in the remote rural areas. Although dairy farming is growing rapidly, the level of farmers' skills and knowledge are evidently low or even non-existent. As a result, cattle diseases occur frequently and the consequences on farmers' financial status are enormous, as sometimes the whole herd of cows could be infected at one time. The demand for help from experts is increasing rapidly. However, experts are scarce and not readily available, especially in rural areas. It takes long time to train novices and even longer for them to establish their experience in practice. An expert system is considered as an effective tool to help dairyman in solving the problems they meet in practice. Therefore, an expert system for dairy cow disease diagnosis is required to substitute human vets in helping farmers in China. Also the specialists need an expert system because the human expert' life is limited; he/she can only obtain limited experience. Furthermore, various personal characteristic and body or mood factors can impact their diagnosis. The computer systems have no such problems. They can accumulate knowledge all the while and always conclude the same result from the same evidences.

3. SYSTEM DESIGN AND DEVELOPMENT

3.1 System architecture

Based on the analysis and assessment of the users' needs, a web-based decision expert system called DCDDS was developed. The system is a typical three-tier web application which uses Internet Information Server (IIS), Microsoft SQL Server 2000, Windows XP as the operating system and Windows XP with Internet Explorer (IE) on the client side. The web pages were programmed using html and C# code running from ASP.NET filter which is a component of IIS. Users (patient or medical worker) enter the system via the web.

3.2 Inference engine

An inference engine of multi-models for the system is designed, which is the brain of the system. As the former described, clinical diagnosis is often complicated because similar clinical signs may arise in animals with different diseases while the same disease may well express itself in different ways in different individuals within an infected population. Hence, the diagnosis of disease is not well suited to the binary logic seen in the structure of many rule-based algorithms, which is often adopted in the lots of diagnosis systems. The adoption of three algorithms includes CBR, subjective Bayesian theory and D-S evidential theory provides a more-flexible method for use in disease diagnosis. Because the different three algorithms can get results with different efficiency and exactness, users can choose one of the three models in specifically occasion to get a reliable diagnosis result. The CBR fit for familiar symptom groups and it has high efficiency of runtime. The subjective Bayesian theory can be adapted with the groups of less and inconspicuous symptoms. It can deduce the possible disease and figure out its probability. However, when the information volume is very low and the use of probability functions is not possible, it is advisable to use other alternative techniques. In this case, a candidate for managing uncertainty in the expert systems is the Dempster–Shafer evidence theory.

3.3 Flowchart

The DCDDS is composed of four subsystems: case management, diagnosis, prevention and cure and medicine management. Users can choose to browse the records about their cows when login in the system. They can also choose to study by looking through the medicine component and the prevention component. Certainly, the main function of the system is to diagnose. If users choose to diagnose, they will be shown the first step to submit the environment information around the sick cow and to select the symptoms visible. Within the set of symptoms the users selected, the system will look for relevant disease. Only when some diseases can explain and match as many symptoms as possible based on the rule, the process is finished. To do this, the system will first look for a disease that can explain all the symptoms. If there is no such disease, it will look for a disease that can explain all but one symptoms, etc., until it finds a possible disease [2]. The users also can select the disease the users supposed based on his experience. With the set of disease the user selected, the system will look for relevant symptoms and output them to let user confirm. When get a disease finally, a proposal about prevention and cure will be given. At the end, the users will be asked if to save the diagnosis result as a case history to be checked in the future.

4. DISCUSSIONS AND CONCLUSIONS

This system has the following characteristics:

It is a web-based expert system that provides an easy access for dairymen. Comparing with most of the exist expert systems which based on C/S mode or setup on a stand alone computer, the DCDDS is easy to upgrade the program. With its online data and knowledge acquisition system, farmers, technicians and experts can easily update the knowledge base and data base with new instance. Users can reach the system only with a PC connected with the internet and they can get help from the expert system with up-to-the-minute version.

The users' feedback form serves as an effective and efficient mechanism to collect specialists and common users' comments, problems and suggestions.

It also provides general information support tools, such as a dictionary, information on medicine, contact information of medicine vendors, etc, which help the users to study online.

ACKNOWLEDGEMENT

This program is supported by Huo Yingdong Young Teacher's Foundation (Contract No. 94032). We also would like to thank many domain experts from China Agriculture University.

REFERENCE

Schreiber, G., Wielinga, B., de Hoog, R., Akkermans, H., van de Velde, W., CommonKADS: a comprehensive methodology for KBS development. IEEE Expert 12, 2837 (1994).

Kramers, M.A., Conijn, C.G.M., Bastiaansen, C. EXSYS, an Expert System for Diagnosing Flowerbulb Diseases, Pests and Non-parasitic Disorders. Agricultural Systems Volume: 58, Issue: 1, September, 1998, pp. 57-85.

A CALL CENTER ORIENTED CONSULTANT SYSTEM FOR FISH DISEASE DIAGNOSIS IN CHINA

Jie Zhang [1,2], Daoliang Li [1,2,*]
[1] *China Agricultural University, College of Information and Electrical Engineering, Beijing 100083, China*
[2] *Key Laboratory of Modern Agriculture System Integration, China Agricultural University, Beijing 100083, China*
[*] *Corresponding Author, Tel: +86-10-62736717, Fax: +86-10-62737679, Email: dliangl@cau.edu.cn or*

Abstract: Fish expert systems have been developed to solve the fish disease diagnosis problems. However, despite these inherent advantages, there are some limitations yet in the application of the systems because of the poor infrastructure and low capacity for the use of computers. A Call Center-based consultant system for fish disease diagnosis has been developed to overcome the difficulties, and the objective of this system is the automation of fish diagnostic process based on call center via the telephone. This paper explains the needs for the consultant system, the fish disease diagnosis process and the development of the system. The system structure and its components are described. It can provide the accurate and timely diagnosis, prevention and treatment all around the clock and everywhere the fish sites are located. The experience and the further development of the system are also discussed in the paper.

Keywords: Call Center, fish disease diagnosis, consultant system, knowledge acquisition

1. INTRODUCTION

Great efforts have been put by researchers, institutes and organizations to solve the fish disease diagnosis problems in China. Especially, a number of

Zhang, J. and Li, D., 2008, in IFIP International Federation for Information Processing, Volume 259; Computer and Computing Technologies in Agriculture, Vol. 2; Daoliang Li; (Boston: Springer), pp. 1447–1451.

agricultural web sites have emerged on the internet providing on-line information in the form of Hyper Text Markup Language (HTML) pages. (Allan et al., 2000) A web-based expert system for fish disease diagnosis, called Fish-Expert, has been developed by Agricultural Information Technology Institute at China Agricultural University (Daoliang et al., 2002).

As most aquaculture sites are scattered in remote rural areas which are weak in infrastructure and capability for the use of computers. Fish farmers can hardly get the service the Fish-Expert provides and the application of the system is restricted to some extent. Nowadays, with the rapid development of Telecommunication Industry, the Telephone Coverage Rate has been increased in Rural Areas in China. Especially, after the Cun Cun Tong program was launched. In early 2004, telephone services were available in 630,000 villages, or around 91 percent of the nation's total. By November 2005 the figure had been raised to 97 percent (http://english.sohu.com /20060314/n227789404.shtml).

Agriculture distance diagnosis has been applied since the 1990s (Holmes et al., 2000). A call center can provide telephone access to information services that are delivered by a human operator (agent), who in turn has access to an information resource (database) (Marco et al., 2004), or by IVR.

This paper presents a call center-based consultant system for fish disease diagnosis. It was a major outcome of Huo Ying Dong Fundation (Contract Number 94032). During the development of the system, fish expert system and the call center were integrated together to overcome the limitations and enhance the functionality of traditional ES. The system can provide the accurate and timely diagnosis, prevention and treatment all around the clock and everywhere the fish sites are located via the telephone. It has been tested and evaluated in Tianjin province of North China. The experience with developing the system, lessons learnt and conclusions are provided at the end of the paper.

2. SYSTEM DESIGN AND DEVELOPMENT

Though rural areas are weak in the infrastructure and capacity for the use of computers, telephone and mobile uses increase rapidly in rural China. And also the provision of call centre service appears as a sound alternative support channel for fish farmer to acquire counseling and support in disease diagnosis in China. So a call center-based consultant system for fish disease diagnosis was developed that allows fish farmers to easily get the fish disease diagnosis information via the telephone.

2.1 System architecture

The overall architecture of the system contains three parts. Users including mobile phone users and telephone users are connected to the call center via the telephone. Then calls are distributed to the IVR (Interactive Voice Response) or the Agents by the ACD (Automatic call distribution). IVR is a technology that automates interaction with telephone callers. IVR solutions use pre-recorded voice prompts and menus to present information and options to fish farmers, and touch-tone telephone keypad entry to gather responses, also helps the caller to be connected to an agent who can assist them. For the fish farmer, he or she can describe the symptoms of their fishes' disease to the call center agent by telephone, and they input all these symptoms into the Expert system for fish disease diagnosis interface, then tell the fish farmers the diagnosis, prevention and the treatment. After many fish farmers using the system, a great quantity of fish disease diagnosis cases were acquired and stored in the case base for the further knowledge acquisition. Some efficient algorithms will be used to find hidden patterns in data and generate sets of decision rules from cases. Then they will be used in the next diagnosis processing.

2.2 System work flow

Telephone callers (or wireless callers) dial, typically, a toll-free telephone number to contact the call center via the PSTN (or PLMN), then the interactive voice response (IVR) system prompts the caller with a series of announcements and menus. The caller chooses the service mode from IVR automatic service and agent service by pressing the right number.

For general fish disease problems, the call center is usually "front-ended" with the interactive voice response system. The IVR tree has some main branches (e.g. on-the-sport behavior query, body part query, etc) represented by the telephone number from 1 to N. Each main branch has its own branches corresponding to the phone number. For instance, after you choosing the on-the-sport behavior query, you still should make a choice from the multiple symptoms. And at last the treatment, prevention and the diagnosis will be given to you. Maybe you can not get a conclusion by one symptom, what you should do then is to press # button for returning to the main menus and choose other symptoms until you get the suggestion.

In some cases, if a caller wants to exit the IVR system at certain points in the "dialogue", he can press any key at any time to disconnect the telephone directly. Callers needn't wait to make a key choice until the end of every performance. The system is allocated to each call only for a very short fraction of time, an IVR experience may have a duration of several minutes.

Therefore, if time outs, the phone will be disconnected by the system automatically.

After choosing to speak directly with an agent, a caller is routed to the agent that has been idle for the longest duration. Otherwise, if all agents are busy, the caller's call is routed to the queue with the shortest expected wait time. The agents interact with the expert system for fish disease diagnosis that provides the fish disease knowledge.

2.3 Knowledge acquisition

Knowledge acquisition is the process of eliciting the expertise of domain expert and of formulating the expertise into a representation that can be used by a knowledge-based system. It is the most difficult stage in the development of expert systems. Many techniques have been developed for knowledge acquisition (KA). Some commonly used approaches are interviews, observations, taking experts through case studies and rule induction by machines.

Plenty of the fish disease diagnosis cases can be required from the call center system, maybe the cases available are insufficient or poorly-specified. The challenge here is to get hold of the information, and turn them into knowledge by making cases usable. A knowledge acquisition tool can be used to produce a set of rule cases with the quantity of the original cases.

In this system, Rough sets theory and neural networks are combined together to acquire knowledge for their complementary features. The rough set theory, proposed by Pawlak in 1982 (Pawlak et al., 1982) adopts the concept of equivalence classes to partition training instances according to some criteria. The rule induction from the original data model is data-driven without any additional assumptions. Neural networks (NN) are a computer 'simulation' of the interconnected neurons in our brains. This method simulates a network of interconnected neurons usually arranged in three layers, where the lowest one receives inputs and the top signals the outputs.

The approach is to use rough set approach as a pre-processing tool for the neural networks. By eliminating the redundant data from database, rough set methods can greatly accelerate the network training time and improve its prediction accuracy. Rough set method was also applied to generating rules from trained neural networks. In the system, neural networks were the main knowledge bases and rough sets were used only as a tool to speedup or simplify the process of using neural networks for mining knowledge from the databases.

3. DISCUSSIONS AND CONCLUSIONS

The system demonstrates the possibility and potential benefits of using the call center to facilitate the expert system in aquaculture. So it is a good idea for integrating call center system with the fish disease diagnosis expert system, the two systems can complement with each other, and meet the needs for the fish farmers and vets in different situations to overcome the limitations of each individual system.

This paper presents the development of a call-center based consultant system for fish disease. This system is a good try for telemedicine techniques application based on the call center in aquaculture. It realized the long-distance fish disease diagnosis without having the necessary IT equipment or low Internet access speed in rural areas in China and was successfully used in the practice. It presents the tendency of the telemedicine techniques application based on call center in Agriculture, we will continue this interesting work unmoved.

ACKNOWLEDGEMENTS

This research is supported by the Huo Ying Dong Fundation (Contract Number: 94032) in China. We would like to thank many domain experts from the Beijing Aquaculture Science Institute, Aquaculture Department of Tianjin Agricultural College, Aquaculture Bureau of Shandong province, for their co-operation and support. Our special thanks should go to Dr Zahng Xiaoshuan for the fish disease diagnosis and Dr Zhang Jian for the call center use. We would also like to acknowledge Mr YouLongyong and Mr Zhuwei for their valuable suggestions on the improvement of the system.

REFERENCES

Allan Leck Jensen, Peter S. Boll, Iver Thysen and B.K. Pathak., PlanteInfor — a web-based system for personalised decision support in crop management, Computers and Electronics in Agriculture, 25 (2000), 271–293.

Daoliang Li, Zetian Fu and Yanqing Duan, Fish-Expert: a web-based expert system for fish disease diagnosis, Expert Systems with Applications, 23 (2002), 311–320.

http://english.sohu.com/20060314/n227789404.shtml

Marco Adria and Shamsud D. Chowdhury, Centralization as a design consideration for the management of call centers, Information and Management, 41 (2004), 497–507.

Pawlak, Z, Rough set, International Journal of Computer and Information Sciences, 341–356, 1982.

A CBR SYSTEM FOR FISH DISEASE DIAGNOSIS

Wei Zhu, Daoliang Li*

Key Laboratory of Modern Precision Agriculture System Integration Research, China Agricultural University, 17 Tsinghua East Road, Beijing, China

* *Corresponding author, Address: P.O. Box 121, EU-China Center for Information & Communication Technologies, China Agricultural University, 17 Tsinghua East Road, Beijing, 100083, P. R. China, Tel: +86-10-72736717, Fax: +86-10-62737679, Email: li_daoliang@yaohoo.com*

Abstract: Fish disease diagnosis is a complicated process and requires high level of expertise. However, there's no accepted general knowledge in fish disease diagnosis. This paper describes a CBR (case-based reasoning) system for fish disease diagnosis. A two-step case retrieve model is proposed in this paper.

Keywords: case-based reasoning, fish disease diagnosis, nearest neighbor

1. INTRODUCTION

Fish disease diagnosis is a complicated process and requires high level of expertise. This problem domain has its own problems, the major one being that the effort to deal with a multitude of diseases for multiple species, and another one being that there's no accepted general knowledge in fish disease diagnosis (Daniel Zeldis, 2000).

An expert system for fish disease diagnosis called Fish-Expert has been developed. Fish-Expert, in which rule-based reasoning is applied, can mimic human fish disease experts (Daoliang Li, 2002). There are two deficiencies with the Fish-Expert, the major one being that rule-based reasoning requires some 'deep' knowledge in order to be truly effective, however there's no accepted general knowledge, another one being that inference engine is too complex to work efficiently and is time-consuming.

Zhu, W. and Li, D., 2008, in IFIP International Federation for Information Processing, Volume 259; Computer and Computing Technologies in Agriculture, Vol. 2; Daoliang Li; (Boston: Springer), pp. 1453–1457.

This paper describes a CBR system for fish disease diagnosis. CBR is a well recognized and established method for building medical expert systems [4]. Instead of relying on general domain knowledge, CBR uses the storage of a large number of previously solved cases (Isabelle Bichindaritz, 2006). Through the interview with these fish experts, plenty of fish disease cases those can be used in CBR have been acquired. And a two-step case retrieve model is proposed. Some experiences with developing CBR system are discussed and the conclusions are provided at the end of the paper.

2. SYSTEM ARCHITECTURE

The overall architecture of the system is show in Figure 1. Through a forms interface, the fish farmers' requirements could be input. And the CBR system will search the fish disease case base of past cases and retrieve similar cases. Details of the similar case will then be available to the fish farmers. All this information would then be automatically passed back to an agent to authorize or change, if necessary. Once a case is completed, its details would be added to the fish disease case base (Isabelle Bichindaritz, 2006).

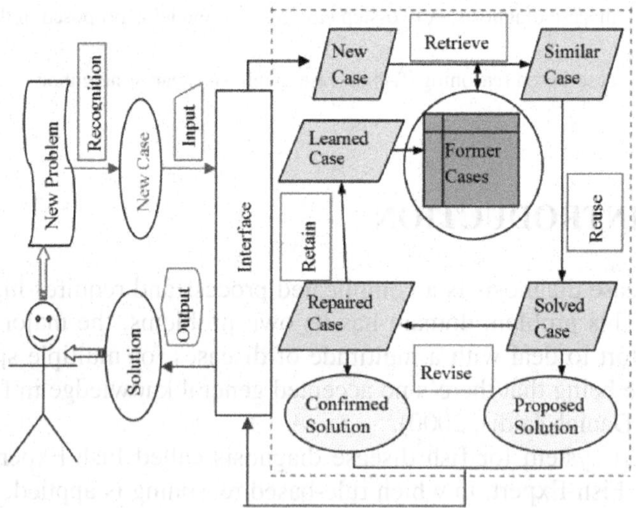

Figure 1. System architecture

3. METHODOLOGY

In the following, how the CBR to be applied in fish disease diagnosis is described. The subtasks are referred to as case representation, case indexing, case retrieval and case reuse.

3.1 Case Representation

A case of fish disease diagnosis is a set of empirical data, gathering the experience of the fish vets involved in a previous situation. The fish vets write the usual description of a case in natural language in a general standardized report. From the analysis of a corpus of available reports, a model has been developed. Within this model, a case of fish disease diagnosis is consisted of (Object, Symptoms, Treatments). As there is no fish disease vocabulary standard available for descriptions, the domain-related items have been chosen by consensus between some fish vets (Anil Varma, 1999). The symptoms are described as (0-1) vector, where 0 means the symptom doesn't appear, and 1 means the symptom appears.

3.2 Case Indexing

A two-step case indexing model is proposed to quickly locate similar previous cases in this paper. In the first step index, a clustering algorithm is used to partition all past cases in the fish disease case base into clusters according to their design specifications. Primary index features are part of the explanation of the new case input by the fish farmers. It initially identifies which case base the new case belongs to. With the indexing mechanism conducted in advance, the case retrieval procedure can be accelerated. In the second step, the observed features themselves are used as secondary features only. The similarity between the new case and the small set of retrieved cases of the first step is calculated using this simple nearest neighbor algorithm (Florian Hartge, 2006).

3.3 Case Retrieval

The retrieve task starts with a partial fish disease symptoms description, and ends when a best matching previous fish disease case has been found. Its subtasks are referred to as Identify Features, Initially Match and Select, executed in that order (Isabelle Bichindaritz, 2006).

3.4 Identify Feature

To identify a new fish disease case whose input symptoms features should be noticed. Unknown symptoms features may be disregarded or requested to be explained by the fish farmer. To understand a new case involves filtering out noisy symptoms descriptors, inferring other relevant problem features, checking whether the feature values make sense within the context, etc (Kyung-Sup Kim, 2001).

3.4.1 Initially Match

Once the fish disease symptoms features have been identified, the Initially Match process starts. Finding a set of matching disease is done by using the symptoms features as indices to the fish disease knowledge base in a direct way. The possible fish disease set is retrieved solely from input symptoms features. A way to assess the degree of similarity is needed, and several similarity measurements have been proposed, based on surface similarities of fish disease symptoms features (Nirmalie Wiratunga, 2004).

3.4.2 Select

The selection process typically generates consequences and expectations from each retrieved case, and attempts to evaluate consequences and justify expectations. The similarity between the new case and the small set of retrieved cases is calculated using this simple nearest neighbor algorithms. From the set of similar cases, a best match case is to be chosen. The best matching case is usually determined by evaluating the similarity of the fish disease symptoms features between the new case and the small set of retrieved cases. The case that has the strongest explanation for being similar to the new problem is chosen (Isabelle Bichindaritz, 2006). In this stage, the small set of retrieved fish disease cases is compared by the client-side applet with the original query and similarity is calculated using this simple nearest neighbor algorithm (Abdus Salam Khan, 2003).

3.5 Case Reuse

The reuse of the retrieved fish disease case solution in the context of the new case focuses on two aspects: (a) the differences among the previous and the new case and (b) what part of a retrieved case can be transferred to the new case (I. Watson, 1999). The modification is done by using the fish disease diagnosis knowledge in this paper.

4. CONCLUSION

In conclusion, CBR is applied as the inference engine to mimic human fish experts. Instead of relying on general fish disease knowledge which is lacking, CBR uses the storage of a large number of previously solved cases. And a two-step case indexing model is proposed to quickly locate similar previous cases. Based on the two-step case indexing model, an efficient case retrieval procedure is developed to find similar cases from the fish disease case base for a new case. Experience has shown that CBR approach has been able to contribute significantly in fish disease diagnosis. Though the system is developed specific for fish disease diagnosis, it can also easily be developed available for other domains.

ACKNOWLEDGEMENTS

The authors would like to thank many fish experts from Beijing Aquaculture Science Institute and Aquaculture department of Tianjin Agricultural College for their fruitful comments and suggestions. The authors acknowledge the Fok Ying Tung Education Foundation (Grant No. 94032) for funding our research.

REFERENCES

Abdus Salam Khan, Achim Hoffmann. Building a case-based diet recommendation system without a knowledge engineer. Artificial Intelligence in Medicine, 2003(27): 155-179

Anil Varma, Nicholas Roddy. ICARUS: design and deployment of a case-based reasoning system for locomotive diagnostics. Engineering Applications of Artificial Intelligence, 1999(12): 681-690

Daniel Zeldis, Shawn Prescott. Fish disease diagnosis program-problems and some solutions. Aquacultural Engineering, 2000(23): 3-1

Daoliang Li, Zetian Fu, Yanqing Duan. Fish-Expert: a web-based expert system for fish disease diagnosis. Expert Systems with Applications, 2002(23): 311-320

Florian Hartge, Thomas Wetter, Walter E. Haefeli. A similarity measure for case based reasoning modeling with temporal abstraction based on cross-correlation. Computer Methods and Programs in Biomedicine, 2006(8): 41-48

I. Watson. Case-based reasoning is a methodology not a technology. Knowledge-Based Systems, 1999(12): 303-308

Isabelle Bichindaritz, Cindy Marling. Case-based reasoning in the health sciences: What's next? Artificial Intelligence in Medicine, 2006(36): 127-135

Isabelle Bichindaritz. Memoire: A framework for semantic interoperability of case-based reasoning systems in biology and medicine. Artificial Intelligence in Medicine, 2006(36): 177-192

Kyung-Sup Kim, Ingoo Han. The cluster-indexing method for case-based reasoning using self-organizing maps and learning vector quantization for bond rating cases. Expert Systems with Applications, 2001(21): 147-156

Nirmalie Wiratunga, Susan Craw, Bruce Taylor, Genevieve Davis. Case-based reasoning for matching Smarthouse technology to people's needs. Knowledge-Based Systems, 2004(17): 139-146

SUMMARY OF PIVOTAL TECHNIQUE OF COTTON-HARVEST ROBOT

Ling Wang[1], Changying Ji[1,*]
[1] *College of Engineering, Nanjing Agricultural University, Nanjing, 210031, China*
[*] *Corresponding author, Address: Department of Agricultural Mechanization, College of Engineering, Nanjing Agricultural University, 40 Dianjiangtai Road, Nanjing, 210031, P. R. China, Tel: 13951994628, Email: chyji@njau.edu.cn*

Abstract: Pivotal techniques of cotton-harvest robots were summarized, including image segmentation, features generation, artificial classifiers and performances evaluation. Solutions based on machine vision and pattern recognition were analyzed to distinguish ripe from under-ripe/over-ripe cottons, and rank cottons according to government grading standards.

Keywords: Cotton, Harvest robot, Classifier

1. INTRODUCTION

Cotton production systems have benefited from precision agriculture (PA) (Searcy et al., 2000). Agricultural robot combines advantages of Cotton-harvest in manual and mechanism way, including ecological benefits, needless of biological pretreatments, high-precision of classification and economic profitability. And it has potentials to resolve these conflicts in either efficiency or quality: America is characterized by these aspects, including large scale, single-variety, middling-long fibers and one-off harvest, which are suitable for cotton harvest in mechanism way. While China is characterized by these factors, containing small-scale, multi-varieties, harvest in batches, short-middling fiber, which are inapplicable in either mechanism or manual way, and the latter has been restricted by plenty of pickers (Wang et al., 2006). So cotton-harvest robot adapts to Chinese

Wang, L. and Ji, C., 2008, in IFIP International Federation for Information Processing, Volume 259; Computer and Computing Technologies in Agriculture, Vol. 2; Daoliang Li; (Boston: Springer), pp. 1459–1463.

situation to a great extent, and cotton-harvest robot towards PA in China is a new trend in practice.

2. PIVOTAL TECHNIQUE

A mobile harvest robot consists of manipulators, end-effectors, mobile wheels, machine visions, control systems etc, and fundamentals in robotic auto-navigation and objects detection are obtained. Development trends of harvest robot lie in fairly accurate, nondestructive, robust and low cost, so classifiers design is the pivotal techniques.

The appearances of cottons, including size, shape, color, texture, orientation, blemishes and impurities, plays an important role of Chinese standards in the purchase of pre-harvest cottons. Various stages for classifiers design as follows, including image segmentation, features generation, artificial classifiers and performances evaluation, are interrelated and depend on the results each other to improve the overall performances.

2.1 Image segmentation

Intensity images segmentation algorithms are generally based on edge, threshold and region. Threshold has become a fundamental approach enjoying a significant degree of popularity, especially in applications where speed is an important factor.

Color image segmentation algorithms are also robust such as clustering and discriminant analysis based on pixels. C-means clustering segmentation algorithm is suitable for recovering compact clusters with computational simplicity (see Figure 1).

Figure 1. Segmentation of cotton based on C-means clustering

2.2 Features generation

Image represented as boundaries is suitable for analysis of morphology, including chain codes, moments and geometrics traditionally, and which are insensitive to position, size, orientation. Image represented as regions is used for evaluation of texture and color, consisting of multiple-order statistics moments, autoregressive parameters, and fractal dimension D (Pentland et al., 1984).

Features via linear transforms can exhibit high "information packing" properties. Some transforms based on problem-dependent use unfixed vectors with high computational complexity, such as Principal Component Analysis (PCA), the optimality of which leads to excellent information packing properties, and the disadvantage of which does not lead to maximum class separability in the lower dimensional subspace. Other transforms are based on fixed images and their sub-optimality with respect to decorrelation and information packing properties is often compensated by its low computational requirements, such as Fast Fourier Transform (FFT) and Discrete Wavelet Transform (DWT), and wavelet-based features selected as morphology or texture exhibit smaller within class variance and stronger between class separation than the Fourier-based ones without shift invariant (Wuncsh et al., 1995), based on which classification may result in low error rates.

2.3 Artificial classifiers

Classifiers are designed as follows to distinguish ripe from under/over-ripe cottons, and rank cottons according to government grading standards.

2.3.1 Bayes classifier

It classifies an unknown pattern in the most probable of the classes by minimizing the classification error probability, and assumes a known class-conditional probability density functions of the feature in each class. In practice, we can assume that the features follow the general multivariate normal density due to its computational tractability and the fact that it models adequately a large number of cases.

2.3.2 Linear discriminant

Discriminant analysis is suboptimal with respect to Bayes classifiers by computing decision surfaces directly by means of alternative costs. Fisher linear discriminant function is an optimal linear classifier by minimizing MSE with fast, fairly accurate and easy to implement. It is useful in a wide variety of applications even if the function to be learnt is mildly nonlinear. The better generalization may be got with a simple linear classifier than with a complicated nonlinear classifier if there is too little data or too much noise.

2.3.3 Non-linear discriminant

The major limitations of Support Vector Machine (SVM) are the high computational burden and the difficult selection methods of the kernel functions. A notable characteristic of SVM is suitable for high dimension features to exhibit good generalization performances.

The major limitations of Artificial Neural Network (ANN) are easy to lead the weights to a local minimum of the MSE. MSE is decomposed into the bias and the variance, and under-fitting produces excessive bias in the outputs whereas over-fitting produces excessive variance. Experiment proves that the error on training set keeps decreasing while the error on test set changes from decreasing to increasing, and the inflexion of which can be used in practice to terminate the iterations. And the size of classifiers should be kept as small as possible to a satisfactory performance.

2.4 Performances evaluation

Different classifiers would result in different generalization error. Resampling based cross-validation is a method for estimating generalization errors, the resulting estimates of generalization error are often used for choosing among various models. K-fold cross-validation is preferred for discontinuous error functions such as the number of misclassified cases. A value of 10 for k is popular for estimating generalization error while leave-one-out cross-validation often works well for estimating generalization error for continuous error functions such as MSE.

3. CONCLUSION

The pivotal techniques of cotton-harvest robots lie in classifiers. Machine vision and pattern recognition can be applied to recognize cottons and grade them while being harvested with a accepted accurate, and algorithms such as autoregressive model, fractal geometry, PCA, FFT, DWT, ANN, and SVM can be applied to enhance robustness of classifiers.

REFERENCES

Ling Wang, Changying Ji. Technical analysis and expectation for cotton harvesting based on agricultural robot. Cotton Science. Vol. 18(2). pp. 124-128, 2006.

Pentland A. "Fractal based decomposition of natural scenes," IEEE Transactions on Pattern Analysis and Machine Intelligence, Vol. 6(6), pp. 661-674, 1984.

Searcy S.W., Beck A.D. Real time assessment of cotton plant height. Proceedings of Fifth International Conference on Precision Agriculture (CD), July 16 19, 2000. Bloomington, MN, USA.

Wuncsh P. Laine A. "Wavelet descriptors for multiresolution recognition of handwritten characters," Pattern Recognition, Vol. 28(8), pp. 1237-1249, 1995.

RANKING FOR PREHARVEST COTTONS BY USING MACHINE VISION

Ling Wang[1], Changying Ji[1,*]
[1] *College of Engineering, Nanjing Agricultural University, Nanjing, 210031, China*
[*] *Corresponding author, Address: Department of Agricultural Mechanization, College of Engineering, Nanjing Agricultural University, 40 Dianjiangtai Road, Nanjing, 210031, P. R. China, Tel: 13951994628, Email: chyji@njau.edu.cn*

Abstract: In order to assess the quality of preharvest cottons objectively, ranking classifiers were designed based on machine vision technologies to grade preharvest cottons on dark background based on their sizes and colors. Experiments showed that the classifiers can classify preharvest cottons into seven grade categories with an accuracy of nearly 91.5%.

Keywords: Preharvest Cottons, Grade, Machine vision, Classifier

1. INTRODUCTION

Over a long period, cottons have been mostly harvested in either manual or machinery way in China. The former is highly subjective, which is not accurate for grading, while the latter is not able to grade at all. It is necessary to come up with an approach to grade the preharvest cottons.

At present, although much research has been done on ginned cottons for grading outside and in by HVI equipment (Poceciun, 1999), little work has been done on un-ginned cottons as preharvest cottons for grading outside. This thesis presented the results of experiments in which classifiers was designed to sample the preharvest cottons and then grade them by employing machine vision and pattern recognition.

Wang, L. and Ji, C., 2008, in IFIP International Federation for Information Processing, Volume 259; Computer and Computing Technologies in Agriculture, Vol. 2; Daoliang Li; (Boston: Springer), pp. 1465–1469.

2. MATERIALS AND METHODS

The samples were acquired by camera afield. A black board was placed behind each sample in the course of photo to segment from backdrop well. The total samples (402) were classified into seven grades ranged from 1 to 7 in manual, each of which contained 10, 28, 91, 67, 67, 69, 70.

3. RESULTS AND DISCUSSIONS

3.1 Images segmentation

Morphology operation, including dilation, erosion, opening, closing, top-hat cutting, and bottom-hat cutting, is a shape-based technique of image processing (Gonzalez, 2002). Opening operation eroded original intensity images and then dilated eroded images with the same big structuring elements. And top-hat cutting images subtracted morphologically opened images from intensity images. Accordingly, closing operation dilated the original intensity image and then eroded the dilated image using the same small structuring element for both operations. And bottom-hat cutting images subtracted intensity images from morphologically closed images.

3.1.1 Cottons with bracteoles segmentation from background

Intensity images with brightened bracteoles were extracted by using intensity images plus adjusted top-hat cutting images, and then transformed into binary images based on Otsu's threshold. So cotton binary images with bracteoles were segmented from their background with noises (Figure 1).

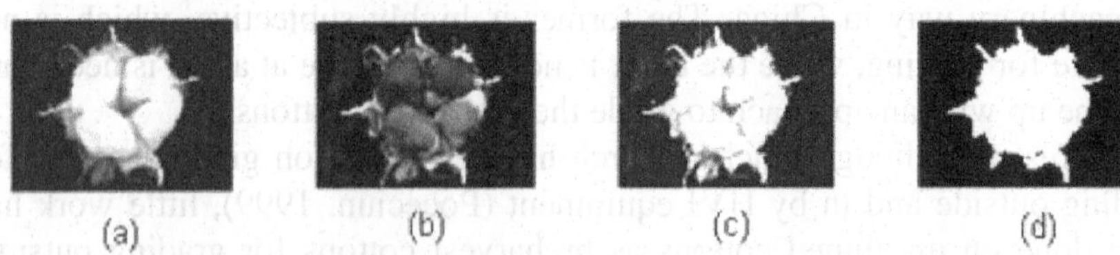

Figure 1. Cotton image segmentation from background

(a) Intensity image

(b) Top-hat cutting image

(c) Intensity image with brightened bracteoles

(d) Cotton binary image with bracteoles

3.1.2 Cottons segmentation from their bracteoles

Intensity images with darkened bracteoles subtracted adjusted bottom-hat cutting images from intensity images, and then transformed into binary images based on Otsu's threshold. So cotton binary images were segmented from their bracteoles in turn because of bits and pieces of bracteoles being far small (Figure 2).

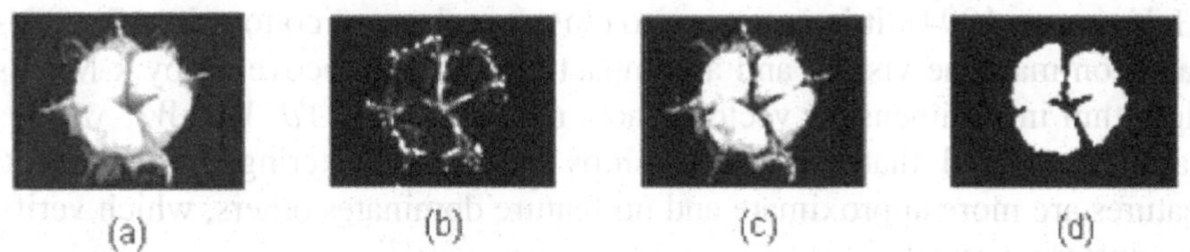

Figure 2. Cotton image segmentation from bracteoles

(a) Color image
(b) Bothat cutting image
(c) Intensity image with darkened bracteoles
(d) Cotton binary image without bracteoles

3.2 Features selection

According to Chinese government standards in letters in the purchase of preharvest cottons (Xiong, 2005), the primary factors for determining acceptability grades prior to purchase are external quality of cottons, including size, colors, textures, and impurities. Size and colors, implying textures and impurities respectively, were selected in this experiment.

Size (*size*) was calculated from the rate of the number of pixels of cotton with bracteoles to bracteoles. Much experiment results showed that hue of cotton is close to 10YR in Munsell Color Ring (Xiong, 1995), accordingly, and their color was only described by using saturation and intensity in HSI color space. So colors contained yellow degree (*Yd/yd*) and white degree (*Wd/wd*), which were calculated from the mean value of saturation and intensity image of cotton with/without bracteoles, respectively. Furthermore, white contrast (*Wc/wc*) was calculated from the standard deviation of intensity image of cotton with/without bracteoles.

Minimum information redundancy among the features is a major goal. According to Chinese government standards, the relationship of *size* and *Yd/yd* is negative, similarly, *size* and *Wc/wc*, but that of *size* and *Wd/wd* is positive. Table 1 showed that *wd* was invalid because of sunlight.

Table1. The relationship between size and colors

	Yd	Wd	Wc	yd	wd	wc
Size	-0.668	0.574	-0.617	-0.332	-0.125	-0.245

3.3 Classifiers design

3.3.1 Grades clustering

Different inspectors would result in different grades. Heinemann studied mushroom whereby the disagreement between inspectors varied from 14% to 34% compared to less than 20% of the machine vision system (Heinemann, 1994). It is necessary to cluster preharvest cottons into 7 grades based on machine vision, and a compact cluster was recovered by k-Means algorithm in 6-dimension vector space, including *size, Yd, Wd, Wc, yd, wc*. Table 2 showed that the relationships between clustering grades and 6 features are more approximate and no feature dominates others, which verify the clustering results.

Table 2. The relationship between grades and features

Grades	Size	Yd	Wd	Wc	yd	wc
Inspector	-.737	0.602	-.264	0.629	0.310	0.507
Clustering	-.822	0.860	-.623	0.649	0.621	0.366

3.3.2 Principal component analysis

The optimality of Principal component analysis (PCA) with respect to the minimum MSE will lead to excellent information packing properties. PCA was performed to generate optimally uncorrelated features, keep the size of classifiers as small as possible and increase the generalization capabilities. In this experiment, the 6 correlated features of *size, Yd, Wd, Wc, yd, wc* were reduced to only 2 orthonormal eigenvectors, i.e. the first and the second principal component (Prin. 1 & 2), which corresponded to nearly 78% in cumulative variation.

3.3.3 Linear discriminant

Fisher linear discriminant function is an optimal linear classifier by minimizing MSE, which had two inputs (i.e. Prin.1&Prin.2) and one output (i.e. clustered grades). Fisher's linear discriminant functions were computed based on within-groups covariance in SPSS (Table 3).

Table 3. Fisher's classification functions coefficients

Principal components	Function						
	1	2	3	4	5	6	7
Const.	-25.723	-7.778	-4.769	-1.922	-5.351	-4.974	-8.767
Prin. 1	-16.441	-7.940	-5.573	-1.220	-.516	4.914	8.680
Prin. 2	1.894	1.735	-1.263	1.489	-4.806	3.142	-2.340

The apparent reclassification results varied from 64% to 98% with the average of 91.5% by re-substitution method, which gave some indication on the consistency of machine vision decision and verified the clustering results (Table 4).

Table 4. Reclassification results by using re-substitution (%)

Grouped	Group						
cases	1	2	3	4	5	6	7
10	90	10	0	0	0	0	0
39	0	64	23	13	0	0	0
53	0	4	90	6	0	0	0
92	0	1	0	98	1	0	0
64	0	0	5	0	95	0	0
89	0	0	0	4	0	96	0
55	0	0	0	0	9	0	91

4. CONCLUSION

This thesis indicates the applications of machine vision technology involving size and colors grading of preharvest cottons. A prototype automated preharvest cottons classifier based on previously developed algorithm for size and colors assessment is successfully developed and tested. Reclassification results of cotton grade categories with an average accuracy of nearly 96%, and the results demonstrate the validity of the clustering results previously. Hence, machine vision systems helps standardizing and quantifying the inspection process of field preharvest cottons by promoting grading consistency and objectivity.

REFERENCES

Gonzalez, R.C. and Woods, R.E. Digital Image Processing, Second Edition. [M]. Beijing: Publishing House of Electronics Industry. 2002, 7:519-569.

Heinemann, P.H., Hughes, R., Morrow, C.T., Sommer III, H.J., Beelman, R.B., & West, P.J. (1994). Grading of mushrooms using a machine vision system. Transactions of the ASAE, 37(5), 1671-1677.

Poceciun V.V., Temkin B., Ethridge M.D., Hequet E. Computerized collection and analysis of HVI data. U.S.A. Orlando, National Cotton Council of America, Proceedings Beltwide Cotton Conference, 1999.

Xiong Z.W. Study on the fiber quality and the classification of color grades of China's cotton. Master Paper of China Agricultural University, 2005:10-25.

Xiong Z.W. Cotton Color Features [J]. China Cotton, 1995, 22 (4): 39-40.

ESTIMATING PIG WEIGHT FROM 2D IMAGES

Yan Yang, Guanghui Teng*

Department of Agricultural and Bioenvironmental Engineering, China Agricultural University, P.O. Box 195, Beijing, P. R. China

* *Corresponding author, Address: Department of Agricultural and Bioenvironmental Engineering, China Agricultural University, P.O. Box 195, Beijing, P. R. China, phone: 86-10-62737583, Email: futong@cau.edu.cn*

Abstract: This paper presents a new method that utilizes the technologies of image processing and computer vision. Firstly, the projected areas of the pig's image captured directly from top view are computed. Secondly, the heights are obtained from side view. Then the pig's weight is estimated by the projected areas and heights. By comparing with the real weight, the mean relative error is 3.2%. The experiment indicates that this hands-off method has great significance in scientific management of the pig's production which does not require large labor and material resources, and also avoid the loss in production resulted from stress.

Keywords: pig; weight; image processing; projected area; height

1. INTRODUCTION

This paper estimates the weight of pigs from images obtained by two cameras which are installed on the top of the ceiling and side of wall. We can measure the projected areas and heights of pigs through computer vision technique (Schofield et al., 1999), and then find a close relationship between the weight and these parameters. It indicates that weights of pigs can be estimated accurately from projected areas and heights. From many experiments we obtained a regressive formulation which is used for the estimation of the pig's weight.

Yang, Y. and Teng, G., 2008, in IFIP International Federation for Information Processing, Volume 259; Computer and Computing Technologies in Agriculture, Vol. 2; Daoliang Li; (Boston: Springer), pp. 1471–1474.

2. IMAGE PROCESSING ALGORITHM

2.1 Segmentation of images

1) Construction of the difference image. A difference image is constructed from the RGB channels, which suppresses the background and emphasizes the foreground objects (Wan et al., 2001):

$$D(x,y) = \frac{g(x,y)+b(x,y) - r(x,y)}{[g(x,y)+b(x,y) + r(x,y)]^{1.5}} \quad (1)$$

where (x,y) is the coordinate of the pixel. Figure 1(a) is an example of the difference image where the pig region is salient.

2) Automatic selection of threshold. Figure 1 (b) is a typical histogram in our experiments. The pixels of the pig region concentrate on the first peak and background concentrate on the right one. (Yu et al., 2001).

3) Binarization and after-processing. A binary connected component analysis is utilized to extract a single-connectivity region. Figure 1 (c,d) is an example of the after-processing results.

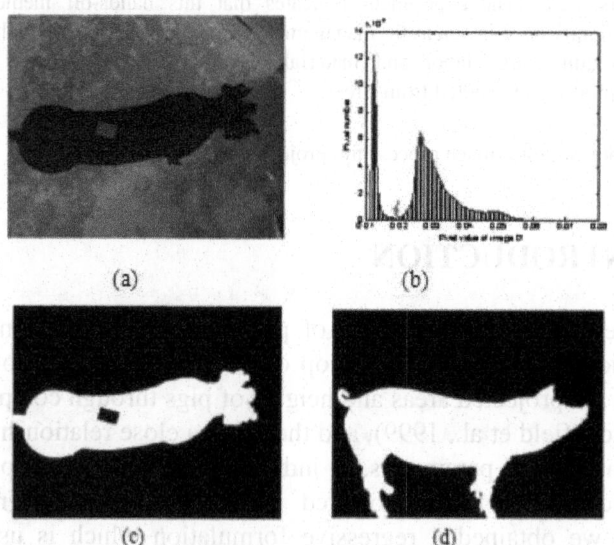

Fig. 1. Segmentation of images

(a) Difference image (b) The histogram of Difference image, threshold is 0.0262
(c) Top image after segmenting (d) Side image after segmenting

2.2 Measurement of projection area and height

The calculation of projected area is to count the pixels of value 1 in the binary image, i.e. the projected area is calculated by the following equation:

$$\frac{A_1}{N_1+N_2}=\frac{A_2}{N_2} \tag{2}$$

In the above equation:

A1 stands for the top projected area of pig to be calculated, cm^2;

A2 stands for the real area of reference cm^2;

N1 stands for the number of pixels in the segmented image of projected area after moving the reference. Pixels;

N2 stands for the number of pixels of the reference. Pixels.

We calculate the pig height from the side image. The segmented region (binary-valued pixels) of the pig is horizontally projected and the height is roughly estimated as the maximum projection value. We can obtain real height of pig by following equation:

$$\frac{L1}{N1}=\frac{L2}{N2} \tag{3}$$

L1 stands for real height of pig, cm;

L2 stands for the real length of the reference cm;

N1 stands for the maximum projection value as above;

N2 stands for the length of the reference in the segmentation image.

3. RESULTS AND DISCUSSIONS

As we know, volume correlates closely to weight. The pig weights were estimated using the following multiple regression equation after many experiments.

$$W=0.003\times S^{1.2811}\times H^{0.6121} \tag{4}$$

where W is the estimated weight (kg) of a pig, A is specified projected image area (cm^2), and H is the estimated height.

On the whole, the mean relative error of this equation in these samples was 3.2%. These errors partly result from the fact that the pigs did not adapt themselves to the electric balance. When driven into the balance, they did not remain stationary. Hence errors might occur when the weight is read on the display of the balance.

4. CONCLUSIONS AND FUTURE WORKS

To estimate the pig weight by a hand-off method, image analysis techniques were used in this work (Minagawa et al., 1994). The results indicate that there is a strong relationship between the weight of a pig and the volume composed by projected area and height. By comparing with the real weights, the mean relative error is 3.2%. More work is required to improve the image quality through carefully fixing the cameras and choosing suitable lighting, and to improve the image analysis algorithm. Measurement of weight using this method enables the stockman to monitor the performance and health of the pigs in time, and to predict and control their market weights and dates.

Future work is to develop a greater database as well as taking into account other factors, such as size and shape (Doeschl et al., 2004) which are considered as useful bases to describe the pig growth, body form and function, and also as possible indicators of muscle volume and lean content, and environmental parameters for industrial purpose. The final goal is to develop a pig management system based on the actual performance of the growing pigs. An automatic image processing system has to be developed, and images should be taken within the pens without moving pigs to special booths. This research will increase the efficiency of pig's production using the pig growth monitoring system, thereby improving pig performance without reducing the pig welfare.

REFERENCES

Schofield, C.P., J.A. Marchant et al., 1999. Monitoring pig growth using a prototype imaging system. Journal of Agric. Engineering. 72, 205-210.

Doeschl, A.B., Green, D.M., Whittemore, C.T., Schofield, C.P., Fisher, A.V. and Knap, P.W. 2004. The relationship between the body shape of living pigs and their carcass morphology and composition. Animal Science. 79: 73-83.

Minagawa, H. and T. Ichikawa. Determining the weight of pigs with image analysis. Trans. ASAE. 37(3):1011-1015.

Wan Q.Q., 1994. Digital image processing, Beijing: electronic industry publish company, 2001.

Yu X.W., Shen Z.R.. 2001, Research of insects' digital image segmentation. Transactions of the Chinese Society of Agricultural Engineering, Vol. 17(3): 137-141.